u

uni
univ
cc
Di
Kw
Pa
Rie
typ

Lecture Notes in Computer Science 10450

Commenced Publication in 1973
Founding and Former Series Editors:
Gerhard Goos, Juris Hartmanis, and Jan van Leeuwen

More information about this series at http://www.springer.com/series/7409

Jaap Kamps · Giannis Tsakonas
Yannis Manolopoulos · Lazaros Iliadis
Ioannis Karydis (Eds.)

Research and Advanced Technology for Digital Libraries

21st International Conference on Theory and Practice
of Digital Libraries, TPDL 2017
Thessaloniki, Greece, September 18–21, 2017
Proceedings

 Springer

Editors
Jaap Kamps
Faculteit der Geesteswetenschappen
Universiteit van Amsterdam
Amsterdam
The Netherlands

Giannis Tsakonas
Library & Information Center
University of Patras
Patras
Greece

Yannis Manolopoulos
Aristotle University of Thessaloniki
Thessaloniki
Greece

Lazaros Iliadis
Civil Engineering
University of Thrace
Kimmeria
Greece

Ioannis Karydis
Informatics
Ionian University
Kerkyra
Greece

ISSN 0302-9743   ISSN 1611-3349   (electronic)
Lecture Notes in Computer Science
ISBN 978-3-319-67007-2   ISBN 978-3-319-67008-9   (eBook)
DOI 10.1007/978-3-319-67008-9

Library of Congress Control Number: 2017952390

LNCS Sublibrary: SL3 – Information Systems and Applications, incl. Internet/Web, and HCI

Printed on acid-free paper

This Springer imprint is published by Springer Nature
The registered company is Springer International Publishing AG
The registered company address is: Gewerbestrasse 11, 6330 Cham, Switzerland

Preface

This volume of proceedings contains the reviewed papers presented at the 21st International Conference on Theory and Practice of Digital Libraries (TPDL), which was held in Thessaloniki, Greece from September 18 to 21, 2017. The conference was organized by the Aristotle University of Thessaloniki and the Democritus University of Thrace. The general theme of the 21st International Conference on Theory and Practice of Digital Libraries was "Part of the Machine: Turning Complex into Scalable" and its aim was to create a dialogue that addressed the challenge of creatively transforming these highly-synthesized environments into solutions that can scale for the benefit of varied communities.

TPDL 2017 received 85 full-paper submissions, up from 50 at TPDL 2016 and 44 at TPDL 2015, making the conference in 2017 very competitive and selective, and requiring the Program Committee to uphold the highest possible academic standards. We introduced a two-layered structure for oral presentations, long and short oral, in order to include an adequate number of interesting papers that expand the field of digital libraries on innovative topics and to strengthen the areas already known. Of the 85 long-paper submissions, only 20 (24%) were accepted for a long oral presentation, and an additional 19 (22%) long papers were accepted for a shorter oral presentation. This makes a grand total of 39 (46%) full papers accepted for the proceedings.

Of the 8 short-paper submissions, only 4 (50%) were accepted, and of the 5 poster/demo submissions, only 2 (40%) were accepted. Selected full-paper submissions were redirected for evaluation as potential short or poster/demo papers, following the recommendations of the reviewers.

Each submission was reviewed by at least three Program Committee members, and two Senior Program Committee members, and the two chairs oversaw the reviewing and often extensive follow-up discussion. Where the discussion was not sufficient to make a decision, the paper went through an extra review by the Program Committee. Each paper was discussed individually, based on the reviews, the meta reviews, and the discussion at a PC meeting, where the final decisions were made.

The conference was honored by three very interesting keynote speeches by Paul Groth on "Machines are People Too", Elton Barker on "Back to the Future: Annotating, Collaborating, and Linking in a Digital Ecosystem" and Dimitrios Tzovaras on "Visualization in the Big Data Era: Data Mining from Networked Information". All three covered important areas of the digital library field.

The program of TPDL 2017 also included a doctoral consortium track and four tutorials on "enriching digital collections using tools for text mining, indexing and visualization", "putting historical data in context: how to use DSpace-GLAM", "innovation search", and "enabling precise identification and citability of dynamic data – recommendations of the RDA working group on data citation". Finally, four workshops were organized in conjunction with the main conference, namely the long-established "European Networked Knowledge Organization Systems (NKOS)" in its 17th year and the newly introduced "(Meta)-data Quality Workshop", "International Workshop on Temporal Dynamics in Digital Libraries", and "Modeling Societal Future (FUTURITY)".

We would like to thank all our colleagues for trusting their papers to the conference, as well as our Program Committee members, both the senior and the regular, for the precise and thorough work they put into reviewing the submissions. A word of gratitude must be addressed to our workshop chairs, Philipp Mayr and Kjetil Nørvåg, our tutorial chairs, Thomas Risse and Gianmaria Silvello, our panel chair, Cristina Ribeiro, our posters/demo chairs, Vangelis Banos and Annika Hinze, and our doctoral consortium chairs, Maja Žumer and Heiko Schuldt, for the substantial effort they put into in running their tracks.

September 2017   Jaap Kamps
Thessaloniki   Giannis Tsakonas
                                              Yannis Manolopoulos
                                                  Lazaros Iliadis
                                                 Ioannis Karydis

Organization

TPDL 2017 was organized by the Aristotle University of Thessaloniki, Greece and the Democritus University of Thrace, Greece.

General Chairs

Yannis Manolopoulos	Aristotle University of Thessaloniki, Greece
Lazaros Iliadis	Democritus University of Thrace, Greece

Program Chairs

Jaap Kamps	University of Amsterdam, The Netherlands
Giannis Tsakonas	University of Patras, Greece

Organizing Chair

Apostolos Papadopoulos	Aristotle University of Thessaloniki, Greece

Publicity/Publication Chair

Ioannis Karydis	Ionian University, Greece

Workshop Chairs

Philipp Mayr	GESIS, Germany
Kjetil Nørvåg	Norwegian University of Science and Technology, Norway

Tutorial Chairs

Thomas Risse	L3S, Germany
Gianmaria Silvello	University of Padua, Italy

Panel Chair

Cristina Ribeiro	University of Porto, Porto

Posters/Demo Chairs

Vangelis Banos	Aristotle University of Thessaloniki, Greece
Annika Hinze	University of Waikato, New Zealand

Doctoral Consortium Chairs

Maja Žumer   University of Ljubljana, Slovenia
Heiko Schuldt   University of Basel, Switzerland

Program Committee

Senior Program Committee

Trond Aalberg   Norwegian University of Science and Technology,
                         Norway
David Bainbridge   University of Waikato, New Zealand
Tobias Blanke   University of Glasgow, UK
Jose Borbinha   IST/INESC-ID, UK
George Buchanan   City University London, UK
Donatella Castelli   CNR-ISTI, Italy
Stavros Christodoulakis Technical University of Crete, Greece
Milena Dobreva   University of Malta, Malta
Nicola Ferro   University of Padua, Italy
Edward Fox   Virginia Polytechnic Institute and State University, USA
Ingo Frommholz   University of Bedfordshire, UK
Norbert Fuhr   University of Duisburg-Essen, Germany
Richard Furuta   Texas A&M University, USA
Marcos Goncalves   Federal University of Minas Gerais, Brazil
Annika Hinze   University of Waikato, New Zealand
Sarantos Kapidakis   Ionian Uninversity, Greece
Laszlo Kovacs   Hungarian Academy of Sciences, Hungary
Clifford Lynch   CNI, USA
Wolfgang Nejdl   L3S and University of Hanover, Germany
Michael Nelson   Old Dominion University, USA
Erich Neuhold   University of Vienna, Austria
Christos Papatheodorou   Ionian University, Greece
Edie Rasmussen   University of British Columbia, Canada
Andreas Rauber   Vienna University of Technology, Austria
Thomas Risse   L3S Research Center, Germany
Laurent Romary   Inria and HUB-ISDL, France
Seamus Ross   University of Toronto, Canada
Heiko Schuldt   University of Basel, Switzerland
Mário J. Silva   Instituto Superior Técnico, Universidade de Lisboa,
                         Portugal
Hussein Suleman   University of Cape Town, South Africa
Costantino Thanos   ISTI-CNR, Italy
Herbert Van De Sompel   Los Alamos National Laboratory, USA

Program Committee

Hamed Alhoori	Northern Illinois University, USA
Robert Allen	Yonsei University, South Korea
Avishek Anand	L3S Research Center, Germany
Vangelis Banos	Aristotle University of Thessaloniki, Greece
Valentina Bartalesi	ISTI-CNR, Italy
Christoph Becker	University of Toronto, Canada
Maria Bielikova	Slovak University of Technology in Bratislava, Slovakia
Pável Calado	Instituto Superior Técnico, Universidade de Lisboa, Portugal
Vittore Casarosa	ISTI-CNR, Italy
Lillian Cassel	Villanova University, USA
Panos Constantopoulos	Athens University of Economics and Business, Greece
Fabio Crestani	University of Lugano (USI), Switzerland
Sally Jo Cunningham	Waikato University, New Zealand
Theodore Dalamagas	IMIS-"Athena" R.C., Greece
Makx Dekkers	Spain
Giorgio Maria Di Nunzio	University of Padua, Italy
Fabien Duchateau	Université Claude Bernard Lyon 1 – LIRIS, France
Maria Economou	University of Glasgow, UK
Schubert Foo	Nanyang Technological University, Singapore
Nuno Freire	INESC-ID, Portugal
Dimitris Gavrilis	Athena Research Centre, Greece
Manolis Gergatsoulis	Ionian University, Greece
C. Lee Giles	Pennsylvania State University, USA
Julio Gonzalo	UNED, Spain
Paula Goodale	University of Sheffield, UK
Sergiu Gordea	Austrian Institute of Technology, Austria
Stefan Gradmann	KU Leuven, Germany
Jane Greenberg	Drexel University, USA
Mark Michael Hall	Edge Hill University, UK
Bernhard Haslhofer	AIT-Austrian Institute of Technology, Austria
Frank Hopfgartner	University of Glasgow, UK
Nikos Houssos	RedLink, Greece
Antoine Isaac	Europeana Foundation, Belgium
Adam Jatowt	Kyoto University, Japan
Nattiya Kanhabua	Aalborg University, Denmark
Roman Kern	Graz University of Technology, Austria
Claus-Peter Klas	GESIS – Leibniz Institute for Social Sciences, Germany
Martin Klein	Los Alamos National Laboratory, USA
Petr Knoth	KMi, The Open University, UK
Stefanos Kollias	National Technical University of Athens, Greece
Ronald Larsen	University of Pittsburgh, USA
Séamus Lawless	Trinity College Dublin, Ireland

Hyowon Lee	Singapore University of Technology and Design, Singapore
Suzanne Little	Dublin University, Ireland
Zinaida Manžuch	Vilnius University, Lithuania
Bruno Martins	IST – Instituto Superior Técnico, Portugal
Philipp Mayr	GESIS, Germany
Cezary Mazurek	Poznań Supercomputing and Networking Center, Poland
Robert H. Mcdonald	Indiana University/Data to Insight Center, USA
Dana Mckay	Swinburne University of Technology, Australia
Andras Micsik	SZTAKI, Hungary
David Nichols	University of Waikato, New Zealand
Jeppe Nicolaisen	University of Copenhagen, Denmark
Ragnar Nordlie	Oslo and Akershus University College, Norway
Moira Norrie	ETH Zurich, Switzerland
Kjetil Nørvåg	Norwegian University of Science and Technology, Norway
Raul Palma	Poznan Supercomputing and Networking Center, Poland
Nils Pharo	Oslo & Akershus University College of Applied Sciences, Norway
Dimitris Plexousakis	Institute of Computer Science, FORTH, Greece
Panayiota Polydoratou	ATEI of Thessaloniki, Greece
Cristina Ribeiro	University of Porto, Portugal
Ian Ruthven	University of Strathclyde, UK
J. Alfredo Sánchez	UDLAP, Mexico
Michalis Sfakakis	Ionian University, Greece
Gianmaria Silvello	University of Padua, Italy
Nicolas Spyratos	University of Paris South, France
Shigeo Sugimoto	University of Tsukuba, Japan
Tamara Sumner	University of Colorado at Boulder, USA
Anastasios Tombros	Queen Mary University of London, UK
Theodora Tsikrika	Information Technologies Institute, CERTH, Greece
Chrisa Tsinaraki	European Union - Joint Research Center (EU-JRC), Belgium
Douglas Tudhope	University of Glamorgan, UK
Yannis Tzitzikas	University of Crete and FORTH-ICS, Greece
Stefanos Vrochidis	Information Technologies Institute, CERTH, Greece
Michele Weigle	Old Dominion University, USA
Marcin Werla	Poznań Supercomputing and Networking Center, Poland
Iris Xie	University of Wisconsin-Milwaukee, USA
Maja Žumer	University of Ljubljana, Slovenia

Additional Reviewers

Agathos, Michail
Alian Nejadi, Mohammad
Cancellieri, Matteo
Carvalho, André
Chuda, Daniela
Fafalios, Pavlos
Giachanou, Anastasia
Kalogeros, Eleftherios
Kamateri, Eleni
Kanellos, Ilias
Kaššák, Ondrej
Kim, Kunho
Kompan, Michal
Kondylakis, Haridimos
Kotzinos, Dimitris
Körner, Martin
Landoni, Monica
Li, Liuqing

Liu, Lu
Marketakis, Yannis
Medina, María Auxilio
Minadakis, Nikos
Mountantonakis, Michalis
Papachristopoulos, Leonidas
Papadakos, Panagiotis
Pride, David
Rocha Da Silva, João
Rörden, Jan
Santos, Rui
Schlarb, Sven
Srba, Ivan
Tzouramanis, Theodoros
Vergoulis, Thanasis
Williams, Kyle
Wu, Jian
Zhang, Xuan

Sponsors

The Coalition for Networked Information (CNI)

Keynotes

Back to the Future: Annotating, Collaborating and Linking in a Digital Ecosystem

Elton Barker

The Open University, UK

Abstract. Classical philology has rarely been a self-enclosed discipline: in order to interpret Greek and Latin texts, it is necessary to place them in context— grounding them in the histories of the time and exploring them in and against those cultural horizons. Using the linking potential of the Web, Pelagios Commons (http://commons.pelagios.org/) has been pioneering a means of digital 'mutual contextualization', whereby any online document—be it a text, map, database or image—can be connected to another simply by virtue of having something in common with it, and then draw on this external content to enrich its own, or in turn be drawn upon by and enrich another. In Pelagios this linking is achieved through the method of annotating places. From having originally been seeded in collaboration with partners who already curated data and had the technical know-how to align datasets, Pelagios Commons now offers any researcher, librarian, museum curator, student or member of the public a simple, intuitive means to encode place information in a document of their choosing.

This presentation will set out and explain this annotation process in the Web-based, Open Source platform, Recogito (http://recogito.pelagios.org/) developed by the Pelagios team. It will go through the steps that the researcher would take in order to geoannotate their material—first identifying the place entity in their document, then resolving that information to a central authority file: i.e. a gazetteer of placenames (e.g. http://pleiades.stoa.org/). It also considers the potential uses of this kind of semantic annotation, outlining the mapping of places in texts, the repurposing of the data in other systems (such as GIS), and the linking to other related resources. Throughout, however, it will be concerned to identify challenges and persistent issues that are not only related to the technical development and use; using Recogito puts a primary demand on defining and conceptualising place. Thus, contrary to much current thinking, this presentation hopes to show how digital tools can enhance the close reading of texts and facilitate a more nuanced understanding of the status and role of places in our historical sources.

Elton Barker is Reader in Classical Studies, having joined The Open University as a Lecturer in July 2009. Before then, he had been a Tutor and Lecturer at Christ Church, Oxford (2004-09), and also lectured at Bristol, Nottingham and Reading. He has been a Junior Research Fellowship at Wolfson College, Cambridge (2002-04) and a Visiting Fellow at Venice International University (2003-04). From 2012-2013 he had a Research Fellowship for Experienced Researchers awarded by the Alexander von Humboldt Foundation for research at the Freie Universität Berlin and the University of Leipzig. He has been awarded a Graduate Teaching Award from Pembroke College

(Cambridge) and twice won awards from the University of Oxford for an Outstanding Contribution to Teaching.

His research interests cross generic and disciplinary boundaries. Since 2008, he has been leading and co-running a series of collaborative projects, which are using digital resources to rethink spatial understanding of the ancient world. The Hestia project investigates the underlying ways in which Herodotus constructs space in book 5 of his Histories. Meanwhile, the Pelagios project has been establishing the Web infrastructure by which data produced and curated by different content providers – from academic projects like the Perseus Classical Library to cultural heritage institutions like the British Museum – can be linked through their common references to places.

Machines are People Too

Paul Groth

Elsevier Labs, Elsevier Inc., USA

Abstract. The theory and practice of digital libraries provides a long history of thought around how to manage knowledge ranging from collection development, to cataloging and resource description. These tools were all designed to make knowledge findable and accessible to people. Even technical progress in information retrieval and question answering are all targeted to helping answer a human's information need.

However, increasingly demand is for data. Data that is needed not for people's consumption but to drive machines. As an example of this demand, there has been explosive growth in job openings for Data Engineers – professionals who prepare data for machine consumption. In this talk, I overview the information needs of machine intelligence and ask the question: Are our knowledge management techniques applicable for serving this new consumer?

Paul Groth is Disruptive Technology Director at Elsevier Labs. He holds a Ph.D. in Computer Science from the University of Southampton (2007) and has done research at the University of Southern California and the Vrije Universiteit Amsterdam. His research focuses on dealing with large amounts of diverse contextualized knowledge with a particular focus on the web and science applications. This includes research in data provenance, data science, data integration and knowledge sharing. He leads architecture development for the Open PHACTS drug discovery data integration platform. Paul was co-chair of the W3C Provenance Working Group that created a standard for provenance interchange. He is co-author of "Provenance: An Introduction to PROV" and "The Semantic Web Primer: 3rd Edition" as well as numerous academic articles. He blogs at http://thinklinks.wordpress.com. You can find him on twitter: @pgroth.

Visualization in the Big Data Era: Data Mining from Networked Information

Dimitrios Tzovaras

Information Technologies Institute, Centre for Research and Technology, Greece

Abstract. Network graphs have long formed a widely adapted and acknowledged practice for the representation of inter- and intra-dependent information streams. Nowadays, they are largely attracting the interest of the research community mainly due to the vastly growing amount (size & complexity) of semantically dependent data produced world-wide as a result of the rapid expansion of data sources.

In this context, the efficient processing of the big amounts of information, also known as Big Data forms a major challenge for both the research community and a wide variety of industrial sectors, involving security, health and financial applications.

In order to address these needs the current presentation describes a proprietary platform built upon state-of-the-art algorithms that are combined to implement a top-down approach for the facilitation of Data & Graph Mining processes, like behavioral clustering, interactive visualizations, etc.

The applicability of this platform has been validated on α series of distinct real-world use cases that involve large amounts of intra-exchanged information and can be thus help as characteristic examples of modern Big Data problems. In particular, they refer to (i) DoS attacks in a real-world mobile networks and (ii) early event detection in social media communities, (iii) traffic management and (iv) DNA sequences analysis.

In all these cases, the large volumes of data are addressed via a Data Minimization approach that starts with an aggregated overview of network at its whole, and gradually the focus is put on smaller data subsets (i.e. approach upon successive levels of abstraction). In parallel, insights on the network's operations are allowed through the detection of behavioral patterns. Similarly, a dynamic hypothesis formulator and the corresponding backend solver can subsequently be exploited through graph traversing and pattern mining. This way, an analyst is provided with the appropriate equipment to set and verify concrete hypotheses through simulation and extract useful conclusions.

Dr. Dimitrios Tzovaras is a Senior Researcher Grade A' (Professor) and Director at CERTH/ITI (the Information Technologies Institute of the Centre for Research and Technology Hellas). He received the Diploma in Electrical Engineering and the Ph.D. in 2D and 3D Image Compression from the Aristotle University of Thessaloniki, Greece in 1992 and 1997, respectively. Prior to his current position, he was a Senior Researcher on the Information Processing Laboratory at the Electrical and Computer Engineering Department of the Aristotle University of Thessaloniki. His main research interests include network and visual analytics for network security, computer security, data fusion, biometric security, virtual reality, machine learning and artificial intelligence. He is author or co-author of over 110 articles in refereed journals and over 300 papers in international conferences.

Since 2004, he has been Associate Editor in the following International journals: Journal of Applied Signal Processing (JASP) and Journal on Advances in Multimedia of EURASIP. Additionally, he is Associate Editor in the IEEE Signal Processing Letters journal (since 2009) and Senior Associate Editor in the IEEE Signal Processing Letters journal (since 2012), while since mid-2012 he has been also Associate Editor in the IEEE Transactions on Image Processing journal. Over the same period, Dr. Tzovaras acted as ad hoc reviewer for a large number of International Journals and Magazines such as IEEE, ACM, Elsevier and EURASIP, as well as International Scientific Conferences (ICIP, EUSIPCO, CVPR, etc.).

Since 1992, Dr. Tzovaras has been involved in more than 100 European projects, funded by the EC and the Greek Ministry of Research and Technology. Within these research projects, he has acted as the Scientific Responsible of the research group of CERTH/ITI, but also as the Coordinator and/or the Technical/Scientific Manager of many of them (coordinator of technical manager in 21 projects – 10 H2020, 1 FP7 ICT IP, 7 FP7 ICT STREP, 3 FP6 IST STREP and 1 Nationally funded project).

Contents

Linked Data

Exploiting Interlinked Research Metadata. 3
    Shirin Ameri, Sahar Vahdati, and Christoph Lange

Preserving Bibliographic Relationships in Mappings from FRBR
to BIBFRAME 2.0 15
    Sofia Zapounidou, Michalis Sfakakis, and Christos Papatheodorou

Exploring Ontology-Enhanced Bibliography Databases
Using Faceted Search 27
    Tadeusz Pankowski

What Should I Cite? Cross-Collection Reference Recommendation
of Patents and Papers 40
    Julian Risch and Ralf Krestel

Corpora

Taxonomic Corpus-Based Concept Summary Generation
for Document Annotation. 49
    *Ikechukwu Nkisi-Orji, Nirmalie Wiratunga, Kit-Ying Hui,
    Rachel Heaven, and Stewart Massie*

RussianFlu-DE: A German Corpus for a Historical Epidemic
with Temporal Annotation 61
    Tran Van Canh, Katja Markert, and Wolfgang Nejdl

A Digital Repository for Physical Samples: Concepts, Solutions
and Management. 74
    *Anusuriya Devaraju, Jens Klump, Victor Tey, Ryan Fraser, Simon Cox,
    and Lesley Wyborn*

Facet Embeddings for Explorative Analytics in Digital Libraries 86
    *Sepideh Mesbah, Kyriakos Fragkeskos, Christoph Lofi,
    Alessandro Bozzon, and Geert-Jan Houben*

Data in Digital Libraries

Automatic Hierarchical Categorization of Research Expertise
Using Minimum Information 103
   Gustavo Oliveira de Siqueira, Sérgio Canuto, Marcos André Gonçalves,
   and Alberto H.F. Laender

Extracting Event-Centric Document Collections from Large-Scale
Web Archives.. 116
   Gerhard Gossen, Elena Demidova, and Thomas Risse

Information Governance Maturity Model Final Development Iteration 128
   Diogo Proença, Ricardo Vieira, and José Borbinha

Challenges of Research Data Management for High
Performance Computing... 140
   Björn Schembera and Thomas Bönisch

Quality in Digital Libraries

How Linked Data can Aid Machine Learning-Based Tasks 155
   Michalis Mountantonakis and Yannis Tzitzikas

Can Plausibility Help to Support High Quality Content
in Digital Libraries? 169
   José María González Pinto and Wolf-Tilo Balke

Classifying Document Types to Enhance Search and Recommendations
in Digital Libraries 181
   Aristotelis Charalampous and Petr Knoth

Understanding the Influence of Hyperparameters on Text Embeddings
for Text Classification Tasks 193
   Nils Witt and Christin Seifert

Digital Humanities

Europeana: What Users Search for and Why 207
   Paul Clough, Timothy Hill, Monica Lestari Paramita,
   and Paula Goodale

Metadata Aggregation: Assessing the Application of IIIF and Sitemaps
Within Cultural Heritage 220
   Nuno Freire, Glen Robson, John B. Howard, Hugo Manguinhas,
   and Antoine Isaac

A Decade of Evaluating Europeana - Constructs, Contexts,
Methods & Criteria   233
  Vivien Petras and Juliane Stiller

On the Uses of Word Sense Change for Research
in the Digital Humanities   246
  Nina Tahmasebi and Thomas Risse

Entities

Multi-aspect Entity-Centric Analysis of Big Social Media Archives   261
  Pavlos Fafalios, Vasileios Iosifidis, Kostas Stefanidis, and Eirini Ntoutsi

A Comparative Study of Language Modeling to Instance-Based Methods,
and Feature Combinations for Authorship Attribution   274
  Olga Fourkioti, Symeon Symeonidis, and Avi Arampatzis

What Others Say About This Work? Scalable Extraction of Citation
Contexts from Research Papers   287
  Petr Knoth, Phil Gooch, and Kris Jack

Semantic Author Name Disambiguation with Word Embeddings   300
  Mark-Christoph Müller

Scholarly Communication

Towards a Knowledge Graph Representing Research Findings
by Semantifying Survey Articles   315
  Said Fathalla, Sahar Vahdati, Sören Auer, and Christoph Lange

Integration of Scholarly Communication Metadata
Using Knowledge Graphs   328
  Afshin Sadeghi, Christoph Lange, Maria-Esther Vidal, and Sören Auer

Analysing Scholarly Communication Metadata of Computer
Science Events   342
  Said Fathalla, Sahar Vahdati, Christoph Lange, and Sören Auer

High-Pass Text Filtering for Citation Matching   355
  *Yannis Foufoulas, Lefteris Stamatogiannakis, Harry Dimitropoulos,
  and Yannis Ioannidis*

Sentiment Analysis

Sentiment Classification over Opinionated Data Streams Through
Informed Model Adaptation   369
  Vasileios Iosifidis, Annina Oelschlager, and Eirini Ntoutsi

Mining Semantic Patterns for Sentiment Analysis of Product Reviews   382
  Sang-Sang Tan and Jin-Cheon Na

A Comparison of Pre-processing Techniques for Twitter
Sentiment Analysis   394
  Dimitrios Effrosynidis, Symeon Symeonidis, and Avi Arampatzis

Employing Twitter Hashtags and Linked Data to Suggest Trending
Resources in a Digital Library   407
  *Ioannis Papadakis, Konstantinos Kyprianos, Apostolos Karalis,
  and Christos Douligeris*

Information Behavior

Social Tagging: Implications from Studying User Behavior
and Institutional Practice   421
  Õnne Mets and Jaagup Kippar

The Ghost in the Museum Website: Investigating the General Public's
Interactions with Museum Websites.   434
  David Walsh, Mark Hall, Paul Clough, and Jonathan Foster

Evaluating the Usefulness of Visual Features for Supporting
Document Triage   446
  Dagmar Kern, Maria Lusky, and Dirk Wacker

Building User Groups Based on a Structural Representation
of User Search Sessions.   459
  Wilko van Hoek and Zeljko Carevic

Information Retrieval

Multiple Random Walks for Personalized Ranking
with Trust and Distrust   473
  Dimitrios Rafailidis and Fabio Crestani

Plagiarism Detection Based on Citing Sentences   485
  Sidik Soleman and Atsushi Fujii

Lexicon Induction for Interpretable Text Classification.   498
  Jérémie Clos and Nirmalie Wiratunga

The Clustering-Based Initialization for Non-negative Matrix Factorization
in the Feature Transformation of the High-Dimensional Text Categorization
System: A Viewpoint of Term Vectors   511
    Le Nguyen Hoai Nam and Ho Bao Quoc

Short Paper

Analysis of Interactive Multimedia Features in Scientific
Publication Platforms.   525
    Camila Wohlmuth da Silva and Nuno Correia

Extending R2RML with Support for RDF Collections and Containers
to Generate MADS-RDF Datasets.   531
    Christophe Debruyne, Lucy McKenna, and Declan O'Sullivan

Building the Brazilian Academic Genealogy Tree   537
    Wellington Dores, Elias Soares, Fabrício Benevenuto,
    and Alberto H.F. Laender

When a Metadata Provider Task Is Successful   544
    Sarantos Kapidakis

Semantic Enrichment of Web Query Interfaces to Enable Dynamic
Deep Linking to Web Information Portals   553
    Arne Martin Klemenz and Klaus Tochtermann

A Complete Year of User Retrieval Sessions in a Social Sciences
Academic Search Engine.   560
    Philipp Mayr and Ameni Kacem

Social Dendro: Social Network Techniques Applied to Research
Data Description.   566
    Nelson Pereira, João Rocha da Silva, and Cristina Ribeiro

Incidental or Influential? - Challenges in Automatically Detecting
Citation Importance Using Publication Full Texts   572
    David Pride and Petr Knoth

User Interactions with Bibliographic Information Visualizations   579
    Athena Salaba and Tanja Merčun

Towards Building Knowledge Resources from Social Media
Using Semantic Roles   585
    Diana Trandabăţ

Poster and Demonstration Paper

Towards Finding Animal Replacement Methods   595
  Nadine Dulisch and Brigitte Mathiak

Environmental Monitoring of Libraries with MonTreAL   599
  Marcel Großmann, Steffen Illig, and Cornelius Matějka

Introducing Solon: A Semantic Platform for Managing Legal Sources   603
  *Marios Koniaris, George Papastefanatos, Marios Meimaris,
  and Giorgos Alexiou*

Towards a Semantic Search Engine for Scientific Articles   608
  *Bastien Latard, Jonathan Weber, Germain Forestier,
  and Michel Hassenforder*

Development of an RDF-Enabled Cataloguing Tool   612
  *Lucy McKenna, Marta Bustillo, Tim Keefe, Christophe Debruyne,
  and Declan O'Sullivan*

Towards Semantic Quality Control of Automatic Subject Indexing   616
  Martin Toepfer and Christin Seifert

Doctoral Consortium Paper

Research Data in Scholarly Practices: Observations of an Interdisciplinary
Horizon2020 Project   623
  Madeleine Dutoit

Research Data in Norway: How Do Expectations, Demands and Solutions
Correspond in the Knowledge Infrastructure for Research Data?   628
  Live Kvale

Top-Down and Bottom-up Approaches to Identify the Users,
the Services and the Interface of a 2.0 Digital Library   632
  Elina Leblanc

Cross-Language Record Linkage Across Humanities Collections
Using Metadata Similarities Among Languages.   640
  Yuting Song

Machine Learning Architectures for Scalable and Reliable Subject
Indexing: Fusion, Knowledge Transfer, and Confidence.   644
  Martin Toepfer

Explaining Pairwise Relationships Between Documents   648
    Nils Witt

Studying Conceptual Models for Publishing Library Data
to the Semantic Web   652
    Sofia Zapounidou

Tutorials

Putting Historical Data in Context: How to Use DSpace-GLAM   659
    Andrea Bollini and Claudio Cortese

Innovation Search   661
    Michail Salampasis

Enabling Precise Identification and Citability of Dynamic Data:
Recommendations of the RDA Working Group on Data Citation   663
    Andreas Rauber

Enriching Digital Collections Using Tools for Text Mining, Indexing
and Visualization   665
    Riza Batista-Navarro, Axel J. Soto, Nhung T.H. Nguyen,
    William Ulate, and Sophia Ananiadou

Workshops

NKOS 2017 – 17th European Networked Knowledge Organization
Systems Workshop   669
    Philipp Mayr, Douglas Tudhope, Koraljka Golub, Christian Wartena,
    and Ernesto William De Luca

MDQual – (Meta)-Data Quality Workshop   671
    Dimitris Gavrilis and Christos Papatheodorou

TDDL 2017 – 1st International Workshop on Temporal Dynamics
in Digital Libraries   673
    Annalina Caputo, Nattiya Kanhabua, Pierpaolo Basile,
    and Séamus Lawless

FUTURITY 2017 – Workshop on Modeling Societal Future   675
    Daniela Gîfu and Diana Trandabăţ

Author Index   677

Linked Data

Exploiting Interlinked Research Metadata

Shirin Ameri[1], Sahar Vahdati[1(✉)], and Christoph Lange[1,2]

[1] Smart Data Analytics (SDA), University of Bonn, Bonn, Germany
{ameri,vahdati,langec}@cs.uni-bonn.de
[2] Fraunhofer, Intelligent Analysis and Information Systems (IAIS),
Sankt Augustin, Germany

Abstract. OpenAIRE, the Open Access Infrastructure for Research in Europe, aggregates metadata about research (projects, publications, people, organizations, etc.) into a central Information Space. OpenAIRE aims at increasing interoperability and reusability of this data collection by exposing it as Linked Open Data (LOD). By following the LOD principles, it is now possible to further increase interoperability and reusability by connecting the OpenAIRE LOD to other datasets about projects, publications, people and organizations. Doing so required us to identify link discovery tools that perform well, as well as candidate datasets that provide comprehensive scholarly communication metadata, and then to specify linking rules. We demonstrate the added value that interlinking provides for end users by implementing visual frontends for looking up publications to cite, and publication statistics, and evaluating their usability on top of interlinked vs. non-interlinked data.

Keywords: Interlinking · Linked open data · Research metadata · Scholarly communication · Semantic publishing

1 Introduction

Linked Open Data (LOD) is a popular approach for maximizing both legal and technical reusability of data, and enabling its connection with further datasets [2]. However, without further work, LOD datasets do not yet provide added value to end users, as they are only accessible for service and application developers familiar with Semantic Web technology and the datasets' vocabularies.

OpenAIRE (OA), the Open Access Infrastructure for Research in Europe [9], aggregates metadata about research (projects, publications, people, organizations, etc.) into a central Information Space. It so far covers more than 13 M publications, 12 M authors and scientific datasets. OA metadata has been exposed as LOD [14], aiming at maximizing its reusability and technical interoperability by:

- providing an infrastructure for data access, retrieval and citation (e.g., a SPARQL endpoint or a LOD API),
- interlinking with popular LOD datasets and services (DBLP, ACM, CiteSeer, DBpedia, etc.),

© Springer International Publishing AG 2017
J. Kamps et al. (Eds.): TPDL 2017, LNCS 10450, pp. 3–14, 2017.
DOI: 10.1007/978-3-319-67008-9_1

– enriching the OpenAIRE Information Space with further information from other LOD datasets.

This work focuses on enriching the OpenAIRE LOD by interlinking, and utilizing this interlinked data to provide added value to users in situations where they need scholarly communication metadata, e.g., when they are looking for a publication to cite, or for all publications of a given author.

2   Related Work

Rajabi has studied the exploitation of educational metadata using interlinking methods [8]. His work objectives closely related to ours; however its application domain is eLearning services and therefore he discusses the benefits of interlinking educational (meta)data in practice. Rajabi et al. provide a comparison of interlinking tools as well as interlinking rules [7] and a method for identification of duplicate links [6]. Hallo et al. follow the same objective as we do, i.e., publishing Open Access metadata as LOD [3]. Their work focuses on providing better search services on top of open journal datasets, but their data could be used as a candidate dataset for our interlinking. Recent work by Purohit et al. addresses the problem of scholarly resource discovery [5]. They also reviewed tools providing such services and present a framework for Resource Discovery for Extreme Scale Collaboration (RDESC)[1] which has common objectives with OA. However, they have not yet initiated interlinking of research metadata and the provision of a comprehensive knowledge graph.

3   Background: OpenAIRE LOD Services

The main motivation for exposing OA as LOD is to provide wider data access, and easier and broader metadata retrieval by enabling interlinking with relevant and popular LOD datasets [14]. Metadata about different types of entities – research results (publications and datasets), persons, projects and organizations – that the OA infrastructure aggregates is being exposed as LOD. OA LOD uses terms from existing vocabularies and, where necessary, defines new terms. Existing ontologies reused include SKOS, CERIF, DCMI Terms, FOAF [14,15]. Two prefixes/namespaces are OA specific: oav: http://lod.openaire.eu/vocab/ for the OA vocabulary, and oad: http://lod.openaire.eu/data/ for OA instance data.

The data has been exposed in three ways: (1) small fragments of RDF, accessible by dereferencing the URI that identifies a particular entity, (2) a downloadable all-in-one dump[2], and (3) a SPARQL endpoint, i.e. a standardized query interface accessible over the Web[3].

It is envisaged to extend the OA LOD by enriching and interlinking it with the following types of data:

---

[1] https://tw.rpi.edu/web/project/RDESC.
[2] http://tinyurl.com/OALOD.
[3] http://lod.openaire.eu/sparql.

– data that has not (yet) been collected by OA's existing mechanisms, e.g., certain types of persistent identifiers of publications or people (e.g., ORCID),
– data that is expensive to collect and/or not included in the OA data model, e.g., data about scientific events, and
– data that is related to open research but out of the scope of the OA infrastructure itself and therefore not targeted to be ever collected, e.g., biographies of persons, or geodata about the locations of organizations.

The primary objectives are (1) providing added value to users, by enabling those who develop user-oriented applications and services to access a richer collection of relevant data than just OA's own, and (2) facilitating internal data management, e.g., by aiding the resolution of duplicates resulting from metadata being harvested from different repositories by linking to external reference points.

4   Interlinking

Interlinking the OA LOD with other LOD datasets required us to do the following preparatory work: (1) analyzing the OA metadata schema to find appropriate entity types and properties on which to interlink, (2) identifying candidate target datasets, and, (3) among existing link discovery tools, finding the one most appropriate for our purpose, before we could finally implement interlinking rules (Fig. 1).

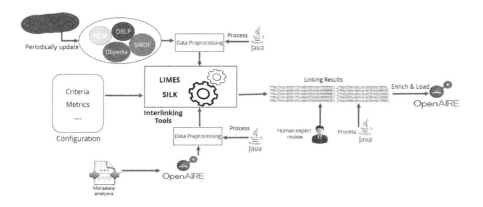

Fig. 1. Interlinking process

4.1   Identifying Properties Suitable for Interlinking

Not all properties of an OA entity are suitable for the purpose of interlinking to other entities, as Rajabi et al. have investigated in the related domain of metadata about educational resources [7]. Following their method, we analyzed all OpenAIRE entities and their properties to discover linkable elements. We filtered out properties that potentially cannot be linked due to their specific values, for example Booleans (Yes/No), format values (PDF, JPEG), or language codes (en, de), and properties whose meaning is local to some source repository according to its policy, for example local identifiers or version numbers. This left us with properties such as 'publication title' and 'author name', 'published year', 'description', 'subject', etc., which have string or integer values. Where initial interlinking tests yielded subjectively satisfactory results, we chose the respective properties for interlinking – i.e. the following:

- **Title** and **Digital Object Identifier** of **Publication**,
- **Full name**, **First name** or **Last name** of **Person**s, and
- **Label** or **Homepage** of **Organization**s.

4.2   Investigating Existing Interlinking Tools

There exist a number of tools for creating semi-automatic links between datasets by running some matching techniques. These linking tools identify similarities between entities and generate links (e.g.owl:sameAs) that connect source and target entities. Rajabi et al. conducted a study that suggests that data publishers can trust interlinking tools to interlink their data to other datasets; accordingly, LIMES and Silk are the most promising frameworks [7]. Simperl et al. have compared various linking tools by addressing aspects such as required input, resulting output, considered domain and matching techniques used [11]. This allowed for a comparison from several perspectives: degree of automation (to what extent the tool needs human input) and human contribution (the way in which users are required to do the interlinking.

In summary, these comparisons point out the two well-known open source interlinking frameworks that we also used: LIMES[4] (Link Discovery Framework for Metric Spaces) and Silk[5] (Link Discovery Framework for the Web of Data). In an evaluation of the two frameworks, the LIMES developers showed that LIMES considerably outperforms Silk in terms of running time, with a comparable quality of the output. Moreover, LIMES can be downloaded as a standalone tool for carrying out link discovery locally and consists of modules that can be extended easily to accommodate new or improved functionality.

Our comparative evaluation of Silk and LIMES, which finally made us choose LIMES based on the quality of the output, is presented in Sect. 6.1.

4.3   Identifying Interlinking Target Datasets

To identify appropriate target datasets to be interlinked with OA, we examined several datasets from the LOD Cloud, in the following steps:

---

[4] http://aksw.org/Projects/LIMES.html.
[5] http://silkframework.org/.

1. **Identifying publication-related datasets in DataHub:** our aim is to find datasets tagged with the same domain as that of OA or a related one. We therefore searched the DataHub portal[6] for datasets tagged with 'publication' or related domains. This search yielded more than 900 datasets.
2. **Checking data endpoint availability:** we filtered the datasets identified previously by checking their SPARQL endpoints' or RDF dumps' availability.
3. **Retrieving datasets specification:** of the remaining datasets (still more than 60), we next retrieved each dataset's specification (size, metadata schema, etc.). From an interlinking point of view, we considered data volume, frequent updates, and matches with the entity types and properties identified previously (Sect. 4.1) as the most important characteristics of a dataset. Moreover, we considered available links to other related datasets desirable.

Table 1 lists the ten most relevant datasets according to these criteria.

Table 1. List of candidate Datasets

Datasets	Size	Endpoint	Dump	Covered OA entity types
DBpedia	1 B	Available	NT	Person, Organization
DBLP	55 M	–	NT	Publication, Person
ACM	12 M	Available	RDF/XML	Publication, Person
CiteSeer	8 M	Available	RDF/XML	Publication, Person
BibBase	200 K	–	RDF/XML	Person, Publication, Organization
IEEE	200 K	Available	RDF/XML	Publication, Person
OpenCitations	3 M	Available	JSON-LD	Person, Publication, Organization
SWDF	242 K	–	RDF/XML	Person, Publication, Organization
BNB	109 M	–	NT, RDF/XML	Person, Publication
COLINDA	149 K	Available	RDF/XML	Publication
GeoNames	93 M	–	RDF/XML	Organization

4.4 Identifying String Matching Algorithms

One of the most important factors in discovering links effectively is choosing the right string matching algorithm. The results of our heuristic experiments shows that both tools supports string matching according to trigrams, Levenshtein [7], Jaro, Jaro-Winkler and cosine (all of them normalized); cf. Table 2. It shows detailed definition of the algorithms. In our initial experiments, Jaro and Levenshtein proved most reliable for identifying equivalent names and titles. Thus, we chose Levenshtein for long string values, i.e., publication titles, and Jaro for short string values, i.e., person names. An example of a metric definition in LIMES is shown below.

---

[6] https://datahub.io/.
[7] https://wikipedia.org/Levenshtein_distance.

Table 2. String matching algorithms

Metric	Description
Trigrams	uses the number of matching triples in both strings as $s = 2 \times \frac{m}{(a \times b)}$ where m is the number of matching trigrams, a is the number of trigrams in string 1, and b is the number of trigrams in string 2 [10]
Levenshtein	is based on the minimum number of insertion, deletion or replacement operations required to transform string 1 into string 2
Jaro	is a measure of characters in common, being no more than half the length of the longer string in distance, with consideration for transpositions; it is best suited for short strings such as person names [12]
Jaro-Winkler	is an optimized version of Jaro designed and best suited for short strings such as person names
Cosine	is the cosine of the angle between string vectors; for equal strings the angle between them will be 0 and the cosine will be 1 [10]

```
<METRIC>
  AND(Jaro(x.foaf:name, y.foaf:name)|0.8, Levenshtein(
  x.dcterms:creator/cerif:name, ^y.dblp:hasAuthor/dblp:title)|0.8)
</METRIC>
```

5   Use Cases

The main objective of OA LOD is to achieve maximum re-usability of OA data for developers of third-party applications and services [14]. Such applications and services may include statistical analyses beyond those in the scope of OA itself, efforts aggregating OpenAIRE and other data such as research data, or tools that support scientific writing and communication, e.g. online collaborative editors. To this end, we aimed to exploit the interlinked metadata of OA LOD in plugins for online collaborative editors to provide recommendations for authors of scientific papers. In the remainder of this section, two example scenarios are discussed in more detail to demonstrate our approach.

5.1   Look-Up Publications to Cite

The process of generating citations is too time consuming using state-of-the-art editors such as Fidus Writer[8]. Citations are created manually either by entering metadata such as author names, publication titles, etc., or copied from an existing BibTeX snippet. An application plugin to simplify the frustrating citing process can support researchers by instantly generating all required and possible citations.

---

[8] https://www.fiduswriter.org.

We implemented this plugin as a modal dialog window (jQuery/UI) [4]. Consider the following example scenario: *Suppose a researcher wants to* ***cite*** *a publication. He cannot remember the full information of that publication but just its partial title, which contains:* ***'opencourseware observatory'.*** Our implementation supports this in the following steps: (1) the user can select the desired type of research output (Publication or Dataset) from a drop-down menu (Fig. 2A), (2) the user can perform a search based on different attributes, e.g., publication title, author name or publication year (Fig. 2B), (3) the user specifies the selection of the corresponding text, i.e., here, 'opencourseware observatory', in the search field (Fig. 2C). (4) From the results suggested, the user selects the desired one to insert into the text (Fig. 2D).

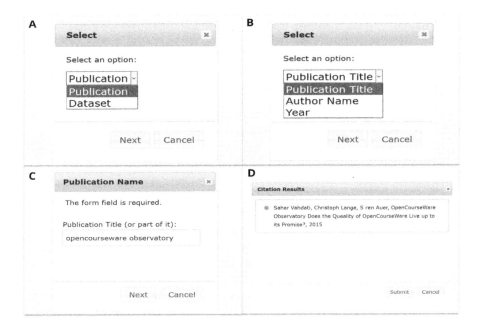

Fig. 2. Looking up Publications to Cite

5.2 Look-Up Author and Statistics

For researchers and publishers, it is important to find publications, authors, journals or conferences related to their research area. However, most of the time it is difficult to find this information in the enormous amount of data on the Web [13]. When users run multiple queries over the most popular data sources for their research fields, the results will not be connected with each other. Thus, our motivation is to develop a plugin that not only retrieves and visualizes data from the OA dataset, but also finds and displays related objects that may be of interest to the user, obtained from interlinking with information from various other online resources, such as DBLP. Furthermore, we explore possibilities for presenting related data in a useful manner (e.g., using statistical analysis); cf. [13]. This plugin provides the following features:

- Perform a search based on author name
- Retrieve and visualize the author's information obtained from OA dataset
- Find further information by following links from a search result to other datasets, e.g., DBLP
- Display statistics for a certain type of information, e.g., an author's number of publications per year, or co-author relationships.

We implemented a modal search dialog, which enables users to run keyword searches (Fig. 3A). By forming the query with a part of an author name and selecting the desired person (Fig. 3B), our plugin yields the following results (Fig. 3C):

- list of publications and year of publication for each author
- list of co-authors
- statistical graphs based on the above results

Moreover, we utilize links to external datasets such as DBLP, SWDF, and enrich our result with information from those datasets (Fig. 3D).[9]

6   Evaluation

6.1   Evaluation of Interlinking Tools

To find the common and individual links created by selected interlinking tools, we wrote a script [1, Appendix C], which compares the contents of results obtained by two tools and returns the number of common links and also the number of links found by one tool but not by the other. In an experiment with considering publications of OA data and publications of DBLP data LIMES was able to match 432 entities, i.e. more than Silk. The number of common records discovered by both Silk and LIMES is 358. 74 links were found by LIMES but not by Silk, and 3 links were found by Silk but not by LIMES.

In addition to the number of discovered links, reliability of the obtained links is also important. Thus, to evaluate the quality and reliability of the links obtained via each tool, we created a reference linkset (gold standard) consisting of 100 publication resource selected from OA and by manual research found 38 links to SWDF. We then ran Silk and LIMES to find only links from these 100 selected OA resources to SWDF and then compared their output to the gold standard. We computed precision, recall and F-measure to check completeness and correctness of the links found; Table 3 shows the results. Precision is the ratio

---

[9] Note that the encoding problem ('Sören Auer' in OA instead of 'Sören Auer' in DBLP) stems from the OA data.

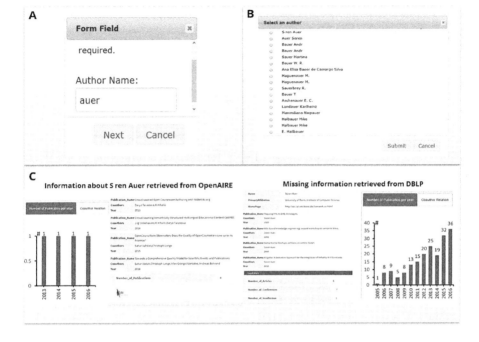

Fig. 3. Author lookup feature's process

of the number of relevant items to the number of retrieved items, i.e.: Precision

$$= \frac{\text{true positive}}{\text{true positive} + \text{false positive}}.$$ In our case, this means

$$\text{Precision} = \frac{(\text{Number of created links} - \text{Number of incorrect links})}{\text{Number of created links}}$$ and indi-

cates the correctness of links discovered. Recall is the ratio of the number of retrieved relevant items to the number of relevant items, i.e.:

$$\text{Recall} = \frac{\text{true positive}}{\text{true positive} + \text{false negative}}.$$ In our case, this means

$$\text{Recall} = \frac{(\text{Number of created links} - \text{Number of incorrect links})}{(\text{Number of correct links} + \text{Number of missing links})}$$ and indi-

cates the completeness of links discovered. F-measure is a combined measure of accuracy defined as the harmonic mean of precision and recall, i.e. $F1 = \frac{2 * precision * recall}{precision + recall}$.

The evaluation revealed 9 missing links and one incorrectly discovered link in Silk and 1 missing in LIMES. This corresponded to a Precision of 1, a Recall of 0.97 and an F-measure of 0.98 for LIMES and a Precision of 0.96, a Recall of 0.76 and an F-measure of 0.84 for Silk. The main advantage for LIMES within this small evaluation is the execution time. However we consider the best practices so far which showed that LIMES outperforms Silk dealing with big data. Therefore, due to the fact that we got more relevant, reliable and accurate results from

Table 3. Evaluation of interlinking tools result against a gold standard

Tool	Number of created links	Number of missing links	Number of incorrect discovered links	Precision	Recall	F-measure
LIMES	37	1	0	1	0.97	0.98
Silk	29	9	1	0.96	0.76	0.85

LIMES compared to Silk, we chose LIMES for further interlinking OpenAIRE with other datasets.

6.2   Evaluation of Interlinking Results

We configured LIMES to generate owl:sameAs links between resources with a similarity of above 95%. However, the question is to what extent resources linked in this way are actually the same. Given the size of the linkset, manually assessing and analyzing each link would have been too time-consuming. We therefore picked a number of sample links from each linkset based on its size, aiming at feasibility of a manual inspection (150 samples of publication links, 200 samples of person links and 25 samples of organization links). We then manually verified the correctness of each link and computed precision as 'number of correct links'/'number of sample links'. In the absence of a gold standard, we did not compute recall.

Table 4. Number of inter-links and precision values obtained between OA and DBLP, SWDF, ACM and DBpedia for publications, persons and organizations.

Links between	Target dataset	Target instances	Generated links	Sample of generated links	Verified links	Precision
Publication	DBLP	164890	2276	150	147	0.98
Publication	SWDF	5009	432	150	150	1.0
Publication	ACM	10378	1082	150	136	0.9
Person	SWDF	11184	2000	200	180	0.9
Person	DBLP	932000	6852	200	111	0.55
Person	DBpedia	23373	1088	200	80	0.40
Organization	SWDF	3212	866	30	30	1.0
Organization	DBpedia	3472	38	30	30	1.0

The number of links obtained between OA and DBLP, SWDF, ACM and DBpedia for publications, persons and organizations is displayed in Table 4 along with the precision for each linkset. We obtained high precision in Publication and Organization interlinking, but not in Person interlinking. This is because initially we carried out Person interlinking by just comparing the names, which was not sufficient, as different persons may have the same name. In future work, we should improve the linking rule for persons taking into account not only their names but also the titles of their publications.

6.3   Usability and Usefulness of Services

We used a custom survey to measure the usability and usefulness of the implemented services discussed in subsection 5.1 (for full details see [1]). 8 participants were first introduced to the idea and the services. We asked them to use the services and figure out the answer of 10 pre-defined questions. Finally, two questionnaires, one for usability and the other for usefulness (10 questions each) were handed out to be filled by them. The questionnaires were designed using System Usability Scale[10]. The results show that most of the participants agreed that our applications are very useful in terms of supporting authors and publishers as well as easy to use an easy to learn. Two of them indicated they needed to learn a bit in the beginning on how the system works. Half of the participants were confident using system and they found it easy to explore. They also mentioned, they would recommend it to experts and use it frequently. Overall, usability of the services is scored as 76.56%.

Satisfaction of the users on usefulness was much higher. Author look up and citation services are selected as a highly useful feature to assist researchers. Three participants were experts of SPARQL queries, however the rest asked for a bit more use-friendly interface both for querying and result representation. 5 of the participants scored the system as easy to use.

7   Conclusion and Future Work

We have presented an approach for interlinking the OpenAIRE research metadata with related Linked Open Datasets, and tools that exploit these new connections to the benefit of end users. After identifying appropriate elements for interlinking, selecting candidate datasets and comparing interlinking tools, we applied the LIMES tool to interlink OpenAIRE concepts to four datasets providing related information (DBLP, DBpedia, ACM, SWDF) and evaluated the precision of the results. We achieved high precision for publications and organizations, whereas the interlinking of persons requires further improvement. Aiming at enhancing the reusability of the interlinked OpenAIRE LOD, we implemented two plugins to assist researchers: a citation lookup service and a tool that looks up statistics about authors. Our usability evaluation suggests that these plugins are easy to use, consistent, adequate for frequent use, and well integrated.

Interlinking OA dataset with other relevant datasets is an ongoing task for the OA LOD team. Deployment of OA interlinking with already examined datasets in the infrastructure of OA is a future work. Based on the current observations,

---

[10] https://wikipedia.org/wiki/System_Usability_Scale.

we also plan to enhance the interlinking results between OA and other candidate datasets related to other fields such as biology and astronomy and provide a more advanced evaluation. We plan to adopt the implemented services into the infrastructure of the OA and have them publicly available with a better design.

Acknowledgments. This work has been partially funded by European Commission grant 643410 (OpenAIRE) and by DFG grant AU 340/9-1. This work has been partially funded by the European Commission with a grant for the H2020 project OpenAIRE2020 (GA no. 643410) and OpenBudgets.eu (GA no. 645833).

References

1. Ameri, S.: Exploiting interlinked research metadata to provide recommendations for authors of scientific papers. MA thesis. University of Bonn (2017). http://eis-bonn.github.io/Theses/2017/Shirin_Ameri/thesis.pdf
2. Bauer, F., Kaltenböck, M.: Linked Open Data: The Essentials (2011)
3. Hallo, M., Luján-Mora, S., Cháez, C.: An approach to publish scientific data of open-access journals using linked open data technologies. In: EDULEARN (2014)
4. JQuery UI 1.12 API Documentation. https://jqueryui.com/dialog/
5. Purohit, S., et al.: Effective tooling for linked data publishing in scientific research. In: International Conference on Semantic Computing (2016)
6. Rajabi, E., Sicilia, M.-A., Sanchez-Alonso, S.: Discovering duplicate and related resources using an interlinking approach: the case of educational datasets. J. Inf. Sci. **41**(3), 329–341 (2015)
7. Rajabi, E., Sicilia, M.-A., Sanchez-Alonso, S.: Interlinking educational resources to web of data through IEEE LOM. Comput. Sci. Inf. Syst. **12**(1), 233–255 (2015)
8. Rajabi, E., et al.: Interlinking educational data to web of data. In: Big Data Optimization: Recent Developments and Challenges (2015)
9. Rettberg, N., Schmidt, B.: OpenAIRE supporting a european open access mandate. Coll. Res. Libr. News **76**(6), 306–310 (2015)
10. Rodichevski, A.: Approximate String Matching Algorithms. http://www.morfoedro.it/doc.php?n=223&lang=en
11. Simperl, E., et al.: Combining human and computation intelligence: the case of data interlinking tools. Int. J. Metadata Semant. Ontol. **7**(2), 77–92 (2012)
12. Stackoverflow: Difference between Jaro-Winkler and Levenshtein distance. http://stackoverflow.com/questions/25540581/difference-betweenjaro-winkler-and-levenshtein-distance
13. Uyar, E., Brehmer, S., Athamnah, M.: Accessing, Analyzing and Linking Data from DBLP with other Internet Resources (2013)
14. Vahdati, S., et al.: LOD Services. Deliverable D8.2. OpenAIRE2020 (2015)
15. Vahdati, S., Karim, F., Huang, J.-Y., Lange, C.: Mapping large scale research metadata to linked data: a performance comparison of HBase, CSV and XML. In: Garoufallou, E., Hartley, R.J., Gaitanou, P. (eds.) MTSR 2015. CCIS, vol. 544, pp. 261–273. Springer, Cham (2015). doi:10.1007/978-3-319-24129-6_23. arXiv:1506.04006 [cs.DB]

Preserving Bibliographic Relationships in Mappings from FRBR to BIBFRAME 2.0

Sofia Zapounidou$^{(\boxtimes)}$ ⓘ, Michalis Sfakakis ⓘ,
and Christos Papatheodorou ⓘ

Department of Archives, Library Science and Museology,
Ionian University, Corfu, Greece
{ll2zapo, sfakakis, papatheodor}@ionio.gr

Abstract. In the environment of the World Wide Web large volumes of library data have been published following different conceptual models. The navigation through these volumes and the data interlinking require the development of mappings between the conceptual models. Library conceptual models provide constructs for the representation of bibliographic families and the relationships between Works. A key requirement for successful mappings between different conceptual models is to preserve such content relationships. This paper studies a set of cases (Work with single Expression, Work with multiple Expressions, translation, adaptation) to examine if and how bibliographic content relationships and families could be preserved in mappings from FRBR to BIBFRAME 2.0. Even though, relationships between Works of the same bibliographic family may be preserved, the progenitor Work is not always represented in BIBFRAME after mappings.

Keywords: BIBFRAME · Bibliographic families · Content relationships · FRBR · Interoperability · Linked data · Representation patterns

1 Introduction

Linked data technologies enable the integration of bibliographic data into the web and allow the web users to navigate through the bibliographic universe. Library linked data initiatives have already been launched in various countries all over the world. Each initiative was developed within the framework of different projects aiming to address different needs. Therefore, entities of the bibliographic universe are perceived, defined and described in different manners. Definitions of these entities may be found either in bibliographic conceptual models (e.g. FRBR, BIBFRAME, etc.) or in the local schemata used by the projects (e.g. Linked Open BNB/British Library, data.deichman. no/Oslo Public Library).

Navigating through the bibliographic universe is often an intricate process due to the relationships, explicitly or implicitly defined, that interlink bibliographic entities. Content relationships may explicitly or implicitly exist between bibliographic entities generating *bibliographic families*. The term *bibliographic family* has been coined by Professor Smiraglia to describe 'a set of related bibliographic works that are somehow derived from a common progenitor' [1]. *Works* or *Expressions* within the same

© Springer International Publishing AG 2017
J. Kamps et al. (Eds.): TPDL 2017, LNCS 10450, pp. 15–26, 2017.
DOI: 10.1007/978-3-319-67008-9_2

bibliographic family may share the same intellectual content and be related to the progenitor through different types of relationships. The identification of *bibliographic families* and the clustering of all related entities are extremely important and one of the main functions that library catalogs need to deliver [2–4].

Library data conceptual models include constructs that enable the description of such content relationships. A key requirement for successful mappings between different conceptual models is to preserve content relationships and hence to approach the model's compatibility degree to the *bibliographic families*, after the mapping and the data transformation [5–8]. Preservation of *bibliographic families,* based on the Smiraglia definition [1, 4], means the preservation of information that two or more *Works* originate from a common progenitor. This study investigates whether and how content relationships could be preserved when transforming data from FRBR to BIBFRAME 2.0 (hereafter referred as BIBFRAME), as well as their *bibliographic families.* We focus on these two data models because FRBR is a major milestone in the evolution of bibliographic data conceptualization; BIBFRAME is being developed by the Library of Congress and is expected to supersede the MARC21 standard.

Due to the models' different conceptualizations, mappings should be refined by revealing content relationships and *bibliographic families.* A content relationship and a *bibliographic family* within the semantics of a library data conceptual model are instantiated following *representation patterns.* Therefore, in order to evaluate whether content relationships and *bibliographic families* are preserved after their transformation from a source to a target model, their *representation patterns* in the source and the target models have to be defined. Then, the target *representation pattern* should be compared with the *representation pattern* resulted from the transformation mappings. *Representation patterns* have been studied by other scholars in terms of identifying good practices for the representation of specific bibliographic cases using a model's semantics [9, 10]. It should be clarified that a *representation pattern* does not express uniquely a bibliographic description case, because there exist alternatives of expressing the same semantics using the terms of a model.

In the next section some definitions are given and the background of our research. In Sect. 3, mappings for selected content relationships and *bibliographic families* using their *representation patterns* are presented. Conversions from FRBR to BIBFRAME are studied following the proposed methodology. Key findings are presented in the discussion and conclusions section. It must be noted that for clarity reasons the names of models' classes/entities and properties are written in the text in italics.

2 Background

In the environment of the different conceptual models for the library data and the volumes of data that have been published to the World Wide Web, the development of automated mechanisms for their transformations and interlinking requires the development of mappings between the conceptual models. Mappings is one way of tackling interoperability problems and enable either the transformation of instances of a source model to instances of a target model or the integration of data that are expressed by the terms of different models.

Successful mappings preserve the semantics of the source model into the target model. Bibliographic relationships are important for navigation in the bibliographic universe and both FRBR and BIBFRAME models include constructs to describe bibliographic entities and the relationships between them. Bibliographic relationships between works have been studied by Tillett in [11]. Tillett created a taxonomy of bibliographic relationships and identified seven types of them: equivalent, derivative, descriptive, whole-part, accompanying, sequential and shared characteristic ones [11]. The equivalent, derivative and descriptive relationships have been characterized by Tillett [12] as "close content relationships that can be viewed as a continuum starting from an original work". The derivative bibliographic relationship is "broad ranging" [13]. Therefore, Smiraglia [14] focused on derivation and identified eight types of derivative bibliographic relationships. He also coined the term *bibliographic family* to express *Works* that somehow derive from a common original *Work*, also known as the *progenitor*. Smiraglia also found that older and/or popular *Works* tend to have large and complex families [1, 14]. Such families formulate information networks consisting of nodes, which are instances of bibliographic entities, and arcs, that interconnect the instances and denote their relationships. Therefore, Smiraglia has extended the concept of *bibliographic families* using the new term *instantiation network* [7].

The preservation of content relationships and *bibliographic families* in the mapping and data transformation process between two library data conceptual models is not a straightforward issue due to semantic and structural heterogeneities between the models [15]. Therefore, *representation patterns* for both FRBR and BIBFRAME need to be identified so as the semantics of the content relationships and *bibliographic families* in the terms of each conceptual model is described. We use the term *representation pattern* for the representation of each relationship/*bibliographic family* in each conceptual model. We define the concept of *representation pattern* for a *bibliographic family* F a graph $G_{fm}(C_{fm}, P_{fm})$, where C_{fm} is a subset of the set C of the classes of a conceptual model M and P_{fm} a subset of the set P of the properties of a conceptual model M, such that for every triple $(C_{fmd}, P_{fmi}, C_{fmr})$ in G_{fm}, C_{fmd} is the domain class and C_{fmr} is the range class of the property P_{fmi}.

The methodology followed in this paper for developing mappings between library data conceptual models is presented below:

1. Description of the bibliographic relationships and *family* (e.g. translation)
2. Definition of their *representation pattern(s)* in each model.
3. Mapping between Source *representation pattern* and Target *representation pattern*. Due to the semantic and structural heterogeneities of FRBR and BIBFRAME, it is important in particular cases to define the conditions that enable proper mapping, e.g. the existence of a specific attribute of a class or a specific value to an attribute.

The mappings are tested using a real example, the Homer's 'Odyssey' *bibliographic family* and some of its members, for the cases where FRBR is the source model and BIBFRAME is the target model.

3   Mapping Content Relationships and *Bibliographic Families*

The paper gradually leads the examination process from simple to more complex *bibliographic families*. The cases studied are *Works* with a single *Expression*, *Works* with multiple *Expressions*, and derivations, namely translations and adaptations. When *representation patterns* are depicted, the nodes symbolize the corresponding classes, while the edges illustrate the properties between the classes of each model. Each node is divided in two smaller boxes: the upper one denotes the class, while the lower one provides its instance. For readability reasons the lower box denoting a class' instance includes a small description and neither the instance's full title and/or related details, nor its complete URI.

3.1   *Work* with a Single *Expression*

The simplest and the most frequent bibliographic case [16] is a *Work* with a single *Expression* and a single *Manifestation*, e.g. a monograph (book) in a language. In FRBR the *Work* entity is an abstract entity that delimits a distinct intellectual creation, as initially intended by its author(s). The *Work is realised through* an *Expression*, a realization of the *Work* in a specific form and set of signs. It must be noted that due to the abstract nature of the *Work* entity, a *Work* is mainly recognized through its various *Expressions*. These *Expressions* are embodied in *Manifestation* entity instances. An exemplar of all identical copies exemplifying a *Manifestation*, is represented by the *Item* entity. The *representation pattern* of this bibliographic description case in the terms of the FRBR model is presented in Fig. 1.

Fig. 1. *Representation pattern* for a *Work* with a single *Expression* in FRBR.

BIBFRAME defines different conceptualizations. A *Creative Work* instance represents both the idea of an intellectual creation and its form of realization. The material embodiment of the *Creative Work* (*bf:Work*) is expressed with the *bf:Instance* class. A copy of the *bf:Instance* held at a library is represented by a *bf:Item* class instance. BIBFRAME does not define different classes for differentiating between the abstract idea of an intellectual creation and therefore *Creative Work* 'seems to be semantically closer to the (union of the) FRBR *Work* and *Expression* entities' [17, 18]. This difference in conceptualizing basic bibliographic entities is likely to prove crucial to prospective transformations of bibliographic data between the two models. The *representation pattern* of this bibliographic description case in terms of BIBFRAME is presented in Fig. 2.

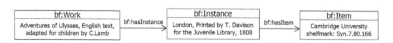

Fig. 2. *Representation pattern* for a *Work* with a single *Expression* in BIBFRAME.

Mapping FRBR entities to BIBFRAME shall ensure preservation of semantics. FRBR uses two entities, namely *Work* and *Expression,* to represent intellectual creation and the signs used for its realization, while BIBFRAME uses only one class, *Creative Work*. Physical embodiment is represented in both models in the same way. FRBR represents embodiments with the *Manifestation* entity and *Manifestation* exemplars with the *Item* entity. Likewise, BIBFRAME defines the *bf:Instance* class for embodiments and *bf:Item* class for *bf:Instance* exemplifications. This mapping is depicted in Fig. 3, which is actually a generalization of the mapping for the *Work* with a single *Expression* example presented in Figs. 1 and 2. The instances of two FRBR classes, namely the *Work* and *Expression* instances, are semantically subsumed by instances of the class *bf:Work* in BIBFRAME. The *Manifestation* entity is mapped to the *bf: Instance* class and the *Item* entity to the *bf:Item* class. Moreover, in Fig. 3 the mapping rules between the core classes of FRBR and BIBFRAME are presented. These rules also refer to the "inherent relationships" [19] among FRBR Group 1 entities.

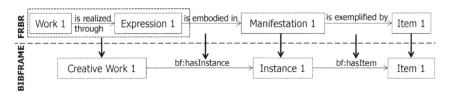

Fig. 3. Mapping from FRBR to BIBFRAME 2.0 *representation pattern* for a *Work* with a single *Expression*.

Specialization by attributes. While BIBFRAME uses the *Creative Work* class to represent both the intellectual content and its realization, it specializes its semantics by a set of 10 subclasses. Accordingly, the *bf:Instance* class has 5 subclasses. The mapping of the *representation patterns* presented in Fig. 3 is generic and involves the high-level classes of the target model. Hence, in order to achieve closest similarity between the source and the target classes and properties, more detailed *representation patterns* regarding the FRBR triple *Work - is realized through - Expression* and the *Manifestation* entity should be generated. Such patterns are generated by exploiting information lying in the attributes of the FRBR *Expression* and *Manifestation* entities. Moreover, controlled vocabularies from the Library of Congress Linked Data Service (http://id.loc.gov/) should be used for the values of the attributes so as the mapping rules to be precisely expressed.

Regarding the mapping of the FRBR triple *Work - is realized through - Expression* to the *bf:Work* class and subclasses, we have identified the *form of expression* attribute. This attribute of the *Expression* entity describes the way a *Work* has been realized, e.g. text, still image, notated music, etc. The LC Content Types Scheme (http://id.loc.gov/vocabulary/contentTypes) may be used for the values of the attribute *form of expression*. Depending on these values, the FRBR triple *Work - is realized through - Expression* shall be mapped to a different *bf:Work* subclass. The attribute *form of expression,* along with the values of the LC Content Types Scheme, enables more precise mappings for all *bf:Work* subclasses. In some cases these values may even

Table 1. The values of the FRBR attribute *form of expression* trigger the mapping of the FRBR *'Work – is realized through – Expression'* triple to different *bf:Work* subclasses.

If	then map (Work – is realized through – Expression) to bf:Work subclass	and Manifestation to bf:Instance subclass
(Expression - form of expression - contentTypes:cartographic image)	bf:Cartography	
(Expression - form of expression - contentTypes:computer dataset)	bf:Dataset	bf:Electronic
(Expression - form of expression - contentTypes:text)	bf:Text	
(Expression - form of expression - contentTypes:tactile text)	bf:Text	bf:Tactile

determine the mapping to a *bf:Instance* subclass. As an example, some mapping rules triggered by this attribute's values are exhibited in Table 1.

Concerning the mapping of the FRBR *Manifestation* entity to the *bf:Instance* class and subclasses, the attribute *form of carrier* has been identified. This attribute of the *Manifestation* class describes the physical carrier in which an *Expression* of a *Work* is embodied. The Carriers Scheme (http://id.loc.gov/vocabulary/carriers), already used in RDA cataloging, may be also used as the vocabulary for the values of the *form of carrier* attribute. These values adjust the mapping of a *Manifestation* instance to a *bf:Instance* subclass. Some examples of mapping rules that are triggered by the *form of carrier* attribute values are presented in Table 2. It must be noted though that the *form of carrier*, along with the values of the Carriers Scheme, enables some mappings but not for all *bf:Instance* subclasses, such as the *bf:Manuscript* subclass.

Table 2. The values of the FRBR attributes *form of carrier* trigger mapping of the FRBR *Manifestation* to different *bf:Instance* subclasses.

If	map *Manifestation* to *bf:Instance* subclass
(Manifestation - form of carrier - (carriers:computer tape reel OR carriers:online resource OR carriers:computer disc))	bf:Electronic
(Manifestation - form of carrier - carriers:volume)	bf:Print

3.2 Work with Multiple Expressions

The mapping rules of the previous section preserve information when transforming *'Work* with single *Expression'* data from FRBR to BIBFRAME. In FRBR the classes *Work* and *Expression* are correlated by the relationship *is realized through*, having an *one to many* cardinality, meaning that for a *Work* several *Expressions* might exist. Indeed, classical works tend to have great bibliographic families. For instance, there are different editions of Homer's 'Odyssey' and many translations in a variety of languages. In Fig. 4 two *Expressions* of *The Essential Homer* by Stanley Lombardo are represented: the English text and the audio narration of the text (sound recording).

Using the rules in Fig. 3, each one of the two triples *Work-is realized through-Expression* depicted in the upper side of Fig. 4 will be mapped to an instance of a *bf:Work* class in BIBFRAME. It is worth noting that the same instance of the *FRBR Work* entity 'Odyssey' participates in two different mappings. However, following the aforementioned rules to transform the FRBR *representation pattern* for the *Work with multiple Expressions* to BIBFRAME, the semantics of the origination of the two instances of the *bf:Work* class from the same *Work* (intellectual idea) are lost. BIBFRAME provides the property *bf:hasExpression* to correlate the two *Expressions*, as depicted in the BIBFRAME side of Fig. 4 for the two 'The Essential Homer' editions. In this case, in order to indicate in the target representation that the *bf:Work* class originated from the same intellectual idea the rules must be extended and connect all pairs between these two *bf:Work* instances with an instance of the *bf:hasExpression* property. The additional semantics incorporated by the *bf:hasExpression* property in the target pattern, preserve the content relationship. Yet, the information that the *bf: Work* instances have the same progenitor (*Work*) is not preserved.

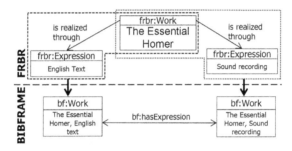

Fig. 4. Mapping from FRBR to BIBFRAME 2.0 *representation pattern* for a *Work* with more than one *Expressions*.

3.3   Derivation Patterns: Translation and Adaptation

The *bibliographic family* of 'Odyssey' has become really great due to derivatives; there are many translations, as well as adaptations, dramatizations, imitations, etc. There are many types of derivation, as described in [15]. In this paper, the case of literal translation is studied. In Fig. 5 an example for the literal translation case is represented using the well-known translation of 'Odyssey' by Alexander Pope. Literal translation is represented at the *Expression* level in FRBR (Fig. 5). Two *Expression* instances of the same *Work* are related to each other with the *has translation* property, where one instance of the *Expression* entity (ancient text edited by D.Chalcocondylis) *has a translation* in another language represented by an instance of a second *Expression* entity (English translation by A.Pope). In BIBFRAME, translation is represented as a relationship between two *Creative Work* instances, as depicted in the BIBFRAME side of Fig. 5.

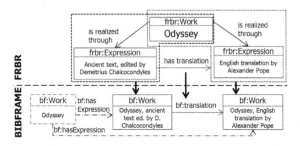

Fig. 5. Mapping from FRBR to BIBFRAME 2.0 *representation pattern* for the translation case. The *bf:Work* with the long dash-dot outline has been added in the mapping to preserve the progenitor *bf:Work* of the *Odyssey bibliographic family*.

As in the case of the *Work with multiple Expressions*, the same FRBR *Work* entity instance of 'Odyssey' participates in two mappings (Fig. 5). Moreover, to transform the FRBR translation *representation pattern* to BIBFRAME, the FRBR *has translation* property has to be utilized in order to correlate the two different *Expressions*. Then, the property will be mapped to the *bf:translation* property. Thus, in the derivation-translation case information regarding the content relationship between the two *Expressions* is preserved in the two *bf:Works*. However, following this mapping the information that the *bf:Work* instances have the same progenitor (*Work*) is not preserved. In order to preserve information about the common progenitor, mappings should be changed. More specifically, an additional *bf:Work* instance will be created (*bf:Work* with the long dash-dot outline in Fig. 5). Then this additional *bf:Work* instance will be linked with the others *bf:Work* instances using the *bf:hasExpression* property (also depicted with a long dash-dot line).

In case the *Expression* of derivation is not known, there will be *Expressions* in different languages of a *Work*. These *Expressions* will not be related with a *has translation* property, but the translation could be implied due to the different values between the *language of expression* attributes of each *Expression* instance. Since there is no explicit representation of the translation relationship, mapping of this representation would be similar to Fig. 4. Ideally, the mapping would be similar to the adaptation case depicted in Fig. 7 where the representation is made with an *Expression*-agnostic *Creative Work* instance related to another *bf:Work* instance through a *bf:-translation* property. In order to achieve such mapping, new rules must be implemented taking into account the existence of differing values for *language of expression* attributes. Differences between the entity *Person/Family/Corporate Body* that *created* the *Work* instance and the *Person/Family/Corporate Body* that *realized* an *Expression* of the same *Work* instance must also be considered.

A derivation that results in a new *Work* is represented in FRBR at the *Work* level with various properties, namely *has adaptation, has a transformation, has an imitation, has a paraphrase, has a dramatization*. By contrast, BIBFRAME utilizes only the *bf: hasDerivative* property at the *bf:Work* level. Hence, all these FRBR properties are mapped to a single property in BIBFRAME.

In FRBR adaptation may be represented by the *has adaptation* property at either the *Work* or *Expression* level. When information regarding which *Expression* has been used for creating an adaptation is not known, then the representation of adaptation is preserved at the *Work* level and hence it is *Expression*-agnostic. The *has adaptation* property is used at the *Expression* level, when there is information about the particular *Expression* used to create both the *Work* and the *Expression* of the new adaptation.

In Fig. 6 an adaptation of 'Odyssey' for children is represented. Charles Lamb used the English translation of George Chapman and then "turned… [Odyssey] into prose, simplified the order of the narrative, abbreviated or combined episodes, and deleted descriptions and whole books in order to … eliminate anything inappropriate for young readers" [20]. As depicted in Fig. 6, the progenitor *Work* 'Odyssey' along with one of its *Expression* instances (English translation by G.Chapman) is mapped to one *bf:Work* instance, while its derivative *Work* 'Adventures of Ulysses' with its *Expression* instance is mapped to a second *bf:Work* instance. The *has adaptation* relationship at the *Expression* level is mapped to the *bf:hasDerivative* property instance that relates the two *bf:Work* instances. In this case both content relationships and the bibliographic family are preserved.

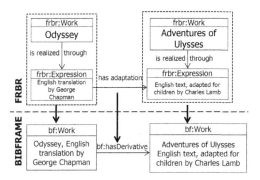

Fig. 6. Mapping from FRBR to BIBFRAME 2.0 *representation pattern* for the adaptation case.

In Fig. 7 an *Expression*-agnostic adaptation at the *Work* level is depicted. The exact *Expression* of 'Odyssey' used by Anne Terry White to create her adaptation for children is not known. Therefore, the progenitor *Work* 'Odyssey' is mapped to a *bf: Work* instance that lacks *Expression*-related information (e.g. language), while the derivative *Work* "Odysseus comes home from the sea" along with its *Expression* is mapped to a second *bf:Work* instance. The *bf:Work* on the left side of the *bf: hasDerivative* property may serve as an abstract *bf:Work* and it cannot have any *bf: Instances* because its *Expression*-related information is not known. In this case both content relationships and the bibliographic family are preserved.

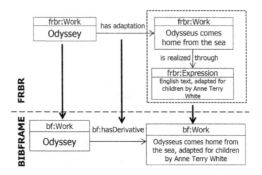

Fig. 7. Mapping from FRBR to BIBFRAME 2.0 *representation pattern* for the derivation-adaptation case. The exact Expression used to produce the adaptation Expression is not known.

4 Discussion and Conclusions

The navigation in an ever-changing overloaded bibliographic universe that preserves the contextual semantics of the bibliographic descriptions largely depends on the control of content relationships and *bibliographic families*. Library conceptual models include constructs to describe and control *bibliographic families*. This paper examines if and how information about content relationships and *bibliographic families* may be preserved under mappings. It focuses on FRBR and BIBFRAME models, and on mappings where FRBR is the source model and BIBFRAME is the target one. The cases of a *Work* with a single *Expression*, as well as *bibliographic family* cases (e.g. *Work* with multiple *Expressions, Works* with derivative relationships) are studied and some interesting findings were derived.

The generic mapping of the simplest case of a *Work* with a single *Expression* may be considered straightforward (Fig. 3). Additionally, more precise mapping rules may be applied combining FRBR attributes and values from controlled vocabularies (Tables 1 and 2). The utilization of controlled vocabularies for mapping purposes and automated exchange of library data demands a shift in working culture and an adoption of new cataloging rules and policies. From now on librarians shall perform cataloging having in mind collaboration and reuse of data, not just indexing their library's collection for local purposes. This may affect cataloging systems, as well as workflows.

An interesting finding of this study is that the relationships between members of a *bibliographic family* may be preserved in BIBFRAME only when FRBR *Expressions* are related by a particular property (*has translation, has adaptation*, etc.). In the case of mapping an FRBR *Work* with multiple *Expressions* to BIBFRAME there is no relationship between the FRBR *Expression* instances. Therefore, the information regarding the common progenitor is lost in BIBFRAME. The mapping has been extended with the insertion of two *bf:hasExpression* property instances (Fig. 4), to preserve the content relationship. Still the common progenitor is not explicitly represented. Information about the progenitor *Work* may be preserved in BIBFRAME following the practice shown in Figs. 5 and 7, where a new *bf:Work* is generated to hold the information of the *Work* entity as progenitor. In both cases an *Expression*-agnostic *bf:Work* instance has been used as the progenitor. Then, this progenitor *bf:Work* is related to the other members of the family with *bf:hasExpression* property instances (Fig. 5) or with another property (*bf:hasDerivative* in Fig. 7), if such exists based on the mappings.

This *Expression*-agnostic *bf:Work* is similar to the *superwork* expressed by Svenonius in [21] and may be used to group all *bf:Works* that are somehow derived by it. *Expression*-agnostic *bf:Works* are not expected to have any *bf:Instances*. At this point, it must be noted that BIBFRAME does not impose cardinalities regarding the triple *bf:Work-bf:hasInstance-bf:Instance*. This may provide flexibility in some implementations of BIBFRAME, but at the same time may cause ambiguity. Totally different mapping rules can be defined when different cardinality constraints exist, if for example a *bf:Work* must or may have one or more *bf:Instances*.

This study uses a limited set of cases and data. More bibliographic relationships need to be studied and findings shall be checked using larger and more complicated datasets. The mappings produced in this study need to be converted through a mapping language in conversion rules. A follow-up study shall compare the transformation based on these rules in contrast to the MARCXML to BIBFRAME Transformation software [22]. Moreover, existing software tools should be selected and adapted to evaluate the degree of preservation of bibliographic relationships after mappings. Interesting findings are also anticipated for testing the opposite mappings, where BIBFRAME is the source model and FRBR is the target one. Updates of the two models are likely to cause changes in mappings. The consolidated FRBR-LRM is expected to be announced in 2017. BIBFRAME model is regularly updated and its second version has already included FRBR conceptualizations to enable mappings, e.g. the *bf:Item* class. There is the possibility that prospective BIBFRAME versions shall include more changes for interoperability reasons.

References

1. Smiraglia, R., Leazer, G.: Derivative bibliographic relationships: the work relationship in a global bibliographic database. J. Am. Soc. Inform. Sci. **50**, 493–504 (1999). doi:10.1002/(SICI)1097-4571(1999)50:6<493::AID-ASI4>3.0.CO;2-U
2. Coyle, K.: Future considerations: the functional library systems record. Libr. Hi Tech. **22**, 166–174 (2004). doi:10.1108/07378830410524594
3. Mimno, D., Crane, G., Jones, A.: Hierarchical catalog records - Implementing a FRBR catalog. D-Lib Mag. **11**, 1–9 (2005). doi:10.1045/october2005-crane
4. Smiraglia, R.P.: The Elements of Knowledge Organization. Springer, Cham (2014)
5. Arastoopor, S., Fattahi, R., Parikosh, M.: Developing user-centered displays for literary works in digital libraries: integrating bibliographic families, FRBR and users. In: Proceedings of the 2nd International Conference of Asian Special Libraries ICoASL. Special Libraries Association - Asian Chapter, Tokyo, pp 83–91 (2011)
6. Merčun, T., Žumer, M., Aalberg, T.: FrbrVis: an information visualization approach to presenting FRBR work families. In: Zaphiris, P., Buchanan, G., Rasmussen, E., Loizides, F. (eds.) TPDL 2012. LNCS, vol. 7489, pp. 504–507. Springer, Heidelberg (2012). doi:10.1007/978-3-642-33290-6_60

7. Smiraglia, R.P., Heuvel, C.V.D.: Classifications and concepts: towards an elementary theory of knowledge interaction. J. Doc. **69**, 360–383 (2013). doi:10.1108/JD-07-2012-0092

8. Merčun, T., Žumer, M., Aalberg, T.: Presenting bibliographic families: Designing an FRBR-based prototype using information visualization. J. Doc **72**, 490–526 (2016). doi:10.1108/JD-01-2015-0001

9. Urban, R.J.: Representation patterns for cultural heritage resources. Proceedings of the ASIST Annual Meeting **50**, 1–4 (2013). doi:10.1002/meet.14505001123

10. Aalberg, T., Vennesland, A., Farrokhnia, M.: A pattern-based framework for best practice implementation of CRM/FRBRoo. In: Morzy, T., Valduriez, P., Bellatreche, L. (eds.) ADBIS 2015. CCIS, vol. 539, pp. 438–447. Springer, Cham (2015). doi:10.1007/978-3-319-23201-0_44

11. Tillett, B.B.: Bibliographic Relationships: Toward a Conceptual Structure of Bibliographic Information used in Cataloging, Doctoral Dissertation. University of California, Los Angeles (1987)

12. Tillett, B.B.: Bibliographic relationships. In: Bean, C.A., Green, R. (eds.) Relationships in the Organization of Knowledge. Information Science and Knowledge Management, vol. 2, pp. 19–35. Springer, Dordrecht (2001). doi:10.1007/978-94-015-9696-1_2

13. Riva, P.: Mapping MARC 21 linking entry fields to FRBR and Tillett's taxonomy of bibliographic relationships. Libr. Resour. Tech. Serv. **48**, 130–143 (2004)

14. Smiraglia, R.: Authority control and the extent of derivative bibliographic relationships. Doctoral Dissertation. The University of Chicago, Chicago (1992)

15. Zapounidou, S., Sfakakis, M., Papatheodorou, C.: Representing and integrating bibliographic information into the Semantic Web: a comparison of four conceptual models. J. Inf. Sci. (2016). doi:10.1177/0165551516650410

16. Bennett, R., Lavoie, B.F., O'Neill, E.T.: The concept of a work in WorldCat: an application of FRBR. Libr. Collect. Acquis. **27**, 45–59 (2003). doi:10.1016/S1464-9055(02)00306-8

17. Zapounidou, S., Sfakakis, M., Papatheodorou, C.: Highlights of library data models in the era of linked open data. In: Garoufallou, E., Greenberg, J. (eds.) MTSR 2013. CCIS, vol. 390, pp. 396–407. Springer, Cham (2013). doi:10.1007/978-3-319-03437-9_38

18. BIBFRAME - Bibliographic Framework Initiative: BIBFRAME Profiles: Introduction and Specification (Draft - 5 May 2014). Library of Congress, Washington, DC. http://www.loc.gov/bibframe/docs/bibframe-profiles.html

19. Tillett, B.: What is FRBR?: A conceptual model for the bibliographic universe. Washington, D.C (2004)

20. The University of Chicago Library. Homer in Print: The Transmission and Reception of Homer's Works, Chapter VII The Children's Homer. https://www.lib.uchicago.edu/e/webexhibits/homerinprint/everyone.html

21. Svenonius, E.: The Intellectual Foundation of Information Organization. MIT Press, Cambridge, Mass, London (2009)

22. lcnetdev/marc2bibframe2. Convert MARC records to BIBFRAME2 RDF. https://github.com/lcnetdev/marc2bibframe2

Exploring Ontology-Enhanced Bibliography Databases Using Faceted Search

Tadeusz Pankowski[✉]

Institute of Control and Information Engineering,
Poznań University of Technology, Poznań, Poland
`tadeusz.pankowski@put.poznan.pl`

Abstract. In this paper, we exploited the application of the Ontology Based Data Access (OBDA) approach, equipped with faceted search utilities, to explore bibliography databases. A bibliography database is enhanced by means of an ontology leading to a bibliography information space. We show that faceted search paradigm to explore such information space is particularly attractive. We describe an implementation of this approach in DAFO system. We focus on formulating faceted queries over the ontology, mapping the ontology to a relational database, and on transforming the query to executable forms. The final version of a faceted query is a SQL query that is executed in a relational database system. The computational results show that the usage of faceted search-oriented way of modeling and retrieving information is very promising.

1 Introduction

Faceted search is commonly used in retrieving data in e-commerce applications [7,17]. In this paper, we adapt this approach to explore bibliography databases. To take advantages of this retrieval paradigm, a bibliography database should be first enriched with an ontology, leading to the creation of an ontology-enhanced bibliography database. The purpose of this extension is to provide the database with concepts, relationships and rules, which both facilitate query formulation and allow for a flexible perceiving the domain information space. The main advantages of the faceted search are: (a) iterative and interactive support for query formulation, usually based on a user-friendly graphical interface; (b) coexistence of many different views over the underlying information space; (c) effective implementation due to the expression power of faceted queries limited to first order monadic positive existential queries (MPEQ) [1,11,16].

Related work. This paper refers both to Ontology-Based Data Access (OBDA) and faceted search. In OBDA, an ontology is used as a global schema and a database is used as a data repository and a mapping is established between the ontology and the database [3,13]. Some issues concerning this approach were discussed in data integration and data exchange contexts [5,8]. However, the problem in such system is a query language. It is unrealistic to require the user to know the database schema in details, so the usage of SQL, SPARQL or XQuery as

© Springer International Publishing AG 2017
J. Kamps et al. (Eds.): TPDL 2017, LNCS 10450, pp. 27–39, 2017.
DOI: 10.1007/978-3-319-67008-9_3

end-user languages is unacceptable. On the other hand, relying only on keyword queries (even with boolean operators) significantly limits search capabilities. Thus, it is quite obvious that users should be provided with graphical-oriented tools. Such solution can be based on Query-By-Example [9,18], which was the inspiration for developing a number of visual query systems and languages [4]. Visual systems provide an intuitive and natural perceiving of the information space, and follow the direct manipulation idea with visual representation of domain and query manipulation. End users recognize the relevant fragments of information space and formulate queries by directly manipulating them. To this family of information retrieval paradigms we can count faceted search. The faceted search combines two classical approaches, namely keyword-based search and manipulation search with narrowing the information space.

Contribution. We discuss the aforementioned issues in the context of our system called DAFO (Data Access with Faceted queries over Ontology) [14,15]. The main contributions of this paper are as follows: (1) we discuss an ontology based on OWL 2 RL to describe an information space relevant to bibliography database; (2) we show how the ontology is used in faceted query formulation; (3) we describe main steps in answering faceted queries: (a) translating to first order faceted queries (FOFQ), (b) rewriting faceted queries using ontology rules, (c) mapping the ontology into relational database; (4) we report some computational experiments which prove that the formulation and evaluation of faceted queries in DAFO is very promising.

Paper outline: The structure of the paper is the following. Ontology-enhanced databases are discussed in Sect. 2. In Sect. 3, a mapping of the ontology into relational database is defined. Faceted search over bibliography ontology and some experimental results are presented in Sect. 4. Section 5 summarizes the paper.

2 Ontology-Enhanced Bibliography Database

2.1 Relational Schema of Bibliography Database

In this section we motivate our research showing the advantages of combining information representation capabilities provided by relational database and an ontology. The discussion will be focused on a bibliography database, BibDb, with the schema in Fig. 1.

The schema was designed based on analysis of DBLP Computer Science Bibliography [6,10]. An instance of the database was prepared by extracting data from DBLP resources (from XML, HTML, and BibTex files), and enriched with data extracted form personal and conference home pages. Some tables have primary keys (denoted by Id) used to identify entities and to establish cross references between tables. Same tables have also unique DBLP identifiers ($DblpKey$) used as references to DBLP bibliography.

The schema in Fig. 1 can be used as a target schema for posting relational queries. However, direct operating on such schema is troublesome. In general, the schema can be large, incomprehensible, and a language for query formulation

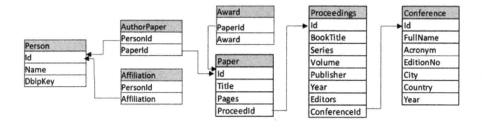

Fig. 1. Diagram of bibliography relational database BibDb.

can make requirements that are difficult to accept. For example, it is unrealistic to demand that the user can write SQL queries.

Formally, a relational database schema is a tuple $\mathcal{S} = (\mathbf{R}, att, pkey, InclDep)$, where $\mathbf{R} = \{R_1, \ldots, R_n\}$ is a set of relation names; att assigns a set of attributes to each $R \in \mathbf{R}$, $att(R) \subseteq \mathsf{Att}$; $pkey$ assigns a primary key to some $R \in \mathbf{R}$, $pkey(R) \in att(R)$; $InclDep$ is a set of *inclusion dependencies* (or *referential constraints*), i.e., expressions of the form $R[A] \subseteq R'[A']$, where $A \in att(R)$, $A' = pkey(R')$, and A is called a *foreign key* referring from R to R'.

In Fig. 1 there is a graphical representation of relational database schema. By Id we denote primary keys, and inclusion dependencies are denoted by arrows.

2.2 Ontology Describing Bibliography Information Space

By a *bibliography information space* we understand a specification of the knowledge relevant to the bibliography domain. In practice, this knowledge covers and enriches relational schema, and is defined by means of an *ontology*. Then we say about *ontology-enhanced* database.

An ontology [2,11] is a pair $\mathcal{O} = (\mathcal{T}, \mathcal{A})$, where $\mathcal{T}$ and $\mathcal{A}$ are, respectively, the *terminological* and *assertional* parts of the ontology. The terminological part is a pair $\mathcal{T} = (\Sigma, \mathcal{R})$, where $\Sigma = \mathsf{UP} \cup \mathsf{BP} \cup \mathsf{Const}$ is the *signature* of the ontology, and specifies a set of *unary predicates* (UP), a set of *binary predicates* (BP), and a set of *constants* (Const). $\mathcal{R}$ is a set of ontology *rules*. The assertional part is a set of *assertions* (facts), i.e., expressions of the form $C(a)$ or $P(a, b)$, where $C \in \mathsf{UP}$, $P \in \mathsf{BP}$, and $a, b \in \mathsf{Const}$. The set UP of unary predicates is divided into *extensional* (UP_E) and *intentional* (UP_I) ones. Similarly, BP is divided into BP_E and BP_I. Extensional predicates are those, which appear in $\mathcal{A}$, while intentional predicates do not appear explicitly in $\mathcal{A}$ but are defined by means of rules in $\mathcal{R}$.

For example, the considered bibliography information space can be defined by means of on ontology BibOn. A fragment of terminological part of BibOn is depicted in Fig. 2. By solid lines we drawn extensional predicates, and intentional predicates are denoted by dashed lines.

Rules in ontologies usually conform to those specified in OWL 2 profiles [12]. We will restrict ourselves to categories of rules given in Table 1. Additionally, following so called *extended knowledge bases* introduced in [11], we divide the

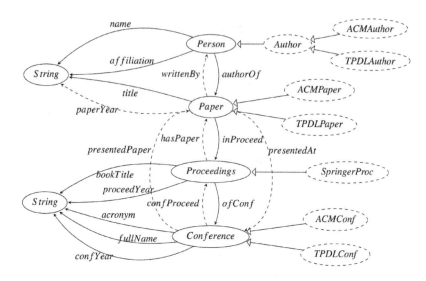

Fig. 2. Terminological part of BibOn ontology.

set of rules into *deductive rules* (Ded-1 – Ded-9) and *integrity constraints* (IC-1 – IC-4). A deductive rule is used to deduce (infer) new assertions (intentional or extensional) from extensional and already deduced intentional ones. Integrity constraints are used to check correctness of the given set of assertions, and not to infer new assertions. Note that all but the last two are in OWL 2 RL [12]. Moreover, functionality (IC-1) and key (IC-2) rules can be used as deductive rules in the case when labeled nulls are allowed (see [8,14,15]).

Deductive rules and integrity constraints referred to in this paper are:

1. *Deductive rules* – specify how some types or properties may be deduced from another. In this paper they are used to infer:
   (a) *inheritance hierarchies* (subtypes and subproperties), e.g., (a) a unary predicate (type) *Author* is a subtype of *Person*, and (b) binary predicate (property) *awardedAt* is a subproperty of *presentedAt*;
   (b) *domains* and *ranges* of binary predicates, e.g., property *authorOf* has *Person* as its domain (although may be not defined for all persons), and *Paper* as the range.
   (c) a *value-driven specialization* – a subtype *C* may be a subset of domain of *P*, for which *P* has a given value *a*, e.g., *SpringerProceed* is a subtype of *Proceedings*, for which *publisher* has value *Springer*;
   (d) a *pattern-driven specialization* – like value-driven specialization but now, *a* is treated as a pattern, and the value of *P* must conform to this pattern, e.g., *ACMConf* is a subtype of *Conference* if *acronym* of the conference contains "*ACM*", i.e., conforms to the pattern "*%ACM%*";
   (e) a *type-driven specialization* – a subtype *C* is a subset of domain of property *P*, for which *P* has value of a given type *D*, e.g., *ACMAuthor* is the type of those authors, who presented their papers at an *ACMConf*;

Table 1. Categories of ontology rules used in bibliography ontology BibOn.

Id	General form of a rule	Name	Representation
Ded-1	$\forall x\ (D(x) \rightarrow C(x))$	Subtype	$subtype(D, C)$
Ded-2	$\forall x, y\ (S(x, y) \rightarrow P(x, y))$	Subproperty	$subprop(S, P)$
Ded-3	$\forall x, y\ (P(x, y) \rightarrow C(x))$	Domain	$dom(P, C)$
Ded-4	$\forall x, y\ (P(x, y) \rightarrow D(y))$	Range	$rng(P, D)$
Ded-5	$\forall x\ (P(x, y) \wedge y = a \rightarrow C(x))$	Specialization (value and pattern driven)	$spec1(P, a, C)$
Ded-6	$\forall x, y\ (P(x, y) \wedge D(y) \rightarrow C(x))$	Specialization (type driven)	$spec2(P, D, C)$
Ded-7	$\forall x, y, z\ (R(x, y) \wedge S(x, z) \wedge z = a \rightarrow P(x, y))$	Property specialization (value and pattern driven)	$spec3(R, S, a, P)$
Ded-8	$\forall x, y, z\ (S(x, z) \wedge T(z, y) \rightarrow P(x, y))$	Chain (composition)	$chain(S, T, P)$
Ded-9	$\forall x, y\ (S(y, x) \rightarrow P(x, y))$	Inversion	$inv(S, P)$
IC-1	$\forall x, y_1, y_2\ (P(x, y_1) \wedge P(x, y_2) \rightarrow y_1 = y_2)$	Functionality	$func(P)$
IC-2	$\forall x_1, x_2, y\ (P(x_1, y) \wedge P(x_2, y) \rightarrow x_1 = x_2)$	Key (functionality of inversion)	$key(P)$
IC-3	$\forall x(C(x) \rightarrow \exists y\ P(x, y)$	Existence	$exists(C, P)$
IC-4	$\forall x(C(x) \rightarrow \exists y\ P(x, y) \wedge y = a$	Has value	$hasVal(C, P, a)$

(f) a *chain* (or *composition*) – a property is a chain (composition) of two other properties, e.g., *paperYear* (year of a paper) is the chain of *inProceed* and *proceedYear* (i.e., year of the proceedings the paper is in);

(g) an *inversion* – a property is the inversion of other property, e.g., *writtenBy* is the inversion of *authorOf*.

2. *Integrity constraints* – are used to check whether a given set of assertions is consistent:

(a) *functionality* – states that a property P is a function, e.g., *name*, *inProceed*;

(b) *key* – states that value of a property P uniquely identifies the domain object, i.e., inversion of P is a function, e.g., *title*, or *hasPaper* (paper uniquely identifies the proceeding in which the paper is included);

(c) *existence* – states that a property P is defined on each element in C, e.g., *authorOf* is defined on each object of *Author* type;

(d) *has value* – states that a property P on each element in C has the same value a (or conforms to pattern a), e.g., the value of *acronym* for each object of *ACMConf* type conforms to the pattern "*%ACM%*".

A first order (FO) formula $\varphi(x)$ is a *monadic positive existential query* (MPEQ), if it has exactly one free variable and is constructed only of: (a) atoms of the form $C(x)$, $P(x_1, x_2)$ and $x = a$; (b) conjunction ($\wedge$), disjunction ($\vee$), and existential quantification ($\exists$). A constant a is an *answer* to $\varphi(x)$ if $\varphi(a)$ is satisfied in $\mathcal{O}$.

Any rule can be treated either as deductive rule or as an integrity constraint. Our choice is motivated by the use of rules to rewrite queries. The following example explains the role of rules and integrity constraints in query rewriting.

Example 1. It seems obvious that a query $q_1(x) = Author(x)$ can be replaced with $q_2(x) = Person(x) \land \exists y(authorOf(x,y))$. We would expect that sets of answers to these queries were equal. However, this is the case only when the data source satisfies the integrity constraint (IC-3) $Author(x) \rightarrow \exists y(authorOf(x,y))$. Indeed, let
$\mathcal{A} = \{Person(a), Person(b), Person(c), Author(a), Author(c), authorOf(a,p)\}$.
Then $q_1(x)(\mathcal{A}) = \{a, c\}$, but $q_2(x)(\mathcal{A}) = \{a\}$.   □

New rules can be dynamically added to the ontology. For example, we can add the binary predicate *authorConf* connecting authors with conferences, which is the chain of *authorOf*, *inProceed*, and *ofConf*.

3   Mapping Ontology to Relational Database

In DAFO, queries formulated over an ontology are evaluated in a relational database. Thus, a mapping of the ontology into relational database must be defined. Two levels of the mapping are distinguished: (1) *metaschema level* – assigns schema elements to ontology predicates, and (2) *schema level* – assigns relational data to ontology assertions.

A mapping on a metaschema level is specified by four functions: *table*, *domCol*, *rngCol* and *inclDep*, defined as follows (some examples are given in Table 2).

1. Extensional unary predicates are mapped to relational names. Formally, if $C \in \mathsf{UP}_E$ then $table(C) = R_C \in \mathbf{R}$, and $domCol(C) = pkey(R_C)$; $R_C \neq R_{C'}$ for $C \neq C'$.
2. Extensional binary predicates are divided in four classes: *functional data* properties (BP_{fd}), *multivalued data* properties (BP_{md}), *functional object* properties (BP_{fo}), and *multivalued object* properties (BP_{mo}). Data properties are binary predicates with *String* as their ranges. Object properties have ranges different from *String*. Then
   - if $P \in \mathsf{BP}_{fd}$, and C is domain of P, then: $table(P) = R_C$, $domCol(P) = pkey(R_C) = Id$, $rngCol(P) = A_P^r \in att(R_C)$, e.g., *name*;
   - if $P \in \mathsf{BP}_{md}$, and C is domain of P, then: $table(P) = R_P \neq R_C$, $domCol(P) = A_P^d \in att(R_P)$, $rngCol(P) = A_P^r \in att(R_P)$, and $R_P.A_P^d \subseteq R_C.Id$, where A_P^d is a foreign key referring from R_P to R_C, e.g., *affiliation*;
   - if $P \in \mathsf{BP}_{fo}$, C is domain of P, D is range of P, then: $table(P) = R_C$, $domCol(P) = pkey(R_C) = Id$, $rngCol(P) = A_P^r \in att(R_C)$, and $R_C.A_P^r \subseteq R_D.Id$, where A_P^r is a foreign key referring from R_C to R_D, e.g., *inProceed*;

Table 2. Mapping predicates of ontology BibOn into database schema BibDb.

Name	Class	table	domCol	rngCol	inclDep
Person	UP_E	*Person*	*Id*		
name	BP_{fd}	*Person*	*Id*	*Name*	
affiliation	BP_{md}	*Affiliation*	*PersonId*	*Affiliation*	*Affiliation*[*PersonId*] $\subseteq Person[Id]$
authorOf	BP_{mo}	*AuthorPaper*	*PersonId*	*PaperId*	*AuthorPaper*[*PersonId*] $\subseteq Person[Id]$ *AuthorPaper*[*PaperId*] $\subseteq Paper[Id]$
inProceed	BP_{fo}	*Paper*	*Id*	*ProceedId*	*Paper*[*ProceedId*] $\subseteq Proceedings[Id]$

- if $P \in \mathsf{BP}_{mo}$, C is domain of P, D is range of P, then: $table(P) \neq R_C$, $table(P) \neq R_D$, $domCol(P) = A_P^d \in att(R_P)$, $rngCol(P) = A_P^r \in att(R_P)$, and $R_P.A_P^d \subseteq R_C.Id$, $R_P.A_P^r \subseteq R_D.Id$, where A_P^d is a foreign key referring from R_P to R_C and A_P^r is a foreign key referring from R_P to R_D, e.g., *authorOf*.

A mapping on schema level is specified by means of the following mapping rules (see source-to-target dependencies studied in [5,8]):

1. For each $C \in \mathsf{UP}_E$: $\forall x \ (C(x) \rightarrow \exists r \ (R_C(r) \wedge r.Id = x))$.
2. For each $P \in \mathsf{BP}_{fd}$, and $dom(P,C) \in \mathcal{R}$:
   $\forall x,y \ (P(x,y) \rightarrow \exists r \ (R_C(r) \wedge r.Id = x \wedge r.A_P^r = y))$.
3. For each $P \in \mathsf{BP}_{md}$, and $dom(P,C) \in \mathcal{R}$:
   $\forall x,y \ (P(x,y) \rightarrow \exists r,s \ (R_P(r) \wedge R_C(s) \wedge r.Id = x \wedge r.A_P^r = y \wedge s.Id = x))$.
4. For each $P \in \mathsf{BP}_{fo}$, $dom(P,C) \in \mathcal{R}$, and $rng(P,D) \in \mathcal{R}$:
   $\forall x,y \ (P(x,y) \rightarrow \exists r,s \ (R_C(r) \wedge R_D(s) \wedge r.Id = x \wedge r.A_P^r = y \wedge s.Id = y))$.
5. For each $P \in \mathsf{BP}_{mo}$, $dom(P,C) \in \mathcal{R}$, and $rng(P,D) \in \mathcal{R}$:
   $\forall x,y \ (P(x,y) \rightarrow \exists r,s,t \ (R_P(r) \wedge R_C(s) \wedge R_D(t) \wedge r.A_P^d = x \wedge s.Id = x \wedge r.A_P^r = y \wedge t.Id = y))$.

Note that the mapping rules above take into account also inclusion dependencies. Thus, it is guaranteed that the set of answers to a faceted query executed against the ontology is equal to the query evaluated in the relational database.

4   Faceted Search over Bibliography Ontology

Faceted search implies a new approach to modeling and perceiving data. It allows to see information objects in a multidimensional information space, like in multidimensional datawarehouse modeling. For example, *conferences* can be perceived in a multidimensional space determined by such dimensions (called *facets*) as *time*, *location*, *authors* of papers, *publishers* of proceedings, etc.

However, in contrast to modeling in datawarehousing, where the distinction between target data (measures) and dimensions is fixed, in the case of faceted-oriented modeling this perception can change from query to query. For example, in one query the target information can be *persons* in a space determined by *conferences*, *papers* and *universities*. In another – *conferences* in a space determined by *persons* and *publishers*, etc.

An ontology, like that in Fig. 2, is in general a complex and large semantic network. Any unary predicate in this ontology can be treated as the target object. Then the others determine the multidimensional information space used to search the expected set of target objects. Thus, in one ontology can coexist many such information spaces.

To explore the ontology and utilize it to formulate queries, we implemented in DAFO an approach based on faceted search. In this implementation we distinguish the following three steps: (a) providing a faceted interface and initializing a faceted query by means of a keyword query, (b) refining the faceted query, (c) transforming the faceted query into an executable form.

4.1   Keyword Queries and Faceted Interfaces

A keyword query in DAFO is a partially ordered set of unary predicate names from the underlying ontology, $kq = (C_0, \ldots, C_N)$, $C_i \in$ UP, for $0 \le i \le N$, where C_0 is the type of expected answers (target objects). Elements in the sequence $(C_1, \ldots, C_N)$ are used to appropriate restricting and pivoting the ontology, and to arrange it in a hierarchy consistent with the ordering of unary predicates in the keyword query. This hierarchy forms a *faceted interface*, which is used by the user to iterative and interactive refinement of the faceted query.

In Figs. 3(a) and (b), there are two faceted interfaces determined by two different keyword queries. In Fig. 3(a), a user is interested in *Papers* presented at TPDL conferences and written by some persons not yet specified. In Fig. 3(b), a user is interested in *Persons* connected to some ACM or TPDL conferences.

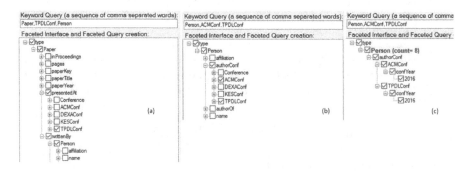

Fig. 3. Two faceted interfaces, (a) and (b), determined by two different keyword queries (checked nodes denote initial faceted queries), and a final form of faceted query (c), created over the interface (b).

In both cases, we have faceted interfaces, where checked elements form the first approximation of a created faceted query.

In both cases, the underlying ontology is pivoted in the way corresponding to the user intention expressed by the keyword query. Thus, the following two objectives are achieved: (1) the presentation of the ontology is restricted to some neighborhood of the given set of keywords (unary predicates); and (2) the ontology is somehow pivoted so that it is presented in the way conforming to the ordering of predicates in the keyword query.

A final faceted query in Fig. 3(c) is a result of operating over the faceted interface in Fig. 3(b).

4.2 Transforming Faceted Queries into First Order Faceted Queries

A faceted query is created over a faceted interface by means of selecting/unselecting nodes, inserting values of binary predicates, and discarding unselected nodes. During creation of faceted queries, the user is informed about the number of answers corresponding to the current form of the query.

The query in Fig. 3(c), has the following meaning:

"Get persons who are authors of papers presented at an ACM conference in year 2016, or at a TPDL conference in year 2016".

The textual form of this query is:

$$\alpha = \{Person\}[(authorConf, \text{any})/\{ACMConf[(confYear, \{``2016"\})], \atop TPDLConf[(confYear, \{``2016"\})]\}]. \quad (1)$$

The formal syntax of a faceted query α is [14,15]:

$$\alpha ::= t \mid t[\beta] \mid \alpha \vee \alpha \atop \beta ::= b \mid b/\alpha \mid \beta \wedge \beta, \quad (2)$$

where: (a) t is a set $\{C_1, \ldots, C_n\}$ of unary predicates; (b) b is a pair (P, any) or $(P, \{a_1, \ldots, a_n\})$, where any denotes *any* constant, and $\{a_1, \ldots, a_n\}$ is a set of allowed constants (possible values of property P).

A faceted query with syntax (2) is transformed to a first order faceted query (FOFQ), which is in the class of MPEQs. The transformation is made by means of the following semantic function $[\![\alpha]\!]_x$:

$$[\![\{A_1, \ldots, A_n\}]\!]_x = A_1(x) \vee \cdots \vee A_n(x) \qquad [\![t[b]]\!]_x = [\![t]\!]_x \wedge \exists y([\![b]\!]_{x,y})$$
$$[\![\{a_1, \ldots a_n\}]\!]_x = (a_1 = x) \vee \cdots \vee (a_n = x) \qquad [\![t[b/\alpha]]\!]_x = [\![t]\!]_x \wedge \exists y([\![b]\!]_{x,y} \wedge [\![\alpha]\!]_y)$$
$$[\![(R, \text{any})]\!]_{x,y} = R(x, y) \qquad [\![t[\beta_1 \wedge \beta_2]]\!]_x = [\![t[\beta_1]]\!]_x \wedge [\![t[\beta_2]]\!]_x$$
$$[\![(R, \{a_1, \ldots, a_n\})]\!]_{x,y} = R(x, y) \wedge [\![\{a_1, \ldots a_n\}]\!]_y \qquad [\![\alpha_1 \vee \alpha_2]\!]_x = [\![\alpha_1]\!]_x \vee [\![\alpha_2]\!]_x.$$

In result, a monadic positive existential query is obtained. For example, for the faceted query (1), we obtain:

$$[\![\alpha]\!]_x = Person(x) \wedge \exists x_1(authorConf(x, x_1) \wedge \atop (ACMConf(x_1) \wedge \exists x_2(confYear(x_1, x_2) \wedge x_2 = ``2016") \vee \atop TPDLConf(x_1) \wedge \exists x_2(confYear(x_1, x_2) \wedge x_2 = ``2016"))). \quad (3)$$

4.3   Rewriting FOFQs into Extensional Form

In FOFQs may occur both extensional and intentional predicates. In the rewriting process all intentional predicates are replaced with extensional ones using deductive rules from the ontology [13]. The rewriting algorithm recursively looks for intentional predicates. If C (or P) is such a predicate, then a rule with C (or P) occurring on its right-hand side is used in the rewriting procedure. The rewriting concerns the entire atom, i.e., $C(x)$ (or $P(x, y)$), and the atom is replaced by the left-hand side of the rule with appropriate substitution of variables. In result, some new intentional predicates can appear in the query, so the process of rewriting must be repeated recursively. If the set of rules is not recursive with respect to intentional predicates, and is complete, i.e., any intentional predicate occurs on the right hand side of some rule, then the rewriting process ends successfully.

For example, the atom containing intentional predicates $authorConf$ (see (3) and Fig. 4(b)) has the following rewriting (Fig. 4(c)) in ontology BibOn:

$$rewrite_{BibOn}(authorConf(x, x_1)) = \exists x_5(authorOf(x, x_5) \wedge Paper(x_5) \\ \wedge \exists x_6(inProceed(x_5, x_6) \wedge Proceedings(x_6) \wedge ofConf(x_6, x_1))). \tag{4}$$

In Fig. 4(a), we give a slightly modified version of the faceted query from Fig. 3(c) (requirements about affiliation of authors are added). The FOFQ before rewriting is presented in Fig. 4(b), and FOFQ after rewriting is in Fig. 4(c). Queries are depicted as syntactic trees, where all variables except x are quantified existentially. In Fig. 4(d) there is a sample set of answers to the query in DAFO.

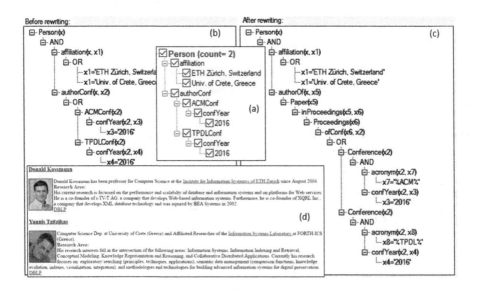

Fig. 4. Sample faceted query in DAFO (a), its presentation as: FOFQ tree before rewriting (b); FOFQ tree after rewriting (c); and answers to it (d).

4.4   Answering Faceted Queries - Experimental Evaluation

FOFQs are translated to SQL queries over relational database using the mapping defined in Sect. 3. A result SQL query is executed in relational databases using a commercial RDBMS (SQL Server, in this case). Advanced optimization capabilities provided by RDBMS guarantee high efficiency. This was verified in computational experiments made in DAFO system with the following setting: (a) a database containing: 3818 papers, 1907 conferences, 1853 proceedings, and 61 persons; (b) computation environment: 2.60 GHz Intel Core i7 processor, and 8GB RAM memory; (a) ontology with 182 elements (predicates and rules). Results of evaluations are given in Table 3 (query q3 is that in Fig. 4). Time costs (in milliseconds) are divided into *total preparing* and *execution* costs. The preparing time highly depends on both the size of query and ontology (in our experiments the ontology and the database were fixed).

Table 3. Evaluation of time costs for preparing and executing faceted queries.

Query	#Nodes after rewriting	Creation [msec]	Rewriting [msec]	Translation [msec]	Total preparing [msec]	Execution [msec]
q1	5	12	23	2	37	22
q2	9	6	58	8	72	29
q3	24	32	122	13	167	45
q4	37	53	210	16	279	57

We can observe that there is a linear relationship between the size of queries (expressed in the number of nodes in its syntactic tree after rewriting) and preparing and execution times. These relationships are presented in Fig. 5.

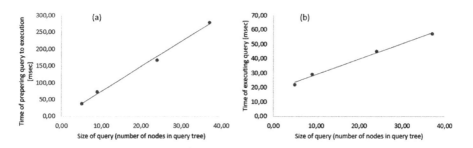

Fig. 5. Time of preparing faceted query to execution (a) and execution time (b) depending on the number of nodes in FOFQ tree after rewriting.

5 Summary

In this paper, we exploited the application of the ontology based data access (OBDA) approach equipped with faceted search utilities to explore bibliography databases. We discussed a way of using ontology to describe bibliography information space. Universality and flexibility of an ontology depends on the choice of deductive and integrity constraint rules. The former are used to deduce new facts and the latter to check correctness of data. Both are significant in rewriting queries. We proposed a way of presenting the ontology to users in conformance with the faceted search methodology. The implemented system DAFO provides users with a graphical interface allowing the user for interactive and iterative creation of faceted queries. We shown how the ontology can be mapped to a relational database. This mapping together with translation queries to SQL enables high efficiency of query execution. The crucial in achieving this efficiency was the usage of a commercial RDBMS with excellent optimization capabilities.

This research has been supported by Polish Ministry of Science and Higher Education under grant 04/45/DSPB/0163.

References

1. Arenas, M., Grau, B.C., Kharlamov, E., Marciuska, S., Zheleznyakov, D.: Faceted search over ontology-enhanced RDF data. In: ACM CIKM 2014, pp. 939–948. ACM (2014)
2. Baader, F., Calvanese, D., McGuinness, D., Nardi, D., Petel-Schneider, P. (eds.): The Description Logic Handbook: Theory, Implementation and Applications. Cambridge University Press, Cambridge (2003)
3. Bak, J., Blinkiewicz, M.: RuQAR: Querying OWL 2 RL ontologies with rule engines and relational databases. In: 9th International Conference on Computational Collective Intelligence (ICCCI 2017), LNAI, Springer (2017). (in print)
4. Catarci, T., Costabile, M.F., Levialdi, S., Batini, C.: Visual query systems for databases: A survey. J. Vis. Lang. Comput. **8**(2), 215–260 (1997)
5. ten Cate, B., Kolaitis, P.G.: Structural characterizations of schema-mapping languages. Commun. ACM **53**(1), 101–110 (2010)
6. DBLP Computer Science Bibliography. http://dblp.org/ (2017)
7. Dumais, S.T.: Faceted search. In: Liu, L., Tamer Özsu, M. (eds.) Encyclopedia of Database Systems, pp. 1103–1109. Springer, Heidelberg (2009)
8. Fagin, R., Kolaitis, P.G., Miller, R.J., Popa, L.: Data exchange: semantics and query answering. Theor. Comput. Sci **336**(1), 89–124 (2005)
9. Krishnamurthi, S., Gray, K.E., Graunke, P.T.: Transformation-by-Example for XML. In: Pontelli, E., Santos Costa, V. (eds.) PADL 2000. LNCS, vol. 1753, pp. 249–262. Springer, Heidelberg (1999). doi:10.1007/3-540-46584-7_17
10. Ley, M.: DBLP - some lessons learned. PVLDB **2**(2), 1493–1500 (2009)
11. Motik, B., Horrocks, I., Sattler, U.: Bridging the gap between OWL and relational databases. J. Web Semant. **7**(2), 74–89 (2009)
12. OWL 2 Web Ontology Language Profiles. www.w3.org/TR/owl2-profiles (2009)
13. Pankowski, T.: Rewriting and executing faceted queries over ontology-enhanced databases. In: 21st Conference on Knowledge-Based and Intelligent Systems (KES 2017). Procedia Computer Science, Elsevier (2017, in print)

14. Pankowski, T., Brzykcy, G.: Data Access Based on Faceted Queries over Ontologies. In: Hartmann, S., Ma, H. (eds.) DEXA 2016. LNCS, vol. 9828, pp. 275–286. Springer, Cham (2016). doi:10.1007/978-3-319-44406-2_21
15. Pankowski, T., Brzykcy, G.: Faceted Query Answering in a Multiagent System of Ontology-Enhanced Databases. In: Jezic, G., Chen-Burger, Y.-H.J., Howlett, R.J., Jain, L.C. (eds.) Agent and Multi-Agent Systems: Technology and Applications. SIST, vol. 58, pp. 3–13. Springer, Cham (2016). doi:10.1007/978-3-319-39883-9_1
16. Papadakos, P., Tzitzikas, Y.: Hippalus: Preference-enriched faceted exploration. In: Workshops of the EDBT/ICDT. CEUR Workshop Proceedings, vol. 1133, pp. 167–172. CEUR-WS.org (2014)
17. Tunkelang, D.: Faceted Search. Morgan & Claypool Publishers, San Rafael (2009)
18. Zloof, M.M.: Query-by-example: A data base language. IBM Syst. J. **16**(4), 324–343 (1977)

What Should I Cite? Cross-Collection Reference Recommendation of Patents and Papers

Julian Risch[✉] and Ralf Krestel

Hasso-Plattner-Institut, Prof.-Dr.-Helmert-Str. 2–3, 14482 Potsdam, Germany
{Julian.Risch,Ralf.Krestel}@hpi.de

Abstract. Research results manifest in large corpora of patents and scientific papers. However, both corpora lack a consistent taxonomy and references across different document types are sparse. Therefore, and because of contrastive, domain-specific language, recommending similar papers for a given patent (or vice versa) is challenging.

We propose a recommender system that leverages topic distributions and keywords to recommend related work despite these challenges. As a case study, we evaluate our approach on patents and papers of two fields: medical and computer science. We find that topic-based recommenders complement word-based recommenders for documents with collection-specific language and increase mean average precision by up to 27%. As a result of our work, publications from both corpora form a joint digital library, which connects academia and industry.

Keywords: Recommender systems · Text mining · Topic modeling

1 Searching for Related Work Across Patents and Papers

More than 1.2 million patents will be granted[1] and more than 1.5 million scientific papers will be published in 2017 according to bibliometric growth models [1]. These large collections form an extensive library of latest research results in an almost unstructured form, thus challenging to mine automatically. Searching for related work in papers is an important task for academic researchers. Similarly, patent applicants search for prior art to prove novelty and to define scope. Prior art denotes publicly available, state-of-the-art information in any form. Therefore, it is not limited to patents but includes also papers. Content-based recommender systems for text documents typically rely on tf-idf-based measures to identify representative keywords of a document and to recommend similar documents. However, linguistic differences of patents and papers are challenging for word-based recommender systems: Although a patent and a paper deal with the same topic, they might use different words to describe their work.

Because patents claim the scope of an invention, they cover as much variation of the invention as possible. As a consequence, patent descriptions use vague

---

[1] http://www.wipo.int/edocs/pubdocs/en/wipo_pub_941_2016.pdf.

© Springer International Publishing AG 2017
J. Kamps et al. (Eds.): TPDL 2017, LNCS 10450, pp. 40–46, 2017.
DOI: 10.1007/978-3-319-67008-9_4

language, such as "electronic imaging apparatus", whereas a paper might call the same invention "digital camera". Moreover, patents have specific linguistic characteristics, such as a higher frequency of words with indefinite, general meaning. Approximately 1% of scientific papers cite at least one patent [3]. Because existing references across patents and papers are sparse, we assume that these references are not suited to train graph-based recommender systems.

With this work, we propose cross-collection topic modeling to bridge the linguistic gap between patents and scientific papers. Based on topic distributions, we identify and recommend topically similar patents and papers even if they do not share keywords. In contrast to manual classification with inconsistent taxonomies, topic modeling is an unsupervised machine learning technique. As a consequence, our approach allocates topics to millions of documents automatically. We present two case studies on datasets consisting of U.S. patents, computer science papers, and medical articles. For an evaluation on these datasets, we use existing references as a gold standard for recommendations and compare the mean average precision (MAP) of topic-based, word-based, and combined recommender systems.

2   Related Work

Mining Patents and Papers. More than 200 research articles address recommender systems for scientific papers. For example, Liu et al. mine citation graphs of computer science literature to predict further citations [5]. Most recently, Momeni et al. evaluate how co-authorship networks support author name disambiguation for common names [8]. Wang et al. identify topics in patents based on latent Dirichlet allocation (LDA) and noun phrase extraction [10]. They compare different institutions with regards to their patents' topic distributions. Krestel et al. propose a recommender system for patents based on topic modeling and document ranking techniques [4]. Although patent mining and paper mining face similar challenges, such as keyword extraction and topic modeling, they form two separate research fields. Especially different document style and the variation of wording limits the capabilities of holistic approaches. For example, Google Scholar[2] provides a search interface for patents and papers, but its word-based approach neglects linguistic contrasts.

Topic Modeling. Wang et al. combine collaborative filtering and topic modeling to recommend scientific papers in a user's field of interest [11]. Given a citation graph, Mei et al. propose a concept of "topical inheritance" and enforce similar topic distributions in cited and citing documents [7]. However, all previous approaches consider only single collections and neglect linguistic contrasts of patents and papers. Extending LDA, Paul et al. model topics across multiple corpora with cross-collection LDA (ccLDA) [9]. Their approach considers collection-specific and collection-independent word distributions per topic but not in the domain of recommender systems or patents and scientific papers.

---

[2] https://scholar.google.com.

Cross-Domain Recommendation. To match query terms and document terms in heterogeneous digital libraries, Mayr et al. propose to manually map terminology from one controlled vocabulary to another [6]. However, this approach requires an enormous manual effort. With our cross-collection topic model, we automate the matching of collection-specific and collection-independent terms. Recently, recommender systems have been proposed to transfer users' rating patterns from one domain to another, such as movies and books [2]. However, such cross-domain recommender systems rely on users' rating histories to transfer knowledge and our task lacks user ratings. To the best of our knowledge, so far no research addresses patents and scientific papers as a joint library of related work. Neither cross-collection topic models nor cross-domain recommender systems have been used to bridge the linguistic gap between both corpora.

3   Jointly Recommending Patents and Papers

To recommend similar patents or papers, we propose two complementing similarity measures based on (i) keywords and (ii) topic distributions. Whereas word-based similarity is fine-grained, topic-based similarity is coarser-grained.

Keyword Similarity. For each document, we extract 10 representative keywords[3] with highest tf-idf scores. The similarity of two documents is calculated based on this keyword vector representation. While keywords are an established relevance measure for document retrieval systems, such as Elasticsearch[4], they are constrained by the exact wording in a document. This limitation emerges as a problem on patents and papers, because they make intensive use of collection-specific language. Even closely related documents may not have any keywords in common.

Topic Distribution Similarity. To reveal documents with similar latent topics across both collections, we adapt the topic model ccLDA and distinguish patent-specific, paper-specific and collection-independent word distributions per topic. In contrast to Paul et al., we distinguish collection-specific and collection-independent word types instead of word tokens. Types with similar frequency in patents and papers are modeled with a single, collection-independent probability, whereas all other types are modeled with multiple, collection-specific probabilities. Because collection-specific and collection-independent word types together constitute a topic, even documents that have no words in common can share the same topic distribution. We train the adapted ccLDA model and estimate topic distribution, collection-specific word distributions, and collection-independent word distribution in 500 iterations[5]. To compare documents based on their topic distribution, we use cosine similarity.

---

[3] Larger keyword vectors increase runtime but do not improve result quality.

[4] https://www.elastic.co/products/elasticsearch.

[5] parameters set as suggested in the original paper: $\beta = 0.01$, $\delta = 0.01$, $\gamma_1 = 1$, $\gamma_2 = 1$.

Table 1. The number of documents and gold standard references per dataset

	#Patents	#Papers	#References
Computer science dataset	3,377	2,443	6,488
Medical dataset	19,419	21,921	70,588

Word-Based and Topic-Based Recommender System. We propose a recommender system that leverages the best combination of these two similarity measures to rank and recommend related work across patents and papers. We transfer the concept of explicit relevance feedback in information retrieval systems, where users evaluate initial query results to control subsequent queries and improve relevance. In our scenario a patent applicant wants to retrieve relevant papers for his patent. Based on this patent, keyword-based recommendations are presented. If the user's information need is not fulfilled, the recommendation approach can be manually switched to topic-based recommendations. We do not rely on an automatically switching hybrid but on the explicit decision of the user.

4   Case Study

Our evaluation task is to recommend related papers for each patent in our datasets. Although this limited case study considers only inter-collection references from patents to papers, our approach works also for references from papers to patents as well as intra-collection references without any adjustments. We evaluate the mean average precision of the top 100 recommendations, MAP@100.

Datasets. The first dataset contains granted U.S. patents and referenced ACM papers. We extract patent abstracts from United States Patent and Trademark Office (USPTO) publications and ACM paper abstracts from a citation network dataset[6]. We assume that referenced documents are related work. Therefore, cross-collection references serve as the gold standard for recommendations in our evaluation. The second dataset is based on medical research projects funded by the National Institutes of Health (NIH). These projects are required to list their patent and paper publications in a public database[7]. We consider projects with at least one patent and one paper. Publications of the same project are assumed to be related work and therefore serve as the gold standard for reference recommendations. Table 1 lists the number of documents and cross-collection references for each dataset.

   With a preliminary experiment for the topic-based approach, we determine the number of topics with the highest MAP@100 per dataset. To this end, we split the dataset by time: We determine the number of topics on the oldest 50% of the documents and use the most recent 50% for the final evaluation of

---

[6] https://bulkdata.uspto.gov/ and https://aminer.org/citation.
[7] https://exporter.nih.gov/.

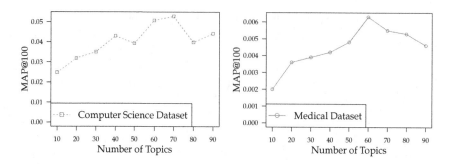

Fig. 1. MAP@100 of the topic-based approach for different numbers of topics

Table 2. MAP@100 comparison of topic-based, word-based, and combined approach

	Topic-based	Word-based	Best Comb
Computer science dataset (70 Topics)	0.0528	0.1332	0.1696
Medical dataset (60 Topics)	0.0068	0.0372	0.0414

MAP@100. According to the results visualized in Fig. 1, we set the number of topics to 70 for the computer science dataset and to 60 for the medical dataset. We find that MAP@100 is consistently approximately one order of magnitude higher for the computer science dataset compared to the medical dataset. We assume, recommendation on the medical dataset is a more difficult task because of larger corpus size.

Recommendation Quality. Table 2 illustrates that the topic-based and the word-based approach are significantly outperformed by the best combination of both recommendation approaches on both evaluation datasets. Especially on the computer science dataset, the best combination achieves a 27% higher MAP@100 than the word-based approach. On the medical dataset, the MAP@100 is 11%

Table 3. Top three recommendations for the patent "Method and apparatus for enhancing data storage efficiency". Relevant recommendations are in bold print.

Word-Based Paper Recommendations
1. Improving locality of reference in a garbage collecting memory management system
2. Garbage collection in a large LISP system
3. Page placement algorithms for large real-indexed caches
Topic-Based Paper Recommendations
1. A real-time garbage collector based on the lifetimes of objects
2. Garbage collection in a large LISP system
3. Design of the opportunistic garbage collector

higher. The experiment results demonstrate also that keywords are superior to topic distributions as a feature for recommending related work. However, their combination achieves the by far best results in our evaluation. Table 3 exemplifies a patent and its top three paper recommendations.

5   Conclusions and Future Work

In order to recommend patent and paper references despite their linguistic differences, we proposed a recommender system based on keywords and topic distributions of a cross-collection topic model. Experiment results demonstrate the effectiveness of this combination on two datasets of publications in the fields of medical and computer science. The combined approach outperforms word-based approaches by up to 27% MAP@100. A promising path for future work is to combine content-based and collaborative recommendation across patents and papers. For example, authors could be compared based on their co-authorship relations or citation history. Furthermore, word-based and topic-based approaches could be combined for an automatic diversification of recommendations.

References

1. Bornmann, L., Mutz, R.: Growth rates of modern science: a bibliometric analysis based on the number of publications and cited references. J. Assoc. Inf. Sci. Technol. (JASIST) **66**(11), 2215–2222 (2015)
2. Gao, S., Luo, H., Chen, D., Li, S., Gallinari, P., Guo, J.: Cross-domain recommendation via cluster-level latent factor model. In: Blockeel, H., Kersting, K., Nijssen, S., Železný, F. (eds.) ECML PKDD 2013. LNCS, vol. 8189, pp. 161–176. Springer, Heidelberg (2013). doi:10.1007/978-3-642-40991-2_11
3. Glänzel, W., Meyer, M.: Patents cited in the scientific literature: an exploratory study of 'reverse' citation relations. Scientometrics **58**(2), 415–428 (2003)
4. Krestel, R., Smyth, P.: Recommending patents based on latent topics. In: Proceedings of the Conference on Recommender Systems (RecSys), pp. 395–398. ACM (2013)
5. Liu, Y., Niculescu-Mizil, A., Gryc, W.: Topic-link LDA: joint models of topic and author community. In: Proceedings of the International Conference on Machine Learning (ICML), pp. 665–672. ACM (2009)
6. Mayr, P., Mutschke, P., Petras, V.: Reducing semantic complexity in distributed digital libraries: treatment of term vagueness and document re-ranking. Libr. Rev. **57**(3), 213–224 (2008)
7. Mei, Q., Cai, D., Zhang, D., Zhai, C.: Topic modeling with network regularization. In: Proceedings of the International Conference on World Wide Web (WWW), pp. 101–110. ACM (2008)
8. Momeni, F., Mayr, P.: Using co-authorship networks for author name disambiguation. In: Proceedings of the Joint Conference on Digital Libraries, pp. 261–262. ACM (2016)
9. Paul, M., Girju, R.: Cross-cultural analysis of blogs and forums with mixed-collection topic models. In: Proceedings of the Conference on Empirical Methods in Natural Language Processing (EMNLP), pp. 1408–1417. ACL (2009)

10. Wang, B., Liu, S., Ding, K., Liu, Z., Xu, J.: Identifying technological topics and institution-topic distribution probability for patent competitive intelligence analysis: a case study in LTE technology. Scientometrics **101**(1), 685–704 (2014)
11. Wang, C., Blei, D.M.: Collaborative topic modeling for recommending scientific articles. In: Proceedings of the International Conference on Knowledge Discovery and Data Mining (SIGKDD), pp. 448–456. ACM (2011)

Corpora

Taxonomic Corpus-Based Concept Summary Generation for Document Annotation

Ikechukwu Nkisi-Orji[1]([✉]), Nirmalie Wiratunga[1], Kit-Ying Hui[1],
Rachel Heaven[2], and Stewart Massie[1]

[1] Robert Gordon University, Aberdeen, UK
{i.o.nkisi-orji,n.wiratunga,k.hui,s.massie}@rgu.ac.uk
[2] British Geological Survey, Nottingham, UK
reh@bgs.ac.uk

Abstract. Semantic annotation is an enabling technology which links documents to concepts that unambiguously describe their content. Annotation improves access to document contents for both humans and software agents. However, the annotation process is a challenging task as annotators often have to select from thousands of potentially relevant concepts from controlled vocabularies. The best approaches to assist in this task rely on reusing the annotations of an annotated corpus. In the absence of a pre-annotated corpus, alternative approaches suffer due to insufficient descriptive texts for concepts in most vocabularies. In this paper, we propose an unsupervised method for recommending document annotations based on generating node descriptors from an external corpus. We exploit knowledge of the taxonomic structure of a thesaurus to ensure that effective descriptors (concept summaries) are generated for concepts. Our evaluation on recommending annotations show that the content that we generate effectively represents the concepts. Also, our approach outperforms those which rely on information from a thesaurus alone and is comparable with supervised approaches.

Keywords: Taxonomy · Text annotation · Information discovery

1 Introduction

Digital library resources that were not born-digital are increasingly being made available for electronic access through mass digitisation efforts. Unlocking the content of such resources to enhance search and browse remains a challenge as facilities for content linking and navigation are often absent. Semantic annotation plays an important role in this regard by mapping the content of documents to unambiguous concepts from controlled vocabularies (or thesauri). The thesaurus models an organisation of knowledge in a domain and when used to annotate documents, is expected improve organisation, access and dissemination [1]. Accordingly, several digital repositories have controlled vocabularies from which authors and annotators select concepts to annotate or tag digital

© Springer International Publishing AG 2017
J. Kamps et al. (Eds.): TPDL 2017, LNCS 10450, pp. 49–60, 2017.
DOI: 10.1007/978-3-319-67008-9_5

content. Popular thesauri for knowledge organisation include the Medical Subject Headings (MeSH) and Library of Congress Subject Headings (LCSH). The selection of concepts for use in annotation largely rely on manual efforts which is tedious, time-consuming, and lacks scalability. Controlled vocabularies can contain several thousand concepts making it difficult to find the right concepts for annotation. Although it may not be possible to fully automate the annotation process as it is quite subjective, the ability to recommend a useful subset of concepts will reduce the burden on annotators.

Semantic annotation can be done at different levels of granularity (e.g. entire documents, sections/chapters, or specific terms) and approaches for recommending annotations differ accordingly. While a high-level understanding of content may be sufficient when annotating an entire document, areas such as named entity recognition, word sense disambiguation, and co-reference resolution are more pertinent to annotating specific terms. This work focuses on the annotation of segments of documents (e.g. chapters and sections) which is especially useful for books and other publications that can cover a range of domain topics. Digital agents can reuse such annotations in bespoke ways such as to meet an information need by dynamically assembling a document using relevant segments of other documents. The strategies for annotating segments of documents can be generalised for annotating entire documents. Accordingly, we treat segments of documents as individual documents. Also, we use the terms thesaurus and controlled vocabulary interchangeably and in either case, assume a taxonomy of domain concepts.

The most effective approaches for recommending annotations rely on reusing the concepts that were assigned to documents in an annotated corpus which have features that are similar to the document being annotated. However, such supervised methods make it difficult to recommend concepts that do not appear in the annotated corpus. Also, an annotated corpus has to be created before use for annotating new documents. Alternative approaches that do not require an annotated corpus often rely on the use of thesaurus-based features (concept terms, synonyms, and descriptions) to recommend annotations. Relying on thesaurus-based features can lead to poor results as controlled vocabularies often lack sufficient textual content that effectively describe concepts. In this work, we use a corpus-based approach for generating descriptive texts for concepts (concept summaries) which are subsequently used for recommending annotations for documents. Our main contribution is the generation of concept summaries from a corpus which are sufficiently descriptive of the concepts in a thesaurus. A key process in our approach is the use of knowledge of semantic relatedness between the concepts of a thesaurus to identify the documents from which concept summaries are extracted. In our evaluation, we use generated concept summaries for recommending annotations and compare its performance to alternative approaches.

The remainder of this paper is organised as follows: Sect. 2 reviews relevant literature; Sect. 3 presents our corpus-based approach for generating concept summaries and recommending document annotations; Sect. 4 is an experimental evaluation which compares our approach to alternative approaches; and Sect. 5 concludes with an outline for future work.

2   Related Work

We categorise the popular approaches for recommending annotations in the literature as either supervised or unsupervised methods as shown in Fig. 1.

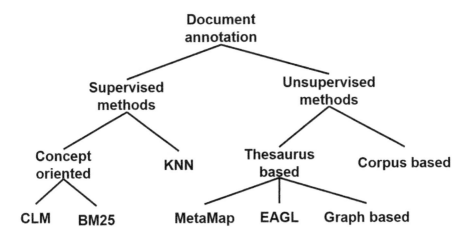

Fig. 1. Document annotation approaches.

The supervised methods reuse the annotations of previously tagged documents that share some similarity to the document being annotated. The intuition is that a new document can inherit some or all the annotations that were assigned to similar previously seen documents. Concept-oriented approaches generate concept summaries by merging all the documents that have been annotated with a concept. Concept summaries are then indexed so that a document to be annotated forms a query for which relevant summaries are retrieved. The corresponding concepts for top ranked summaries are recommended as annotations. Popular approaches for retrieving relevant concept summaries are CLM and BM25 [2]. CLM uses a language model for retrieval while BM25 uses the Okapi BM25 ranking function [3]. K-nearest neighbour (KNN) is a state-of-the-art supervised approach and is used in systems such as the Medical Text Indexer (MTI) [4,5]. Instead of generating concept summaries, KNN adds each annotated document to an index. When a new document is to be annotated, it uses a document ranking function to retrieve K most similar documents. The annotations of retrieved documents form candidate concepts to be recommended. One variant of KNN ranks candidate concepts by combining the relevance scores of documents for which they form annotations [6,7]. Another variant passes the features of candidate concepts to a machine classifier which determines which concepts to put forward for annotation [8]. Some features that are used by a classifier include the proportion of retrieved documents that were annotated with the concept and if the concept appears in the title or content of a document. Experimental results show that KNN or hybrids of it are most effective in recommending annotations [2,8]. However, supervised approaches cannot be used when a corpus of annotated documents does not exist. It is also difficult to effectively recommend concepts that appear sparingly or are absent from the annotated corpus.

Unsupervised methods do not require an annotated corpus when recommending annotations. They rely on thesaurus-based indicators or resources that are generated from external sources. MetaMap parses the document to be annotated to identify exact and partial mappings to concept terms. Identified mappings form candidate concepts for annotation and using several linguistic principles, a ranking of concepts is generated [9]. Some considerations for ranking candidate concepts include the number of times it appears in a document and if it was a partial or complete match. Similar to the supervised concept-oriented approaches, EAGL generates concept summaries so that the concepts whose summaries are most similar to a document are recommended. The concept summaries are generated by merging the textual features of each concept (e.g. synonyms and descriptions). Concept summaries are indexed and a variety of retrieval approaches can be used to retrieve summaries (e.g. vector space model). Experiments show that EAGL performs better than MetaMap [2]. However, both approaches perform poorly when compared with the supervised methods. In MetaMap, the inability to disambiguate terms in documents was cited as one reason for poor performance [9]. EAGL is fast and efficient but controlled vocabularies often lack sufficient textual content to generate effective concept summaries.

In [10], documents are annotated with DBpedia concepts using a graph-based approach. First, the key terms in a document are identified and linked to corresponding DBpedia concepts. Titles of DBpedia entries form concept terms while corresponding textual contents provide textual context for disambiguating terms in documents. The DBpedia graph structure is then analysed to identify central nodes which connect the concepts that were linked to the document. These central concepts are used to annotate the document. Although the results are promising, this approach is suitable if the intent is to annotate with DBpedia or a similar knowledge resource with rich textual content. When using a different thesaurus, candidate concepts may not have equivalent DBpedia entries. This is especially true in specialised domains whose concepts may not have DBpedia entries. An analysis of the geoscience-related concepts used in this work showed that over 50% of the concepts have no corresponding DBpedia entries. Also, DBpedia often conflate concepts (e.g. "Rocks" and "Rock type" point to the same article). It may be desirable to maintain subtle differences in specialised domains. In this work, we adopt an approach that is similar to EAGL but augment thesaurus-based concept summaries with node descriptors from an external corpus. The use of external corpus such as Wikipedia has helped in generating additional useful information to aid the alignment/matching of concepts from different taxonomies [11].

3   Corpus-Based Concept Summaries

We use an external corpus to generate concept summaries for the concepts in a thesaurus. When generating a concept's summary, other concepts in its neighbourhood are used for disambiguation forming a semantic filter which ensures that the summary generated is relevant to a concept. This relies on a taxonomic structure of the thesaurus to measure semantic relatedness between concepts. A high-level overview of the process for generating a concept summary is presented in Fig. 2. We summarise the process in the following steps:

1. The concept term (textual label) is issued as a query to retrieve documents from a corpus. We refer to this concept term as *query concept.*
2. The documents retrieved in step 1 are mapped to the thesaurus to identify the concepts expressed in them. We refer to the set of concepts that are identified in a document as *document concepts.*
3. Each document in step 2 is re-ranked based on the semantic overlap between the query concept and document concepts. We use a semantic relatedness algorithm to measure semantic overlap.
4. The query-biased snippets of top ranked documents in step 3 are extracted and merged to form a concept summary.
5. Steps 1–4 is repeated for all the concepts in the thesaurus generating a corpus of concept summaries.

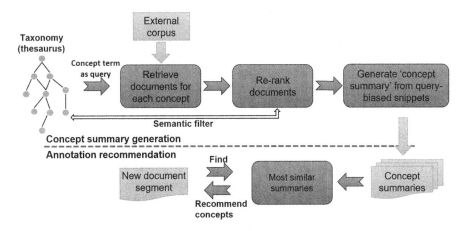

Fig. 2. Overview of concept summary generation which is used for annotation recommendation.

Generated concept summaries are indexed for use in recommending annotations. In order to annotate a document, the most similar concept summaries are retrieved using the BM25 ranking function which is a state-of-the-art term vector based model. Concepts are recommended in the order of ranking of their summaries. The remainder of this section describes the steps above in more detail.

3.1   Discover Candidate Source Documents

Our intent is to generate concept summaries from documents which are most relevant to the concepts of a thesaurus. First, we identify a set of documents that are potentially relevant to each concept. Accordingly, each concept term is issued as query to a corpus that is assumed to contain documents which are relevant to the thesaurus. The documents that are retrieved for the query concept form candidate sources for generating its summary. Concept terms are often very short making it difficult to appropriately represent an information need. Due to reasons such as the presence of polysemous terms (e.g. rock: music or stone?), some of the documents that are retrieved for a concept may not be relevant. Therefore, we introduce a semantic re-ranking step to identify a subset of retrieved documents that we are more relevance to the concept.

3.2   Semantic Re-Rank of Documents

Semantic re-rank measures the degree to which a document's concepts cluster about the query concept. The intuition is that a document's relevance increases as its concepts cluster closer to query concepts on the taxonomy. To identify document concepts, we match concept terms from the thesaurus to a keyword index of documents. Both concept terms and the keyword index are stemmed to maximise match discovery. Considering that there may be polysemous terms in the keyword index and the likely introduction of errors by conflating words through stemming, we impose the requirement that a document should also contain a *semantic context* of the concept before it is deemed present. The semantic context of a concept is the set of all concepts that are directly linked to it in the thesaurus [12]. For example, the semantic context of "rock" in a geological thesaurus may include "igneous rock" and "sedimentary rock". A document that describes the music genre "rock" is unlikely to contain those semantic contexts. The outcome of mapping documents to the thesaurus is a bag-of-concepts representation for each document.

Next, we estimate the semantic closeness of a document's concepts to the query concept by cumulating pairwise semantic relatedness measures. We use the Wu and Palmer algorithm [13] as shown in Eq. 1 to measure relatedness between concepts which correlates well with human judgments of relevance [14]. The algorithm preserves the specificity cost and specialisation cost properties which are important when comparing the nodes of a taxonomy. Specificity cost property requires that relatedness between neighbouring concepts increase with greater taxonomic depth while specialisation cost property requires that further specialisation implies reduced relatedness [15]. Wu and Palmer requires finding the most specific common subsumer (MSCS) of a concept pair being compared which is the most distant concept from the root node that subsumes them.

$$rel(c_i, c_j) = \frac{2 * n(c_i, c_j)}{n(c_i) + n(c_j) + 2 * n(c_i, c_j)} \qquad (1)$$

where c_i and c_j are concepts being compared, $n(c_i)$ is minimum node count from c_i to the MSCS, $n(c_j)$ is minimum node count from c_j to the MSCS, and $n(c_i, c_j)$ is minimum node count from the MSCS to the root node.

In other words, the query concept forms a central node on the taxonomy from which document concepts are measured. Let x denote query concept and C_d denote the concepts of document d. A cumulation of pairwise semantic relatedness measures between x and C_d as shown in Eq. 2 determine the semantic relevance score of d.

$$semScore(d, x) = \sum_{x, c_i \in C_d} rel(x, c_i) \tag{2}$$

Afterwards, the documents that were retrieved for a query concept are sorted by semantic relevance scores.

3.3   Generate Concept Summaries

The final step of concept summary generation is the extraction of relevant content from top ranked documents. We generate and extract document *snippets* which are short textual summaries in search result listings for the purpose of determining relevance prior to viewing entire documents. We use a dynamic snippet generation approach that scores the sentences of a document with respect to a query and retrieves the most relevant sentences. This query-biased snippet generation approach has been shown to be effective in extracting useful document summaries and is adopted by several search engines [16]. We use the BM25 ranking function to identify relevant sentences for snippet generation. Snippets of top K documents are then merged to create a concept's summary.

4   Evaluation

We compare our approach for recommending annotations using corpus-based concept summaries (CCS) to alternative approaches in the literature.

4.1   Dataset and Experiment Setup

The evaluation dataset is from 1,948 document sections in 30 geological documents (mostly geology memoirs) which were manually annotated by domain experts in a project aimed at enhancing access content. These documents are book-like containing multiple sections[1]. Figure 3 is an example of a document section that is annotated with two concepts. The entries "value" and "scheme" refer to concepts and their source thesaurus respectively.

We selected 3 controlled vocabularies that were used to annotate the documents – BGS Geoscience Thesaurus (THESAURUS)[2], BGS Geochronology

---

[1] An example of documents used http://pubs.bgs.ac.uk/publications.html?pubID= B01745.
[2] http://www.bgs.ac.uk/discoverymetadata/13603129.html.

```
<section>
    <sectioninfo>
        <indexterm scheme="CHRONOSTRAT" value="KU"> </indexterm>
        <indexterm scheme="AMF" value="5283"> </indexterm>
    </sectioninfo>
    <title>MID-CRETACEOUS TO END CRETACEOUS REGIONAL SHELF SUBSIDENCE</title>
    <para>Crustal extension had effectively ceased by mid-Cretaceous times (e.
    established over the Manx region, with structural demarcation between bloc
    Ireland implies deposition of Chalk across the entire Manx region. The max
    </para>
</section>
```

Fig. 3. Example of an annotated document section.

(CHRONOSTRAT)[3] and BGS Lexicon of Named Rock Units (LEXICON)[4]. Concepts from these vocabularies were used 701 times (276 unique concepts) to annotate 397 document sections making an average of 1.8 concepts per document section. We use these concepts (110 from THESAURUS, 122 from LEXICON, and 44 from CHRONOSTRAT) and subset of document sections for our evaluation. We randomly select 2/3 of the dataset for training in supervised approaches and for parameter tuning, and we report results on remaining 1/3 for all approaches.

Wikipedia was used as the corpus for generating concept summaries and the vector space model (with BM25 ranking) for retrieving articles for query concepts. Specifically, we extracted a subset of Wikipedia (286,766 articles) that were tagged with one or more terms from the "Earth sciences" sub-category hierarchy. Articles from the Earth sciences category align with our evaluation dataset. We generated concept summaries from 5-sentence snippets of top 10 ranked articles as determined using the training dataset.

Alternative approaches for recommending annotations which we compare are:

SUP_{BM25}: Supervised approach which generates concept summaries using the content of all documents that were annotated with a concept. BM25 ranking function ($k1 = 1.2$, $b = 0.75$) is used to identify summaries that are most similar to an unseen document from indexed concept summaries. The concepts are recommended in the order of the relevance scores of retrieved summaries.

SUP_{LM}: Language model approach that is similar to SUP_{BM25}. We use a language model based on Dirichlet similarity ($\mu = 3500f$) to retrieve summaries from the concept summary index. Dirichlet similarity uses Dirichlet priors for Bayesian smoothing and is a popular language model retrieval approach.

SUP_{KNN}: KNN approach that indexes annotated documents and notes the annotating concepts. When a new document is to be annotated, the most similar documents are retrieved from the index. The concepts that were used to annotate top K ($K = 10$) retrieved documents are ranked by summing

---

[3] http://data.bgs.ac.uk/doc/Geochronology.html.
[4] http://data.bgs.ac.uk/doc/Lexicon.html.

the relevance scores of their respective documents. Although some previous works have used a language model for retrieval, BM25 ranking gave the best results which we report.

$UNSUP_{EAGL}$: Unsupervised approach that generates concept summaries from concept terms, synonyms, definitions, and other textual content in a thesaurus. Concept summary indexing and the process for recommending annotations are similar to SUP_{BM25}.

CCS_{Lite}: A variant of our approach (CCS) which does not re-rank documents that are retrieved for a concept from the corpus. This enables us to evaluate the impact of semantic re-rank in CCS.

All document indexing and ranking functions were implemented on Elasticsearch using its Java API[5]. We use mean average precision (MAP), recall and F1 measures to compare approaches. MAP combines the precision and ranking quality of recommended annotations in a single performance measure making it easier to compare different systems (see Eq. 3). Precision is the proportion of recommended annotations that are correct. Recall is the proportion of correct annotations that are included in recommended annotations. We measure recall at 5, 7 and 10 top recommended concepts for annotation. F1 measure is the harmonic mean of precision and recall.

$$MAP = \frac{\sum_{d=1}^{|D|} AP(d)@n}{|D|} \tag{3}$$

D is the set of all documents being annotated, $AP(d)@n = \sum_{k=1}^{n} P(k)/min(m,n)$ is the average precision (AP) of recommended concepts for document d, $P(k)$ is the precision at position k of ordered concepts recommended, m is the number of relevant annotations for d, n is the maximum number of recommended concepts being evaluated. We set $n = 10$ for all approaches.

4.2   Results and Discussion

The results of different methods for recommending annotation are presented in Table 1. CCS outperformed $UNSUP_{EAGL}$ on all the evaluation metrics used. Also, results of CCS are not very far from those of supervised approaches. CSS outperformed CCS_{Lite} highlighting the utility of re-ranking documents before generating concept summaries. Unsurprisingly, the difference between CCS and CCS_{Lite} is minimal given that we used a subset of Wikipedia that is mostly relevant to the domain. We expect the impact of semantic re-ranking to be more pronounced when using a more diverse corpus where there is greater possibility of encountering polysemous terms.

The F1 measures are low for all the methods because there are only few concepts that annotate each document. As an illustration, consider a document that

---

[5] Elasticsearch   Java   API   http://www.elastic.co/guide/en/elasticsearch/client/java-api/5.2.

Table 1. Mean average precision (MAP), recall (R) and F-measure (F1) of the approaches that we compared for recommending document annotations.

	MAP	F1@5	R@5	R@7	R@10
SUP$_{BM25}$	0.2967	**0.2489**	**0.4935**	0.5176	**0.5691**
SUP$_{LM}$	0.2632	0.2192	0.4508	0.4874	0.5177
SUP$_{KNN}$	**0.3093**	0.2412	0.4767	**0.5192**	0.5387
UNSUP_{EAGL}	0.2221	0.1258	0.3749	0.4007	0.4394
CCS	0.2647	0.2045	0.4419	0.4860	0.5345
CCS$_{Lite}$	0.2469	0.2074	0.4409	0.4754	0.5157

is annotated with one concept which is correctly included in the top 5 recommendations. The precision will be 0.2 (1/5) and F1 0.33. This appears to be low even though the correct annotation was recommended. Recall value is more relevant in this case since it shows the proportion of the correct annotations that an approach was able to discover. The choice of concepts for annotating documents is quite subjective and attaining high recall values remain a challenge [8].

The results that are obtained for the other approaches in our evaluation mostly agree with previous comparisons [2]. The performance of UNSUP$_{EAGL}$ was weak due to insufficient textual content in the controlled vocabularies. CHRONOSTRAT and THESAURUS describe only concept terms and synonyms (or alternative spellings), while LEXICON includes some descriptive text. As expected, the supervised approaches performed strongly. SUP$_{BM25}$ was overall best in retrieving the right annotations as shown in recall values. SUP$_{KNN}$ ranked correct concepts slightly better as MAP values indicate.

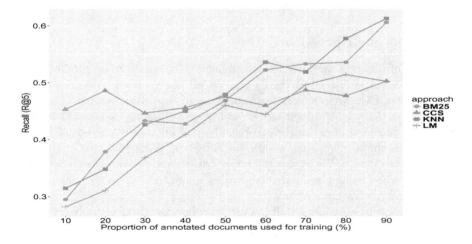

Fig. 4. Performances (R@5) with varying proportion of dataset used for training.

In Fig. 4, we show the performance of different approaches as the proportion of training and test dataset vary. The training dataset simulates the proportion of documents that were annotated prior to recommending annotations for test documents. A random function was used to split the documents and recall (R@5) in the figure show performances on the test dataset. The performance of CCS remained fairly similar and was only outperformed by the supervised approaches when over 50% of the dataset was used for training.

Although the supervised approaches may be better at recommending annotations, unsupervised approaches remain relevant for generating an initial set of annotated corpus. Supervised and unsupervised approaches are usually combined to form hybrid document annotation systems.

5   Conclusion

In this work, we introduced a corpus-based approach for generating descriptive textual content (concept summaries) for the concepts of a thesaurus. We used semantic knowledge from the thesaurus to identify the best documents for generating concept summaries. Concept summaries were then used to recommend annotations for documents. Our goal was to overcome the limitations of unsupervised thesaurus-based approaches which suffer from insufficient descriptive texts for effective use in recommending document annotations. Evaluation using a manually annotated corpus showed that this objective was achieved and that our results were somewhat comparable with the supervised approaches.

Future work will explore alternative ways of using concept summaries for recommending annotations. For example, the graph-based approach in [10] can be applied using concept summaries as DBpedia articles. Also, an approach that is based on explicit semantic analysis (ESA) will be explored. Instead of recommending annotations based on the similarity of term vectors, the ESA approach utilises concept vectors. Finally, we have assumed an independence of document sections by treating them as separate documents. In reality, the rest of the document from which a section is extracted can provide useful information for determining the right concepts for annotating the section.

Acknowledgement. This work is partly funded by the British Geological Survey (BGS) through the BGS University Funding Initiative (BUFI). We are grateful for the valuable comments of our reviewers.

References

1. Berlanga, R., Nebot, V., Pérez, M.: Tailored semantic annotation for semantic search. In: Web Semantics Science, Services and Agents on the World Wide Web (2014)
2. Trieschnigg, D., Pezik, P., Lee, V., De Jong, F., Kraaij, W., Rebholz-Schuhmann, D.: MeSH Up: effective MeSH text classification for improved document retrieval. Bioinformatics **25**(11), 1412–1418 (2009)

3. Robertson, S.E., Walker, S., Beaulieu, M., Gatford, M., Payne, A.: Okapi at TREC-4, pp. 73–96. NIST Special Publication SP (1996)
4. Große-Bölting, G., Nishioka, C., Scherp, A.: A comparison of different strategies for automated semantic document annotation. In: Proceedings of the 8th International Conference on Knowledge Capture, vol. 8. ACM (2015)
5. Aronson, A.R., Mork, J.G., Gay, C.W., Humphrey, S.M., Rogers, W.J.: The NLM indexing initiative's Medical Text Indexer. Medinfo 11(Pt 1), 268–72 (2004)
6. Giannopoulos, G., Bikakis, N., Dalamagas, T., Sellis, T.: GoNTogle: a tool for semantic annotation and search. In: Aroyo, L., Antoniou, G., Hyvönen, E., Teije, A., Stuckenschmidt, H., Cabral, L., Tudorache, T. (eds.) ESWC 2010. LNCS, vol. 6089, pp. 376–380. Springer, Heidelberg (2010). doi:10.1007/978-3-642-13489-0_27
7. Huang, M., Névéol, A., Lu, Z.: Recommending MeSH terms for annotating biomedical articles. J. Am. Med. Inform. Assoc. 18(5), 660–667 (2011)
8. Dramé, K., Mougin, F., Diallo, G.: Large scale biomedical texts classification: A kNN and an ESA-based approaches. J. Biomed. Semant. 7(1), 40 (2016)
9. Aronson, A.R.: Effective mapping of biomedical text to the UMLS Metathesaurus: The MetaMap program. In: Proceedings of the AMIA Symposium, American Medical Informatics Association, vol. 17 (2001)
10. Hulpus, I., Hayes, C., Karnstedt, M., Greene, D.: Unsupervised graph-based topic labelling using DBpedia. In: Proceedings of the Sixth ACM International Conference on Web Search and Data Mining, pp. 465–474. ACM (2013)
11. Hertling, S., Paulheim, H.: Wikimatch: using wikipedia for ontology matching. In: Proceedings of the 7th International Conference on Ontology Matching, vol. 946, pp. 37–48. CEUR-WS.org (2012)
12. Fernández, M., Cantador, I., López, V., Vallet, D., Castells, P., Motta, E.: Semantically enhanced information retrieval: An ontology-based approach. Web Semantics Science, Services and Agents on the World Wide Web, vol. 9(4), pp. 434–452 (2011)
13. Wu, Z., Palmer, M.: Verbs semantics and lexical selection. In: Proceedings of the 32nd Annual Meeting on Association for Computational Linguistics, pp. 133–138. Association for Computational Linguistics (1994)
14. Hliaoutakis, A., Varelas, G., Voutsakis, E., Petrakis, E.G., Milios, E.: Information retrieval by semantic similarity. Int. J. Semant. Web Inf. Syst. 2(3), 55–73 (2006)
15. Knappe, R., Bulskov, H., Andreasen, T.: Perspectives on ontology-based querying. Int. J. Intell. Syst. 22(7), 739–761 (2007)
16. Leal Bando, L., Scholer, F., Turpin, A.: Query-biased summary generation assisted by query expansion. J. Assoc. Inf. Sci. Technol. 66(5), 961–979 (2015)

RussianFlu-DE: A German Corpus for a Historical Epidemic with Temporal Annotation

Tran Van Canh[1(✉)], Katja Markert[2], and Wolfgang Nejdl[1]

[1] L3S, Hannover, Germany
{ctran,nejdl}@l3s.de
[2] Institute of Computational Linguistics, Heidelberg University, Heidelberg, Germany
markert@cl.uni-heidelberg.de

Abstract. Temporally annotated corpora about historic events can be crucial to digital humanities research: they allow to extract and date events as well as reactions to them, and to construct timelines of events and of language use, among other applications. However, producing a precise corpus of a particular event in history is very challenging due to the lack of noise-free digitalized data. This paper introduces RussianFlu-DE, a temporally annotated corpus of 639 articles extracted from noisy OCR text of newspaper issues in German. All articles are about the Russian flu epidemic that took place during 1889–1893. We describe the development of RussianFlu-DE, including methods to clean different types of noise in the OCR text, and our tool for extracting Russian flu related articles. In addition, the task of temporal annotation using the TIMEX2 schema is discussed and the characteristics of the corpus compared to other corpora are presented. To show how our contribution supports epidemiology, we present some preliminary yet interesting results obtained from analyzing the articles in RussianFlu-DE. The corpus and associated tools for exploration are publicly available.

Keywords: Corpus in German · Russian flu epidemic · TIMEX2 · Temporal annotation

1 Introduction

Analyzing past events such as wars or epidemic diseases has received significant attention as knowledge obtained from such events paves the way for solving current or future similar issues. This is especially important for epidemiology due to many infectious diseases being discovered every year in many parts of the world. By analyzing historical accounts of earlier epidemics researchers can find useful insights into, for example, disease transmission methods, development stages, and community vulnerability factors. Such knowledge helps authorities prepare effective interventions, e.g., closing churches, schools, and public events, controlling methods of transport, or requiring masks to be worn in certain places

© Springer International Publishing AG 2017
J. Kamps et al. (Eds.): TPDL 2017, LNCS 10450, pp. 61–73, 2017.
DOI: 10.1007/978-3-319-67008-9_6

for reducing risks from a particular infectious disease [1,2]. However, studying any epidemic happening in the past always faces a difficult challenge because reliable datasets to conduct analysis are only rarely found. One of the main reasons is that creating a corpus of a particular epidemic event requires a lot of effort, from identifying data sources to extracting appropriate content. It is even more challenging when dealing with events that happened pre-digitization due to the lack of data and the noisy nature of the content found. Because of this, research studies on historical epidemic events are typically carried out based on documents manually collected from various repositories, with size, coverage and representativeness limited by the large amount of human effort required [3–5].

Our first goal in this work is therefore to target a historical epidemic event and create a reliable corpus for it as a main contribution for others to use. Specifically, the largest nineteenth-century epidemic of influenza, called "the Russian flu epidemic" is of our interest. The epidemic reached Europe from the East in November and December of 1889 and spread over the whole globe in the space of a few months. It was one of the first epidemics of influenza that occurred during the period of the rapid development of bacteriology. In addition, it was the first epidemic that was publicly and intensively narrated in the developing daily press, especially those published in German located in Germany and Austria [6]. However, as stated in [1], very limited information about the epidemiology of this influenza has been found in materials published in English. Motivated by these observations and by the fact that no epidemic corpus in German has been publicly available so far, we create RussianFlu-DE corpus[1], which contains 639 German news articles (stories) extracted from noisy OCR text of newspapers published during 1889–1893.

In addition, reliable extraction of interesting knowledge regarding epidemiology from a text corpus often needs to refer to the time at which reported events take place. Temporal information supports the development of techniques for timeline creation and tracking the progress of events over geographic areas. Apart from these epidemiological applications, temporal information plays an important role in many natural language processing and understanding tasks. Therefore, the extraction and normalization of temporal expressions from documents are crucial preprocessing steps in these research areas. An important application is to evaluate the quality of temporal taggers. Thus, as a second contribution, we provide a temporally annotated version of the corpus using the TIMEX2 annotation schema [7]. Finally, to show examples of how useful RussianFlu-DE is for epidemiology, we present some preliminary yet interesting results obtained from analyzing the corpus.

The remainder of the paper is structured as follows. In Sect. 2, we discuss the related work for this paper. In Sect. 3, we detail our tasks to produce RussianFlu-DE. Section 4 describes the methods used to annotate temporal expressions in the corpus. We then present in Sect. 5 some exploratory results obtained from the corpus before concluding the paper in Sect. 6.

---

[1] RussianFlu-DE is accessible on our project website: http://russianfluweb.l3s. uni-hannover.de.

2   Related Work

In this section, we first discuss recent studies on the Russian flu epidemic and then present relevant work for the creation of a temporally annotated corpus.

Studies on the Russian flu epidemic. As mentioned in the previous section, there is limited information about the epidemiology of the Russian flu epidemic 1889–1893. In [6], the authors conducted an analysis to examine the impact of the epidemic in 14 cities in Europe. Their results showed that the epidemic spread quickly from Saint Petersburg, Russia, to other parts of Europe with a speed of around 400 km/week and reached the American continent only 70 days after the original peak in Saint Petersburg. In addition, some detailed information about case fatality ratio and the median basic reproduction was also given. However, their work was based on reports of only two local daily newspapers in Poznań, Poland, which implies some uncertainty due to the lack of data coverage. Valleron et al. [8] presented a case study on the transmissibility and geographic spread of the Russian flu. A similar approach was followed by Valtat et al. [1] to examine the age distribution of the affected people and the mortality rate of this flu event. In a recent study, Ewing et al. [9] collected contemporary reports and explored a digital humanities approach to interpret information dissemination regarding this epidemic. The limitations common to all these studies are the heterogeneity and lack of coverage of data used.

Development of temporally annotated corpora. Since the last decade, there has been significant effort in the area of temporal annotation of text documents. Annotation standards such as TIDES TIMEX2 [7] and TimeML [10] were defined and temporal taggers e.g., DANTE [11] and HeidelTime [12], were developed. Furthermore, research challenges such as the Automatic Content Extraction (ACE) time expression and normalization (TERN) were organized where temporal taggers were evaluated. In 2010, temporal tagging was one task in the TempEval-2 challenge [13]. However, the research focus was mainly on English documents. A few temporally annotated corpora have been published, e.g., ACE EN Train corpora[2], TimeBank [14], TempEval EN [13], and WikiWars [11]. Only recently, a German WikiWars corpus consisting of Wikipedia articles in German about famous wars in history was developed [15]. Nevertheless, no historical epidemic corpus is available so far.

3   Corpus Creation

In this section, we detail the corpus creation tasks. In Sect. 3.1 we describe our methods for collecting and cleaning data. A tool for extracting Russian flu related stories from OCR text is then introduced in Sect. 3.2.

---

[2] See corpora LDC2005T07 and LDC2006T06 on http://www.ldc.upenn.edu.

3.1   Data Collection and Noise Reduction

Data used in this work was collected from the Austrian Newspapers Online
(ANNO)[3] repository. ANNO contains almost all issues from many newspapers
in Austria and Germany during the time the Russian flu epidemic took place.
The data are accessible, both in scanned PDF and OCR formats. These are
appropriate for our goal in terms of extracting Russian flu related stories from
noisy OCR text and checking against the scanned PDF content for validity.
To establish the data collection, the keywords listed in Table 1 were used to
search the ANNO repository[4]. The search query was constrained to the time
interval from 1889 to 1893. Empirically, these keywords are likely to appear in
texts talking about diseases in general and about the Russian flu epidemic in
particular, therefore resulting in a high recall collection. After preprocessing the
search results we obtained 4,806 issues, which become the candidates to extract
stories about the Russian flu. ANNO search always returns a whole issue of a
newspaper and fully automatic extraction of individual stories is not possible.

Table 1. Keywords used to collect newspaper issues containing stories about the
influenza epidemic. We aimed for high recall.

ID	Keyword	Variation	ID	Keyword	Variation
1	Influenza	Jnfluenza, Jnsolvenza	4	Grippe	
2	Epidemie		5	erkrankt	ertrankt
3	Influenza-Epidemie	Influenzaepidemie	6	Pathologie	

Due to the low quality of the scanned images of newspaper issues, a lot of
noise is present in the corresponding OCR texts. A very frequent type of error is
the so-called antistring, where the OCR output of a sequence of words consists of
individual characters with spaces inbetween. For example, Fig. 1 shows a scanned
image of three short messages about the Russian flu in London, Prague, and
Munich, which was published on December 14, 1889 by Die Presse, and the
corresponding OCR text. We can observe that a string "I f l u e n z a i n P r a
g n o c h n i c h t" was produced instead of five words "Influenza in Prag noch
nicht". Besides, several misrecognized words exist in the OCR text.

Our goal was to correct OCR errors but at the same time keep the language as it was so that the derived corpus pertains its historical perspective.
As modern German is rather different in writing and usage of many words due
to language evolution, text normalization models that were trained on modern
German datasets could not be applied [16]. To cope with these issues, we
adopted a snapshot of the Google-2-gram dataset for German[5] from 1885 to 1895.

---

[3] The Austrian National Library: http://anno.onb.ac.at.

[4] Some misspelt variations of keywords were used due to possible OCR errors.

[5] The Google-2-gram dataset is publicly available at: http://storage.googleapis.com/
books/ngrams/books/datasetsv2.html.

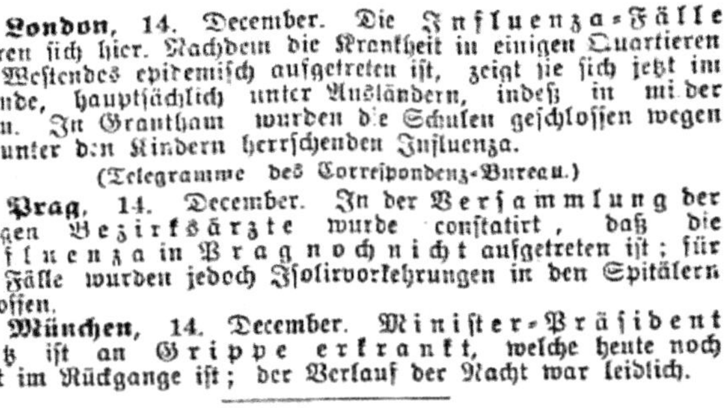

Londdon, 14. December. Die Influenza-Fälle
mehren sich hier. Nachdem die Krankheit in einigen Quartieren
des Westendes epitemisch aufgetreten ist, zeigt sie sich jetzt im
Ostende, hauptsächlich unter Ausländern, indeß in milder
Form. In Grantham wurden die Schulen geschlossen wegen
der unter den Kindern herrschenden Influenza.
(Telegramme des Correspondenz-Bureau.)
Prag, 14. December. In der Versammlung der
hiesigen Bezirksärzte wurde constatirt, daß die
Influenza in Prag noch nicht aufgetreten ist; für
alle Fälle wurden jedoch Isolirvorkehrungen in den Spitälern
getroffen.
München, 14. December. Minister-Präsident
Lutz ist an Grippe erkrankt, welche heute noch
nicht im Rückgange ist; der Verlauf der Nacht war leidlich.

Lo ndon, 14. December. DieInfluenza-Fälle
mehren sich hier. Nachdem die Krankheit in einigen Quartieren
dcs Wcstcndes epidemisch aufgetreten ist, zeigt sie sich jetzt im
Ostende, hauptsächlich unter Ausländem, indeß in nii der
Form. In Grantham wurden d eSchule» geschlossen wegen,
der unter den Kindem herrschenden Influenza.
(Telegramme desCo rrefpondcnz -Bureeu.) j
Prag, 14. December. In der Versammlung der 1
hiesigen Bezirksärzte wu rde constatirt, daß die f
Inf l u e n za in P ra g no ch ni cht aufgetreten ist; für! l
alle Fälle wu rden jedoch Jsolirvorkehrungen in den Spitälem l
getroffen.
München, 14. December.Minister-Präsident
Lutz ist an Grippe erkrankt, welche heute no ch
nichtim Rückgänge ist; der Verlauf der Nacht war leidlich.

Fig. 1. An image of three short messages about the Russian flu in an issue published on December 14, 1889, by Die Presse and the corresponding recognized OCR text.

We used the dataset to train a bigram-based model for word segmentation (to recover words from antistrings), and for spell checking (to correct misrecognized words).

We applied our model to the 639 Russian flu stories in our corpus (see Sect. 3.2 for details on story extraction). Updates were done for 5,121 antistrings and 79,114 misrecognized words. The final corpus contains over 453,000 words. To validate our noise correction, two German native speakers proofread a random sample of 11 stories in our corpus before and after correction and identified word errors. The word error rate (WER) over all 11 texts before correction was around 18.9%, reduced to 5.5% after correction. Although our proofreaders also found some errors introduced by our model, the quality of the text has greatly improved. Table 2 shows details per story.

3.2 Tool for Extracting Russian Flu Stories

Text block classification. Manually reading thousands of newspaper issues to extract relevant stories about the Russian flu is too time consuming. In addition, a difficult challenge when extracting complete stories is that recognized OCR text blocks are very often not aligned in the same order as they were in the original image of an issue. Our approach was to automatically pre-classify OCR text blocks to identify the ones that are more likely to be part of flu-related

Table 2. Sample results of noise cleaning process. B and A prefixed columns indicate the numbers obtained before and after running our noise cleaning model, respectively.

ID	Filename	B-Errors	B-Words	B-WER	A-Errors	A-Words	A-WER
01	apr18891206.story.1	80	567	14.1%	20	573	3.4%
02	apr18891209.story.1	135	820	16.4%	50	815	6.1%
03	apr18891210.story.1	101	460	21.9%	24	403	5.9%
04	apr18891210.story.2	102	539	18.9%	33	528	6.2%
05	apr18891211.story.1	36	253	14.2%	6	236	2.5%
06	apr18891211.story.2	93	541	17.1%	30	553	5.4%
07	apr18891211.story.3	114	721	15.8%	22	721	3.0%
08	apr18891212.story.1	118	653	18.0%	31	637	4.8%
09	apr18891214.story.1	94	446	21.0%	36	443	8.1%
10	apr18891214.story.2	21	119	17.6%	6	105	5.7%
11	apr18891215.story.1	239	717	33.3%	66	726	9.0%
	Average	**103**	**531**	**18.9%**	**30**	**521**	**5.5%**

stories. For this, we adopted the KL-divergence based technique developed by Schneider [17] to build a classifier. We manually extracted and labeled 245 OCR text paragraphs, which were then used to train the model and obtained a recall of 81.5% and a precision of 68.6% when cross-validating on the training set. The output of the classifier can be used to help annotators start working on an issue by looking at suggested text blocks first, from which they then select paragraphs that are part of the same story.

Extraction tool. We implemented a Web-based tool for annotators to help build our RussianFlu-DE corpus collaboratively. The high-recall classifier described above is an underlying component of the tool that suggests to the annotators to navigate to the text paragraphs that are probably part of a Russian flu story. The main GUI of our tool is shown in Fig. 2. As an important component for the annotator, the bottom-left area displays information on text blocks derived from classifier output and the annotator's recent decisions (e.g., which text blocks were selected to complete an article). For more convenience, one just needs to click on an entry in this list to navigate to the corresponding content that is shown afterwards in the area on the right[6].

Four students worked through all 18,768 classifier-suggested text blocks of all 4,806 newspaper issues returned by the original ANNO keyword search to identify Russian-flu related stories in those blocks and surrounding blocks. The extracted 657 stories were subsequently additionally verified by another two native-speaker annotators resulting in a final 639 Russian flu stories from 42 newspapers, identified with 85.7% agreement between annotators (i.e., 18 stories

---

[6] Detailed guidelines for the tool are available on: http://russianfluweb.l3s.uni-hannover.de.

Fig. 2. Main GUI of our tool for extracting newspaper articles about the Russian flu from OCR text.

Table 3. List of 20 newspapers from which most of the articles about Russian flu in our corpus were extracted.

ID	Newspaper	Stories	ID	Newspaper	Sto.
01	Die Presse	72	11	Linzer Volksblatt	24
02	Neue Freie Presse	56	12	Bludenzer Anzeiger	20
03	(Linzer) Tages-Post	54	13	Vorarlberger Landes-Zeitung	17
04	Bregenzer/Vorarlberger Tagblatt	50	14	Mährisches Tagblatt	16
05	Deutsches Volksblatt	43	15	Prager Abendblatt	15
06	Wiener Zeitung	40	16	Salzburger Chronik	11
07	(Neuigkeits) Welt Blatt	38	17	Badener Bezirks-Blatt	9
08	Das Vaterland	34	18	Neue Warte am Inn	9
09	Prager Tagblatt	33	19	Volksblatt für Stadt und Land	8
10	Bukowinaer Rundschau	25	20	Teplitz-Schönauer Anzeiger	6

were removed, 548 in common, 91 in partial agreement). Table 3 shows a list of 20 newspapers from which most of the articles in the corpus were extracted. Each article was then converted into an SGML file, the format of the ACE TERN corpora containing DOC, DOCID, DOCTYPE, DATETIME, and TEXT tags. The document creation time was set to the publication date. The format complies with widely used tools for temporal annotation tasks, which we address in the next section.

4   Temporal Annotation

This section describes our work to produce a temporally annotated version of RussianFlu-DE. We first give an overview over the TIMEX2 schema (Sect. 4.1), which we used for annotating temporal expressions. We also explain our two-stage strategy for annotating the corpus. Then, in Sect. 4.2, we present some statistics computed on RussianFlu-DE compared to other corpora.

4.1 TIMEX2 Schema and Annotation Strategy

For the annotation of temporal expressions, we followed the authors of the Wiki-Wars corpus [11]. Particularly, we used TIMEX2 as annotation schema to annotate the temporal expressions in our corpus. The TIMEX2 annotation guidelines [7] describe how to determine the extents of temporal expressions and their normalizations. Note that, in addition to date and time expressions, such as "December 10, 1889" and "9:30 p.m.", temporal expressions describing durations and sets are to be annotated as well [15]. Examples for expressions of the types duration and set are "24 months" and "daily", respectively. The normalization of temporal expressions is based on the ISO 8601 standard for temporal information. In particular, the following five attributes can be used to normalize a temporal expression: VAL (value), MOD (modifier), SET (set identification), ANCHOR_VAL (anchor value), and ANCHOR_DIR (anchor direction).

The most important attribute of a TIMEX2 annotation is VAL, which holds the normalized value of a temporal expression. Table 4 gives values of VAL for the four examples described above. The SET attribute is used to identify set expressions. In addition, the modifier MOD is used to provide additional specifications not captured by VAL. For instance, for expressions such as "the end of December", MOD is set to END. Finally, ANCHOR_VAL and ANCHOR_DIR are used to anchor a duration to a specific date, using the value information of the date and specifying whether the duration starts or ends on this date.

Table 4. Normalized values (VAL) of temporal expressions of different types. We here assume that 9:30.p.m in the second example refers to 9:30.p.m on December 10, 1889.

Temporal expression	VAL attribute	Temporal expression	VAL attribute
December 10, 1889	1889-12-10	24 months	P24M
9:30.p.m	1889-12-10T21:30	daily	XXXX-XX-XX

Similar to Strötgen et al. [15,18], we used the Heideltime temporal tagger [12] as a first-pass annotation tool. HeidelTime is a multilingual, rule-based temporal tagger that was developed to have strict separation between the source code and the resources (rules, extraction patterns, normalization information). Because of this, HeidelTime supports several languages [12,13]. The output of HeidelTime was then imported to the annotation tool Callisto[7] for manual correction of the annotations. As in [11,15], this two-stage annotation procedure is

---

[7] http://callisto.mitre.org.

motivated by the fact that "annotator blindness", i.e., annotators missing temporal expressions, is reduced to a minimum. Furthermore, the annotation effort is reduced significantly since the annotator does not have to create TIMEX2 tags for the expressions already identified by the tagger. At the second stage of annotation, the stories were examined for temporal expressions missed by HeidelTime and existing HeidelTime annotations were manually corrected. This task was performed by two rounds of 2 annotators, who each worked separately on half the collection. Overall, relative to the final 7,492 temporal annotations, the annotators contributed 4.4% new temporal annotations and corrected 7.9% of HeidelTime annotations (see Table 5).

Table 5. Updates made by annotators in two rounds on the result of the HeidelTime temporal tagger. The percentage is relative to the final result.

First round of annotators		Second round of annotators	
Extraction	Normalization	Extraction	Normalization
311 (added)	561 (edited)	16 (added)	31 (edited)
4.2%	7.5%	0.2%	0.4%

Finally, the annotated files, which contain inline annotations, were transformed into the ACE APF XML format, a stand-off markup format used by the ACE evaluations. Thus, the RussianFlu-DE corpus is available in the same two formats as the WikiWars and WikiWarsDE corpora. Therefore, evaluation tools of the ACE TERN evaluations can be used with our corpus as well.

4.2 Corpus Statistics

In this section, we present some statistics regarding the length of stories and the number of temporal expressions in our RussianFlu-DE corpus compared to other corpora.

The RussianFlu-DE corpus contains 639 stories related to the Russian flu with a total of more than 453,000 tokens and 7,492 temporal expressions. In Table 6, we compare our corpus to other publicly available, temporally tagged corpora. While ACE 04 EN Train remains the largest corpus in terms of number of documents, RussianFlu-DE is the largest one regarding the number of tokens. Except for the two WikiWars copora that naturally have long narrative documents, RussianFlu-DE has significantly longer documents compared to others. RussianFlu-DE contains around 7,500 temporal expressions. Thus, the corpus is second only to the ACE 04 EN Train corpus. The density of temporal expressions indicated by Tokens/TIMEX in RussianFlu-DE is similar to the other corpora except for the two WikiWars corpora and the ACE 04 EN Train corpus. Finally, RussianFlu-DE contains 11.7 temporal expressions per document on average, which is slightly higher than the others except for the two WikiWars corpora that have an order of magnitude more temporal expressions per document.

Table 6. Statistics computed on the RussianFlu-DE corpus in comparison to other publicly available corpora.

ID	Corpus	Docs	Tokens	Tokens Doc	TIMEX	Tokens TIMEX	TIMEX Doc
1	ACE 04 EN Train	863	306,463	355.1	8,938	34.3	10.4
2	ACE 05 EN Train	599	318,785	532.1	5,469	58.3	9.1
3	TimeBank 1.2	183	78,444	428.6	1,414	55.5	7.7
4	TempEval2 EN Train	162	53,450	329.9	1,052	50.8	6.5
5	TempEval2 EN Eval	9	4,849	538.7	81	59.9	9.0
6	WikiWars	22	119,468	5,430.3	2,671	44.7	121.4
7	WikiWarsDE	22	95,604	4,345.6	2,240	42.7	101.8
8	*RussianFlu-DE*	*639*	*453,288*	*709.3*	*7,492*	*60.5*	*11.7*

5   Preliminary Exploratory Results

In addition to the RussianFlu-DE corpus, we developed associate tools and made them available so that research communities can use them to query for information and conduct explorative studies based on the corpus. We present in this section some functionalites of our tools and preliminary yet interesting results.

The corpus timeline provides statistics on the number of stories in the corpus across time and news outlet. In addition, it provides an interactive visualization

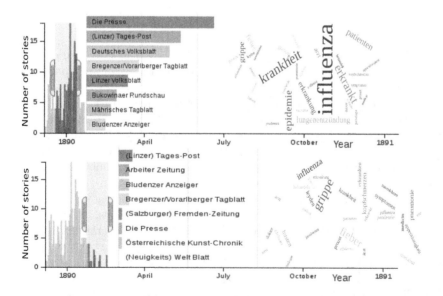

Fig. 3. Press attention on the Russian flu and topic changes over time.

Fig. 4. A pattern of frequent collocations extracted from RussianFlu-DE corpus.

from shallow semantic analysis, such as word usage and word collocations over time. As an example shown in Fig. 3, during the peak time in late December 1889 and January 1890, extensive news about the influenza was published. Newspapers were trying to narrate the outbreak as fast as possible. Words that appear significantly in the stories include *influenza, epidemic, krankheit (disease)*, and *erkrankt (sick)*. A short time after this peak period, i.e., in February and March 1890, fewer reports were published about the outbreak of the flu and communities started discussing its treatment more. Names of doctors appear in the news (e.g., Leyden, Proust) together with words describing symptoms such as *fieber (fever), kopfschmerzen (headache)*, and *appetitlosigkeit (anorexia)*.

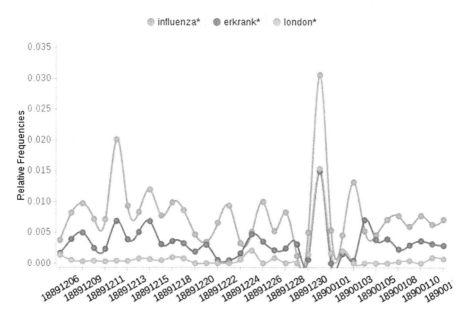

Fig. 5. High correlation between the frequencies of *influenza* and *erkrankt*, and the peak time of the flu in *London*.

Rather than such an overall view, by exploring word collocations in the whole corpus, one can find even more interesting information. For example, Fig. 4 shows a frequent pattern of word collocation describing the influenza. This pattern actually provides useful insights, both on how the media narrates the flu and the flu itself. The words *heute (today)* and *gestern (yesterday)* indicate that news about the flu was updated every day; and the word *jänner (January)* implies that the flu outbreaks happened during winter.

Figure 5 presents the co-occurrences of three words *influenza, erkrankt,* and *london* over time. It shows a strong correlation between the occurrence frequencies of *influenza* and *erkrankt*. In addition, one can observe that the peak time of the flu in London was from late December 1889 to early January 1890 as indicated in [4,19]. This suggests that the temporal distribution of terms can give us more insights into the geographical spread of the epidemic.

6   Conclusions

We have described RussianFlu-DE, a corpus of German articles about the Russian flu of 1889–1893. After discussing the methods for data collection and cleaning, we introduced our tool for extracting relevant articles from OCR text. In addition, a temporally annotated version of the corpus was produced. We further presented some interesting insights that we achieved from analyzing articles in the corpus. In future, we will (i) extend the corpus with articles from German-language newspapers published in countries other than Austria, (ii) use our temporal corpus annotation for information extraction (such as number of deaths) and (iii) investigate information spread as well as sentiment towards events by aligning articles on the same events from different newspapers.

Acknowledgments. This work is supported by the German Research Foundation (DFG) for the project "Tracking the Russian Flu in U.S. and German Medical and Popular Reports, 1889–1893" on Grant No. NE 638/13-1. We also thank you the Austrian National Library for help in data collection.

References

1. Valtat, S., Cori, A., Carrat, F., Valleron, A.J.: Age distribution of cases and deaths during the 1889 influenza pandemic. Vaccine **29**(Supplement 2), B6–B10 (2011)
2. Collinson, S., Heffernan, J.M.: Modelling the effects of media during an influenza epidemic. BMC Public Health **14**(1), 376 (2014)
3. Ewing, E.T., Gad, S., Ramakrishnan, N.: Gaining insights into epidemics by mining historical newspapers. Computer **46**(6), 68–72 (2013)
4. LeGoff, J.M.: Diffusion of influenza during the winter of 1889–1890 in Switzerland. Genus **67**(2), 77–99 (2011)
5. Yan, Q., Tang, S., Gabriele, S., Wu, J.: Media coverage and hospital notifications: correlation analysis and optimal media impact duration to manage a pandemic. J. Theor. Biol. **390**, 1–13 (2016)

6. Kempiska-Mirosawska, B., Woniak-Kosek, A.: The influenza epidemic of 1889–90 in selected european cities - a picture based on the reports of two Poznań daily newspapers from the second half of the nineteenth century. Med. Sci. Monit. **19**, 1131–1141 (2013)
7. Ferro, L., Gerber, L., Mani, I., Sundheim, B., Wilson, G.: TIDES standard for the annotation of temporal expressions. Technical report, MITRE Corporation (2005)
8. Valleron, A.J., Cori, A., Valtat, S., Meurisse, S., Carrat, F., Boëlle, P.Y.: Transmissibility and geographic spread of the 1889 influenza pandemic. Proc. Natl. Acad. Sci. **107**(19), 8778–8781 (2010)
9. Ewing, E.T., Veronica, K., Sinclair, E.N.: Look out for la grippe: using digital humanities tools to interpret information dissemination during the Russian flu, 1889–90. Med. Hist. **60**(01), 129–131 (2016)
10. James, P., Robert, K., Jessica, L., Roser, S.: Temporal and event information in natural language text. Lang. Resour. Eval. **39**(2), 123–164 (2005)
11. Mazur, P., Dale, R.: WikiWars: a new corpus for research on temporal expressions. In: Proceedings of the 2010 Conference on Empirical Methods in NLP, pp. 913–922. Association for Computational Linguistics (2010)
12. Strötgen, J., Gertz, M.: Multilingual and cross-domain temporal tagging. Lang. Resour. Eval. **47**(2), 269–298 (2013)
13. Verhagen, M., Saurí, R., Caselli, T., Pustejovsky, J.: Semeval-2010 task 13: Tempeval-2. In: Proceedings of the 5th International Workshop on Semantic Evaluation, SemEval 2010, PA, USA, pp. 57–62. Association for Computational Linguistics (2010)
14. Pustejovsky, J., Hanks, P., Sauri, R., See, A., Gaizauskas, R., Setzer, A., Radev, D., Sundheim, B., Day, D., Ferro, L., et al.: The TimeBank corpus. Proc. Corpus Linguist. **2003**, 647–656 (2003)
15. Strötgen, J., Gertz, M.: WikiWarsDE: a German corpus of narratives annotated with temporal expressions. In: Proceedings of the Conference of the German Society for Computational Linguistics and Language Technology (GSCL 2011), pp. 129–134 (2011)
16. Bollmann, M.: (Semi-)automatic normalization of historical texts using distance measures and the norma tool. In: Proceedings of ACRH-2 Workshop, pp. 3–14 (2012)
17. Schneider, K.M.: A new feature selection score for multinomial Naive Bayes text classification based on KL-divergence. In: Proceedings of the ACL 2004 (2004)
18. Strötgen, J., Armiti, A., Canh, T.V., Zell, J., Gertz, M.: Time for more languages: temporal tagging of Arabic, Italian, Spanish, and Vietnamese. ACM Trans. Asian Lang. Inf. Process. **13**(1), 1:1–1:21 (2014)
19. Honigsbaum, M.: The great dread: cultural and psychological impacts and responses to the Russian influenza in the United kingdom, 1889–1893. Soc. Hist. Med. **23**(2), 299–319 (2010)

A Digital Repository for Physical Samples: Concepts, Solutions and Management

Anusuriya Devaraju[1]([✉]), Jens Klump[1], Victor Tey[1], Ryan Fraser[1],
Simon Cox[2], and Lesley Wyborn[3]

[1] CSIRO Mineral Resources, P.O. Box 1130, Bentley, WA 6102, Australia
{anusuriya.devaraju,jens.klump,victor.tey,ryan.fraser}@csiro.au
[2] CSIRO Land and Water, Private Bag 10, Clayton South, VIC 3169, Australia
simon.cox@csiro.au
[3] The Australian National University, 56 Mills Road, Acton 2600, Australia
lesley.wyborn@anu.edu.au

Abstract. Physical samples are important resources for sample-based data reuse. They may be utilized in the reproduction of scientific findings, depending on their availability and accessibility. Although several solutions have been developed to curate and publish digital collections (e.g., publications and datasets), considerably less attention has been paid to providing access to physical samples, and linking them to data, reports, and other resources on the Internet. Some progress has been made to bring physical samples into the digital world; for example, through the web-identifier schemes, sample metadata standards and catalogues, and specimen digitization. Existing studies based on the above examples are either project or domain-specific. Also, a particular challenge exists in providing citable and resolvable identifiers for physical samples outside the context of an individual project or a sample data repository. Within the Commonwealth Scientific and Industrial Research Organisation (CSIRO), further work is needed in order to connect the various types of physical samples collected by different entities (individual researchers, projects and laboratories) to the Web, and enable their discovery. We address this need through the development a digital repository of physical samples. This paper presents technical and non-technical components of the repository. They were applied to unambiguously identify the various physical samples and to systematically provide continuous online access to their metadata and data.

Keywords: Physical sample · Specimen · Persistent identifier · IGSN · Sample data curation · Institutional repository

1 Introduction

Physical samples (also called physical specimens) are information sources that come from the Earth's environment. Sampling activities are conducted for scientific research and monitoring purposes. For example, core samples are collected

© Springer International Publishing AG 2017
J. Kamps et al. (Eds.): TPDL 2017, LNCS 10450, pp. 74–85, 2017.
DOI: 10.1007/978-3-319-67008-9_7

to investigate the physical and chemical nature of rocks, soil specimens are gathered to calibrate soil-water measuring instruments, and specimens of plants and insects serve as reference materials to understand the biodiversity of a specific area. The reproducibility of scientific findings and the interpretation of sample-based data requires access to the physical samples and sample metadata, respectively [13]. In addressing the importance of samples, organizations have issued policies and regulations. For example, the NSF Data Sharing Policy asserts that [1], "investigators are expected to share with other researchers, [...] the primary *data, samples, physical collections* and other supporting materials created or gathered in the course of work under NSF grants" [16, p. 66]. The Australian Antarctic program (AAp) Data Policy (2015) indicates that chief investigators "are responsible for ensuring that all *data and samples* generated as part of their research are adequately managed for long-term re-use" [15].

Motivation. Physical samples are important research assets of CSIRO. Within the organization, the community of sample users includes individual researchers, projects, and laboratories, all of which collect or generate samples as a part of their field studies or analytical processes. Examples of samples include groundwater, drill cores, seabed cores, soil archives, sediment, biological specimens, and synthetic materials. The organization curates a large number of legacy samples as well as thousands of samples that will be collected in the future. As in most organizations, there are two main challenges to identifying and discovering samples [6,12]. First, sample collectors may follow their own naming convention to identify samples; therefore, sample names can often be ambiguous. For instance, different collectors may refer to different samples using the same name, or the same sample could be named differently based on analytical procedures performed. This naming ambiguity is also applicable to sample-related physical resources, such as sample collections[1] and sampling features[2]. Herein, we will refer to physical samples, sample collections, and sampling features as 'physical resources' in this paper. The second challenge concerns the discovery of samples. Sample descriptions are often only available to the sample owner. They may not be easily discoverable by other users due to the lack of online catalogues that offer access to the sample metadata.

Contributions. In this paper, we address the challenges through the development of a digital repository, which supports the effective management and discovery of physical resources and their metadata across the Web. The key features of the repository are (a) globally unique and persistent identification of the resources, (b) technical solutions (e.g., tools, data stores, web services and a web portal) that are domain independent, extensible, and easily accessible by members of the organization, and (c) interoperability with sample data repositories managed by other institutions. Given the diverse research communities in

---

[1] A collection may be a group of arbitrary specimens or an aggregation of specimens, e.g., rock chips.

[2] A sampling feature is an entity that is designed to observe some domain features. This may refer to the 'locations' where a sample was collected from such as drillholes, wells, sections, and soil pits.

CSIRO, and the changes in staff over the years, exclusively technological solutions are inadequate. Therefore, another important aspect we considered when developing the digital repository is the non-technical solutions, such as an organizational and governance framework that supports the operation of the technical solutions and the governance of the physical resource identification used in the organization.

Outline. This paper provides an overall view of the digital repository, its related work Sect. 2, and its technical Sect. 3 and non-technical solutions Sect. 4. In addition, we describe the applications of the digital repository in the context of different sample repositories in the organization. We summarize the paper Sect. 5 by detailing its contributions.

2   Related Work

This section provides an overview of IGSN and summarizes related work.

2.1   International Geo Sample Number (IGSN)

Persistent identifiers, such as Digital Object Identifiers (DOI), have proven successful in providing long-term access to digital resources by maintaining the link between a digital resource and its location on the Web [8]. Similarly, assigning globally unique identifiers to physical samples will facilitate unambiguous and systematic access to the samples [6]. IGSN[3] is a persistent, globally unique code for the identification of physical samples and sample collections. The use of the IGSN is not limited to the geosciences but is also relevant to other sciences dealing with specimens, such as biology and oceanography. The IGSN initiative is represented by organizations in various parts of the world, including North America, Europe, Asia, Africa and Oceania.

Figure 1 illustrates the system architecture of the IGSN registration. The Implementation Organization of the IGSN (IGSN e.V.) governs and promotes standard methods for identifying and citing physical samples, and operates the international (top-level) IGSN registration service [10]. The international registration service is modelled after DataCite and utilizes the Handle.net System[4], which is a global persistent identifier resolver service [7]. An allocating agent is a member institution that is authorized by the IGSN e.V. to register the IGSN within an allocated namespace. CSIRO is one of three IGSN allocating agents in Australia alongside Geoscience Australia and Curtin University. In the CSIRO implementation, a client (i.e., individual users or laboratories) may send IGSN registrations to the agent's service based on the *description schema* developed by the respective allocating agent. Then, the agent service forwards the registrations to the international registration service based on the *registration schema*[5].

---

[3] http://www.igsn.org/.

[4] https://www.handle.net/.

[5] http://schema.igsn.org/registration/.

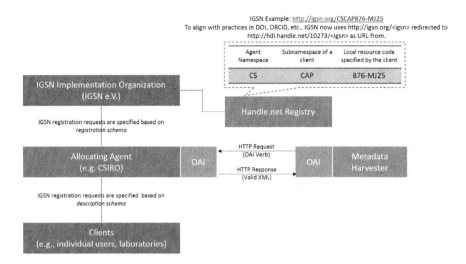

Fig. 1. The hierarchical architecture of the IGSN registration. A namespace refers to the prefix of an allocating agent, e.g., the IGSN e.V. allocated the prefix *CS* to CSIRO. A subnamespace uniquely represents a client. For example, *CAP* is the subnamespace of the Capricorn Distal Footprints project.

The registration schema only covers registration information (e.g., sample number, registrant and log), and excludes sample descriptions to allow greater flexibility in describing samples for different use cases. This separation of registration and description of objects differs from the practice in DOI registration where the registration agents send a standardized set of metadata to the DOI registry as part of the registration process. The Handle.net resolves each individual IGSN handle (e.g., 10273/CSCAP876-MJ25) to a landing page[6] for the resource identified by the handle. Landing pages include more detailed (domain-specific) information of the registered resources and are maintained by the respective client.

2.2   Related Work

Sample Registration Systems. Several organizations have introduced IGSNs to publish their physical samples information. Among these are the System for Earth Sample Registration (SESAR)[11], the Integrated Ocean Drilling Program (IODP) [2] and the International Continental Scientific Drilling Program [3]. The IGSN was developed as SESAR[7] in precursor work at Lamont-Doherty Earth Observatory (LDEO). SESAR was developed with the requirements of individual investigators' geochemical research in mind and have several technical limitations [6]. This work is an expansion of precursor work in SESAR. The existing systems were developed for fairly specific use cases in single research domains.

---

[6] http://capdf.csiro.au/igsn/CSCAP876-MJ25.

[7] http://www.geosamples.org/mysesar.

In contrast, the solutions we developed in CSIRO are domain-independent, i.e., they support representation and registration of various specimen types. Following the IGSN recommended practice, we facilitate the specimen discovery through the meta-data harvesting capabilities across the IGSN communities in Australia.

Specimen Metadata Information Model. There are several metadata schemas representing physical samples. However, some of them are domain specific (e.g., Darwin Core (DwC) [17]), while others have specific design considerations (e.g., modelling sampling features and observation procedures). We developed a comparison between the description metadata schema and the existing schemas in [6]. In this implementation, the sample description metadata schema supports the registration of physical resources through the CSIRO allocating agent service, and the dissemination of resource records through the Open Archives Initiative Protocol for Metadata Harvesting (OAI-PMH)[9] implementation[8]. The description schema adapts some concepts from the DataCite Metadata Schema (v4.0) [5]. It is closely aligned with ISO 19156:2011 (Geographic information - Observatiofs and Measurements (O&M)) [4]. The DataCite Metadata Schema includes the core metadata elements for identifying and describing digital resources, whereas ISO 19156:2011 defines a common set of sampling feature types including *SpatialSamplingFeatures* and *Specimen*. Although the description schema shares some similarities with the two schemas, it differs from them in several aspects. First, it represents the common properties of the three physical resources – physical sample, sample collection and sampling features. We represent new metadata elements, and modify existing elements from the DataCite schema (e.g., cardinality and restrictions) to meet the requirements of the resources. For particular metadata elements (e.g., material, specimen and identifier types), we develop and set controlled vocabularies (expressed as Simple Knowledge Organization System (SKOS)[9] concepts) as their acceptable values. For more information about the description schema and its contributions, see Sect. 3.

Metadata Harvesting and Dissemination. There are several ways to harvest sample metadata catalogues, common examples are the OGC Catalogue Services for the Web (CSW)[10] and OAI-PMH[11]. In the OAI-PMH framework, a *service provider* deploys a client application (harvester) that requests metadata from one or more *data providers*. A data provider operates the metadata catalogue of a repository, which serves the OAI-PMH requests (see Fig. 1). We developed an OAI-PMH provider service to disseminate the sample metadata records in our digital repository in two metadata profiles such as Dublin Core, CSIRO-IGSN Description Schema. We also developed an OAI-PMH service provider which harvests metadata records from our metadata store through the data provider service, and from other allocating agents, e.g., Geoscience Australia.

---

[8] https://igsn.csiro.au/igsn30/api/service/30/oai.

[9] https://www.w3.org/2004/02/skos/.

[10] http://www.opengeospatial.org/standards/cat.

[11] https://www.openarchives.org/pmh/.

3   Solutions

Figure 2 illustrates the architecture of the digital repository that supports sample registration and discovery in CSIRO. Its components are listed as follows:

a. **Clients:** The allocating agent registration service handles requests from two types of clients – sample data curation systems and individual researchers. Individual researchers may register their samples with IGSNs via a web form[12], while sample data curation systems request IGSNs programmatically. Current sample data curation systems in CSIRO are the Capricorn Distal Footprints project, Repository of the Australian Resources Research Centre (ARRC), and Reflectance Spectra Reference Libraries. Table 1 summarizes the local sample systems, material types, and IGSNs registered.

b. **CSIRO allocating agent service:** The allocating agent registration service[13] is a Representational State Transfer (RESTful) web service endpoint that enables clients to register IGSNs of physical resources, to request sub-namespaces, and to retrieve resource metadata programmatically. The IGSN registration requests sent by the clients must be encoded in XML conforming to the CSIRO IGSN description metadata model.[14] The agent registration service mints IGSNs from the international registration service on behalf of the clients.

c. **CSIRO-IGSN Description metadata model:** The description metadata model represents the common concepts associated with physical resources such as identification, collection, curation, and related resources. It is designed to be general enough to catalogue different specimen types in the organization. The metadata schema serves as the basis for IGSN registration through the CSIRO agent registration service and to disseminate resource metadata through the OAI-PMH data provider service. Key features of the schema are that it supports batch registration of resources as our use cases may involve large batches of IGSN registrations, and it has minimal restrictions on which elements are required, e.g., resource identification, types and curation details. Some of the metadata elements are required to obtain IGSNs from the international registration service (e.g., *resourceIdentifier* and *landingPage*), while the others are relevant when discovering the resources through the web portal (e.g., *materiaTypes* and *curationDetails*). In addition, the schema offers flexibility to express both geographic and non-geographic location information (toponym), and time instants and intervals based on the W3C Date and Time Formats[15], which is a simpler profile of ISO 8601[16]. Physical samples are often relocated from one repository to another, therefore the schema captures the provenance of sample curation. It also represents several relation types

---

[12] https://igsn.csiro.au/igsn30/.

[13] https://igsn.csiro.au/igsn30/api/.

[14] The XML schema and its graphical representation are available at https://igsn.csiro. au/schemas/3.0/.

[15] https://www.w3.org/TR/NOTE-datetime.

[16] https://www.iso.org/standard/40874.html.

to associate a registered resource with its related resources, such as subsamples, digital resources (datasets, reports, images) and a reference resource[17]. It leverages existing and new controlled vocabularies that we developed in order to provide standardized information about the metadata elements and to ensure consistent metadata entry by clients. Digitization of specimens is beyond the scope of the project, although the digital images of specimens could be linked to their specimens registered in our system through the description metadata model.

d. **Controlled vocabularies.** To align with existing standards, we incorporated existing vocabularies into the description metadata schema, e.g., OGC definitions of nil reasons[18], material and specimen types defined by the CUAHSI's Observations Data Model (ODM2)[19], the contributor types from the CSIRO Linked Data Registry[20] and EPSG Geodetic Parameter Dataset[21]. We also developed the missing SKOS-based vocabularies that are necessary to connect the registered resources to the Web of data, e.g., registration types, identifier types, and relation types. The new vocabularies were identified with their corresponding persistent URIs to ensure machine actionability to the vocabularies. We use the Research Vocabularies Australia (RVA)[22] system to maintain the new vocabularies.

e. **Metadata store:** Metadata are stored in a PostgreSQL database modelled after the description metadata model. The metadata store captures resource metadata and client information (e.g., subnamespaces).

f. **Metadata provider and harvester:** We implemented an OAI-PMH provider service[23] to disseminate the metadata of registered resources in the metadata store. We also developed an OAI-PMH harvester, which is based on the PANGAEA Framework for Metadata Portals (panFMP) [14]. It harvests sample metadata from our own repository and other allocating agents.

g. **National IGSN web portal:** panFMP is entirely web-service based and does not supply its own graphical user interface, therefore its index is queried through a web portal. The web portal[24] provides a common access to sample metadata harvested from OAI-PMH services operated by different allocating agents in Australia.

---

[17] A physical sample is usually compared with a reference sample.
[18] http://www.opengis.net/def/nil/OGC/0/.
[19] http://vocabulary.odm2.org/.
[20] http://registry.it.csiro.au/.
[21] https://epsg.io.
[22] http://www.ands.org.au/online-services/research-vocabularies-australia.
[23] https://igsn.csiro.au/igsn30/api/service/30/oai.
[24] https://igsn2.csiro.au/portal.

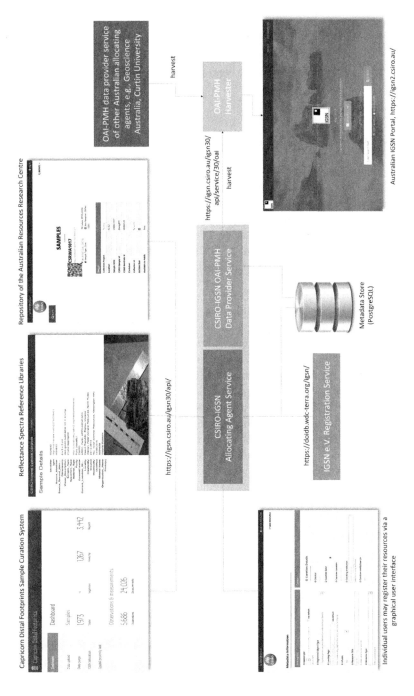

Fig. 2. Architecture of the CSIRO IGSN implementation.

Table 1. Local sample repositories and their IGSN registrations (as on 30.03.2017)

Repositories	Material types	IGSN Registered
Repository of the ARRC	rock, mineral, soil	25652
Capricorn distal footprints	rock, vegetation, water, regolith	4232
Reflectance spectra reference libraries	mineral, rock, synthetic material	94

4   Management

Use cases for samples in CSIRO range from individual researchers managing samples and their data manually (e.g., using spreadsheets) to projects and laboratories using sample curation systems. It is important to become familiar with how the users handle samples so that IGSN can be seamlessly integrated into their workflows and the existing workflows can be improved through technical solutions. To facilitate the integration of IGSN into our workflows, the following non-technical aspects were considered in this project.

a. **What can be identified with IGSNs?** To accommodate the needs of existing and potential sample applications, we allow IGSNs to be used to identify not only physical samples but also sample collections and sampling features. The IGSN Technical Documentation[25] makes recommendations concerning the format (identifier length) and semantic content of IGSN. In our implementation, we do not restrict the total length of an IGSN to allow repositories incorporate their existing identifiers into a globally unique IGSN. The IGSN identifiers are formed from a combination of the prefix of the allocating agent, client and the local sample identifier specified by the client.

b. **Identifier governance:** We established rules for assigning subnamespaces to different groups, collections and laboratories. The use of subnamespaces allows us to decouple the allocation of specific IGSNs in different parts of CSIRO, making it easier to ensure the global uniqueness of registered identifiers through a hierarchical delegation pattern [1].

c. **Integration of identifiers into new and existing systems:** For new sampling campaigns, we recommend that IGSN to be adopted at an early stage of the activity to ensure the consistent use of the identifiers throughout the sample life cycle, i.e., from collection and processing to curation. For existing sample curation systems, the local sample identifiers were extended using the IGSN namespace as a prefix to the local identifiers, thus making them globally unique. A similar method applies when individual researchers register their samples with IGSNs through the web form (Fig. 2). It is also possible to request the system to automatically generate unique identifiers for their samples. The web system also hosts the landing pages of registered resources, thus reducing the technical burden for users.

---

[25] http://igsn.github.io/syntax.

d. **Linking physical resources with their datasets:** For projects that have their own sample data curation systems, we recommend them to use the description metadata schema to associate a registered sample with the persistent URI of its datasets. For users who do not have a data curation system, we recommend to publish the specimen datasets via the CSIRO Data Access Portal, and then use the DOI generated by the data portal to link the dataset to its corresponding sample.

e. **Outreach:** We organized several events (presentations, meetings, and workshops) to introduce IGSN and to identify its potential application in the organization. The technical solutions are documented along with their source code, with examples available on a public repository[26]. This is important to reach the wider community, who may later adopt the solutions.

f. **National collaboration:** CSIRO has a joint project with the other allocating agents (Geoscience Australia and Curtin University) and is funded by the Australian Research Data Services program to implement IGSN for the Australian geoscience community. This collaboration effort involves representatives from academia, research and government agencies, and is essential to coordinating both IGSN-related activities and tool development at the national level, and to promote its implementation and governance in other sciences.

5   Conclusions

This paper described a successful implementation of persistent identifiers for physical resources (physical samples, sample collections, and sampling features) in a large organization. We developed a digital repository for the physical resources and specified its technical and non-technical components underlying the repository. The solutions developed have been applied to unambiguously identify physical resources from various studies, and to connect their metadata and data systematically to the Web. This improves the discovery of resources, and consequently facilitates their reuse and reproducibility.

The digital repository handles IGSN registrations from local sample data curation systems as well as from individual researchers. The hierarchical namespace delegation pattern is well suited for a large organization in which individual users, projects, and laboratories may all have different requirements for identifying and publishing their physical resources. The description metadata model is generic and extensible, and therefore suitable for representing the common properties of resources from different use cases. The digital repository harvests sample metadata from different sources, which can be aggregated to create new applications, for example, the Australian IGSN portal. Following the successful IGSN implementation in CSIRO, we are now collaborating with the John De Laeter Centre for Isotope Research at Curtin University to apply components developed in the context of their Digital Mineral Library. We reached out to some of the potential collections that could benefit from the CSIRO IGSN

---

[26] https://github.com/AuScope.

implementation, for example, the Australian National Soil Archive and the Australian National Insect Collection.

Acknowledgments. The IGSN implementation in CSIRO is part of the Research Data Services (RDS) project funded by the Department of Education as part of their Education Investment Fund (EIF) Super Science Initiative. The Capricorn Distal Footprints was funded by the Science and Industry Endowment Fund as part of The Distal Footprints of Giant Ore Systems: UNCOVER Australia Project (RP04-063).

References

1. Bechtold, S.: Governance in namespaces. Loyola Los Angeles Law Rev. **36**(3), 1239–1320 (2003). doi:10.2139/ssrn.413681
2. Behnken, A., Wallrabe-Adams, H.J., Röhl, U., Krysiak, F.: Application of the IGSN for improved data - sample - drill core linkage. In: EGU General Assembly Conference Abstracts. EGU General Assembly Conference Abstracts, vol. 18, p. 16688, April 2016
3. Conze, R., Lorenz, H., Ulbricht, D., Elger, K., Gorgas, T.: Utilizing the international geo sample number concept in continental scientific drilling during ICDP expedition COSC-1. Data Sci. **16**, 2 (2017). doi:10.5334/dsj-2017-002
4. Cox, S.J.D.: Geographic Information - Observations and Measurements (OGC Abstract Specification Topic 20) (same as ISO 19156:2011) (2011)
5. DataCite Metadata Working Group: DataCite Metadata Schema 4.0. Technical report, DataCite e.V., Hannover, Germany, May 2016. doi:10.5438/0012
6. Devaraju, A., Klump, J.F., Cox, S.J.D., Golodoniuc, P.: Representing and publishing physical sample descriptions. Comput. Geosci. **96**, 1–10 (2016). doi:10.1016/j.cageo.2016.07.018
7. Klump, J., Cox, S.J.D., Wyborn, L.A.I.: Connecting geology with the internet of things. In: Towards Unified Global Research. Melbourne, VIC, Australia, October 2014. http://eresearchau.files.wordpress.com/2014/07/eresau2014_submission_80.pdf
8. Klump, J., Huber, R.: 20 years of persistent identifiers - which systems are here to stay? Data Sci. J. **16**, 9 (2017). doi:10.5334/dsj-2017-009
9. Lagoze, C., Van de Sompel, H., Nelson, M., Warner, S.: Implementation guidelines for the open archives initiative protocol for metadata harvesting. Technical report (2002). http://www.openarchives.org/OAI/2.0/guidelines.htm
10. Lehnert, K.A., Klump, J., Arko, R.A., Bristol, S., Buczkowski, B., Chan, C., Chan, S., Conze, R., Cox, S.J., Habermann, T., Hangsterfer, A., Hsu, L., Milan, A., Miller, S.P., Noren, A.J., Richard, S.M., Valentine, D.W., Whitenack, T., Wyborn, L.A., Zaslavsky, I.: IGSN e.V.: registration and identification services for physical samples in the digital universe. In: American Geophysical Union, Fall Meeting 2011 (2011)
11. Lehnert, K., Carbotte, S., Ryan, W., Ferrini, V., Block, K., Arko, R., Chan, C.: IEDA: integrated earth data applications to support access, attribution, analysis, and preservation of observational data from the ocean, earth, and polar sciences. Geophysical Research Abstracts 13 (2011)

12. Lehnert, K.A., Vinayagamoorthy, S., Djapic, B., Klump, J.: The Digital Sample: Metadata, unique identification, and links to data and publications. EOS, Transactions, American Geophysical Union 87 (52, Fall Meet. Suppl.), Abstract IN53C-07 (2006). http://abstractsearch.agu.org/meetings/2006/FM/sections/IN/sessions/IN53C/abstracts/IN53C-07.html
13. McNutt, M., Lehnert, K.A., Hanson, B., Nosek, B.A., Ellison, A.M., King, J.L.: Liberating field science samples and data. Science **351**(6277), 1024–1026 (2016). http://science.sciencemag.org/content/351/6277/1024
14. Schindler, U., Diepenbroek, M.: Generic XML-based framework for metadata portals. Comput. Geosci. **34**(12), 1947–1955 (2008). doi:10.1016/j.cageo.2008.02.023
15. The Australian Antarctic program (AAp): The Australian Antarctic program data policy 2014 (applied to projects approved between 2 April 2013 and 21 June 2015). Online (June 2015). https://data.aad.gov.au/aadc/about/data_policy_2014.cfm
16. The National Science Foundation: Proposal and award policies and procedures guide (part ii - award & administration guide). Online February 2014. https://www.nsf.gov/pubs/policydocs/pappguide/nsf14001/aagprint.pdf
17. Wieczorek, J., Bloom, D., Guralnick, R., Blum, S., Döring, M., Giovanni, R., Robertson, T., Vieglais, D.: Darwin core: an evolving community-developed biodiversity data standard. PLoS ONE **7**(1), e29715 (2011). doi:10.1371/journal.pone.0029715

Facet Embeddings for Explorative Analytics in Digital Libraries

Sepideh Mesbah[✉], Kyriakos Fragkeskos, Christoph Lofi, Alessandro Bozzon, and Geert-Jan Houben

Delft University of Technology, Mekelweg 4, 2628 CD Delft, The Netherlands
{s.mesbah,k.fragkeskos,c.lofi,a.bozzon,g.j.p.m.houben}@tudelft.nl

Abstract. With the increasing amount of scientific publications in digital libraries, it is crucial to capture *"deep meta-data"* to facilitate more effective search and discovery, like search by topics, research methods, or data sets used in a publication. Such meta-data can also help to better understand and visualize the evolution of research topics or research venues over time. The automatic generation of meaningful deep meta-data from natural-language documents is challenged by the unstructured and often ambiguous nature of publications' content.

In this paper, we propose a domain-aware topic modeling technique called *Facet Embedding* which can generate such deep meta-data in an efficient way. We automatically extract a set of terms according to the *key facets* relevant to a specific domain (i.e. scientific objective, used data sets, methods, or software, obtained results), relying only on limited manual training. We then cluster and subsume similar facet terms according to their semantic similarity into facet topics. To showcase the effectiveness and performance of our approach, we present the results of a quantitative and qualitative analysis performed on ten different conference series in a Digital Library setting, focusing on the effectiveness for document search, but also for visualizing scientific trends.

1 Introduction

In light of the increasing amount of scientific publications, there is a growing need for methods that facilitate the exploration and analysis of a given research field in a digital library collection [1]. Existing approaches rely on word-frequency analysis [2], co-citation analysis [3], co-occurrence word analysis [4], and probabilistic methods like Latent Dirichlet Allocation (LDA) [5]. While popular, these approaches suffer from one major shortcoming: by offering a generic solution, they fail to capture the intrinsic semantics of text related to a specific domain of knowledge. For instance, probabilistic methods like LDA are designed to be generic and widely applicable; however, they often miss out on topics that are relevant from a user's point of view.

To support richer retrieval experience, we advocate extracting *"deep meta-data"* from scientific publication, i.e. meta-data able to represent domain-specific properties and aspects (*facets*) in which a document can be considered and understood within its (research) domain.

© Springer International Publishing AG 2017
J. Kamps et al. (Eds.): TPDL 2017, LNCS 10450, pp. 86–99, 2017.
DOI: 10.1007/978-3-319-67008-9_8

Let us consider, for instance, the domain of *data processing and data science*, which is gaining popularity due to the availability of great amount of digital data, and progress in machine learning. In this domain, researchers and practitioners need to develop an understanding of the properties of available *datasets*; of existing data processing *methods* for the collection, enrichment and analysis of data; and of their respective implementations as *software* packages. The availability of deep meta-data about the facets (*datasets*, *methods*, and *software*) would enable rich queries like: *Which methods are commonly applied to a given dataset?*; *Discover state of the art methods for point of interest recommendation that have been applied to geo-located social media data with high accuracy results*. To the best of our knowledge, no state-of-the-art system is currently able to provide answers to the previous queries.

This paper presents an approach for generating domain-aware *"deep meta-data"* from collections of scientific publications. We focus on the data processing domain, and address the main facets described in the DMS ontology [6], namely *datasets, methods, software, objectives*, and *results*. We build upon a basic distant supervision approach for sentence classification and named entity extraction [7], and extend it with *facet embeddings* to automate the creation of *Facet Topics*, i.e. clusters of semantically similar facet terms which allow for easier querying and visualization. Our contributions are as follows:

- We introduce and formalize the concept of *facet topics*, which subsume a set of facet terms into higher level topics more suitable for exploration, visualization, and topic centered queries.
- We describe a novel approach for facet topic identification through *facet embeddings*. The approach combines distant supervision learning on rhetorical mentions for facet-specific sentence classification; semantic annotation and linking for facet terms extraction; and semantic clustering.
- We quantitatively and qualitatively assess the performance of our approach, and compare to established techniques like LDA topic modeling.
- We showcase our approach with a study exploring and visualizing trends and changes within the domain of data processing research, based on deep meta-data extracted from 11,589 research publications.

2   Related Work

The information overload in digital libraries is a crucial problem for researchers. Online digital libraries like the ACM Digital Library (DL), IEEE Xplore, CiteSeer etc., provide search options for finding relevant publications by using *"shallow"* meta-data such as the title, the authors, keywords or other simple statistical measures like the number of citations and download counts. However they are not designed to support the analysis of *"deep"* meta-data such as the topic, or methods and algorithms used in scientific publications.

There has been a large body of research focused on *deep* analysis of publications in scientific domains such as Software Engineering [1], Bio-informatics [8],

Digital Library evaluation [9], or Computers science [10]; for different purposes, such as finding topic trends in a domain [1,10] and evolution of scientific communities popularity [11]. Common approaches rely on methods such as word-frequency analysis [2], co-citation analysis [3,10], co-word analysis [4], and probabilistic methods like latent Dirichlet allocation [5]. In contrast to existing literature which is either specially tailored to a domain or fully generic, our work combines the strength of both approaches by being partially domain-aware: after defining domain-aware facets using (limited) expert feedback, our approach automatically extracts topics by analyzing the co-occurrence of named entities related to the facets, thus is scalable within a domain while still taking advantage of domain-specific knowledge and peculiarities.

While most current research [1,2,11] limits the analysis of a publication's content to its title, abstract, references, and authors, we extract facet terms from the full text of scientific publications, in order to obtain more descriptive and accurate topics. In addition, our method is not only based on selecting the most frequent keywords (e.g. nouns, verbs set and proper nouns) [2], and, differently from probabilistic methods like Latent Dirichlet Allocation [5], it considers the semantics of terms for topic identification.

Some existing methods for domain-specific concept extraction and categorization are based on noun phrase chunking [11,12] and use a bootstrapping approach to identify scientific concepts in publications. More recent research [13] used both corpus-level statistics and local syntactic patterns of scientific publications to identify and cluster similar concepts. Our method follows a distant supervision approach, a simple feature model (bags-of-words), and does not require prior knowledge about grammatical [12] and part-of-speech characteristics of facet terms. However, we do require a brief training phase for adapting our approach to a new domain.

3   Problem Description and Modeling

The goal of our work is to annotate n documents $D = \{d_1, ..., d_n\}$ of a domain-specific (scientific) corpus with faceted semantic meta-data. This meta-data goes alongside already available structured meta-data like for example author names, publication year, or citations. In particular, we aim at annotating documents with both *facet terms* and *facet topics*, as discussed in the following:

Facets and Facet Sets: The central elements of our approach are *facets*. Facets represent a perceived aspect relevant to user's understanding of documents in corpus D. When adapting our method to a given corpus, a *facet set* has to be defined which is used for describing documents in D, denoted as $F = \{f_1, f_2, ..., f_n\}$. Defining a good facet set requires some domain expertise. In the study presented in this work, we used specific facet set designed based on [6], namely the F_{DMS} facet set covering facets for a document corpus focused on data processing research. This facet set covers the five facets dataset, methods, software, objective, and result. We denote this as $F_{DMS} = \{DST, MET, SFT, OBJ, RES\}$.

Facets Terms: For each document $d \in D$ and facet $f \in F$, we extract a set of *facet terms* FT_f^d. A facet term $ft \in FT_f^d$ represents a term (usually a named entity, but also short phrases are possible) found in the full text of document d, and which can be clearly associated with facet f. We denote the set of all facet terms related to a given facet f found in any document of D as FT_f. Typical examples of facet terms for the method facet $MET \in F_{DMS}$ in our document collection are "Latent Dirichlet Allocation", "Support Vector Machine", or "Description Logic".

Facets Topics: Facet Terms are directly extracted from the full text of documents, and describe a document at a rather low level. In order to also allow for high-level analytics and queries, we introduce the concept of *facet topics*. Facet topics group multiple semantically related facet terms into a larger subsuming topic. In our use case scenario, when focusing on the methods facet, facet topics intuitively relate to research topics. For example, the terms "Support Vector Machine" and "Random Forest" can be subsumed by the facet topic "Machine Learning". The set of all facet topics for a given facet f is denoted as $FTP_f = \{t_1, t_2, \ldots t_k\}$, and each facet topic t is a subset of all facet terms, i.e. $t \in FTP_f : t \subseteq FT_f$. Furthermore, each term can be attributed to a topic, i.e. $FT_f = \bigcup_{t \in FTP_f} t$, and topics of a given facet are disjoint, i.e. $t_i, t_j \in FTP_f, t_i \neq t_j : t_i \cap t_j = \emptyset$ (however, there might be an overlap between topics of different facets, see next section). Terms in a facet topic show strong semantic cohesion.

4 Facet Term Extraction and Facet Topic Identification

In this section, we present our approach for *facet terms* and *facet topics* extraction from a collection of scientific publications, extending our previous work [7] by introducing additional steps for facet topic identification. An overview of our approach is shown in Fig. 1. Our approach is domain-aware in the sense that it requires some limited efforts to adjust it to a new domain (like deciding on facet sets), but is not inherently limited to a specific domain. In the following, we focus on the *data processing* domain, and address the five main *facets* (i.e. datasets, methods, software, results, and objectives) identified in the DMS ontology [6].

The process can base summarized as: First, we identify rhetorical mentions of a *facet* in the full text of documents. In this work, for the sake of simplicity, rhetorical mentions are identified at sentence level (i.e., each sentence is classified whether it contains a rhetorical mention of a given facet or not). Future works will introduce dynamic boundaries, to capture the exact extent of a mention.

After a rhetorical mention was found, we extract potential *facet terms* from it. These terms are filtered and, when applicable, linked to pre-existing knowledge bases. Finally, all filtered facet term candidates finally form the document-specific facet term sets FT_f^d.

The identification of rhetorical mentions is obtained through a workflow inspired by distant supervision, a training methodology for machine learning algorithms that relies on very large, but noisy, training sets. The training sets are

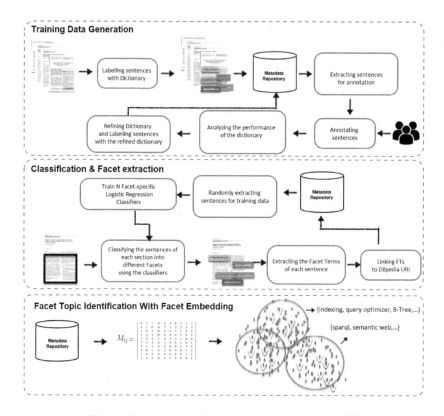

Fig. 1. Domain-aware Facet Modeling Workflow [7]

generated by means of a simpler classifier, for instance a mix of expert-provided dictionaries and rules, refined with manual annotations. Intuitively, the training noisiness can be canceled out by the huge size of the semi-manually generated training data. The method requires significantly less manual effort, while at the same time retaining the performance of supervised classifiers. Furthermore, this approach is more easily adapted to different application domains and changing language norms and conventions (more details in [7]).

Data Preparation: Scientific publications, typically available in PDF, are processed using state-of-art extraction engines, e.g. GeneRation Of BIbliographic Data (GROBID) [14]. GROBID extracts a structured full-text representation as Text Encoding Initiative(TEI)-encoded documents, thus providing easy and reliable access paragraphs and sentences.

Test and Training Data Generation: We created training and benchmarking datasets for evaluating our rhetorical mention classifier by relying on a phrase dictionary for each facet (as described in [7]), automatically labeling all sentences in the document corpus if they contain a mention of relevant for a facet or not. Then, we randomly select a balanced set of 100 mentions of each facet.

As the dictionary-based classifier is not fully reliable, we manually inspect and reclassify the selected sentences using feedback from two expert annotators, and rebalance the sentence set as needed. The inter-annotator agreement using the Cohen's kappa measure averaged over all classes was .58. Using this approach, we can create a reliable manually annotated and balanced test dataset quicker and cheaper compared to annotating whole publications or random sentences, as the pattern-classifier usually delivers good candidate sentences.

Machine-Learning-based Rhetorical Detection: As a next step in our distant supervision workflow, we train a simple binary Logistic regression classifier for each of the (facet) classes using simple TF-IDF features for each sentence. This simple implementation serves as a proof of concept of our overall approach, and can of course be replaced by more sophisticated features and classifiers.

As a test set, we use the aforementioned test set of 100 sentences for each facet. The *method* classifier showed the best performance with respect to its F-measure(0.71). From this, we conclude that our approach is indeed suitable for extracting *DMS* facet terms in a meaningful and descriptive fashion. However, there are still some false positives which cannot easily be recognized using simple statistic means, thus inviting further deeper semantic filtering in future works.

Facet Extraction, Linking, and Filtering: We extract *facet terms* from the labeled rhetorical mentions identified in the previous section, filtering out those terms which are most likely not referring to one of the *facet*, and retaining the others as an extracted term of the class matching the sentence label.

Facet extraction has been performed using the TextRazor API. TextRazor returns the detected *facet terms*, possibly decorated with links to the DBpedia or Freebase knowledge bases. As we get all *facet terms* of a sentence, the result list contains many *facet terms* which are not specifically related to any of the five *facets* (e.g. terms like "software", "database"). To filter the *facet terms*, we decided on a simple filtering heuristic assuming *facet terms* to be not relevant if they come from "common" English language (like software, database), while relevant terms are from domain-specific language or specific acronyms (e.g. SVM, GROBID). In our current prototype, we implement this heuristics by looking-up each term in *Wordnet*. Terms that can be looked-up are removed as we consider them common language. While this simple approach works for the "data science" domain, when extending our approach to a wider range of domains, this implementation should be replaced by more sophisticated heuristics, e.g., based on corpus statistics.

Facet Topic Identification With Facet Embedding After extracting all facet terms, we now strive to discover meaningful *facet topics*. Here, a central goal is to subsume facet terms based on their semantic similarity. We implement a measurement for semantic similarity of terms by *Facet Embeddings*, which exploit co-occurrence of facet terms. For each $t_i, t_j \in FT_f$, we count how often these terms co-occur within the same document: $co_{t_i,t_j} = |\{d \in D : t_i \in FT_f^d \wedge t_j \in FT_f^d\}|$.

This results in a large (sparse) co-occurrence matrix. We reduce the dimensionality of the matrix using truncated Singular Value Decomposition. This step ensures the removal of less informative terms, while increasing the performance and usability of our approach (a smaller matrix is computationally cheaper to process). Using the reduced matrix, we now obtained an embedding of each facet term of a given facet (i.e., each term is represented as row vector in the reduced co-occurrence matrix).

Finally, we now cluster all facet terms of a given facet in order to discover facet topics using K-means clustering, using Euclidean distance between the row vectors of two given terms as a distance measure. In order to find the optimal number k of clusters, we rely on Silhouette analysis, measuring the closeness of each point in a cluster to the points in its neighboring clusters. In addition to the Silhouette analysis, we also manually inspected the resulting clusters, but found that also from an qualitative point of view, the number of clusters determined by the Silhouette analysis is indeed the most satisfying one.

As a last processing step, we have two expert annotators label each facet topic in an iterative process until full agreement between the annotators was reached (see Sect. 5 for more details).

We also implemented an alternative version of facet embeddings, relying on neuronal word embeddings (in our case word2vec [15]) instead of co-occurrence in rhetorical mentions. However, initial qualitative inspection of the results indicate that the distance measure between the term embeddings is an inferior representation of perceived similarity of facet terms from our experts' point of view. A deeper analysis of these results will be the subject of a later study.

5   Evaluation and Experimentation

In this section, we analyze the performance of our facet topic modeling workflow. We analyze and discuss the quality of facet terms extracted from the classified sentences. Next we qualitatively evaluate the quality of the topics extracted using Facet Embeddings. Finally we present some examples of information needs of researcher that can be fulfilled using our approach.

Corpus Analysis: We focused on 11,589 papers from ten conference series: The Joint conference on Digital Libraries (JCDL); the International Conference on Theory and Practice of Digital Libraries (TPDL); the International Conference on Research and Development in Information Retrieval (SIGIR); the Text Retrieval Conference (TREC); the European Conference on Research and Advanced Technology on Digital Libraries (ECDL); the International Conference on Software Engineering (ICSE); the Extended Semantic Web Conference (ESWC); the International Conference On Web and Social Media (ICWSM); the International Conference on Very Large Databases (VLDB); and the International World Wide Web Conference (WWW).

Table 1. Quantitative analysis of the rhetorical sentences and facet terms extracted from ten conference series. Legend: PUB (publications), SNT (sentences), OBJ (objective), DST (dataset), MET (method), SFT (software), RES (results)

Conf.	Size		Rhetorical sentences						Unique facet terms				
	Years	#PUB	#SNT	#OBJ	#DST	#MET	#SFT	#RES	#OBJ	#DST	#MET	#SFT	#RES
ESWC	2005–2016	626	84439	12725	13528	26337	9614	22245	4197	4910	6987	4557	6416
ICWSM	2007–2016	810	34987	6096	4277	8936	1830	13848	2830	2241	3658	1538	4499
VLDB	1975–2007	1884	272380	30360	56647	77123	13317	94933	8008	13207	15319	6262	17532
WWW	2001–2016	2067	322801	47134	40449	97760	21347	116111	10902	10917	17783	8863	19822
ECDL	1997–2010	820	65470	12008	8079	18638	8130	18615	4634	3650	5894	4125	5376
ICSE	1976–2016	1834	182029	29850	16284	57494	26042	52359	8169	5841	12503	8776	11728
JCDL	2001–2016	1416	99747	19290	13002	27786	9692	29977	6524	5240	7754	5037	7979
SIGIR	1971–2016	412	39688	5080	4813	13214	2050	14531	2144	2377	4126	1588	4068
TPDL	2011–2016	276	23176	4660	3342	6032	2489	6653	2168	1871	2625	1719	2503
TREC	1999–2015	1444	122456	11828	14760	39121	8825	47922	6616	3085	4095	3286	7668

Due to changes in publication platforms and PDF format, the corpus does not contain all publications of each conference.[1] We believe the absence of few publications not to have an impact on the significance of our findings, but might still be reflected in the shown diagrams and results. Table 1 provides basic statistics for the analyzed corpus, including: the range of years, the number of publications, the number of extracted rhetorical sentences and mentions, and the distinct facet terms extracted from rhetorical sentences. *Method* and *results* facets are the most frequent, followed by *objectives*.

Quality of extracted topics: We investigated or domain-aware facet embedding compared to the domain-independent technique Latent Dirichlet Allocation (LDA) by asking two domain experts to label the topics derived by each method, while assessing which are more meaningful. For the sake of presentation, we set the maximum number of topics to $T = 30$, and performed the Silhouette analysis to find the number of optimal topics, which resulted in 27 topics.

In order to allow for a more informative comparison, we applied both approaches to the full text of publications, and also to only pre-classified sentences (because LDA is usually applied to full texts. Thus, in one case we use our facet embedding without restricting to classified facet sentences, but we also consider a case where LDA is applied to the set of all sentences which belong to a given facet). For the sake of brevity, we consider only the *method* facet when performing a facet pre-classification of sentences. The *method* classifier has shown the best performance with respect to its F-measure. Our analysis shows comparable results with the other facets.

Full Text without Facet Classification: For full text experiments, the corpus has been pre-processed by removing stop words, and representing each document as a bag-of-words. We use the LDA implementation provided by the `scikit-learn`

---

[1] For instance, around 100 JCDL papers for 2014 are not included in the analysis, as the proceedings were, only for that year, published by ieee.org.

library. For compatibility, we trained for 27 topics. For evaluating facet embeddings without any domain specific pre-classification on full texts, we are assuming that there is only a single facet, and each sentence of a document is classified as such (note: this is not how we usually intend our method to work).

Consider only Sentences classified as Method facet: In this experiment, we perform the *facet topic* extraction as described in Sect. 4, including facet-based sentence classification, facet term extraction, and facet embedding, but limited to only the *Method* facet. As a comparison, we also perform LDA on only those sentences classified as methods (therefore also giving LDA the chance to take advantage of the domain-aware training).

Results: A manual inspection on the resulting topics show that those identified by LDA are very hard to label and are perceived as semantically less meaningful by our experts, while the topics based on Facet Embeddings produced coherent and interpretable topics which were perceived as understandable and useful. In Table 2, we provide an example of 3 randomly selected topics for each aforementioned experimental setup. It can be observed that topics generated from sentences pre-classified as *method* show better semantic cohesion than those generated from full texts. Furthermore, we provide the full result of labeling all 27 topics for the method facet in Table 3. The top-40 term can be found in the companion website[2]

Table 2. Example top terms extracted using the generic (LDA) and domain-aware (FE) topics, using either full texts or only those sentences related to the *method* facet

Full text	LDA	reference, abstracts, linking, sofm, similarity annotations, backup, linkservice, annotation, digital query, data, user, web, information
	FE	sparql, semanticweb, linkeddata, rdf, dbpedia, sql, relationaldatabase, tuple, queryoptimization, datawarehouse, socialnetwork, facebook, randomwalk, pagerank, powerlaw
Facet	LDA	documents, used, classification, libraries, digital measure, performance, given, recommendation, used, social, twitter, media, popular, past
	FE	searchalgorithm, timecomplexity, datastructure, dynamicprogramming, sparql, semanticweb, linkeddata, dbpedia, rdfs, socialmedia, lda, latentdirichletallocation, topicmodel, socialnetwork

Application Example: Scientific Publication Retrieval: In this section we show scenarios of how computer science researchers could use our approach for their work. Furthermore, all faceted deep meta-data used in those scenarios has been published as an RDF knowledge base according to the DMS ontology, accessible from a SPARQL endpoint on the companion website.

---

[2] http://www.wis.ewi.tudelft.nl/tpdl2017.

Table 3. Top five *method* terms for each facet topic. Topic labels have been assigned manually by two xperts.

Topic name	Top five terms
Social Media Analytics: Text-based	social media, lda, latent dirichlet allocation, topic model, social network
Semantic Web: Knowledge Engineering & Representation	sparql, semantic web, linked data, dbpedia, rdfs
Semantic Web: Logic	description logic, dl, abox, tbox, semanticweb
Misc Topics: Web Information Systems	information retrieval, data structure, dataset, natural language, electronic media
Databases: Query Processing	tuple, hash join, sort, relational database, hash table
Databases: Modelling	data model, sql, query language, query optimization, tuple
Web Technologies	side, client, server, javascript, web application
Digital Libraries	digital library, information retrieval, xml, user interface, computer science
Machine Learning	machine learning, support vector machine, supervised learning, dataset, information retrieval
Web Engineering: P2P & Distributed Systems	peer, to, ip address, rdf, webservice
Social Graph Algorithms	greedy algorithm, approximation algorithm, optimization problem, social network, electronic media
Social Graph Analysis	pagerank, random walk, social network, webpage, adjacency matrix
XML Databases	xml, xpath, xquery, xmlschema, sql
Software Engineering: Testing & Formal Methods	source code, test case, control flow, test suite, program analysis
Software Engineering: Systems	software development, software engineering, software development process, software system, case study
Web Engineering: System Modelling	use case, web service, model checking, case study, semantic web
Web Engineering: Client-Side	web page, user interface, web browser, web content, javascript
Information Retrieval: QA, NLP, and Complex Queries	trec, question answering, document retrieval, information retrieval, query expansion
Information Retrieval: Evaluation	adhoc, trec, query expansion, information retrieval, relevance feedback
Information Retrieval: Ranking	query expansion, language model, relevance feedback, trec, information retrieval
Information Retrieval: Mining	score, f1, supervised learning, crf, bic
Microsoft Technology	microsoft, microsoft sqlserver, sql, xml, microsoft word
Databases: Indexing	tree, trees, data structure, access method, search algorithm
Databases: Transaction Mangement	concurrency control, lru, serializability, aries, tion
Databases: Algorithms	search algorithm, time complexity, data structure, dynamic programming, dataset
Recommendation	collaborative filtering, recommender system, gradient descent, singular value decomposition, social network
System Engineering: Architecture	operating system, programming language, file system, data structure, software engineering

Table 4. Examples of papers applying methods (MET) to given datasets(DTS)

Paper title	Dataset and method facet
Personalized Interactive Faceted Search [16]	IMDB(DST), Faceted search(MET)
refeREE: An Open Framework for Practical Testing of Recommender Systems using ResearchIndex [17]	IMDB(DST), Recommender system(MET)
The Party is Over Here: Structure and Content in the 2010 Election [18]	Facebook(DST), Sentiment analysis(MET)

Find publications that applied method X on a given dataset: Table 4 shows the result of an example query for methods which have been applied to movie dataset (i.e. IMDB and MovieLens) or Social media data (i.e. Facebook). For instance, [17] is a paper containing both the facet terms "Recommender system" labeled as *method*, and "IMDB" labeled as *dataset*.

Retrieve the most used methods of a given conference series: To answer this question, we use the number of papers for each *method* facet topic shown in Table 3 for a given conference. Results are shown in Fig. 2. The value in each cell denotes the values normalized by the number of publications in each conference overall. The figure also demonstrate the quality of our approach: topics like "Machine Learning" and "Information Systems" are popular for all considered conferences. "Database" topics are mostly popular in the VLDB conference series, while the topic "Digital Library" appears in JCDL and TPDL. Clearly, the extracted facet topics match the research focus of each conference. Also,

	ECDL	JCDL	TPDL	ICSE	VLDB	SIGIR	TREC	ICWSM	WWW	ESWC
Databases: Algorithms	0.0214	0.0249	0.0217	0.0176	0.1075	0.0362	0.0161	0.0288	0.0369	0.0264
Databases: Indexing	0.0045	0.0031	0.0007	0.0031	0.0758	0.0079	0.0015	0.0057	0.0069	0.0033
Databases: Modelling	0.0291	0.0116	0.0245	0.0292	0.1081	0.0225	0.0083	0.0076	0.0159	0.0279
Databases: Query Processing	0.0086	0.0056	0.0035	0.0117	0.1465	0.0150	0.0054	0.0057	0.0116	0.0160
Databases: Transaction Management	0.0024	0.0020	0.0035	0.0073	0.0728	0.0049	0.0012	0.0014	0.0043	0.0038
Digital Libraries	0.1909	0.1577	0.1058	0.0089	0.0063	0.0154	0.0076	0.0104	0.0116	0.0122
Information Retrieval: Evaluation	0.0122	0.0049	0.0014	0.0093	0.0046	0.0300	0.0797	0.0042	0.0061	0.0083
Information Retrieval: Mining	0.0003	0.0043	0.0014	0.0004	0.0016	0.0000	0.0017	0.0057	0.0052	0.0026
Information Retrieval: QA and NLP and and Complex Queries	0.0410	0.0282	0.0182	0.0085	0.0096	0.0780	0.2392	0.0212	0.0164	0.0191
Information Retrieval: Ranking	0.0395	0.0439	0.0357	0.0093	0.0098	0.1054	0.2548	0.0434	0.0299	0.0217
Machine Learning	0.0689	0.1415	0.1458	0.0542	0.0355	0.1592	0.1131	0.1619	0.1296	0.1070
Microsoft Technology	0.0217	0.0137	0.0070	0.0203	0.0160	0.0026	0.0073	0.0019	0.0143	0.0086
Misc Topics: Web Information Systems	0.2591	0.2549	0.2614	0.2067	0.1806	0.2513	0.1304	0.2011	0.1790	0.1215
Recommendation	0.0125	0.0280	0.0154	0.0080	0.0058	0.0450	0.0117	0.0477	0.0461	0.0191
Semantic Web: Knowledge Engineering & Representation	0.0178	0.0228	0.0596	0.0071	0.0114	0.0150	0.0100	0.0151	0.0449	0.2395
Semantic Web: Logic	0.0068	0.0011	0.0028	0.0026	0.0021	0.0009	0.0010	0.0024	0.0144	0.0753
Social Graph Algorithms	0.0042	0.0038	0.0077	0.0050	0.0152	0.0141	0.0041	0.0198	0.0417	0.0060
Social Graph Analysis	0.0255	0.0390	0.0294	0.0089	0.0154	0.0454	0.0327	0.0760	0.0664	0.0298
Social Media Analytics: Text-based	0.0065	0.0069	0.0308	0.0083	0.0052	0.0454	0.0151	0.2587	0.0561	0.0205
Software Engineering: Systems	0.0237	0.0181	0.0182	0.1705	0.0131	0.0123	0.0078	0.0109	0.0131	0.0176
Software Engineering: Testing & Formal Methods	0.0098	0.0060	0.0070	0.2099	0.0107	0.0093	0.0066	0.0071	0.0196	0.0136
System Engineering: Architecture	0.0175	0.0078	0.0112	0.0397	0.0216	0.0097	0.0034	0.0024	0.0097	0.0036
Web Engineering: Client-Side	0.0748	0.0658	0.0736	0.0283	0.0198	0.0498	0.0202	0.0387	0.0821	0.0350
Web Engineering: P2P & Distributed Systems	0.0116	0.0069	0.0021	0.0024	0.0099	0.0062	0.0034	0.0019	0.0134	0.0100
Web Engineering: System Modelling	0.0359	0.0307	0.0736	0.0892	0.0194	0.0093	0.0071	0.0109	0.0486	0.1063
Web Technologies	0.0062	0.0125	0.0112	0.0207	0.0048	0.0022	0.0027	0.0033	0.0389	0.0102
XML Databases	0.0478	0.0314	0.0266	0.0131	0.0707	0.0071	0.0080	0.0061	0.0371	0.0353

Fig. 2. Heatmap showing the relation between *research methods* and conferences. The values are normalized based on the numbers of papers in each conference.

other popular topics can be explored: conferences like JCDL or TPDL favor methods like Machine Learning, Digital Libraries, Web Information Systems, and Information Retrieval.

What are the trends for methods?: In order to answer this question, we visualize the number of publications covering a *method* facet topic (as listed in Table 3) over the course of the last 10 years. The results are shown in Fig. 3, giving an intuition about the quality of our approach: e.g., methods related to machine learning, software testing, or social media analytics gained great popularity in the last 10 years; while, as expected, topics related to core databases techniques or XML processing are becoming less popular.

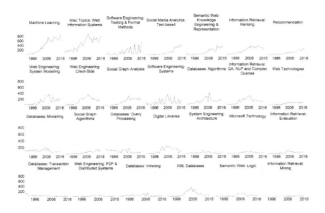

Fig. 3. The trends of *research methods* over years. The y axis shows the contribution of each topic in a certain year by means of the number of method-occurrence

6   Summary and Outlook

This paper presents the design and evaluation of a novel method for domain-aware topic identification for collections of scientific publications. Our method aims at improving the ability of digital libraries systems to support the retrieval, exploration, and visualization of documents based on topics of interest. In contrast to previous work, is taking advantage of some domain-specific insights which vastly improves the quality of the resulting topics, while still being adoptable to other domains by minimal efforts.

Our proposed method relies on a combination of sentence classification, semantic annotation, and semantic clustering to identify *Facet Topics*, i.e. clusters of semantically related *terms* that are tied to an *facet* relevant to an user's understanding of a document collection. The method specializes on the extraction of facet-specific information through the classification of rhetorical mentions in sentences. A lightweight distant supervision approach with low training costs (compared to traditional supervised learning) and acceptable performance, allows for simple adaptation to different domains. Facet terms are extracted from candidate sentences using state-of-the-art semantic annotation tools, and are filtered according to their informativeness. *Facet Topics* are identified using a novel *Facet Embedding* technique.

We applied this novel method to a corpus of 11,589 publications on *data processing* from 10 conference series, and extracted metadata related to the 5 facets of the DMS [6] ontology for data processing pipelines. An extensive set of quantitative and qualitative analysis shows that, despite its simple design, our methods allows for topic identification performance superior to state-of-the-art topic modeling methods like LDA.

While promising, results leave ample space for future improvements. We are interested in investigating the performance of more complex machine learning classifiers (e.g. based on word-embeddings), possibly applied to group of related sentences. We also plan to investigate new techniques for facet terms extractions, and study the performance of our approach with larger amount of *Facet Topics*. Finally, we plan to expand our analysis to additional domains, and investigate new facets of interest.

References

1. Mathew, G., Agarwal, A., Menzies, T.: Trends in topics at SE conferences (1993–2013). arXiv preprint arXiv:1608.08100 (2016)
2. Shubankar, K., Singh, A., Pudi, V.: A frequent keyword-set based algorithm for topic modeling and clustering of research papers. In: 3rd Conference on Data Mining and Optimization (DMO), 2011, IEEE, pp. 96–102 (2011)
3. Chen, C.: CiteSpace II: Detecting and visualizing emerging trends and transient patterns in scientific literature. J. Am. Soc. Inform. Sci. Technol. **57**(3), 359–377 (2006)
4. Isenberg, P., Isenberg, T., Sedlmair, M., Chen, J., Möller, T.: Visualization as seen through its research paper keywords. IEEE Trans. Visual Comput. Graphics **23**(1), 771–780 (2017)
5. Griffiths, T.L., Steyvers, M.: Finding scientific topics. Proc. Natl. Acad. Sci. **101**(suppl 1), 5228–5235 (2004)
6. Mesbah, S., Bozzon, A., Lofi, C., Houben, G.J.: Describing data processing pipelines in scientific publications for big data injection. In: WSDM Workshop on Scholary Web Mining (SWM). Cambridge, UK (2017)
7. Mesbah, S., Fragkeskos, K., Lofi, C., Bozzon, A., Houben, G.-J.: Semantic Annotation of Data Processing Pipelines in Scientific Publications. In: Blomqvist, E., Maynard, D., Gangemi, A., Hoekstra, R., Hitzler, P., Hartig, O. (eds.) ESWC 2017. LNCS, vol. 10249, pp. 321–336. Springer, Cham (2017). doi:10.1007/978-3-319-58068-5_20
8. Song, M., Heo, G.E., Kim, S.Y.: Analyzing topic evolution in bioinformatics: investigation of dynamics of the field with conference data in DBLP. Scientometrics **101**(1), 397–428 (2014)
9. Afiontzi, E., Kazadeis, G., Papachristopoulos, L., Sfakakis, M., Tsakonas, G., Papatheodorou, C.: Charting the digital library evaluation domain with a semantically enhanced mining methodology. In: Proceedings of the 13th ACM/IEEE-CS Joint Conference On Digital Libraries, pp. 125–134. ACM (2013)

10. Hoonlor, A., Szymanski, B.K., Zaki, M.J.: Trends in computer science research. Commun. ACM **56**(10), 74–83 (2013)
11. Gupta, S., Manning, C.D.: Analyzing the Dynamics of Research by Extracting Key Aspects of Scientific Papers
12. Tsai, C.T., Kundu, G., Roth, D.: Concept-based analysis of scientific literature. In: Proceedings of the 22nd ACM International Conference On Conference On Information & Knowledge Management - CIKM 2013, pp. 1733–1738 (2013)
13. Siddiqui, T., Ren, X., Parameswaran, A., Han, J.: FacetGist: Collective extraction of document facets in large technical corpora. In: Proceedings CIKM 2016 (2016)
14. Lopez, P.: GROBID: Combining Automatic Bibliographic Data Recognition and Term Extraction for Scholarship Publications. In: Agosti, M., Borbinha, J., Kapidakis, S., Papatheodorou, C., Tsakonas, G. (eds.) ECDL 2009. LNCS, vol. 5714, pp. 473–474. Springer, Heidelberg (2009). doi:10.1007/978-3-642-04346-8_62
15. Mikolov, T., Sutskever, I., Chen, K., Corrado, G.S., Dean, J.: Distributed representations of words and phrases and their compositionality. Adv. Neural Inf. Process. Syst. **21**, 3111–3119 (2013)
16. Koren, J., Zhang, Y., Liu, X.: Personalized interactive faceted search. In: Proceeding of the 17th International Conference On World Wide Web - WWW 2008, pp. 477–485 (2008)
17. Cosley, D., Lawrence, S.: REFEREE: An open framework for practical testing of recommender systems using ResearchIndex. In: Proceedings of the 28th VLDB Conference, pp. 35–46 (2002)
18. Livne, A., Simmons, M.P., Adar, E., Adamic, L.a.: The Party is Over Here: Structure and Content in the 2010 Election. vol. 161(3), pp. 201–208 (2010)

Data in Digital Libraries

Automatic Hierarchical Categorization of Research Expertise Using Minimum Information

Gustavo Oliveira de Siqueira, Sérgio Canuto, Marcos André Gonçalves, and Alberto H.F. Laender[✉]

Department of Computer Science,
Universidade Federal de Minas Gerais, Belo Horizonte, MG, Brazil
{gosiqueira,sergiodaniel,mgoncalv,laender}@dcc.ufmg.br

Abstract. Throughout the history of science, different knowledge areas have collaborated to overcome major research challenges. The task of associating a researcher with such areas makes a series of tasks feasible such as the organization of digital repositories, expertise recommendation and the formation of research groups for complex problems. In this paper we propose a simple yet effective automatic classification model that is capable of categorizing research expertise according to a hierarchical knowledge area classification scheme. Our proposal relies on discriminative evidence provided by the title of academic works, which is the minimum information capable of relating a researcher to its knowledge area. We also evaluate the use of learning-to-rank as an effective mean to rank experts with minimum information. Our experiments show that using supervised machine learning methods trained with manually labeled information, it is possible to produce effective classification and ranking models.

Keywords: Research expertise categorization · Classification schemes · Supervised classification · Learning-to-rank

1 Introduction

Throughout the evolution of science, scientific problems have become more and more complex over time. Their solution currently requires the combination of multiple expertises for the formation of multidisciplinary research groups to work on those complex problems. One basic premise for this to work is that one may be able to identify the main areas of expertise of scholars/researchers. In fact, the effective and reliable association of a scholar with a knowledge area makes a series of tasks feasible such as: (i) organization of digital repositories according to a knowledge area categorization scheme; (ii) expertise recommendation for specific industrial or scientific problems; and (iii) the formation of research groups for solving very complex problems.

© Springer International Publishing AG 2017
J. Kamps et al. (Eds.): TPDL 2017, LNCS 10450, pp. 103–115, 2017.
DOI: 10.1007/978-3-319-67008-9_9

There are currently several sources of information that can be used to identify a researcher's expertise, such as: (i) digital libraries containing information about a researcher's scientific production over time (e.g., DBLP[1] and ACM Digital Library[2]); (ii) metadata and, in several cases, the full text of an electronic thesis or dissertation (ETD) available in ETD repositories (e.g., NDLTD[3]); and (iii) curricula vitae of researchers made freely available on the Web or in specific official repositories (e.g., the Brazilian Lattes Platform[4]). However, in most of these sources, the researcher's areas of expertise are not explicitly identified and can only be implicitly inferred from the available content in these repositories. This requires some type of text mining treatment such as unsupervised topic extraction [1,6,17,22], automated supervised classification [18,21,23] and learning-to-rank methods [14,15].

In this paper, we focus on supervised techniques as they have historically produced better results, with the drawback of requiring labeled data. More specifically, we exploit a hierarchical classification[5] scheme to establish an automatic categorization model to solve the presented problem as discussed by Ribeiro-Neto et al. [18] and Waltinger et al. [23]. We exploit hierarchical classification in order to classify experts in a finer granularity level. However, hierarchical categorization is still a hard research problem faced by the text mining community. For completeness, we also evaluate the problem of ranking experts according to the knowledge areas.

Particularly, we use the knowledge area hierarchical classification scheme proposed by the Brazilian National Council for Scientific and Technological Development (CNPq), which provides a simple mechanism to systematize and characterize information about researchers and research groups. This classification scheme is organized into the following four levels[6]: *major area* (e.g., Earth and Exact Sciences), *area* (e.g., Computer Science), *subarea* (e.g.,Theory of Computation) and *specialty* (e.g., Formal Languages and Automata). The third and fourth levels of this classification scheme are not used in this paper due to the fact that a researcher might be associated with more than one subarea or specialty, which would characterize a multi-category classification problem [20]. Table 1 shows an excerpt of the two first levels of the CNPq knowledge area classification scheme, which covers nine major areas including altogether 76 specific areas.

Another important source of information used in this paper is the Lattes Platform. Maintained by CNPq, this platform is an internationally renowned initiative in Brazil [11] that provides a repository of researchers' curricula and research groups, all integrated into a single system. The available curricula present a great amount of information about the researchers that can be used for many purposes. In this paper, we focus on exploiting only the title of a researcher's PhD

---

[1] http://dblp.uni-trier.de.

[2] http://dl.acm.org.

[3] http://www.ndltd.org.

[4] http://lattes.cnpq.br.

[5] In this paper we use the terms categorization and classification interchangeably.

[6] http://bit.ly/1JM2j1k.

Table 1. Excerpt of the CNPq knowledge area classification scheme

1.00.00.00-3 Exact and Earth Sciences	**5.00.00.00-4 Agrarian Sciences**
1.01.00.00-8 Mathematics	5.01.00.00-9 Agronomy
1.02.00.00-2 Probability and Statistics	5.02.00.00-3 Forest Engineering
...	...
1.08.00.00-0 Oceanography	5.07.00.00-6 Food Science and Technology
2.00.00.00-6 Biological Sciences	**6.00.00.00-7 Applied and Social Sciences**
2.01.00.00-0 General Biology	6.01.00.00-1 Right
2.02.00.00-5 Genetics	6.02.00.00-6 Administration
...	...
2.13.00.00-3 Parasitology	6.13.00.00-4 Tourism
3.00.00.00-9 Engineering	**7.00.00.00-0 Humanities**
3.01.00.00-3 Civil Engineering	7.01.00.00-4 Philosophy
3.02.00.00-8 Mining Engineering	7.02.00.00-9 Sociology
...	...
3.13.00.00-6 Biomedical Engineering	7.10.00.00-3 Theology
4.00.00.00-1 Health Sciences	**8.00.00.00-2 Linguistics, Letters and Arts**
4.01.00.00-6 Medicine	8.01.00.00-7 Linguistics
4.02.00.00-0 Dentistry	8.02.00.00-1 Letters
...	8.03.00.00-6 Arts
4.09.00.00-2 Physical Education	
	9.00.00.00-5 Others

dissertation or Master's thesis found in her Lattes curriculum, since, in the extreme case, it is the only available (and reliable) information about the researcher when considering, for example, metadata from institutions or curricula vitae. The title of a dissertation or thesis is also a specially important source of information when considering new researchers, since sometimes there is little or no other available evidence about their research interest.

Our goal is to test the limits of some of the current state-of-the-art classifiers and learning-to-rank methods to generate categorization and ranking models to categorize researchers according to a knowledge area hierarchical classification scheme using only the title of their academic works. This is not a trivial task, given the difficulty in training machine learning models to obtain satisfactory results using just a small piece of text and, consequently, a reduced set of features. As any given additional information available by the researcher would probably only improve the results, our investigation would provide a lower bound on the results that can be obtained in this difficult scenario.

To summarize, our goal here is to investigate the benefits of applying supervised machine learning techniques to the tasks of categorizing and ranking research expertise using a knowledge area single-label hierarchical classification scheme. Thus, our main contributions in this paper are:

- An investigation on the limits of solving a combination of two hard problems: hierarchical categorization and categorization of very short texts;
- A comparative analysis of three supervised classification techniques applied to solve the aforementioned combined problem;
- An evaluation of the state-of-the-art ranking technique with recently proposed similarity features to solve the task of ranking experts using very short texts.

Our experimental results show an accuracy of up to 75% and 83% when categorizing researchers according to, respectively, the first and second levels of the CNPq knowledge area hierarchical classification scheme using only the title of their academic works and considering a model trained with Support Vector Machines (SVM). In addition, this classifier is more effective for this particular task than those based on Naive Bayes and Random Forests models. Moreover, the precision in the top positions of our ranking models achieve up to 97% and 88% considering the first and second levels of the hierarchy, respectively. These results provide evidence towards the potential benefits of using state-of-the-art feature representations and learning-to-rank techniques on the hard problem of expert search with minimum information.

The remainder of this paper is organized as follows. Section 2 addresses related work. Section 3 describes the dataset used in our experiments. Section 4 presents the methodology applied to the generation of the machine learning models and describes the results of the experiments performed to evaluate them. Finally, Sect. 5 presents our final considerations and provides directions for future work.

2   Related Work

The closest related tasks associated with automatic categorization and research expertise ranking in the literature are automatic expert profile construction [12,13,17,24], automatic categorization of text documents in digital libraries [1,3,19,20,23] and expert discovery [14,15].

Most of the previous efforts related to our work address the problem of automatically categorizing academic publications from digital libraries. The most effective techniques exploit the supervised learning paradigm to classify documents according to a set of previously defined knowledge areas, usually structured as a specific taxonomy [20,23]. Based on a set of training documents, these strategies are capable of achieving effective results using Support Vector Machines (SVM) to address the high sparsity and dimensionality of textual data derived from academic documents. In order to minimize the manual effort to label training documents, some previous works exploit unsupervised and semi-supervised techniques. They use topic models to categorize documents according to automatically generated taxonomies [3], provide alternative topic representations [1] or rely on linguistic patterns for taxonomy learning [19]. Despite related to our work because the categorization process is based on a specific taxonomy, here we focus on exploiting the minimum necessary discriminative information to categorize research expertise instead of classifying individual documents.

The problem of categorizing expertise is also associated with the task of automatic expert profile construction, which uses associations between an expert and her registered documents to model the expertise [12,13,17,24]. Specifically, after collecting all documents related to an expert, some methods [12,24] classify them using a supervised machine learning approach trained with manually labeled documents from other experts. Alternatively, MacDonald and Ounis [13] model an expert as a set of documents, computing the similarity between her documents and those from a knowledge area. Although automatic methods minimize the manual labor of updating the expert profile, its application in organizational contexts is limited because of the lack of textual documents related to an expert [12].

In addition to classification, the machine learning task of ranking experts has also been recently addressed in the literature [14,15]. Existing approaches rely on information taken from academic works, their citations and the profile information of experts. In this scenario, the use of learning-to-rank techniques presents an effective strategy to combine these different kinds of information [14]. Moreover, the use of such techniques have also been successfully employed to manipulate location-sensitive information [15].

Unlike previous work, we focus on hierarchically categorizing and ranking research expertise using minimum information. Considering the categorization task, both hierarchical categorization and categorization using only short texts are by themselves hard problems [5] and their combination makes this joint problem even harder. The ranking of experts using limited information is also challenging. In order to alleviate this problem, we apply a recently proposed approach to transform sparse textual features [26]. This approach generates a low-dimensional and informative feature space that is more suitable for the task of ranking academic experts using short texts. We evaluate both tasks, classification and ranking with minimum information, using the multi-area dataset described in the next section.

3   Our Dataset

To train a general model to categorize research expertise according to the CNPq knowledge area hierarchical classification scheme, we used the titles of labeled PhD theses and Master's dissertations found on curricula stored on the Lattes Platform. For this, we collected the curricula (XML versions) of 221,119 researchers holding a PhD degree. The respective excerpts of the collected XML documents including data from a thesis or dissertation were parsed and stored into a CSV file with each row containing the following columns: *title*, *major area* and *area*.

For the sake of completeness, we have removed from our dataset the titles of all theses and dissertations without a major area or area associated to them, thus resulting in a dataset that included the title of 49,508 PhD theses and 150,690 Master's dissertations. We have also cleaned the dataset to remove specific errors, such as incompatible major areas and areas associated to a same

Table 2. Distribution of the number of titles per major knowledge area

Major area	Nr. of Areas	Nr. of Titles
Agrarian Sciences	7	22,773
Biological Sciences	13	33,330
Health Sciences	9	25,983
Exact and Earth Sciences	8	33,085
Humanities	10	33,992
Applied and Social Sciences	13	17,398
Engineering	13	22,766
Linguistics, Letters and Arts	3	12,283

thesis or dissertation. For the purpose of this paper, we have also disregarded the major area *Others* due to its lack of a well delimited area grouping in our dataset. Thus, our final dataset comprises data from a total of 199,610 distinct theses and dissertations. We represent our final dataset using the traditional bag-of-words model [9] with the TF-IDF weighting scheme. Table 2 shows the distribution of the curricula vitae in our dataset according to the eight major areas considered for categorizing the researchers in terms of their main research interests, as well as the number of specific areas within each major area.

4   Experiments

In order to evaluate our proposal, we used distinct supervised machine learning methods to generate specific categorizing models and set up a set of experiments to compare them by means of quality metrics aimed to assess their effectiveness when using minimum information, thus allowing us to identify the most promising learning paradigm in our context.

4.1   Model Generation

As aforementioned, our goal is to investigate the benefits of applying supervised machine learning techniques to the task of categorizing and ranking research expertise using a given knowledge area classification scheme. The basic idea is to use such algorithms to "learn" a *good classification* or *ranking function* based on a set of textual features (bag-of-words). We here evaluate three classification techniques that follow completely different learning paradigms, namely Naive Bayes, Support Vector Machine (SVM) and Random Forests (RF) [9]. We also evaluate the task of ranking research expertise with minimum information. For this, we use the state-of-the-art ranking strategy BROOF-L2R [7] and transform textual features into meta-level features designed to improve the ranking results [26].

Hierarchical Classification Models. Considering the evaluated classification models, the Naive Bayes is the most simple and scalable approach, which applies the Bayes' theorem to estimate the category of texts from probability estimates of individual words with the "naive" assumption of independence between every pair of words. RF and SVM are two of the most successful classification methods, being considered by many [8,9] as the top-notch supervised algorithms. The RF approach is based on an ensemble of decision trees, which not only makes the strategy highly parallelizable, but also grants effective non-linear capabilities. Unlike RF, SVM is an inherently binary linear classification approach. Particularly, SVM uses a maximum-margin optimization method that tries to find a hyper-plane that best separates training examples (placed in a hyperspace) belonging to two different categories. The limitations of only discriminating between two linearly separable categories can be surpassed by using non-linear kernels to transform the feature space and building one classifier per category, where each category is fitted against all the other ones (one-vs-all).

Our approach for the hierarchical categorization of researchers involves not only training a classifier to discriminate such researchers among the major areas, but also eight more specific classification models to categorize them within the subareas. In other words, we first apply the general model to identify a researcher's major area (e.g., Exact and Earth Sciences) and, once this is determined, we apply a specific model trained to identify her specific area (e.g., Computer Science)

Learning-to-Rank Approach. In addition to classification, we also exploit the effectiveness of ranking the research expertise in different areas, which can be seen as "queries" in our problem. More specifically, we use the learning-to-rank framework to learn a ranking function from relevant and not relevant items from each area, and them use this function to rank items of unknown relevance.

However, different from the classification task, effective learning-to-rank approaches usually rely on a low-dimensional meta-feature space containing primarily similarity features that explicitly measure the proximity between queries (in our case the knowledge areas) and items [16]. Moreover, the fine-grained features in a high-dimensional feature space of words (the bag-of-words representation) usually used in classification may not be sufficiently expressive for effective learning-to-rank [26].

In order to overcome the challenges related to learning-to-rank using the Lattes categorized data, we propose an effective approach to rank research expertise that first transforms the bag-of-words representation of instances and categories into the recently proposed low-dimensional meta-feature space of similarity features designed for learning-to-rank [26]. In our scenario, these features provide the similarity relationship between an item and a research expertise. Particularly, the research expertise is represented using the centroid of its relevant items, as well as the closest relevant items with respect to a specific item. After this pre-processing transformation step, we use the generated compact meta-feature space as input to the state of the art learning-to-rank approach BROOF-L2R [7].

4.2 Evaluation, Algorithms and Procedures

The classification models were compared using two standard text categorization metrics: micro averaged F_1 (MicroF$_1$) and macro averaged F_1 (MacroF$_1$) [25]. While MicroF$_1$ measures the classification effectiveness over all decisions (i.e., the pooled contingency tables of all classes), MacroF$_1$ measures the classification effectiveness for each individual class and averages them.

The ranking results were measured with two widely used ranking evaluation metrics: Precision at position k (P@k) [2] and Normalized Discounted Cumulative Gain (NDCG) [10]. Both measures evaluate the effectiveness of the top-ranked results, which are the most relevant to a human searching for an expert. All experiments were executed using a 5-fold cross-validation procedure (which selects 4/5 of the dataset as training data and the remaining as testing data). The parameters were set via cross-validation on the training set and the effectiveness of distinct algorithms were measured in the test partition. In order to evaluate the classification effectiveness, we used the *scikit-learn*[7] implementations of RF and Multinomial Naive Bayes, and the LIBLINEAR[8] implementation of SVM. To evaluate the ranking effectiveness, we used the BROOF-L2R implementation provided by their authors [7]. The free parameters of these classifiers include the cost C for SVM and the number of features N considered in the split of a node on the RF-based approaches. These free parameters were set using a 5-fold cross-validation within the training set. The regularization parameter C of SVM was chosen among 11 values from 2^{-5} to 2^{15} and the parameter N of RF was selected among 10%, 20% and 30% of the number of features. For RF, each tree was grown without pruning, as suggested by Breiman [4]. Considering that the results obtained with 200, 300 and 500 trees were statistically tied (with 95% confidence), we adopted 200 trees due to its lower cost. In all ranking experiments, we adopted the number of iterations of the BROOF-L2R algorithm as 100 iterations, which is also the parameter adopted by its authors [7].

4.3 Experimental Results

Classification Results. Table 3 reports the MicroF$_1$ (MicF$_1$) and MacroF$_1$ (MacF$_1$) values for the classification of the theses and dissertations in our dataset using the three aforementioned methods. We evaluated our model considering the two upper levels of the CNPq knowledge area classification scheme, *major area* and *area*. In addition, we grouped our results according to the scheme described in Table 1. We would like to emphasize the following aspects of our results.

---

[7] http://scikit-learn.org.

[8] https://www.csie.ntu.edu.tw/~cjli/liblinear.

Table 3. Average MacroF$_1$ and MicroF$_1$ of the three classification models on each major area

		SVM	NB	RF
Major areas	MicF$_1$	**75.53 ± 0.25**	72.45 ± 0.08	71.04 ± 0.09
	MacF$_1$	**74.66 ± 0.31**	71.23 ± 0.13	69.15 ± 0.16
Agrarian Sciences	MicF$_1$	**82.84 ± 0.18**	80.96 ± 0.16	75.27 ± 0.18
	MacF$_1$	**73.07 ± 0.29**	70.74 ± 0.27	58.59 ± 0.26
Biological Sciences	MicF$_1$	**62.12 ± 0.63**	59.11 ± 0.32	55.68 ± 0.28
	MacF$_1$	**53.00 ± 0.47**	49.39 ± 0.30	44.82 ± 0.28
Health Sciences	MicF$_1$	**76.47 ± 0.34**	71.23 ± 0.20	67.38 ± 0.17
	MacF$_1$	**65.17 ± 2.02**	58.40 ± 0.20	49.19 ± 0.17
Exact and Earth Sciences	MicF$_1$	**83.47 ± 0.13**	81.99 ± 0.12	78.69 ± 0.13
	MacF$_1$	**76.67 ± 0.13**	75.22 ± 0.20	67.98 ± 0.21
Humanities	MicF$_1$	**68.93 ± 0.29**	61.55 ± 0.25	59.74 ± 0.28
	MacF$_1$	**57.93 ± 0.31**	50.35 ± 0.33	45.60 ± 0.31
Applied Social Sciences	MicF$_1$	**74.88 ± 0.20**	68.08 ± 0.20	66.38 ± 0.21
	MacF$_1$	**51.30 ± 0.95**	37.05 ± 0.18	35.52 ± 0.17
Engineering	MicF$_1$	**67.22 ± 0.11**	65.29 ± 0.24	62.41 ± 0.27
	MacF$_1$	**53.49 ± 0.45**	45.56 ± 0.25	41.63 ± 0.25
Linguistics, Letters and Arts	MicF$_1$	**78.47 ± 0.14**	75.14 ± 0.06	73.52 ± 0.13
	MacF$_1$	**79.70 ± 0.13**	74.36 ± 0.08	72.26 ± 0.18

First, the SVM model significantly outperforms all other evaluated models. The primary reason for the effective SVM results is its remarkable capability of learning in high dimensional feature spaces. This is due to the fact that the SVM classifier measures the complexity of hypotheses based on the margin with which it separates data, not on the number of features. The SVM method is also insensitive to the high sparsity of textual data, since it just "adds" the evidence of each word present in a document to classify it. NB also shares the same "additive" nature of SVM, having achieved the second best set of results in our experiments. The method that presented the worst results was RF, which uses complex non-linear patterns extracted by association rules that relate the words of a document to its category. We argue that, due to its complexity, RF generates models that may not generalize well in the case of highly sparse domains as it is the case of short texts.

Second, most of the generated models provide evidence towards the initial hypothesis that it is possible to categorize researchers' expertise by exploiting the information from the titles of their theses or dissertations. Particularly, the models using only this minimum information achieve up to 83% and 79% on MicroF$_1$ and MacroF$_1$, respectively. Moreover, the effectiveness of the results using SVM are superior to 70% in the major areas and most of the areas.

Finally, despite the excellent overall performance, we attribute the low performance (around 51%) in some major areas (e.g., Applied Social Sciences and Engineering) to two different factors. First, the distribution of labeled examples among areas are very imbalanced. Particularly, for some specific areas from Engineering, such as Mining Engineering and Biomedical Engineering, our dataset has less than 10 labeled examples, which makes it difficult to learn effective models for these areas. Second, the high vocabulary overlap among these areas and their fuzzy delimitations can undermine the classification effectiveness.

Ranking Results. We now turn our attention to the classification-related task of ranking according to expertise. Table 4 reports the precision and NDCG values for the top 10 best ranked results using the BROOF-L2R method. Particularly, we evaluate the ranking of theses and dissertations from our dataset according to the expertise of their authors in each area.

Like in the classification task, the overall results show the effectiveness of our ranking strategy. This provides evidence for the benefits of using learning techniques to generate ranking of experts by exploiting only the information from the titles of their theses or dissertations. Particularly, the ranking results considering as queries only the major areas (i.e., the first level of the hierarchy described in Table 1) presented the most effective results for both NDCG@10 and P@10. These results were already expected, since each major area contains thousands of positive training and testing examples (see Table 2). This led to the learning of very effective ranking functions using BROOF-L2R, as well as to plenty of possible positive test examples that can assume the top positions in the ranking.

Coincidentally, our best results occurred on major areas with many positive examples for its areas. Specifically, two major areas (Linguistics, Letters and Arts, and Exact and Earth Sciences) achieved the best results among all major areas due to the fact that both of them have many examples for their specific areas, which led to good ranking functions for each one of them. Therefore, in

Table 4. Average NDCG@10 and P@10 of BROOF-L2R on each major area

	P@10	NDCG@10
Major Areas	0.97 ± 0.03	0.97 ± 0.03
Agrarian Sciences	0.71 ± 0.17	0.72 ± 0.15
Biological Sciences	0.68 ± 0.12	0.68 ± 0.12
Health Sciences	0.73 ± 0.21	0.77 ± 0.20
Exact and Earth Sciences	0.88 ± 0.20	0.89 ± 0.19
Humanities	0.73 ± 0.18	0.75 ± 0.18
Applied Social Sciences	0.53 ± 0.16	0.54 ± 0.15
Engineering	0.59 ± 0.15	0.60 ± 0.14
Linguistics, Letters and Arts	0.77 ± 0.27	0.77 ± 0.25

these cases we obtained a high average ranking effectiveness in all areas (which can be seen as queries) of a major area. Likewise, our worst results occurred on major areas that had only few positive examples (less than ten) in some of their specific areas, which led to poor ranking functions for each one of them. In fact, the worst performing major areas (Engineering and Applied Social Sciences) include specific areas with only three positive examples. Despite these specific cases, the task of ranking specialists using minimum information achieves effective results where there is enough training examples. This provide evidence to our claim that it is possible to effectively rank experts using only short texts.

5   Conclusions and Future Work

To conclude, in this paper we addressed two distinct problems: (i) determining a researcher's expertise area by automatically categorizing the title of her PhD dissertation or Master's thesis according to a hierarchical scheme using an automatic classification model, and (ii) ranking experts with respect to knowledge areas using as well the same piece of information.

The results obtained using supervised classification methods were in general very good, specially given the restriction of using minimum information. We also performed a comparative analysis of three state-of-the-art supervised classification methods to determine the best one for the proposed task, being SVM the one that significantly outperformed the other two. As for the classification task, the state-of-the-art learning-to-rank method using recently proposed ranking features produced excellent results in general, even considering the same minimum information restriction.

As future work, we intend to: (i) expand the study to other datasets using the models learned with the CNPq knowledge area hierarchical classification scheme to categorize researchers not present in the Lattes Platform (Transfer Learning); (ii) explore deeper levels of the CNPq hierarchy and other hierarchical categorization strategies (e.g., fuzzy); and (iii) propose an expert recommendation system based on our results.

Acknowledgements. This work was partially funded by projects InWeb (grant MCT/CNPq 573871/2008-6) and MASWeb (grant FAPEMIG/PRONEX APQ-01400-14), and by the authors' individual grants from CAPES, CNPq and FAPEMIG.

References

1. Aletras, N., Baldwin, T., Lau, J.H., Stevenson, M.: Representing topics labels for exploring digital libraries. In: Proceedings of the 14th ACM/IEEE-CS Joint Conference on Digital Libraries, pp. 239–248 (2014)
2. Baeza-Yates, R.A., Ribeiro-Neto, B.: Modern Information Retrieval. Addison-Wesley Longman Publishing Co. Inc., Boston (1999)
3. Bakalov, A., McCallum, A., Wallach, H., Mimno, D.: Topic models for taxonomies. In: Proceedings of the 12th ACM/IEEE-CS Joint Conference on Digital Libraries, pp. 237–240 (2012)

4. Breiman, L.: Random forests. Mach. Learn. **45**(1), 5–32 (2001)
5. Chen, M., Jin, X., Shen, D.: Short text classification improved by learning multi-granularity topics. In: Proceedings of the Twenty-Second International Joint Conference on Artificial Intelligence, vol. 3, pp. 1776–1781 (2011)
6. Chen, Y., Fox, E.A.: Using ACM DL paper metadata as an auxiliary source for building educational collections. In: Proceedings of the 14th ACM/IEEE-CS Joint Conference on Digital Libraries, pp. 137–140 (2014)
7. de Sá, C.C., Gonçalves, M.A., Sousa, D.X., Salles, T.: Generalized BROOF-L2R: a general framework for learning to rank based on boosting and random forests. In: Proceedings of the 39th International ACM SIGIR Conference on Research and Development in Information Retrieval, pp. 95–104 (2016)
8. Fernández-Delgado, M., Cernadas, E., Barro, S., Amorim, D.: Do we need hundreds of classifiers to solve real world classification problems? J. Mach. Learn. Res. **15**(1), 3133–3181 (2014)
9. Hastie, T., Tibshirani, R., Friedman, J.: The Elements of Statistical Learning. Springer Series in Statistics. Springer New York Inc., New York (2001)
10. Järvelin, K., Kekäläinen, J.: Cumulated gain-based evaluation of IR techniques. ACM Trans. Inf. Syst. **20**(4), 422–446 (2002)
11. Lane, J.: Let's make science metrics more scientific. Nature **464**(7288), 488–489 (2010)
12. Li, M., Liu, L., Li, C.-B.: An approach to expert recommendation based on fuzzy linguistic method and fuzzy text classification in knowledge management systems. Exp. Syst. Appl. **38**(7), 8586–8596 (2011)
13. Macdonald, C., Ounis, I.: Voting techniques for expert search. Knowl. Inf. Syst. **16**(3), 259–280 (2008)
14. Moreira, C., Calado, P., Martins, B.: Learning to rank for expert search in digital libraries of Academic publications. In: Antunes, L., Pinto, H.S. (eds.) Progress in Artificial Intelligence, pp. 431–445. Springer, Heidelberg (2011)
15. Niu, W., Liu, Z., Caverlee, J.: On local expert discovery via geo-located crowds, queries, and candidates. ACM Trans. Spatial Algorithms Syst. **2**(4), 14:1–14:24 (2016)
16. Qin, T., Liu, T.-Y., Xu, J., Li, H.: Letor: a benchmark collection for research on learning to rank for information retrieval. Inf. Retr. **13**(4), 346–374 (2010)
17. Ribeiro, I.S., Santos, R.L.T., Gonçalves, M.A., Laender, A.H.F.: On tag recommendation for expertise profiling: a case study in the scientific domain. In: Proceedings of the Eighth ACM International Conference on Web Search and Data Mining, pp. 189–198 (2015)
18. Ribeiro-Neto, B.A., Laender, A.H.F., de Lima, L.R.S.: An experimental study in automatically categorizing medical documents. JASIST **52**(5), 391–401 (2001)
19. Sanchez, D., Moreno, A.: Bringing taxonomic structure to large digital libraries. Int'l. J. Metadata Semant. Ontol. **2**(2), 112–122 (2007)
20. Seymour, E., Damle, R., Sette, A., Peters, B.: Cost sensitive hierarchical document classification to triage PubMed abstracts for manual curation. BMC Bioinform. **12**(1), 482 (2011)
21. Silla Jr., C.N., Freitas, A.A.: A survey of hierarchical classification across different application domains. Data Mining Knowl. Disc. **22**(1–2), 31–72 (2011)
22. Srinivasan, V., Fox, E.: Progress towards automated ETD cataloging. In: Proceedings of the 19th International Symposium on Electronic theses, dissertations: Data and dissertations (2016)

23. Waltinger, U., Mehler, A., Lösch, M., Horstmann, W.: Hierarchical classification of OAI metadata using the DDC taxonomy. In: Bernardi, R., Anderson, S., Bjrn, C., Frdrique, G., Zaihrayeu, S. (eds.) Advanced Language Technologies for Digital Libraries, pp. 29–40. Springer, Heidelberg (2011)
24. Yang, K.-W., Huh, S.-Y.: Automatic expert identification using a text categorization technique in knowledge management systems. Expert Syst. Appl. **34**(2), 1445–1455 (2008)
25. Yang, Y.: An evaluation of statistical approaches to text categorization. Inf. Retr. J. **1**(1–2), 69–90 (1999)
26. Yang, Y., Gopal, S.: Multilabel classification with meta-level features in a learning-to-rank framework. Mach. Learn. **88**(1), 47–68 (2012)

Extracting Event-Centric Document Collections from Large-Scale Web Archives

Gerhard Gossen[1]([⊠]), Elena Demidova[1], and Thomas Risse[2]

[1] L3S Research Center, Leibniz Universität, Hanover, Germany
{gossen,demidova}@l3s.de
[2] University Library J.C. Senckenberg, Frankfurt, Germany
t.risse@ub.uni-frankfurt.de

Abstract. Web archives are typically very broad in scope and extremely large in scale. This makes data analysis appear daunting, especially for non-computer scientists. These collections constitute an increasingly important source for researchers in the social sciences, the historical sciences and journalists interested in studying past events. However, there are currently no access methods that help users to efficiently access information, in particular about specific events, beyond the retrieval of individual disconnected documents. Therefore we propose a novel method to extract event-centric document collections from large scale Web archives. This method relies on a specialized focused extraction algorithm. Our experiments on the German Web archive (covering a time period of 19 years) demonstrate that our method enables the extraction of event-centric collections for different event types.

1 Introduction

Web archives created by the Internet Archive[1] (IA), national libraries and other archiving services contain large amounts of information collected for a time period of over twenty years [6]. These archives constitute a valuable source for research in many disciplines, including the digital humanities, the historical sciences and journalism by offering a unique possibility to look into past events and their representation on the Web. They can enable a better understanding of past events and offer a lot of novel research directions for these disciplines.

Most Web archive services aim to capture the entire Web (IA) or national top-level domains (national libraries) and are therefore very broad in their scope. Consequently they are also very diverse regarding the topics they contain and the time intervals they cover. Due to the large size and the broad scope it is difficult for interested researchers to locate relevant information in the archives as search facilities are very limited compared to the live Web.

In previous work [14, 26] we have argued that these users are typically interested in studying smaller and more focused event-centric collections of documents contained in a Web archive. Such collections can reflect specific events such as elections, sports tournaments or natural disasters, for example the Fukushima

---

[1] https://archive.org

© Springer International Publishing AG 2017
J. Kamps et al. (Eds.): TPDL 2017, LNCS 10450, pp. 116–127, 2017.
DOI: 10.1007/978-3-319-67008-9_10

nuclear disaster in 2011, the German federal election in 2009 or the FIFA World Cup 2006, especially in regard to their media coverage and public reactions.

Archive services such as Archive-IT[2] collect documents around specific events. These *special collections* are however defined and crawled on an individual basis, such that users are restricted to the collections that exist and their selected scope. Other existing access methods to temporal Web collections do not support creating ad-hoc collections, often forcing users to create their own corpora manually. Currently, access to large-scale Web archives is limited to browsing of individual Web pages through browser-based tools such as the Wayback machine[3], or initial support for keyword-based access[4]. However, these access methods are not sufficient for several reasons. First, the Wayback machine requires the user to already know the URL of the document. Second, full-text indexing of large-scale archived collections incurs high processing and storage costs. Third, such indexes only allow retrieval of individual disconnected documents. Instead, automatic methods are needed that can extract collections of documents related to a particular event of user interest. These collections need to preserve the original link structure to achieve a high degree of authenticity and enable the application of analytical methods on the relevant parts of the Web archive [14].

In this paper, we present a starting point for tackling the novel problem of extracting topically and temporally coherent, interlinked event-centric document collections from large-scale and broad scope Web archives. The key contributions of this paper are: (1) a definition of a *Collection Specification* that describes the temporal and topical scope of the collection to be extracted and gives the user intuitive but powerful options to control the data collection process; and (2) a *focused crawling-based extraction method* for Web archives to create event-centric collections without requiring any full-text indexes. We evaluate our approach in a local environment using file system crawling. However, our approach can easily be used across Web archives using existing access methods. We make our source code and evaluation data available to encourage further research[5].

2   Related Work

Our method is related to crawling methods for creating Web Archives (e.g. [17,24]), as well as to methods for temporal information retrieval [5].

The collection of Web documents from the live Web for retrieval and archiving purposes is usually performed using Web crawlers. Crawling methods that aim to create broad scope collections for search and archiving purposes intend to capture as much of the Web as possible. An example of a web-scale archiving crawler currently used by the Internet Archive is Heritrix [20]. In contrast, *focused crawling* [4] aims to only collect pages that are related to a specific topic. Focused crawlers [1,22] learn a model of the topic and follow links only if they are

---

[2] https://archive-it.org/
[3] http://netpreserve.org/openwayback
[4] https://blog.archive.org/2016/10/24/beta-wayback-machine-now-with-site-search/
[5] https://github.com/gerhardgossen/archive-recrawling

Table 1. Examples of temporal event characteristics.

Event	Type	Duration	Lead time	Cool-down time
Olympic games	recurring	2 weeks	weeks	days
Federal election	recurring	1 day	months	weeks
Fukushima accident	non-recurring	1 week	—	months
Snowden leaks	non-recurring	1 day	—	years

expected to match that topic, e.g. based on the page containing the link. This follows the obervation that relevant documents will preferentially link to other relevant documents ("topical locality" [1]). Extensions of this model use ontologies to incorporate semantic knowledge into the matching process [9,10], 'tunnel' between disjoint page clusters [3,25] or learn navigation structures necessary to find relevant pages [7,17]. In time-aware focused crawling [24] the document or event time is used as the primary focusing criterion. In event-based crawling [11] events are described using an event model that incorporates event location and date. Here Web page relevance is computed as a weighted average of content, location and date similarity. As location extraction increases the overall complexity of the process, we focus on the content and time-based features. Freshness as a specific aspect of temporal relevance has been addressed in the context of joint crawling of the Web and Social media sites [12] where URLs present in Social media posts are used as entry points to recently published content on the Web. In summary, most existing approaches to focused Web crawling consider the topical and temporal relevance in isolation and do not address the problem of jointly finding temporally and topically relevant content. Furthermore, whereas existing approaches operate on the live Web, we are the first to apply focused crawling techniques to existing Web archives.

The notion of temporal relevance has also been explored in the area of temporal information retrieval. Existing ranking methods have been extended to rank documents based on their creation time [5] or to diversify search results over relevant time periods [2]. Contemporary search engines also rank documents based on their freshness (estimated based on their crawling history) [8]. Similarly, time information has been combined with the hypertext link graph to detect the most relevant documents for a given query [21]. These approaches depend on full-text or graph indexes and therefore have a high up-front computational and index storage cost. Moreover, these approaches only allow retrieval of individual disconnected documents and do not preserve the link structure. In contrast, our method allows on demand extraction of interlinked event-centric collections without requiring any additional indexes on the archive.

3 Event-Centric Collections

Events are typically characterized through a certain date or a time interval such as the date of an accident or the duration of a tournament. Here the event time interval is clearly defined. Nevertheless, event-related documents also appear outside of this time interval. For planned and in particular regularly recurring events such as sports competitions or elections, relevant documents often appear in advance of the actual begin of the event during the event *lead time*, and are still published after the event completion during the *cool-down time*. For unexpected non-recurring events such as natural disasters, event-related documents are published from the start of the event onward, i.e. there is no lead time and the relevant documents appear during the cool-down time of the event. The duration of the lead time as well as the duration of the cool-down time depend on the specific event (see Table 1).

Given an event of user interest and a large-scale broad-scope Web archive, our goal is to generate an interlinked collection of documents relevant to this event. The scope of the target collection is defined in the **Collection Specification** :

Definition 1 (Collection Specification). The Collection Specification defines the topical and the temporal scope of an event-centric collection using:

○ Topical Scope:
  – one or more topical reference documents (e.g. from the Web);
  – zero or more representative *keywords*.
○ Temporal Scope:
  – time span of the event (including the start and end dates) $T_e = [t_e^s, t_e^e]$;
  – time duration of the lead time (T_l) and the cool-down time (T_r).

The Collection Specification may be extended to include additional scopes, for example domain black and white lists as used by existing crawlers.

Given the Collection Specification, our goal is to create a collection containing the Web documents temporally and topically relevant to this specification. In the following we propose a focused extraction method that prioritizes URLs during the crawling process according to the Collection Specification and generates interlinked event-centric collections.

4   Event-Centric Collection Extraction

Our goal is to efficiently extract an event-centric interlinked collection of a manageable size from a large scale Web archive. A naïve approach is to iterate through all documents and check their relevance with respect to the Collection Specification using an automatic method. However, this is computationally expensive and does not scale to Web archives spanning tens or hundreds of terabytes. While a full-text index could reduce the iteration cost, it requires high up-front computational and index storage resources and extensive post-filtering of the many near-identical document versions contained in the Web archive [16]. Furthermore, such an index can only be used to retrieve individual documents, where we want to extract interlinked document collections.

We propose an alternative approach that uses the hypertext characteristics of the archived documents by adapting focused Web crawling. A Web crawler collects documents by recursively following the links from a Web document to

---

Algorithm 1. Event-centric Collection Extraction

Input: Collection Specification CS, $targetSize$
Output: Document collection c, excluded URLs $missing$
  $q \leftarrow$ priorityQueue(seedUrls(CS)); $c \leftarrow \{\}$; $missing \leftarrow \{\}$
  while not isEmpty(q) **and** $|c| < targetSize$ **do**
    $url \leftarrow pop(q)$
    $v \leftarrow resolveSnapshots(url, CS)$ {Find all snapshots of url in c}
    if $v = \emptyset$ **then**
      $missing \leftarrow missing \cup \{url\}$
    else
      $v_i \leftarrow selectSnapshot(CS, v)$
      $c \leftarrow c \cup \{v_i\}$
      $out \leftarrow extractOutlinks(v_i) - seenUrls$ {$seenUrls = c \cup missing$}
      $insert(q, out, relevance(v_i))$ {Insert outlinks into queue according to relevance}
    end if
  end while

---

other documents, starting from an initial set of *seed URLs*. A focused Web crawler improves the relevance of the resulting collection by following only links to the documents predicted to be relevant. We therefore extend the Collection Specification to include the seed URLs required for the crawling process:

Definition 2 (Crawl-based Collection Specification). A Crawl-based Collection Specification contains a Collection Specification (Definition 1) and a non-empty set of URLs, which are contained in the archive and refer to relevant documents.

The Crawl-based Collection Specification is created by the user. Semi-automatic approaches include the use of Web search engines to select seed URLs [13].

We adapt the focused crawling algorithm as shown in Algorithm 1 by including steps to resolve snapshots and select the best among them. URLs extracted from collected documents are prioritized in the crawler queue during the focused crawl using the relevance function defined in Sect. 5.

5 Relevance Estimation

We need to prioritize the URLs during the focused crawl to effectively extract event-centric collections based on a *relevance function*. We use a linear combination of the temporal and topical relevance (TTR) to estimate the relevance of a Web document d with respect to the Collection Specification CS:

$$\text{TTR}(d, CS) = \alpha \times \text{TopicR}(d, CS) + (1 - \alpha) \times \text{TempR}(d, CS), \tag{1}$$

where $TempR$ and $TopicR$ are the temporal and topical relevance of d to CS, and $\alpha \in [0,1]$ is the parameter to trade off between the topical and temporal relevance. $\alpha = 1$ results in a standard topically focused crawler, whereas values closer to 0 increase the weight of the temporal dimension. In our setting we consider $TempR$ and $TopicR$ to be equally important, therefore we use $\alpha = 0.5$, but we will in future work investigate the influence of this parameter in detail.

5.1   Temporal Relevance

As described in Sect. 3, event-related documents are published not only during the event time interval, but also before and after. Consequently, we need to estimate the relevance of a document based on the Collection Specification and a time point associated with the Web document (e.g. the creation, last modification or capture date). We define this *Temporal Relevance Function* as follows:

Definition 3 (Temporal Relevance Function). Given a time point t_d associated with the Web document d and the event time interval $T_e = [t_e^s, t_e^e]$, the function $f(t_d, t_e) \rightarrow [0, 1]$ is a temporal relevance function iff (a) $f(t_d, t_e) = 1 \Rightarrow t_d \in t_e$ and (b) f is monotonically non-decreasing in $(-\infty, t_e^s)$ and monotonically non-increasing in $(t_e^e, +\infty)$.

We assume that in general the relevance of documents decreases rapidly as the distance to the event increases and therefore define a temporal relevance function based on the exponential decay function (similar to [18]):

$$\text{TempR}(t_d, t_e) = \begin{cases} 1 & \text{if } t_e^s \leq t_d \leq t_e^e, \\ e^{-\Delta t / \gamma_l} & \text{if } t_d < t_e^s, \\ e^{-\Delta t / \gamma_r} & \text{if } t_d > t_e^e, \end{cases} \tag{2}$$

where Δt is the time difference between the document time point t_d and the nearest end of the reference time interval T_e, and γ_l and γ_r are *time decay factors*. The time decay factors determine how fast the value of this function decreases by giving the Δt at which the relevance has dropped to 0.5. We use the expected duration of the lead and the cool-down time as the time decay factors γ_l and γ_r. For events with no lead time (e.g. accidents) we set $\gamma_l = 0$.

The document time point can be estimated using the date discussed in the document. This would give the most accurate relevance value, especially for documents that describe the event after some time has passed (e.g. at the one year anniversary), but is computationally expensive and highly heuristic. Therefore we extract the document publication time, which is often explicitly contained in the document metadata or content. If no publication time is available, we use the crawl time as a fallback.

5.2   Topical Relevance Estimation

The topical relevance of Web documents with respect to the Collection Specification is estimated by computing the similarity of the textual content of Web documents to the topical scope of the Collection Specification (similar to [23]).

The topical scope is specified primarily through a set of *reference documents* that describe the event (e.g. as Wikipedia pages or newspaper articles). When these documents have an ambiguous topic or the scope should be narrowed down further, keywords can be provided to clarify the topical intent. Together this allows an intuitive yet powerful topical specification.

We represent the topical scope as a term vector, called the *reference vector*, to enable automatic relevance estimation with respect to the topic. To construct the reference vector we tokenize and stem the text of the reference documents and remove stop words using the language-specific analyzers of Apache Lucene[6]. As previous work has shown bigrams to be effective for crawl focusing [19], we use term unigrams and bigrams. Each term is weighted using its frequency (TF) and its inverse document frequency (IDF). IDF scores are based on the frequencies of the last 25 years of the Google Books NGram datasets[7].

The weights of terms explicitly given as Collection Specification keywords are boosted. This helps to shift the reference vector towards the expected interpretation. To perform boosting, we check the overlap of each term with the user-defined keywords, as terms (in the case of bigrams) can contain multiple tokens. Based on whether there is a full or partial overlap, we assign a *term weight* tw_t to the term t in the document vector. In our evaluation, we experimentally set the values for full, partial and no overlap to 2, 1.5 and 1, respectively.

Finally, the topical relevance of a document is the cosine similarity between the reference vector and a document vector computed using the same method.

6   Web Archive and Platform

Our Web Archive contains all Web pages from the .de top-level domain as captured by the Internet Archive until 2013. In this paper we only consider HTML documents with a HTTP status code of 200. This archive has a size of about 30 TB and contains 4.05 billion captures of 1 billion URLs, covering a time period from December 1994 to September 2013.

We manually defined 28 events to be extracted from the Web archive, focusing on events that are likely to be represented in the archive: The selected events fall within the time period of the archive and have a strong connection to Germany, either because the event happened in Germany or was in the focus of public attention. We balanced singular events like the Fukushima nuclear accident and recurring events like federal elections. To create the Collection Specification for each event we selected one or more pages from the German Wikipedia that provide the topical scope of the event. We also defined a start and end date, as well as an estimate for the duration of the event lead and cool-down time. The outgoing links of the Wikipedia pages were extracted and used as seed URLs.

All experiments were conducted on a Hadoop cluster. This cluster has 25 worker and 2 master nodes with in total 296 CPU cores. The worker nodes provide in total 1.37 TB of RAM and 1 PB of hard disk capacity. All data is stored in the standard ARC/WARC formats and available to all worker nodes.

6.1   Crawler Implementation

As mentioned in Sect. 4, the architecture of the archive crawler can be simpler than that of a standard Web crawler because it can access the data of the Web

---

[6] http://lucene.apache.org/core/

[7] Code available at: https://github.com/gerhardgossen/dictionary-creator/

archive locally. As our data is stored as WARC files in a Hadoop filesystem, we implemented the crawler as a multi-thread process running on Hadoop YARN.

WARC files are unordered collections of documents, therefore a lookup table is necessary to find the location of the document snapshots for a given URL. By using Apache HBase for this table we can look up URLs in 1–5 ms. While typically CDX files are used as a lookup method for WARC files, our preliminary experiments showed that this method is considerably faster.

The crawler queue is stored in a file-based queue based on the Mercator architecture [15], which offers prioritisation of URLs and is fast enough for our purposes. Each retrieved document is analysed according to the relevance function described in Sect. 5. The URLs of all outgoing links of that document are inserted into the crawler queue according to the calculated relevance score.

As the Web archive covers a long time period, many documents have been crawled multiple times. To choose among the available versions, we observe that later versions typically have the same content but may have changes in e.g. navigation menus and thus do not represent the document in its original form. Therefore we use the following heuristic: If multiple versions are available that were crawled during the event timespan, we pick the earliest. Otherwise, we use the version that was crawled closest to the event timespan. Future work will investigate further methods to select the most relevant version(s).

7   Evaluation

The goal of the evaluation is to assess the precision of the proposed extraction method in light of different event types and to better understand the influence of this method on the quality of the resulting event-centric collections. We compare our combined relevance function with two baselines that use state-of-the-art relevance functions, each taking only one relevance dimension into account, topical (C-F, cf. [23]) or temporal (T-F, cf. [24]). We also use an unfocused crawl that does not use any relevance estimates as an additional baseline.

7.1   Extraction Evaluation

Our focused crawling approach allows us to adjust the effort invested into the extraction by changing the number of documents processed. By increasing this number to the size of the archive we could clearly guarantee that this method finds all the relevant documents, as long as they are reachable through links. However, the proposed approach should be able to extract most of the relevant documents early on, so that the extraction can be stopped when not sufficiently many relevant documents are discovered anymore or when the user is satisfied with the collection. We therefore look at the accumulated relevance (i.e. the sum of the relevance values of the extracted documents) of the collected results as a function of crawl runtime. Additionally, we look at the number of documents that the crawler attempts to capture but are missing from the archive.

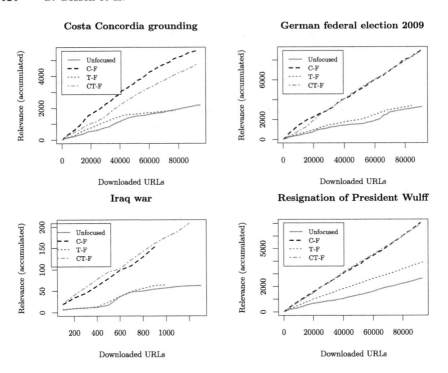

Fig. 1. Accumulated relevance of different event collections.

The relevance of the extracted documents is computed with the C-F relevance function. This is possible because we estimate the relevance of a document during the crawl using the content of a linking document and evaluate using the content of the actual document. A small annotation experiment (omitted for space) showed that this relevance measure correlates with the actual relevance.

For each of the 28 events we started a crawl using each of the configurations described above. Each crawl ran until it had retrieved 100,000 documents or until the crawler queue was empty. Figure 1 shows the accumulated relevance of document collections for selected events in relation to the number of documents crawled. This function should ideally start with a strong incline, meaning that the crawler fetches many relevant documents early on, flattening into a plateau when no relevant documents are available anymore. We see that for all topics the C-F and CT-F functions outperform the T-F function and the unfocused baseline both in terms of average relevance of documents retrieved at any given point and total relevance. The C-F function often performs slightly better than the CT-F function, although closer analysis shows that the differences between both functions often result from discovering some highly relevant hosts earlier.

The relevance focused strategies manage to uncover more potentially relevant URLs even if they are not contained in the locally available Web archive. This is shown by the number of URLs that each focusing method considers (see Table 2), where we see an increase in discovered URLs for these methods. Based on this

Table 2. URLs considered for each event crawl for different relevance strategies.

Topic	CT-F	Unfocused	Ratio
Costa Concordia grounding	239,628	142,851	1.67
German federal election 2009	283,311	161,934	1.74
Iraq War	1,862	2,192	0.84
Pope Election 2013	2,057	1,624	1.26
Stuttgart 21 protests	2,070	1,513	1.36
Resignation of President Wulff	213,039	149,706	1.42

result, the development of methods for cross-archive collection extraction is an interesting direction for future research.

7.2 Effect of the Temporal Scope Parameters

In the Collection Specification we require that the user specifies *lead* and *cool-down times* for the event (cf. Sect. 5.1) to adapt the temporal relevance function to different event types. We crawled each event using a exponential decay function with a fixed decay and compared it to the crawl using the specified lead and cool-down times. Table 3 (left columns) shows the relevance improvement of the time-sensitive relevance functions over the corresponding baseline. We see that the event-specific parameters cause an improvement for most of the events. On average this improvement is moderate, but statistically significant.

Table 3. Effect of temporal scope and keyword parameters. Each row shows improvement ratio of the accumulated relevance for a topic with event-specific time parameters (left) or keywords (right). The last line contains the average improvement over all topics. All values are statistically significant at $p = 0.01$.

Event	Time		Keywords		Event	Time		Keywords	
	T-F	CT-F	C-F	CT-F		T-F	CT-F	C-F	CT-F
Book by Thilo Sarrazin	0.98	0.99	1.28	1.07	Iraq war	0.92	1.19	1.05	1.13
Eruption of Eyjafjallajökull	0.99	1.20	0.83	0.88	Launch of LHC	1.09	0.72	1.21	0.99
European Stability Mechanism	1.16	4.07	1.02	1.04	Costa Concordia grounding	1.14	1.49	0.92	0.98
European floods 2013	1.12	1.12	1.39	1.49	Loveparade disaster	0.84	1.25	0.81	0.97
Eurovision Song Contest 2010	1.00	1.73	1.06	0.68	NSU process	1.01	1.24	1.05	1.05
Football World Cup 2006	0.58	1.27	1.23	1.10	Olympia 2004	0.94	1.03	1.20	1.34
Football World Cup 2010	1.59	1.09	1.11	1.10	Olympia 2008	1.27	1.48	1.39	1.50
Fukushima nuclear disaster	1.17	1.73	1.03	1.02	Olympia 2012	1.18	1.11	1.16	1.12
German federal election 2002	1.21	1.48	1.35	1.02	Olympia 2010	1.02	1.37	1.24	1.65
German federal election 2005	1.33	1.41	1.14	0.89	Pope Election 2005	1.17	1.08	1.10	1.09
German federal election 2009	1.27	1.84	1.03	0.96	Pope Election 2013	1.07	1.50	0.99	0.95
German federal election 2013	1.12	2.17	0.84	0.92	Resig. of Pres. Wulff	1.03	1.05	1.00	1.03
Guttenberg plagiarism affair	0.96	1.01	1.24	1.19	Snowden leaks	1.46	1.43	1.18	1.19
Average						1.10	1.44	1.10	1.08

7.3   Effect of Keywords in the Specification

We use the keywords in the Collection Specification to clarify the topical intent (cf. Sect. 5.2). To measure the impact, we crawled using the same reference documents with and without keywords to describe the topical scope. Table 3 (right columns) shows the relevance improvement of the T-F and CT-F relevance functions compared to the corresponding baseline. We see that the addition of keywords leads on average to a statistically significant improvement. Some events such as the floods in Europe during 2013 can be better focused using keywords, whereas for other events adding keywords leads to a small loss in effectiveness. Further research is needed to better understand the influence of keywords.

8   Conclusions and Outlook

In this work we presented a novel method to create interlinked event-centric collections from large-scale Web archives. The key of this method is to adapt focused Web crawling to previously collected Web archives and to select documents by iteratively following links from relevant documents. We proposed relevance estimation functions that take the temporal and topical aspects of the documents into account and evaluated them as part of the focused extraction process. Specifically, we demonstrated that the relevance function CT-F can improve on topical content selection methods by taking temporal information into account. This holds especially for events that occur repeatedly in similar form, such as Olympic games or elections, where the different instances are hard to distinguish using only topical information. We showed that our re-crawling method can retrieve event-centric collections from large-scale Web archives, especially using the CT-F relevance function, and discussed how the method deals with the challenges inherent to Web archives.

Our method presents a first step towards the extraction of event-centric collections. Further research is needed to understand the influence of extraction methods, relevance functions and parameters in regard to different events, time periods and Web archives. For Web archives that have full-text indexes, methods based on full-text search should be investigated. Furthermore, cross-archive collection extraction is an interesting direction for future research. We therefore provide our source code and evaluation data to encourage similar efforts [8].

Acknowledgments. This work was partially funded by the ERC under ALEXAN-DRIA (ERC 339233), H2020 under SoBigData (RIA 654024) and BMBF under Data4UrbanMobility (02K15A040).

References

1. Aggarwal, C., Al-Garawi, F., Yu, P.S.: Intelligent crawling on the world wide web with arbitrary predicates. In: World Wide Web Conference, pp. 96–105 (2001)

---

[8] https://github.com/gerhardgossen/archive-recrawling

2. Berberich, K., Bedathur, S.: Temporal diversification of search results. In: Workshop on Time-aware Information Access (TAIA 2013) (2013)
3. Bergmark, D., Lagoze, C., Sbityakov, A.: Focused crawls, tunneling, and digital libraries. In: Agosti, M., Thanos, C. (eds.) ECDL 2002. LNCS, vol. 2458, pp. 91–106. Springer, Heidelberg (2002). doi:10.1007/3-540-45747-X_7
4. Chakrabarti, S., van den Berg, M., Dom, B.: Focused crawling: a new approach to topic-specific web resource discovery. Comput. Netw. 31(11–16) (1999)
5. Costa, M., Couto, F., Silva, M.: Learning temporal-dependent ranking models. In: SIGIR 2014 (2014)
6. Costa, M., Gomes, D., Silva, M.J.: The evolution of web archiving. IJDL (2016)
7. Diligenti, M., Coetzee, F., Lawrence, S., Giles, C.L., Gori, M.: Focused crawling using context graphs. In: VLDB (2000)
8. Dong, A., Chang, Y., Zheng, Z., Mishne, G., Bai, J., Zhang, R., Buchner, K., Liao, C., Diaz, F.: Towards recency ranking in web search. In: WSDM 2010 (2010)
9. Dong, H., Hussain, F.K.: SOF: a semi-supervised ontology-learning-based focused crawler. Concurrency Computat. Prac. Experience 25(12) (2013)
10. Ehrig, M., Maedche, A.: Ontology-focused crawling of web documents. In: ACM SAC (2003)
11. Farag, M.M.G., Lee, S., Fox, E.A.: Focused crawler for events. IJDL (2017)
12. Gossen, G., Demidova, E., Risse, T.: iCrawl: Improving the freshness of web collections by integrating social web and focused web crawling. In: JCDL 2015 (2015)
13. Gossen, G., Demidova, E., Risse, T.: The iCrawl Wizard – supporting interactive focused crawl specification. In: ECIR 2015 (2015)
14. Gossen, G., Demidova, E., Risse, T.: Analyzing web archives through topic and event focused sub-collections. In: WebSci 2016. pp. 291–295, May 2016
15. Heydon, A., Najork, M.: Mercator: a scalable, extensible web crawler. World Wide Web 2(4), 219–229 (1999)
16. Jackson, A., Lin, J., Milligan, I., Ruest, N.: Desiderata for exploratory search interfaces to web archives in support of scholarly activities. In: JCDL2016 (2016)
17. Jiang, J., Song, X., Yu, N., Lin, C.Y.: Focus: Learning to crawl web forums. IEEE TKDE 25(6) (2013)
18. Kanhabua, N., Nørvåg, K.: A comparison of time-aware ranking methods. In: SIGIR 2011 (2011)
19. Laranjeira, B., Moreira, V., Villavicencio, A., Ramisch, C., Finatto, M.J.: Comparing the quality of focused crawlers and of the translation resources obtained from them. In: LREC 2014 (2014)
20. Mohr, G., Kimpton, M., Stack, M., Ranitovic, I.: Introduction to Heritrix, an archival quality web crawler. In: 4th International Web Archiving Workshop (2004)
21. Nguyen, T.N., Kanhabua, N., Niederée, C., Zhu, X.: A time-aware random walk model for finding important documents in web archives. In: SIGIR 2015 (2015)
22. Pant, G., Srinivasan, P.: Learning to crawl: Comparing classification schemes. ACM Trans. Inf. Syst. 23(4) (2005)
23. Pant, G., Srinivasan, P., Menczer, F.: Crawling the web. In: Web Dynamics (2004)
24. Pereira, P., Macedo, J., Craveiro, O., Madeira, H.: Time-aware focused web crawling. In: Rijke, M., Kenter, T., Vries, A.P., Zhai, C.X., Jong, F., Radinsky, K., Hofmann, K. (eds.) ECIR 2014. LNCS, vol. 8416, pp. 534–539. Springer, Cham (2014). doi:10.1007/978-3-319-06028-6_53
25. Qin, J., Zhou, Y., Chau, M.: Building domain-specific web collections for scientific digital libraries. In: JCDL 2004 (2004)
26. Risse, T., Demidova, E., Gossen, G.: What do you want to collect from the web? In: Proceedings of the Building Web Observatories Workshop (BWOW) 2014 (2014)

Information Governance Maturity Model Final Development Iteration

Diogo Proença[1,2(✉)], Ricardo Vieira[1,2], and José Borbinha[1,2]

[1] Instituto Superior Técnico, Universidade de Lisboa, Lisbon, Portugal
{diogo.proenca,rjcv,jlb}@tecnico.ulisboa.pt
[2] INESC-ID - Instituto de Engenharia de Sistemas e Computadores Investigação e Desenvolvimento, Lisbon, Portugal

Abstract. Information Governance (IG) as defined by Gartner is the "specification of decision rights and an accountability framework to encourage desirable behavior in the valuation, creation, storage, use, archival and deletion of information. Includes the processes, roles, standards and metrics that ensure the effective and efficient use of information in enabling an organization to achieve its goals".

Organizations that wish to comply with IG best practices, can seek support on the existing best practices, standards and other relevant references not only in the core domain but also in relevant peripheral domains. Thus, despite the existence of these references, organizations still are unable, in many scenarios, to determine in a straightforward manner two fundamental business-related concerns: (1) to which extent do their current processes comply with such standards; and, if not, (2) which goals do they need to achieve in order to be compliant.

In this paper, we present the third and last iteration of an IG maturity model based on existing reference documents. The development process is based on existing maturity model development methods that allow for a systematic approach to maturity model development backed up by a well-known and proved scientific research method called Design Science Research.

Keywords: Information governance · Maturity model · Measurement

1 Introduction

A maturity model defines a pathway of improvement for organizational aspects and is classified by a maturity level. The maturity levels often range from zero to five, where zero consists on the lack of maturity and five consists of a fully mature and self-optimizing process. Maturity models can be used for assessing and/or achieving compliance since they allow the measurement of a maturity level and, by identifying the gap between the current and pursued level, allow the planning of efforts, priorities and objectives in order to achieve the goals proposed.

The use of maturity models is widely used and accepted, both in the industry and the academia [1]. There are numerous maturity models, virtually one for each of the most trending topics in such areas as Information Technology or Management. Maturity

© Springer International Publishing AG 2017
J. Kamps et al. (Eds.): TPDL 2017, LNCS 10450, pp. 128–139, 2017.
DOI: 10.1007/978-3-319-67008-9_11

Models are widely used and accepted because of their simplicity and effectiveness. They depict the current maturity level of a specific aspect of an organization, for example IT, Outsourcing or Project Management, in a meaningful way, so that stakeholders can clearly identify strengths and improvement points and prioritize what they can do in order to reach higher maturity levels, showing the outcomes that will result from that effort which enables stakeholders to decide if the outcomes justify the effort needed to go to higher levels and results in a better business and budget planning.

The objective of this paper is to develop an artifact (the maturity model) by using a research approach to contribute to the body of knowledge. Therefore, Design Science Research (DSR) [19] was chosen as it combines two perspectives, the practical and scientific dimensions. The maturity model focuses on the IG body of knowledge to define IG maturity levels.

The paper is structured in six sections. First, fundamental terms and concepts will be detailed and will be followed by the outline of the research methodology in Sect. 3. Further on, Sect. 4 presents the findings from a literature review. Section 5 elaborates the main insights of the iterative maturity model development and the maturity model itself. Next, the evaluation of the maturity model is presented in Sect. 6. Lastly, this paper presents conclusions from this work and details research limitations.

2   Foundation

To ensure a common understanding, we explain in this section the key terms and concepts, such as, "Maturity" and "Maturity Model".

To evaluate maturity, organizational assessment models are used, which are also known as stages-of-growth models, stage models, or stage theories [12].

The concept of maturity is a state in which, when optimized to a particular organizational context, is not advisable to proceed with any further action. It is not an end, because it is a mobile and dynamic goal [7]. It is a state in which, given certain conditions, it is agreed not to continue any further action. Several authors have defined maturity, however many of the current definitions fit into the context in which each a particular maturity model was developed.

In [6] maturity is defined as a specific process to explicitly define, manage, measure and control the evolutionary growth of an entity. In turn, in [8] maturity is defined as a state in which an organization is perfectly able to achieve the goals it sets itself. In [9] it is suggested that maturity is associated with an evaluation criterion or the state of being complete, perfect and ready and in [10] as being a concept which progresses from an initial state to a final state (which is more advanced), that is, higher levels of maturity. Similarly, in [11] maturity is related with the evolutionary progress in demonstrating a particular capacity or the pursuit of a certain goal, from an initial state to a final desirable state. Still, in [11] it is emphasized the fact that this state of perfection can be achieved in various ways. The distinction between organizations with more or less mature systems relates not only to the results of the indicators used, but also with the fact that mature organizations measure different indicators when comparing to organizations which are less mature. While the concept of maturity relates to one or more items identified as relevant, the concept of capability is concerned only with each of these items. In [12] maturity models are defined as a series of sequential levels, which together form an anticipated or desired logical path from an initial state to a final state of maturity. These models have their origin in the area of quality. The Organizational Project Management Maturity Model (OPM3) defines a maturity model as a structured set of elements that describe the characteristics of a process or product [13]. In [14] maturity models are defined as tools used to evaluate the maturity capabilities of certain elements and select the appropriate actions to bring the elements to a higher level of maturity. Conceptually, these represent stages of growth of a capability at qualitative or quantitative level of the element in growth, in order to evaluate their progress relative to the defined maturity levels.

Some definitions found involve organizational concepts commonly used, such as the definition of [15] in which the authors consider a maturity model as a "… a framework of evaluation that allows an organization to compare their projects and against the best practices or the practices of their competitors, while defining a structured path for improvement." This definition is deeply embedded in the concept of benchmarking. In other definitions there appears the concern of associating a maturity model to the concept of continuous improvement.

In [16], the maturity models are particularly important for identifying strengths and weaknesses of the organizational context to which they are applied, and the collection of information through methodologies associated with benchmarking. In [17] it was concluded that the great advantage of maturity models is that they show that maturity must evolve through different dimensions and, once reached a maturity level, sometime is needed for it to be actually sustained. In [18] it was concluded that project performance in organizations with higher maturity levels was significantly increased. Currently, the lack of a generic and global standards for maturity models has been identified as the cause of poor dissemination of this concept.

3   Research Methodology

The development of maturity models in the IT and IG domains is not new and has been quite popular in recent years. As an example, in [20], the authors have identified more than 100 maturity models, and in [21] even more are identified. However, one major issue can be identified in most these maturity models, which is the lack of disclosure of the development process used to develop them. This leads to a weakness in this research area, which is the lack of contributions regarding how to develop these models. Despite this fact, we have identified some development methods and procedures for maturity models, such as, the general design principles from Roglinger et al. [12], the DSR perspective on maturity models by Mettler [11], the development guidelines from Maier et al. [22], and the procedure model based on DSR [24] from Becker et al. [23], which are quite popular among scholars based on their respective citation counts. To develop the maturity model presented in this paper we decided to apply the development proce- dure of Becker et al. [23] as it is based on DSR and as result it offers a sound methodo- logical foundation, which is suitable for application in the research approach. This development procedure gives a stringent and consistent approach to the DSR guidelines of Hevner et al. [24].

As depicted in the procedure model in Fig. 1 the first steps focus on the problem identification. In this step the research problem is identified and detailed, the practical relevance of the problem is specified and the value of the artifact is justified. This step is followed by the comparison with existing maturity models. This second step is based on the problem identification of the first step and analysis of existing maturity model in the IG domain, which leads to the identification of weaknesses in these models. We conducted a literature analysis, which was based on an extensive online search to find existing maturity models focused on the IG domain. Thus, the analysis of the maturity models was performed according to their functionality.

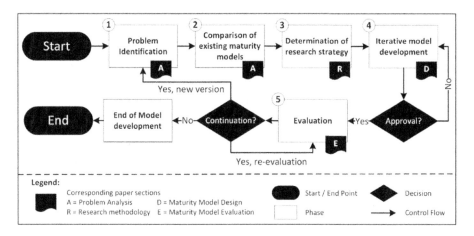

Fig. 1. Maturity Model Development Procedure Model of the research approach based on Becker et al. [23]

The next step deals with the determination of the research strategy outlined in this section of the paper. This is followed by the iterative maturity model development. In this step, we used model adoption techniques, such as, configuration, instantiation, aggregation, specialization and analogy [25] to incorporate the ISO14721, ISO16363 and ISO20652 in the maturity model. This allowed us to create a rigorous maturity model regarding both the structure and content. In the last step, evaluation, we combined the steps of Becker et al. [23], conception of transfer and evaluation, implementation of transfer media, and evaluation, into step 5. All steps will be conducted, but to match the structure of this paper we made this change.

4   Problem Analysis

This section presents the several maturity models from the Information Management, Records Management, IG and Digital Preservation domains that can influence the development of the maturity model proposed in this paper. Each Maturity Model is presented starting with the maturity model name, attributes and maturity levels. These attributes further detail the maturity model by decomposing certain aspects of the maturity model domain. The synthesis of the analyzed maturity models is presented in Table 1.

Table 1. Synthesis of the analyzed maturity models

Maturity model	Attributes		Maturity levels
	Name	Number	
Asset Management Maturity Model [2]	Dimensions/Category	4	Initial; Repeatable; Defined; Managed; Optimizing
Digital Asset Management (DAM) Maturity Model [3]	Categories/Dimensions	4/15	Ad-Hoc; Incipient; Formative; Operational; Optimal
Information Governance Maturity Model [4]	Principles	8	Sub-standard; In Development; Essential; Proactive; Transformational
Digital Preservation Capability Maturity Model (DPCMM) [26]	Domains/Components	3/15	Nominal; Minimal; Intermediate; Advanced; Optimal
Brown Digital Preservation Maturity Model [27]	Process Perspective	10	No Awareness; Awareness; Roadmap; Basic Process; Managed Process; Optimized Process
Preservica Digital Preservation Maturity Model [28]	–	–	Safe Storage; Storage Management; Storage Validation; Information Organization; Information Processes; Information Preservation

5   Maturity Model Design

In accordance to the maturity model development procedure of Becker et al. [23] a new maturity model should be developed, if no existing or the advancement of an existing one can address the identified problem. So, based on the findings of our literature analysis there is no maturity model which acceptably fulfills our needs. Therefore, we decided to develop a new maturity model.

The newly developed maturity model, presented in Fig. 2, adopts established structural elements, domains and functions of the best practice of maturity models analyzed in Sect. 4 and is based in relevant references form the Digital Preservation and Archival Science domains, namely ISO 14721, ISO 16363 and ISO 20652. These artifacts were then extended and adjusted to fit the purpose of assessing the maturity of IG using the guidance from these ISO standards. As outlined within our research methodology, we applied an iterative process for the development of this maturity model. In total, we needed three iterations, which are described in the following:

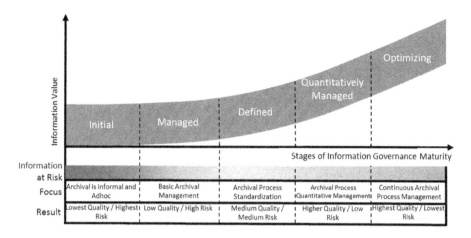

Fig. 2. Information Governance Maturity Model – Maturity Curve

First Iteration: As a first step, we identified the basic characteristics and structure of the model. As a starting point, we proposed five maturity levels – Initial, Managed, Defined, Quantitatively Managed, Optimizing – as this approach is evident in several reputable maturity models, such as, the CMMI [6]. In this initial iteration, we focused in just one dimension of the maturity model, the processes dimension. For each criterion of the maturity model we modeled what is the manifestation of that criterion at the different maturity levels. The first iteration was published through the E-ARK project in Deliverable 7.1 and was communicated to the scientific community through [29].

Second Iteration: The aim of the second iteration was to build on the success of the results of the first iteration. Thus, the maturity model was extended to contemplate all the dimensions of the maturity model. We continued with the approach of the first iteration and modeled each of the criteria at each maturity level. We then conducted a trial assessment using the maturity model, which revealed some issues that will be solved in the third iteration. This second iteration and the results of the trial assessment were published through the E-ARK project in Deliverable 7.2 and were communicated to the scientific community through [29–31].

Third Iteration: After the trial assessment using the maturity model one relevant issue was identified. The trial revealed that there was a difficulty in understanding the differences in each possible answer for the assessment questions. As an example, participants could understand what a "documented procedure" is but it was difficult for them to understand what is a "defined procedure" or even an "ad-hoc assessed procedure". This led to a revision of the assessment questionnaire and an overhaul of the maturity model to accommodate the changes to the assessment questionnaire. The maturity levels definition remained the same, however there are major changes in the overall structure of the criteria. Now instead of modelling each criterion at each maturity level we opted by identifying capabilities for each maturity level and dimension, which resulted in an easily understandable maturity model that is presented in Figs. 3, 4 and 5. This third and final iteration was published through the E-ARK project in Deliverables 7.5 and 7.6.

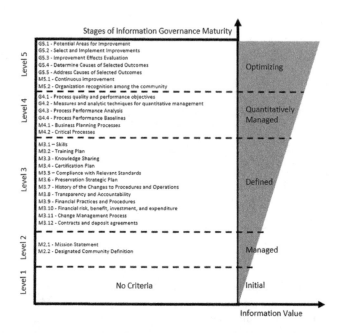

Fig. 3. Information Governance Maturity Model – Management Dimension Maturity Levels

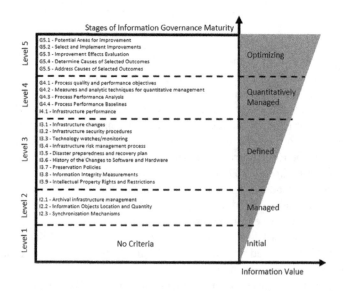

Fig. 4. Information Governance Maturity Model – Infrastructure Dimension Maturity Levels

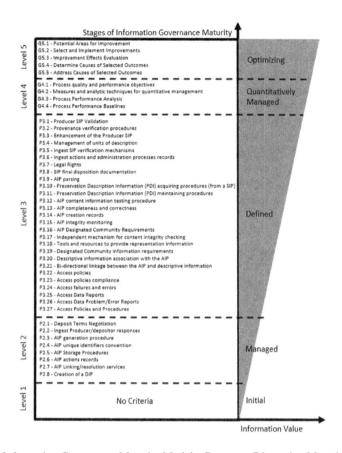

Fig. 5. Information Governance Maturity Model – Processes Dimension Maturity Levels

At maturity level 1, the organization needs to be aware that IG is needed as a relevant function of the organization.

At maturity level 2 IG meets its goals. However, there is no standardization of procedures, which can lead to two people doing different tasks to achieve the same goal and in turn can result in the inability to repeat tasks that were previously performed. Moreover, at this maturity level there is no assignment of responsibilities.

Then at maturity level 3, the organization has a standardized list of procedures with responsibilities assigned. There are also tools and methods that support IG, which are agreed upon and become a standard across the organization. Procedures at this maturity level are well defined and include its purpose, inputs, entry criteria, activities, roles, verification steps, outputs and exit criteria.

At maturity level 4 the organization establishes quantitative objectives for quality and performance of all functions related with IG. Specific measures of performance are collected and are analyzed using statistical and other quantitative techniques. There are also performance baselines and models that help in setting quality objectives. A key difference between maturity levels 3 and 4 is the predictability of performance as predictions are based on the statistical analysis of fine-grain information.

Finally, at Maturity Level 5 the organization continually improves its IG functions based on quantitative analysis of the business objectives and performance baselines. It uses quantitative techniques to understand variations in procedures and the causes of outcomes. It also focuses on continually improving performance using incremental and innovative procedures. Additionally, the quality and performance objectives are established and continuously revised to reflect changing business objective and the organization's performance. A key difference between maturity level 4 and 5 is the focus on improving and managing the organization performance, which at this level is concerned in analyzing performance using data collected from multiple sources. This data helps identify gaps and weak points in performance that are then used to generate a measurable improvement.

To improve from level X to level X + 1, the organization must comply with all the criteria from level X, which makes this maturity model follow a "stages" approach. What an organization can expect from progressing through the maturity levels is that their IG practice will become increasingly managed, defined and optimized.

A maturity table consists of a table that crosses maturity levels with the maturity dimensions and characterizes each dimension in each level. Figure 2 presents the maturity table. The mapping to the assessment criteria for each dimension and maturity level is later detailed in Figs. 3, 4 and 5. The main goal of the IG Maturity Model is to improve the value of information in an organization. Information value will increase when going from a lower to a higher maturity level, as depicted in Fig. 2. Moreover, the lack of procedures and policies in lower levels results in the organization's information being at risk and this risk reduces as policies and procedures become implemented, defined, documented and assessed.

The IG maturity model, consists of three dimensions, Management, Process and Infrastructure. These dimensions provide different viewpoints of IG which help to decompose the maturity model and enable easy understanding. For each dimension we have a set of levels, from one to five, where one show the initial phase of maturity of a dimension and level five shows that the dimension is fully mature, self-aware and optimizing. These levels and their meaning were adapted from the levels defined for CMMI. [6] The management dimension "refers to all the activities that are used to coordinate, direct, and control an organization." [5] The criteria for assessing the maturity of this dimension is depicted in Fig. 3.

The infrastructure dimension "refers to the entire system of facilities, equipment, and services that an organization needs in order to function." [5] The criteria for assessing the maturity of this dimension is depicted in Fig. 4.

Finally, the processes dimension contains the "set of activities that are interrelated or that interact with one another. Processes use resources to transform inputs into outputs." [5] The criteria for assessing the maturity of this dimension is depicted in Fig. 5.

6   Maturity Model Evaluation

This section details the assessment strategy used in the development of the maturity model proposed in this paper. For the purpose of this maturity model we opted for the self-assessment method as it provides a way for organizations to assess their IG practice while maintaining a low cost to the organizations.

Table 2 depicts a comparison between the E-ARK pilots for the initial assessment and final assessment. Pilot 1 is the one which achieved the best overall results, especially the infrastructure dimension achieved the best results. Pilot 3 achieved the second-best results. Pilot 5 also shows a high-level maturity across the dimensions measured in the assessment. However, as in pilot 2, there are still some important enhancements to perform to the infrastructure capability. The other four pilots showed similar results among the dimensions. With some exceptions for pilot 4, where it shows higher maturity levels for the infra-structure dimension. Another exception are pilots 6 and 7 which show higher maturity levels for the processes dimension in the final assessment results.

Table 2.   Initial and Final Self-assessment Results of the E-ARK Pilots

Dimension	Initial Assessment								Final Assessment							
	P1	P2	P3	P4	P5	P6	P7	∅	P1	P2	P3	P4	P5	P6	P7	∅
Management	4	2	4	2	4	1	1	**2.6**	4	2	4	2	4	1	1	**2.6**
Processes	4	1	3	1	3	2	2	**2.3**	4	2	4	1	3	3	4	**3**
Infrastructure	5	2	3	4	2	1	2	**2.7**	5	2	4	4	2	2	3	**3.1**
∅ (Average)	**4.3**	**1.7**	**3.3**	**2.3**	**3**	**1.3**	**1.7**	**2.5**	**4.3**	**2**	**4**	**2.3**	**3**	**2**	**2.7**	**2.9**

The results of the E-ARK project helped the pilots improve their maturity level and as result improved archival practice as can be seen by analyzing the results of the final assessment depicted in Table 2. The final results show several improvements in the overall maturity levels for all pilots. One aspect to take into consideration is that E-ARK outputs focus on the processes dimension as such this is the dimension where the most improvements are as illustrated in Table 2.

7   Conclusion

This paper presented the third and last iteration of a maturity model for IG, as well as, a state of the art on maturity models surrounding IG found in literature. Based on that state of the art and other references from the archival domain, namely the ISO16363, ISO14721 and ISO 20652 we developed a maturity model consisting of three dimensions and five levels.

This paper also presents how the assessment of the E-ARK pilots was performed, as well as, the analysis of the results for the pilots. As can be seen, the self-assessment questionnaire enabled a detailed analysis and comparison of the pilots and proved useful in identifying weak points and strengths of the pilots. Using the results it is then possible for pilots to identify points of improvement which can then lead to the creation of an improvement path for the pilots. Additionally, the self-assessment questionnaire is now available online at http://earkmaturitysurvey.dlmforum.eu. Organizations can use it to assess their current IG Maturity and based on the results plan for improvement.

Despite this there is still room for improvement of the questionnaire, we are now finishing a detailed guide on how to fill the questionnaire and analyze the results which will be available online as a companion to the self-assessment questionnaire.

To extend the research component, we suggest evaluating (and refining) the maturity model within different industry sectors to gather an insight of what IG methods and procedures different industries are using and how far in the maturity scale they are.

Acknowledgements. This work was supported by national funds through Fundação para a Ciência e a Tecnologia (FCT) with reference UID/CEC/50021/2013.

References

1. Shang, S., Lin, S.: Understanding the effectiveness of capability maturity model integration by examining the knowledge management of software development process. In: Total Quality Management & Business Excellence, vol. 20(5) (2009)
2. Lei, T., Ligtvoet, A., Volker, L., Herder, P.: Evaluating asset management maturity in the netherlands: a compact benchmark of eight different asset management organizations. In: Proceedings of the 6th World Congress of Engineering Asset Management (2011)
3. Real Story Group, DAM Foundation, The DAM Maturity Model, http://dammaturity model.org/
4. ARMA International, Generally Accepted Recordkeeping Principles - Information Governance Maturity Model, http://www.arma.org/principles
5. ISO 9001:2008: Quality management systems – Requirements (2008)
6. CMMI Product Team, CMMI for services, version 1.3. Software Engineering Institute. Carnegie Mellon University, Tech. Rep. CMU/SEI-2010-TR-034 (2010)
7. Tonini, A., Carvalho, M., Spínola, M.: Contribuição dos modelos de qualidade e maturidade na melhoria dos processos de software. Produção **18**(2), 275–286 (2008)
8. Anderson, E., Jessen, S.: Project maturity in organizations. Int. J. Project Manage. Account. **21**, 457–461 (2003)
9. Fitterer, R., Rohner, P.: Towards assessing the networkability of health care providers: a maturity model approach. Inf. Syst. E-bus. Manage. **8**, 309–333 (2010)
10. Sen, A., Ramammurthy, K., Sinha, A.: A model of data warehousing process maturity. IEEE Trans. Softw. Eng. (2011)
11. Mettler, T.: A Design Science Research Perspective on Maturity Models in Information Systems. Institute of Information Management, University of St. Gallen, St. Gallen (2009)
12. Röglinger, M., Pöppelbuß, J.: What makes a useful maturity model? a framework for general design principles for maturity models and its demonstration in business process management. In Proceedings of the 19th European Conference on Information Systems, Helsinki, Finland, June 2011
13. OPM3, Organizational Project Management Maturity Model. Project Management Institute, Newtown Square, Pennsylvania, USA (2003)
14. Kohlegger, M., Maier, R., Thalmann, S.: Understanding maturity models: results of a structured content analysis. In Proceedings of the I-KNOW 2009 and I-SEMANTICS 2009, 2–4 September 2009, Graz, Austria (2009)

15. Korbel, A., Benedict, R.: Application of the project management maturity model to drive organisational improvement in a state owned corporation. In: Proceedings of 2007 AIPM Conference, Tasmania, Australia, 7–10 October (2007)
16. Koshgoftar, M., Osman, O.: Comparison between maturity models. In: Proceedings of the 2nd IEEE International Conference on Computer Science and Information Technology, vol. 5, pp. 297–301 (2009)
17. Prado, D.: Gerenciamento de Programas e Projetos nas Organizações. Nova Lima, Minas Gerais (2004)
18. Jamaluddin, R., Chin, C., Lee, C.: Understanding the requirements for project management maturity models: awareness of the ICT industry in Malaysia. In: Proceedings of the 2010 IEEE IEEM, pp. 1573–1577 (2010)
19. Peffers, K., Tuunanen, T., Rothenberger, M., Chatterjee, S.: A design science research methodology for information systems research. J. Manage. Inf. Syst. **24**, 45–77 (2008)
20. Mettler, T., Rohner, P., Winter, R.: Towards a classification of maturity models in information systems. In: D'Atri, A., De Marco, M., Braccini, A.M., Cabiddu, F.: Management of the Interconnected World. Physica-Verlag, Heidelberg (2010)
21. Poeppelbuss, J., Niehaves, B., Simons, A., Becker, J.: Maturity models in information systems research: literature search and analysis. Commun. Assoc. Inf. Syst. **29** (2011)
22. Maier, A., Moultrie, J., Clarkson, P.: Assessing organizational capabilities: reviewing and guiding the development of maturity grids. IEEE Trans. Eng. Manage. **5**(1) 2012
23. Becker, J., Knackstedt, R., Pöppelbuß, J.: Developing maturity models for IT management: a procedure model and its application. Bus. Inf. Syst. Eng. **3**, 213–222 (2009)
24. Hevner, A., Ram, S., March, S., Park, J.: Design science in information systems research. MISQ **28**, 75–105 (2004)
25. Vom Brocke, J.: Design principles for reference modeling-reusing information models by means of aggregation, specialization, instantiation, and analogy. In: Fettke, P., Loos, P. (eds.) Reference modeling for business systems analysis. Idea Group Inc., Hershey (2007)
26. Brown, A.: Practical Digital Preservation - A How-to Guide for Organizations of Any Size. Facet Publishing (2013)
27. Dollar, C.M., Ashley, L.J.: Assessing digital preservation capability using a maturity model process improvement approach, Technical Report, February 2013
28. Preservica, Digital Preservation Maturity Model, White Paper (2014)
29. Proença, D., Vieira, R., Borbinha, J.: A maturity model for information governance. In: 7th DLM Forum Triennial Conference, Lisbon (2014)
30. Proença, D., Vieira, R., Borbinha, J.: A maturity model for information governance. In: 20th International Conference on Theory and Practice of Digital Libraries (TPDL 2016), Hannover (2016)
31. Proença, D., Vieira, R., Borbinha, J.: Towards a systematic information governance maturity assessment. In: 13th International Conference on Digital Preservation (iPres 2016), Bern (2016)

Challenges of Research Data Management for High Performance Computing

Björn Schembera[✉] and Thomas Bönisch

High Performance Computing Center Stuttgart (HLRS),
University of Stuttgart, Nobelstr. 19, 70569 Stuttgart, Germany
{schembera,boenisch}@hlrs.de

Abstract. This paper targets the challenges of research data management with a focus on High Performance Computing (HPC) and simulation data. Main challenges are discussed: The Big Data qualities of HPC research data, technical data management, organizational and administrative challenges. Emerging from these challenges, requirements for a feasible HPC research data management are derived and an alternative data life cycle is proposed. The requirement analysis includes recommendations which are based on a modified OAIS architecture: To meet the HPC requirements of a scalable system, metadata and data must not be stored together. Metadata keys are defined and organizational actions are recommended. Moreover, this paper contributes by introducing the role of a Scientific Data Manager, who is responsible for the institution's data management and taking stewardship of the data.

Keywords: Research data management · HPC · Simulation · Big data · Archive · OAIS · Metadata · Data life cycle

1 Introduction

Today's science can be considered as data-driven. Research data is all scientific data generated or recorded from experiments, studies or simulations. In contrast to theory and classical experiments, simulations produce huge amounts of big research data [19], usually in size of Petabytes (PB). High performance computing (HPC) is one of the driving forces behind big research data enabling large-scale simulations in climate research, engineering or particle physics just to name a few. For researchers it is crucial to keep their data for review or later resumption of work. However, the ability to store and especially manage research data is lagging behind the ability to generate data [15] in HPC: For example, the sheer volume of the data is a specific problem, but not the only critical one.

The following work presents the challenges of research data management of simulation data in the scope of HPC. The first challenge is the problem of Big Data: volume and variety. As a second challenge, insufficient data management concepts will be discussed. Moreover, research data management is not only a technical but also a organizational problem in HPC: A lack of data management plans, regulations and incentives.

© Springer International Publishing AG 2017
J. Kamps et al. (Eds.): TPDL 2017, LNCS 10450, pp. 140–151, 2017.
DOI: 10.1007/978-3-319-67008-9_12

Derived from these challenges, requirements for a feasible research data management in HPC are specified in Sect. 3. This requirement analysis is the main contribution of this paper and includes data management requirements such as metadata, persistent identifiers and data security. Since Open Access will become a key requirement in the future, it will be discussed in a separate subsection. For HPC, scalability requirements are important: Research data management has to cope with the volume and has to provide efficient indexing mechanisms for feasible search of millions of data objects. Since research data management is not only a technical problem, one contribution is the introduction of a new role: The Scientific Data Officer (SDO) that is in charge of the research data management efforts of an institution. All these efforts lead to an improved data life cycle being able to reduce "dark" data. Related work is presented in Sect. 3.7.

2 Challenges

2.1 Big Research Data: Volume and Variety

The data volume produced on an HPC system strongly depends on the amount of main memory of a supercomputer. A DoE study [12] estimates the data volume factor to be 1:35 in worst case: For each Byte of main memory of the compute system, 35 Byte of data to archive is created per year, so the amount of data scales linear with the amount of main memory of the HPC system. This means every time a new HPC system is deployed, an increase in data production (due to a more fine-grained resolution or due to larger scales) has to be expected. The growth over time of the research or during the overall system lifetime data volume must be expected to be exponential, the study concludes. A follow-up study reminds that a storage technology gap exists for Exascale [13]. Figure 1 shows a sample trend (2010–2017) of the data stored in a tape archive at the HLRS related to the main memory of the corresponding HPC flagship systems.

The number of files does not depend on the amount of main memory but on the number of cores and processors, however no estimate can be given. A reason for this is that it strongly depends on the behavior of the researchers how many files are written or if file aggregation is used. Nevertheless, the study states that the number of files to archive per system will grow exponentially from millions to billions during the next 5 years.

Regarding variety, research data is strongly diverse. Most data is formatted according to the researchers bias. For example, as comma-separated values, tables or stored in files of different formats [22]. How this data is organized and managed is another challenge that will be discussed in the following subsection.

2.2 Research Data Management

Research data management in HPC is often handled by the directory structures and an appropriate naming of files and directories, such as

```
/group/project/user/simulation/run/description.format
```

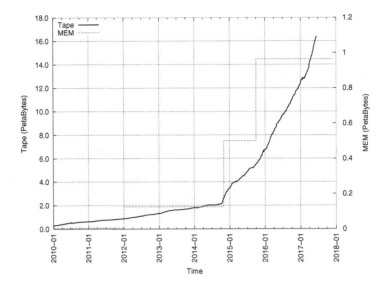

Fig. 1. Data stored (left y-axis/continuous line) in the HLRS HPC center in Stuttgart, Germany in relation to the main memory of a system (right y-axis/dashed line). A dependency is easy to see. The ratio is higher than 1:10 in this center. As of June 2017, 16 PB of data in approximately 9.2 million files of almost 200 users are held on tape.

In doing so, searching, finding and retrieving research data becomes a burden [1,15]: A third person is unable to find data if the person does not have information on who ran the simulation, which group was leading which project or which project acronym is used. Additionally, the directory structures are highly dynamic, for example when a parameter has to be varied unexpectedly.

Moreover in most cases, no explicit metadata is attached to these files. This means that the description what the files contain and how they can be interpreted is nowhere (formally) written down. If metadata is attached to the files, besides file names, it may reside in text files, spreadsheets or encoded in the output files [15]. There are only little common formats for storing and especially describing research data [22], such as NetCDF. It strongly depends on the community whether these possibilities are used, for example in earth sciences with NetCDF [18] or with HDF5 at NASA [19].

2.3 Organizational Workflows

The challenges discussed above have been rather technical issues of scalability. However, research data management is not only a technical management task. What is lacking nowadays for example are incentives for researchers to perform the additional work of tagging their data with metadata and storing the data in a research data management system.

Moreover, plans how to organize data are often missing; specifications that define roles, responsibilities, timelines and descriptions of the data are lacking.

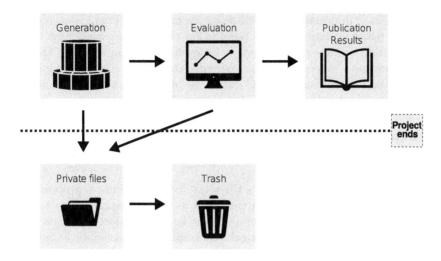

Fig. 2. Today's data life cycle: After generation, evaluation of data and latter publication of the results, the data is put to the private files. There it is deleted or forgotten.

It is often unclear who is taking the stewardship of big research data [19]. Typically, one person of a research group is pushed into a vague role of handling the institution's data. Researchers are acting on their own because there is no assistance offered. This is due to the fact that either the expertise is not available or distributed through the research institution. This is also a question of missing training: How will a researcher get in touch with the tools, practices and regulations of data management? According to a study conducted in the UK [3], lacking skills are one main challenge of data management in general.

Open Access is the paradigm to make research data publicly available and is another challenge in HPC context. Open Access is still not commonly accepted - this is also a problem of regulations and conventions by institutions as well as a legal challenge. However, Open Access will become a top priority in the future [3]. A European Union report argues that in the future, all generated research data has to be made available to the public [7]. In addition, the German Research Foundation (DFG) sets as a vision that publicly funded research data should be publicly available for a long-term period [5]. This prevents duplicate work and saves resources.

2.4 Research Data Life Cycle

Today's HPC research data life cycle is shown in Fig. 2 and can be described as follows: While computing the simulation jobs, data is continuously written to a parallel filesystem, such as Lustre. After the generation of the data, evaluation follows. After the end of the project, the knowledge gained through the results is published. Researchers see the end of data management when the publication is written: Scientific results are communicated and published in papers, where condensed information and statistics are presented. There is no room for referencing research data, so after publication the data is either forgotten in the private files, on tape storage systems or even deleted. In this way, the data becomes dark data, since reusing or reviewing the data is hardly possible [10]. This is not only a problem in HPC but in all big data sciences [11] and runs contrary to good scientific practice, as for example the DFG recommends [5].

3   Requirements for HPC Research Data Management

3.1   Scalability Requirements

To cope with the above challenges, a three-layered architecture is proposed as a technical foundation. It is based on the OAIS model and Askhojs approach, as described in Sect. 3.7 and consists of a storage layer, an object layer and a user layer. The two bottom layers are critical with respect to the scalability challenges of volume and variety.

The storage layer is logically the lowest component and has to handle exponential data growth. It must be possible to easily extend the storage space. This requirement is not specific to HPC but gets critical here [13]. Tape media is the only media that meets the HPC requirements implying low total cost of ownership, large data volume and low error rate [8]. The storage layer of a research data management system needs the possibility to be distributed if the amount of data is too large to store in one single data center[1]. Classic DBMS like DB2 are not suitable for storing the actual data since they are unable to store data in the size of TeraBytes. The storage layer has to be a cost-effective mass storage system on which the object management can be built.

The management of the data objects is located on the object layer which is logically on top of the storage layer. The object layer makes the data stored in the storage layer search- and findable. In HPC this is an important requirement since millions of files are not unusual. According to a report, metadata performance is already critical but will become crucial in the Exascale [13]. Searching must be performed in a feasible time that can only be achieved if metadata and data are not stored together albeit violating the OAIS paradigm of containers holding both data and metadata. Storing only the reference to data residing on another layer of the system, queries can be performed as a first step for retrieving the actual data. Askhojs layered architecture is preferable for HPC due to reasons of flexibility and the distributed character of the data management system. However, Askhojs design to bring OAIS to the cloud is not suitable for the HPC use case since it would not scale for big data volumes and growth. The connection to a cloud storage service would be too poor to transfer huge amounts of data in a feasible time (i.e., retrieving 50TB via a 1GB/s cloud link would take approx. 5 days). Moreover, due to a lack of integration in the HPC workflow and security considerations, cloud services as a backend are not preferable.

---

[1] For example, the data produced by the CERN/LHC experiments is distributed to data centers all over Europe. See: https://home.cern/about/computing/worldwide-lhc-computing-grid, last accessed Nov 28th 2016.

3.2   Data Management Requirements

Metadata is the main concept to handle data [9,23]. Metadata is "data about data" and describes data from a logical point of view as well es from its attributes in a structured form. Enriched with metadata, data becomes a valuable object. For a feasible HPC research data management, the following parts have to be implemented: First, a reasonable metadata scheme that identifies generic as well as domain-specific characteristics of HPC simulation data has to be defined (for example in XML). Following the OAIS metadata scheme (details in Sect. 3.7), structural, administrative and descriptive metadata has to be included. Second, a suitable storage for metadata must be located in the object layer and has to be reliable, safe and performant, for example on mirrored SSDs. Third, efficient index mechanism to search and explore billions of files is needed, which is also a matter of scalability. Following metadata keys for HPC are mandatory:

Descriptive Metadata describes the data content. Important descriptive keys for searching in an HPC context are *Authors, Name, Filename, Creation Date, Access Date, Change Date, Keywords, System, Compiler, Compiler Flags, Batch System, Size, Algorithm, Context, Publications*. There must be an additional field for *domain-specific metadata*. For example in CFD, this could include the Reynolds number, the cases and the exact turbulence model used.

Preservation Metadata or administration metadata is all the metadata ensuring the long-term preservation. Reasonable keys can be derived from the OAIS metadata model and include a *Persistent identifier* (PID) key for the location of the data, such as Handle[2] or ePIC[3]. A PID allows data citation, since data can be uniquely identified. *Provenance information* incorporates information about the origin and the changes in form of list. The *Context information* key holds information that links the data object to others and is a list of persistent identifiers of other data objects. A *fixity* metadata key contains information that ensures the integrity of a data object. This is a checksum of the data object, such as a MD5 hash. An *access rights* key carries information on the access right to the data object. This also includes time limits and embargoes. Additionally to the management of data, preservation metadata plays a crucial role for fulfilling the requirements of data security, Open Access due to the usage of PIDs as well as possibilities of distributing the data over several data centers.

Content Metadata is mostly held by the data itself. However, it may be useful to store it additionally in the metadata store. For example, a key as *Format* should be stored here.

User Interface and Workflow Integration. A user interface that allows browsing, searching, injecting, manipulating of data objects in the research data management system builds the user layer. Possibilities to enter metadata and link it to data is mandatory. The user interface must be able to run in a distributed environment: Since the location of the data stored may differ from the location

---

[2] http://handle.net/, last accessed Nov 26th 2016.
[3] http://www.pidconsortium.eu/, last accessed Nov 26th 2016.

from where data is accessed or managed, location transparency is required that allows a single system view. The system must be accessible via a low-level client on the HPC frontends to perform metadata tagging of created files or retrieving data back for further analysis. All the above points have to be integrated in the HPC workflow seamlessly. Tagging of metadata has to be possible both at creation time of the files (like in the iCurate system discussed in Sect. 3.7) and at the time the files are injected into the system.

3.3   Security Requirements

Data security must be guaranteed by the data management system [23]. First, this means bitstream preservation which is the preservation of bits on a physical layer and also includes the ability to retain the bits if technology changes. Two copies of the data on tape are recommended, preferably in physically distinct locations. Second, this also has to include the fixity of the data objects, that is its integrity. Checksums can guarantee that the data has not been altered. Fixity checks have to be included. Third, this also includes end-to-end data integrity [13]. Fourth, encryption must be possible. These requirements are general requirements and not bound to HPC, but have to take into account the characteristics of big research data, such as how to feasibly perform integrity checks on Petabytes of data. This topic overlaps with the scalability requirement discussed in Subsect. 3.1 and affects components in both the storage and the object layer.

3.4   Open Access Requirements

Open Access is crucial for HPC as a data-driven science and will be raised as a key requirement in the future, as discussed in Sect. 2.3. This is also a technical challenge: On the user layer, an interface has to be provided that is accessible for the world to retrieve data publicly for the concrete use case. Transfer technologies have to be found that are able to move big research data. While publishing or moving the data, metadata annotations have to remain. This can be achieved by including a PID, as described in Sect. 3.2. Moreover, incentives have to be pushed and regulations have to be implemented that make researchers publishing their research data as Open Access.

3.5   Organizational Requirements

Incentives. There are extrinsic and intrinsic reasons why scientists and institutions should participate in research data management efforts. Extrinsic reasons are external influences, regulations or institution-wide standards. For example in Germany, the DFG advises to store scientific research data for at least 10 years [5]. Intrinsic reasons aim for the stakeholders themselves: Reasons for the researcher like the higher reputation when publishing the research data or an easier way to reuse the data after years. These incentives have to be analyzed case-by-case, pushed and incorporated by all future data management approaches.

Data Management Plans (DMP) are formal documents that specify plans and numbers on data management and fulfill the requirement of organizational security [14]. This documentation has to name how research data is handled during and after a research project. Research proposals already require management plans, for example those of the European Union [6] or of the National Science Foundation [20]. A DMP has therefore to be mandatory for all HPC research data management efforts and must be raised as a general requirement when handling research data. All persons involved have to negotiate on a DMP[4].

The DMP has to include a part *Data Description and Metadata*, where the data (and their provenance) should be described as well as the metadata keys used. It has to be specified how, when and where the data is produced. This is of importance for HPC: For example, the data has to be treated differently if it resides on a parallel scratch filesystem than if it resides on tape.

A *Timeline* has to be sketched out. It has to be specified how data is managed during all phases of the HPC workflow. The DMP has to define what tools are used in each step to transfer data back and forth and keep track of the data.

Moreover, *Organizational Topics* have to be covered by the DMP. In the document, the responsibilities have to be defined, that means a SDO has to be named. Legal issues have to be addressed. Access rights have to be defined: Who will be able to access the data at which time? Will the data be made publicly available as Open Access? This part also has to include how data management costs will be covered.

Qualification. Another organizational requirement is qualification [19]. Courses, trainings and integration into curricula have to be provided by and in institutions that apply data management. Persons that later can act as multipliers and end-users have to get data management skills. Only by increasing training activities, scientists will benefit from research data management efforts.

Scientific Data Officer. A person within a project, a department or an institution has to be defined taking the responsibility for data management and the stewardship of the data. This person assumes the role of the Scientific Data Officer (SDO). The SDO has the same position as the security officer with respect to data: Being aware of the trends in research data management, HPC-related tools and infrastructures and knowledge of in-house data storage facilities. The person has to be trained in respective courses. Within the research group or institute, this person has to act as a multiplier and transfer knowledge to the group in talks or on request. Moreover, the data officer has the stewardship of orphaned data that is preliminary dark data still existing when a person leaves the group. If the institution runs own data management systems, technical administration and support for these systems may also be in charge of the SDO.

---

[4] There are online tools available for specifying a DMP, such as: https://dmponline. dcc.ac.uk/, last accessed Nov 25th, 2016.

3.6   An Alternative Data Life Cycle

All the requirements for a feasible research data management for HPC should in the end lead to an improved, ideal data life cycle, as depicted in Fig. 3. Research data management are combined measures of both systems and organization and they have to take effect in all the steps if research data is involved. When data is generated, it has to be enriched with metadata and put to the data management system along with the metadata and a PID. Data with according metadata gained for the evaluation has to be archived as well. Only with data and metadata, the results can be re-evaluated, or checked after the project end. A well-integrated user-interface must enable Open Access to the data. A DMP has to define all the processes, data movements and timelines that occur within the data management process and the SDO has to take responsibilities and coordinate all actions. The SDO also has to take stewardship of all orphaned data. The data life-cycle does not end when the project is finished: The data and metadata remains active in the research data management system for reuse or Open Access.

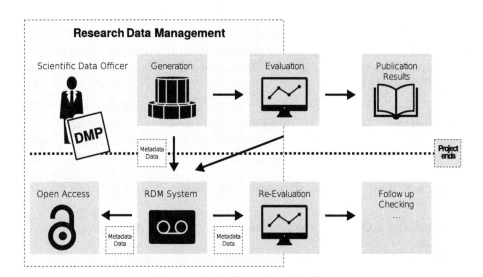

Fig. 3. Ideal data life cycle: Data management affects the generation, evaluation and archiving steps in an HPC ecosystem. Data management is not only a technical, but also a organizational task and is led by the Scientific Data Officer (SDO). A Data Management Plan (DMP) defines all processes and timelines.

3.7   Related Work

Archiving. The Open Archive Information System [21] offers a reference model for archive systems which defines the interaction between humans and machines and moreover proposes a metadata model. As a specification, OAIS is focused on conceptual work and not implementation. The model comprehends three roles:

The producer of data, the consumer and the management. These three parties have to negotiate on preservation planning, administration and data management. Procedures for data ingest and access have to be defined and archival storage must be planned as a reliable, long-term storage. As a framework, OAIS proposes concepts and it is always up to the specific use case in which the archive is used. Emerging from the field of record preservation in the cloud, Askhoj et al. [2] propose to map OAIS to a layered architecture and bring the archive to the cloud in order to combine a well-established concept for archiving with the flexibility and scalability of cloud systems. A PaaS-layer, handling objects as binary strings and ensuring bitstream preservation. The SaaS layer handles digital objects that emerge as objects when they are packaged at the next layer, the packaging layer. A fourth layer called Archives and Records Management layer incorporates all management capabilities and a user interface. In contrast to the OAIS specification, Askhoj et al. argue that it is more beneficial to split data and metadata. They introduce persistent identifiers to reference the data. iCurate is a data management system that is not based on OAIS but has a strong focus on the management layer and metadata [17]: Annotation, retrieval, and validation of metadata should be possible. The system is adapted to the HPC workflow, that means users can specify already in the Portable Batch System (PBS) file some metadata which is added to the output files when the job is complete. The iCurate system can then also harvest the output files for technical metadata to be automatically added. The main contribution of iCurate is the automating of workflows for the annotation of metadata.

Metadata. It is specific to OAIS, that data and metadata is bonded together to a data object. According to the OAIS reference model, there are three major metadata categories. *Content Information* or structural metadata refers to the data object itself with associated, necessary representation information, such as information on the format of the data. Without this representation information, data would not be machine-readable any more and hence become worthless. *Preservation Description Information* or administrative metadata consists of five subcategories. First, Reference Information in general is a unique identifier to locate the data. Second, Provenance Information holds information about the origin of data as well as the history of changes. Third, Context Information describes the relations between digital objects. Fourth, Fixity Information protects the content from unauthorized alteration and may be realized by a checksum. Lastly, Access Rights are access policies. *Descriptive Information* can consist of a whole set of attributes describing properties of the object that emerge to a higher level description of the data. Other existing schemes like DataCite [4] introduce general elements such as *identifier* or *format* to tag and identify data and can be used as a framework to add more specific metadata attributes. The Climate and Environmental Retrieval and Archive system (CERA) data model is a domain-specific metadata model developed for earth sciences at the DKRZ [16].

4 Conclusion

The paper presented the challenges of research data management for HPC. In HPC, simulation data is big in volume and variety. Data reproducibility to diminish the problem of volume is not given for HPC since machines and compilers are renewed every 3 to 5 years. Data management becomes a burden since current solutions disrespect HPC characteristics such as having millions to billions of huge files. Moreover, organizational challenges such as Open Access policies get critical in HPC: Besides legal issues for example, publishing Petabytes of data is not a trivial task. Nowadays, the data life cycle produces a lot of dark data becoming worthless.

Derived from these challenges, requirements for a feasible research data management have been outlined: On the side of technical management, those were metadata, persistent identifiers, data security and workflow integration. To cope with Open Access, research data management has to incorporate the idea and include an appropriate user interface. Scalability requirements aim for providing technologies that can deal with the huge data volume and variety of files. This can be accomplished by a three-layered architecture, separating data storage and metadata storage due to performance considerations. Since research data management is not only a technical task, organizational requirements have been defined, such as qualification and planning documents. Moreover, the role of the SDO has been introduced as one contribution of this paper: Only with a skilled person taking stewardship of research data activities in an institution, research data management can be successful in the interplay of human and machine. In the end, the requirements lead to a data life cycle for HPC where research data management affects all the stages where data is involved and does not end when the project is over.

Acknowledgments. We would like to thank *Wanda Spahn* for proofreading.

References

1. Arora, R.: Data management: state-of-the-practice at open-science data centers. In: Khan, S.U., Zomaya, A.Y. (eds.) Handbook on Data Centers, pp. 1095–1108. Springer, New York (2015). doi:10.1007/978-1-4939-2092-1_37
2. Askhoj, J., Sugimoto, S., Nagamori, M.: Preserving records in the cloud. Rec. Manage. J. **21**(3), 175–187 (2011). https://doi.org/10.1108/09565691111186858
3. Cox, A.M., Pinfield, S.: Research data management and libraries: current activities and future priorities. J. Librarian. Inf. Sci. **46**(4), 299–316 (2014). http://dx.doi.org/10.1177/0961000613492542
4. DataCite: (2016). http://schema.datacite.org/. Accessed 6 Dec 2016
5. DFG: Safeguarding good scientific practice (2013). http://www.dfg.de/download/pdf/dfg_im_profil/reden_stellungnahmen/download/empfehlung_wiss_praxis_1310.pdf. Accessed 6 Dec 2016
6. EU: H2020 programme guidelines on FAIR data management in Horizon 2020 (2016). http://ec.europa.eu/research/participants/data/ref/h2020/grants_manual/hi/oa_pilot/h2020-hi-oa-data-mgt_en.pdf. Accessed 6 Dec 2016

7. EU: European Cloud Initiative - Building a competitive data and knowledge economy in Europe (2016). http://ec.europa.eu/newsroom/dae/document.cfm? doc_id=15266. Accessed 6 Dec 2016

8. Faulhaber, P.: Investing in the future of tape technology. Presentation, HPSS User Forum, New York City (2015)

9. Gray, J., Liu, D.T., Nieto-Santisteban, M., Szalay, A., DeWitt, D.J., Heber, G.: Scientific data management in the coming decade. SIGMOD Rec. **34**(4), 34–41 (2005). http://doi.acm.org/10.1145/1107499.1107503

10. Heidorn, P.B.: Shedding light on the dark data in the long tail of science. Libr. Trends **57**(2), 280–299 (2008). http://doi.org/10.1353/lib.0.0036

11. Helly, J., Staudigel, H., Koppers, A.: Scalable models of data sharing in earth sciences. Geochem. Geophy. Geosyst. **4**(1) (2003). http://dx.doi.org/10.1029/2002GC000318

12. Hick, J.: HPSS in the Extreme Scale Era: Report to DOE Office of Science on HPSS in 2018–2022. Lawrence Berkeley National Laboratory (2010)

13. Hick, J.: The Fifth Workshop on HPC best practices: File systems and archives. Lawrence Berkeley National Laboratory. LBNL Paper LBNL-5262E (2013)

14. Jensen, U.: Datenmanagementpläne. In: Büttner, S., Hobohm, H.-C., Müller, L. (eds.) Handbuch Forschungsdatenmanagement. Bad Honnef: Bock u. Herchen (2011)

15. Jones, S.N., Strong, C.R., Parker-Wood, A., Holloway, A., Long, D.D.E.: Easing the burdens of HPC file management. In: Proceedings of the Sixth Workshop on Parallel Data Storage, PDSW 2011, NY, USA, pp. 25–30 (2011). http://doi.acm.org/10.1145/2159352.2159359

16. Lautenschlager, M., Toussaint, F., Thiemann, H., Reinke, M.: The CERA-2 data model (1998). https://www.pik-potsdam.de/cera/Descriptions/Publications/Papers/9807_DKRZ_TechRep.15/cera2.pdf

17. Liang, S., Holmes, V., Antoniou, G., Higgins, J.: iCurate: a research data management system. In: Bikakis, A., Zheng, X. (eds.) MIWAI 2015. LNCS, vol. 9426, pp. 39–47. Springer, Cham (2015). doi:10.1007/978-3-319-26181-2_4

18. Malik, T.: Geobase: indexing NetCDF files for large-scale data analysis. In: Big Data Management, Technologies, and Applications, pp. 295–313. IGI Global (2014). http://doi.org/10.4018/978-1-4666-4699-5.ch012

19. Mattmann, C.A.: Computing: a vision for data science. Nature **493**(7433), 473–475 (2013). http://dx.doi.org/10.1038/493473a

20. NSF: Grant proposal guide chapter ii.c.2.j (2014). https://www.nsf.gov/pubs/policydocs/pappguide/nsf15001/gpg_2.jsp#dmp. Accessed 6 Dec 2016

21. OAIS: Reference model for an Open Archival Information System. Technical report, CCSDS 650.0-M-2 (Magenta Book) Issue 2 (2012)

22. Parker-Wood, A., Long, D.D.E., Madden, B.A., Adams, I.F., McThrow, M., Wildani, A.: Examining extended and scientific metadata for scalable index designs. In: Proceedings of the 6th International Systems and Storage Conference, SYSTOR 2013, NY, USA, pp. 4:1–4:6 (2013). http://doi.acm.org/10.1145/2485732.2485754

23. Potthoff, J., van Wezel, J., Razum, M., Walk, M.: Anforderungen eines nachhaltigen, disziplinübergreifenden Forschungsdaten-Repositoriums. In: DFN-Forum Kommunikationstechnologien, pp. 11–20 (2014)

Quality in Digital Libraries

How Linked Data can Aid Machine Learning-Based Tasks

Michalis Mountantonakis[1,2($\boxtimes$)] and Yannis Tzitzikas[1,2]

[1] Institute of Computer Science, FORTH-ICS, Heraklion, Greece
{mountant,tzitzik}@ics.forth.gr
[2] Computer Science Department, University of Crete, Heraklion, Greece

Abstract. The discovery of useful data for a given problem is of primary importance since data scientists usually spend a lot of time for discovering, collecting and preparing data before using them for various reasons, e.g., for applying or testing machine learning algorithms. In this paper we propose a general method for discovering, creating and selecting, in an easy way, valuable features describing a set of entities for leveraging them in a machine learning context. We demonstrate the feasibility of this approach by introducing a tool (research prototype), called LODsyndesis$_{\mathcal{ML}}$, which is based on Linked Data technologies, that (a) discovers automatically datasets where the entities of interest occur, (b) shows to the user a big number of useful features for these entities, and (c) creates automatically the selected features by sending SPARQL queries. We evaluate this approach by exploiting data from several sources, including *British National Library*, for creating datasets in order to predict whether a book or a movie is popular or non-popular. Our evaluation contains a 5-fold cross validation and we introduce comparative results for a number of different features and models. The evaluation showed that the additional features did improve the accuracy of prediction.

Keywords: Linked Data · Machine Learning · Feature Discovery & Selection · Automatic classification · Prediction

1 Introduction

It has been written that "Data scientists spend 50%–80% of their time in collecting and preparing unruly digital data, before it can be explored for useful nuggets"[1], thereby, it is beneficial to investigate novel methods for reducing the aforementioned cost. The objective of this paper is to propose a method, that is based on Linked Data, for discovering, creating and selecting, in an easy way, valuable features describing a set of entities for being used in any *Machine Learning (ML)* problem.

---

[1] http://www.nytimes.com/2014/08/18/technology/for-big-data-scientists-hurdle-to-insights-is-janitor-work.html.

© Springer International Publishing AG 2017
J. Kamps et al. (Eds.): TPDL 2017, LNCS 10450, pp. 155–168, 2017.
DOI: 10.1007/978-3-319-67008-9_13

Linked Data [4] refers to a method of publishing structured data while its ultimate objective is linking and integration. It is based on Semantic Web technologies, such as HTTP, URI and RDF, which enables the information to be read automatically by computers and data from different sources to be connected and queried. It differs from other traditional data formats predominantly due to the following reasons: Firstly, data linking facilitates the discovery of datasets containing information about a specific entity (or a set of entities). Secondly, datasets can be integrated more easily through the existence of common entities and common schema elements, which is desirable for exploiting the complementarity of information. For instance, one dataset can contain information about the authors of a book and another about user reviews for that book, thereby, the integration of such datasets offer more features for the entities. Thirdly, complex features can be derived by exploiting SPARQL [17] queries (e.g., "number of awards for each book") and graph metrics (e.g., average degree of an entity).

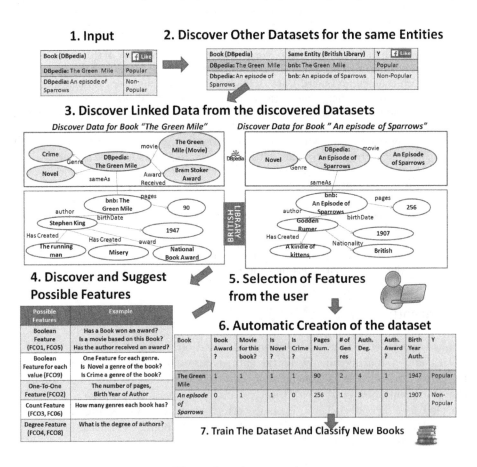

Fig. 1. Running example

A lot of datasets are published in RDF format, i.e., LODStats [6] provides statistics about approximately ten thousand discovered linked datasets.

In this work we show how the wealth of Linked Data and the ML machinery can be jointly exploited for improving the quality of automated methods for various time consuming and/or tedious tasks, which are important also in the area of digital libraries, like automatic semantic annotation or classification, completion of missing values, clustering, or computing recommendations. Specifically, we focus on exploiting Linked Data for discovering and creating features for a set of entities. We introduce a process where (i) we discover datasets and URIs containing information for a set of entities by exploiting *LODsyndesis* [12], (ii) we provide the user with a large number of possible features that can be created for these entities (including features for direct and indirect related entities of any path) and (iii) we produce automatically a dataset for the features selected by the user. For testing whether this enriched dataset can improve ML tasks, we report experimental results over two datasets (from [19]) for predicting the popularity of a set of movies and books. Figure 1 illustrates the running example, where we create features for classifying whether a book is *Popular* or *Non-Popular*, containing data discovered from *DBpedia* [10] and *British National Library* [16]. We evaluate this approach by performing a 5-fold cross validation for estimating the performance of different models for the produced datasets. The evaluation showed that the additional features did improve the accuracy of prediction.

The rest of this paper is organized as follows: Sect. 2 discusses background and related approaches, Sect. 3 states the problem and describes the functionality of the proposed tool (research prototype), Sect. 4 discusses the steps of the process, Sect. 5 reports the results of the evaluation and discusses the effectiveness of the proposed features, and finally, Sect. 6 concludes the paper.

2   Background and Related Work

Background. The Resource Description Framework (RDF) [2] is a graph-based data model. RDF uses `Triples` in order to relate Uniform Resource Identifiers (URIs) or anonymous resources (`blank nodes`) where both of them denote a `Resource`, with other `URIs`, `blank nodes` or constants (`Literals`). Let $\mathcal{U}$ be the set of all URIs, $\mathcal{B}$ the set of all `blank nodes`, and $\mathcal{L}$ the set of all Literals. In Linked Data each statement (or triple) is of the form subject-predicate-object where a subject corresponds to an entity (e.g., a book, a person, etc.), a predicate (or property) to a characteristic of an entity (e.g., genre of a book) and an object to the value of the predicate for a specific subject, e.g., in the following triple ⟨*The Green Mile, hasAuthor, Stephen King*⟩, *The Green Mile* is the subject, *hasAuthor* the predicate and *Stephen King* the object. Let S be the set of all `subjects`, P the set of all `properties`, and O the set of all `objects`. Formally, a *triple* is any element of $\mathcal{T} = S \times P \times O$, where $S = \mathcal{U} \cup \mathcal{B}$, $P = \mathcal{U}$ and $O = \mathcal{U} \cup \mathcal{L} \cup \mathcal{B}$, while an *RDF graph* (or dataset) is any finite subset of $\mathcal{T}$. The linking of datasets is realized by the existence of common URIs, referring to schema elements (defined through RDF Schema and OWL [2]), instances, as well as by equivalence relationships expressed via the `owl:sameAs` predicate.

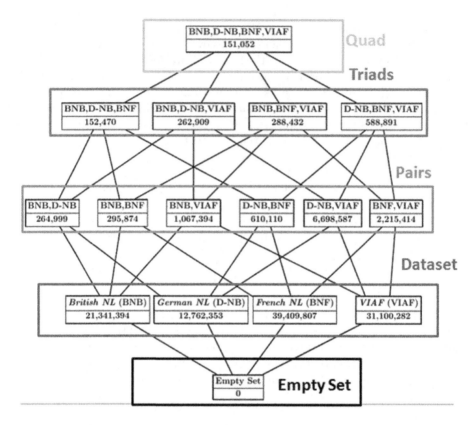

Fig. 2. Lattice of four digital library datasets (common real world objects)

LODsyndesis provides query services and measurements that are useful for several important tasks like (a) object co-reference, (b) dataset discovery, (c) visualization, and (d) connectivity assessment and monitoring [12]. Its public website also provides measurements that concern the commonalities of Linked Datasets, i.e., it provides the number of common real world objects between any set of datasets, that is the number of classes of equivalence of URIs after having computed the symmetric and transitive closure of the set of owl:sameAs relationships from all datasets. Such measurements can be visualized as lattices and Fig. 2 shows a lattice for four digital libraries datasets (i.e., *British National Library, German National Library, French National Library* and *VIAF*). It is evident that these four datasets share 151,052 real world objects. All these equivalent URIs (e.g., among these four datasets) can be found by using the object co-reference service offered by *LODsyndesis*. For instance, one can find all the equivalent URIs among any set of sources, e.g., give me all the equivalent URIs among *British, German* and *French National Libraries*, or all the equivalent URIs for Jules Verne (e.g. http://bnb.data.bl.uk/id/person/VerneJules1828-1905).

Related Work. There are several proposals for using Linked Data for generating features. LiDDM [14] is a tool that retrieves data from Linked Data cloud by sending queries. For finding possible features the users can either construct their own queries or use an automatic SPARQL query builder that shows to the users all the possible predicates that can be used (from a specific SPARQL endpoint). It offers also operators for integrating and filtering data from two or more sources. The authors in [5] presented a modular framework for constructing semantic features from Linked Data, where the user specifies the SPARQL queries that should be used for generating the features. Another work that uses SPARQL queries is described in [13], where the user can submit queries which are combined with SPARQL aggregates (e.g., count). Comparing to our approach, the previous tools presuppose that the user is familiar with SPARQL, and they do not assist the user in discovering automatically datasets containing information for the same entities. The closest tool to our approach is FeGeLOD [15] which combines data from several datasets by traversing `owl:sameAs` paths and generates automatically six different categories of features. *RapidMiner Semantic Web Extension* tool [18] (which is the extension of FeGeLOD) supports the same features while it integrates the data that are derived from multiple sources. Instead we show the provenance of the data without integrating them, i.e., if a feature is provided by two or more sources, the user can decide which source to select for creating this feature. Moreover, we also discover datasets containing the same entities by exploiting *LODsyndesis* [12], where the class of equivalence for each entity has already been pre-computed for more than 300 datasets, whereas the aforementioned tool finds relevant data by traversing links on-the-fly. Finally, we also provide other kinds of features, such as degree of an entity, boolean features for each value of a predicate, as well as features for "sub-entities", i.e., entities correlated with the entities that one wants to classify (e.g., actors of a movie).

3 Linked Data-Based Feature Creation Operators

Let E be the set of entities for which we want to generate features. Below we will show how we can derive a set of features $(f_1, ..., f_k)$ where each f_i is a feature and $f_i(e)$ denotes the value of that feature for an entity $e \in E$. Each $f_i(e)$ is actually derived by the data that are related to e. Specifically we have identified the following nine (9) frequently occurring *Linked Data-based Feature Creation Operators*, for short *FCOs*. In their definition, shown in Table 1, P denotes the set of properties, p, p_1 and p_2 are properties and hereafter $\mathcal{T}$ denotes the triples for the entities that are indexed by *LODsyndesis*.

In our running example of Fig. 1, FCO1 can be used for representing whether a book has been nominated for winning an award or not. FCO2 suits to properties that are functional (one-to-one), e.g. person's birth country, number of pages of the book, and its value can be numerical or categorical. FCO3 counts the values of a property, e.g. the number of genres of a book. FCO4 measures the number of distinct triples that involve e, in our running example the degree of the author of "The Green Mile" book is 3, while the degree of the author of "An episode of

Table 1. Feature creation operators

id	Operator defining f_i	Type	$f_i(e)$
1	p.exists	boolean	$f_i(e) = 1$ if (e, p, o) or $(o, p, e) \in \mathcal{T}$, otherwise $f_i(e) = 0$
2	p.value	num/categ	$f_i(e) = \{\ v\ \mid\ (e, p, v) \in \mathcal{T}\}$
3	p.valuesCard	int	$f_i(e) = \lvert\{\ v\ \mid\ (e, p, v) \in \mathcal{T}\}\rvert$
4	degree	double	$f_i(e) = \lvert\{(s, p, o) \in \mathcal{T}\ \mid\ s = e\ \text{ or }\ o = e\}\rvert$
5	p1.p2.exists	boolean	$f_i(e) = 1$ if $\exists\ o2$ s.t. $\{(e, p1, o1), (o1, p2, o2)\} \subseteq \mathcal{T}$
6	p1.p2.count	int	$f_i(e) = \lvert\{\ o2\ \mid\ (e, p1, o1), (o1, p2, o2) \in \mathcal{T}\}\rvert$
7	p1.p2.value.maxFreq	num/categ	$f_i(e) = $ most frequent $o2$ in $\{\ o2\ \mid\ (e, p1, o1), (o1, p2, o2) \in \mathcal{T}\}$
8	average degree	double	$f_i(e) = \dfrac{\lvert triples(C)\rvert}{\lvert C\rvert}$ s.t. $C = \{\ c\ \mid\ (e, p, c) \in \mathcal{T}\}$ and $triples(C) = \{(s, p, o) \in \mathcal{T}\ \mid\ s \in C\text{ or }o \in C\}$
9	p.values.AsFeatures	boolean	for each $v \in \{\ v\ \mid\ (e, p, v) \in \mathcal{T}\}$ we get the feature $f_{iv}(e) = 1$ if (e, p, v) or $(v, p, e) \in \mathcal{T}$, otherwise $f_{iv}(e) = 0$

Sparrows" is 2. FCO5-FCO9 correspond to features related to "sub-entities" or "related" entities to e. Specifically, FCO5 corresponds to one characteristic of a "sub(related)-entity" of e, e.g. whether at least one actor of a movie has won an award in the past or not. FCO6 counts the distinct values of one characteristic of the "sub-entities", e.g. the total number of movies where the actors of a movie have played. FCO7 finds the most frequently occurring characteristic of these entities, e.g. the country where most of the actors of a movie were born. FCO8 measures the average number of distinct triples for a set of "sub-entities", e.g., the average number of triples for the actors of a movie. The last one, FCO9, does not create one feature but a set of features, e.g. one boolean feature for each genre that a book can possibly belong to. In our running example, we take all genres of both books and for each genre (e.g., novel) we create a distinct boolean feature (both books belong to the genre Novel, but only "The Green Mile" book belongs to the genre Crime). Generally, the operators FCO1-FCO4 and FCO9 concern a single entity (e.g., a book, a person, a country, etc.) while operators FCO5-FCO8 a set of entities (e.g., all actors of a movie). Consequently, for the "sub-entities" that are connected through a functional property (one-to-one) with the entities that we want to classify, operators FCO1-FCO4 and FCO9 are used instead of operators FCO5-FCO8. The user can explore direct or indirect "sub-entities", e.g., authors of a book, countries of authors of a book and so forth, for any formulated path, while the list of operators can be easily extended by adding more operators.

Additional Functionality of LODsyndesis$_{\mathcal{ML}}$.

Here we introduce some useful (for the user) metadata and restrictions for feature selection.

"Completeness" of a Property for a given set of entities. We compute the percentage of instances for which a given property exists, e.g., the percentage of books for which we have information about the number of their pages. If E'_p is the set of entities being subject or object of triples with predicate p, i.e. $E'_p = \{e \in E \mid (e\ p\ o) \text{ or } (o\ p\ e) \in \mathcal{T}\}$, then the percentage is given by $\frac{|E'_p|}{|E|}$.

Multiplicity and Range of a Property. Here we find the multiplicity of a specific property, i.e., whether it is a one-to-one or one-to-many relation. We define the set of one-to-one properties as $P_{1-1} = \{p \mid (e\ p\ o_i) \in \mathcal{T} \text{ and } \nexists\ (e\ p\ o_{ii}) \in \mathcal{T}, o_i \neq o_{ii}, \forall\ e \in E\}$. The rest properties, i.e., one-to-many, are defined as $P_{Many} = P \setminus P_{1-1}$, while we denote as $range(p) \in \{String, Numeric, \mathcal{U}\}$ a property's range, i.e., whether it is a set of Strings, Numeric Values or URIs.

Restrictions derived from metadata. Table 2 shows the restrictions which are derived by taking into account the "completeness", the multiplicity and the range of a property. It is worth mentioning that the "completeness" of a property can also be exploited for discovering missing values for the entities. In addition, the users can define their own restrictions, e.g., they can exclude properties that belong to popular ontologies such as *rdf*, *rdfs*, *foaf* and *owl*.

Table 2. Restrictions of features with respect to the characteristics of a property

Feature operators	Can be applied for
Boolean (FCO1, FCO5)	All properties having $\frac{|E'_p|}{|E|} < 1$
Boolean for each Value (FCO9)	All properties $p \in P_{Many}$, $range(p) \neq Numeric$
One-to-one Relationship (FCO2)	All properties $p \in P_{1-1}$
Count (FCO3, FCO6)	All properties $p \in P_{Many}$
Degree (FCO4, FCO8)	All properties having $range(p) = \mathcal{U}$

4   The Steps of the Proposed Approach

Here we describe the tool (research prototype) LODsyndesis$_{\mathcal{ML}}$ that we have designed and implemented. It is worth noting that LODsyndesis$_{\mathcal{ML}}$ discovers and creates features by exploiting Linked Data for any domain. Even a user that is not familiar with Semantic Web technologies and SPARQL can use it for creating features for feeding a Machine Learning problem. The process is shown in Fig. 3 and is described in brief below. First, it takes as input a file containing a set of URIs that refer to particular entities, i.e., movies, books and so forth. In case of knowing the entities but not their URIs, one can exploit an entity identification tool like DBpedia Spotlight [11] and XLink [7] for detecting automatically a URI for a specific entity. Then, it connects to *LODsyndesis* for discovering automatically datasets containing information for the same entities and shows to the user the available datasets. Afterwards, it discovers and shows

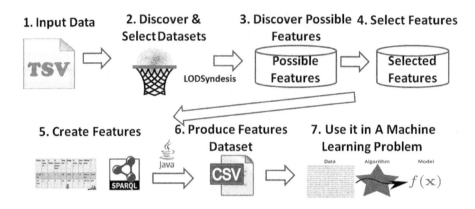

Fig. 3. Process of LODsyndesis$_{\mathcal{ML}}$

to the user possible features that can characterize the entities (or related "sub-entities") of the dataset and the user selects which features to create. The next step is to create the features and to produce the output dataset to be used in any ML problem. Below, we describe in more detail the whole process, while additional information and a demo can be found in http://www.ics.forth.gr/isl/LODsyndesis.

1. Input: The input of LODsyndesis$_{\mathcal{ML}}$ is a file in tab separated value (tsv) format containing URIs describing entities and possibly their class, e.g., URIs for a book and if each book is Popular or Non-Popular.

2. Discover Data by using LODsyndesis: LODsyndesis$_{\mathcal{ML}}$ reads the tsv file and connects to *LODsyndesis* [12] in order to discover (a) datasets containing information for the same entities and (b) the URIs for these entities for each dataset (the indexes of LODsyndesis have already pre-computed the closure of *owl:sameAs* relationships for 300 datasets). Then, the user selects the desired datasets. Concerning the running example of Fig. 1, we observe that we found two different datasets containing information for the books of that example .

3. Discover Possible Features: LODsyndesis$_{\mathcal{ML}}$ sends SPARQL [17] queries for a sample of the aforementioned entities to the SPARQL endpoints of the selected datasets. Afterwards, a number of possible features and their provenance are discovered and returned to the user. Therefore, in this step we do not create any feature, we just discover possible features and we apply the restrictions described in Sect. 3. The result is a table where each row corresponds to a possible feature derived from a specific source while each column consists of a checkbox for a specific feature category. The order that the features appear in the rows is descending with respect to the "completeness" of each property. Particularly, when a property occurs for all the entities, it is placed first in the list, while those with the smallest number of occurrences are placed at the end of the list. Moreover, the user can view the metadata described in Sect. 3. Afterwards, the user can select the desired features (by taking into account their provenance) and can also explore features for (direct or indirect) "sub-entities" of any formulated path and create more features.

4. Feature Selection and 5. Feature Creation: The user selects the desired features and clicks on a button for initiating the dataset creation. Then, the tool sends SPARQL queries for creating the features. For each feature operators category, it sends $|E|$ in number SPARQL queries (one query per entity e for each operator). It is worth noting that for values that are neither numeric nor boolean, it performs a mapping for converting them to numeric. Concerning missing values, we just put a unique constant value. However, for improving datasets' quality, several transformations could be applied after this step, like those proposed in [3] for removing erroneous and inconsistent data or filling missing values. In this paper we do not focus on this task and the data used in the experiments have not been transformed or cleaned by using such techniques.

6. Production of Features' Dataset and 7. Exploitation of the Produced Dataset in a ML problem: The user is informed that the process is completed and that two csv files have been produced: one for the categorical and one for the continuous features. Then, the produced datasets can be given as an input for a ML problem (e.g., classification of books).

5   Evaluation

The datasets, which are used in our experiments (derived from [19]), contain the URIs of movies and books from *DBpedia* [10] and the corresponding classification value, i.e., *Popular* or *Non-Popular* according to the number of Facebook users' likes. We use 1,570 entities for Movies Dataset and 1,076 entities for Books Dataset. The initial datasets are loaded and then more data are discovered by using LODsyndesis$_{ML}$ from the following sources: *British National Library* [16], *Wikidata* [20] and *DBpedia*[10]. In particular, we exploit LODsyndesis$_{ML}$ for discovering, selecting and creating a number of different features for predicting the class of these entities. Afterwards, *MATLAB* [1] is used for performing (a) a 5-fold cross validation for model selection and (b) a comparison of a number of different models for measuring accuracy, which is defined as: $accuracy = \frac{True\ Positive+True\ Negative}{True\ Positive+True\ Negative+False\ Positive+False\ Negative}$ [21]. For each dataset, we repeat the 5-fold cross validation process 15 times for different sizes of the test set, i.e. 10%, 20%, & 30%. Each time a chi-square test of independence [23] is performed (for excluding variables that are independent of the class variable) for 4 different values of significance level (or threshold) a: 0.01, 0.05, 0.1, 1. For each value of threshold a we test 10 different models: (a) 2 *Naive Bayes* models (Empirical & Uniform), (b) 3 *Random Forest* models with 50 trees and different min leaf sizes: 1, 3 & 5, (c) 3 *K-Nearest Neighbours* models with K: 3, 5 &15, (d) a *linear SVM model* and (e) the *trivial* model. In each iteration the best model is obtained for the training set (by using cross validation). Finally, the accuracy of the best model is estimated on the test set.

Creation of Features. In Fig. 4 we can observe how the number of possible features increases when (a) more datasets are added and (b) features of

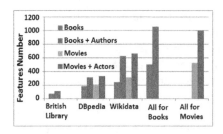

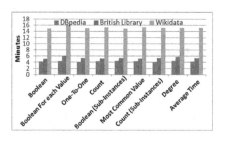

Fig. 4. Features number per dataset for books & movies

Fig. 5. Generation time for each feature operators category and dataset

"sub-entities" are created, i.e., approximately the possible features are doubled when we explore a "sub-entity" (e.g. the authors of a book). Moreover, Fig. 5 shows the time for generating a feature for each different category (and each dataset) for 1,076 books. As we can see, the generation time depends highly on the dataset to which we send SPARQL queries, e.g., DBpedia's response time is much shorter than Wikidata's. Concerning the generation time of a specific feature operators category, the degree operators (FCO4, FCO8) and the boolean for each value of a predicate (FCO9) need more time to be generated while the remaining ones need approximately the same time on average for being generated. Finally, the execution time for retrieving the similar entities from *LODsyndesis* was 105 s.

Results for Movies Dataset. Figure 6 shows the selected features (and their category) for the dataset of movies (features belonging in the additional categories that we propose in this paper are underlined). In total we sent 39,250 queries and we created 159 features (147 categorical and 12 continuous). In Fig. 7 we can see a plot with the accuracy of each test size (using the best model selected by the cross validation process) and we can observe that the accuracy is much higher comparing to a trivial case while the highest variation occurred for test size equal to 0.1. In all iterations, the best model was a *Random Forest*

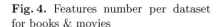

Fig. 6. Selected features for books & movies with their provenance

Table 3. Accuracy for each feature operators category (Movies & Books test size 0.2)

Feature Operators Category	Average (Movies)	Max (Movies)	Average (Books)	Max (Books)
All Features (FCO1-FCO9)	0.871	0.906	0.730	0.762
Continuous Features	0.861	0.896	0.709	0.739
New Features (FCO4-FCO9)	0.835	0.865	0.650	0.675
Existing Features (FCO1-FCO3)	0.827	0.855	0.694	0.716
Count (FCO3,FCO6)	0.830	0.862	0.706	0.709
1-1 Relationship (FCO2)	0.791	0.808	0.570	0.607
Categorical Features	0.760	0.818	0.673	0.694
Boolean (FCO1, FCO5, FCO9)	0.750	0.774	0.634	0.656
Degree (FCO4,FCO8)	0.741	0.780	0.608	0.627
Most Frequent Value (FCO7)	0.698	0.758	0.560	0.595
Trivial Case	0.495	0.532	0.508	0.551

(with different parameters in many cases). Figure 8 shows the average accuracy for each model (and each threshold a) in the cross validation process for test size 0.2. We observe that *Random Forest* models achieved higher accuracy (mainly when min leaf size equals 1, i.e., Random Forest 1) comparing to the other models. The next ones with the highest accuracy is the *linear SVM*, followed by the two *Naive Bayes* models and finally the *K-NN* ones. However, all these models are better comparing to the trivial one whose accuracy is approximately 0.5. Table 3 shows the average and maximum accuracy for each different features' category in descending order with respect to their average accuracy. The continuous ones (mainly the count features, i.e., FCO3 and FCO6) seem to be the most predictive while all the categories achieved high accuracy comparing to a trivial case. Moreover, the average accuracy of features that other approaches also support (i.e., FCO1-FCO3) was 0.827 while for the additional features that we

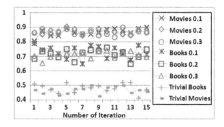

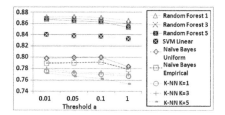

Fig. 7. Accuracy in each iteration & Test size for dataset books and movies

Fig. 8. Average accuracy of models in cross validation for movies with test size 0.2

propose (e, FCO4-FCO9 in Table 3) the average accuracy was 0.835. By combining all the categories of features, the average accuracy was 0.871, which means that the additional features improved the accuracy in this particular problem.

Results for Books Dataset. Figure 6 shows the selected features (and their categories) for the books dataset. In total we sent 21,520 queries and we created 190 features (180 categorical and 10 continuous). In Fig. 7 we can see a plot with the accuracy of each test size and we observe that the accuracy is much higher comparing to the trivial case while the highest variation occurred for test size equal to 0.1. In 42 iterations, the best model was a *Random Forest* (with different parameters in many cases) while in 3 cases the best model was a *linear SVM* while the variations of *K-NN* algorithms were more effective than the *Naive Bayes* ones. As we can observe in Table 3, the combination of all features gave the maximum accuracy, while the continuous features, and especially the count features (FCO3, FCO6), were more predictive comparing to the remaining ones. Moreover, for the feature operators FCO1-FCO3 the average accuracy was 0.694 while for the feature operators FCO4-FCO9 was 0.65. However, by combining both types of features the average accuracy improved, i.e., 0.73, therefore the additional features improved the accuracy for books' dataset, too.

6 Concluding Remarks

We have shown how we can exploit the wealth of Linked Data and the ML machinery for improving the quality of automatic classification. We presented a tool, called LODsyndesis$_{\mathcal{ML}}$, which exploits Linked Data (and the related technologies) for discovering automatically features for any set of entities. We categorized the features and we detailed the process for producing them. For evaluating the benefits of our approach, we used two datasets and the results showed that the additional features did improve the accuracy of predictions, while the most effective model for both datasets was a *Random Forest* one. As future work, we plan to extend our tool for supporting more operators and transformations for improving the quality of the produced dataset. Furthermore, we plan to evaluate our tool in other tasks, e.g., completion of missing values, to support users' SPARQL endpoints and to connect our tool with SPARQL-LD [8] for incorporating also data stored in files (i.e. not hosted by SPARQL endpoints). Finally, it would be interesting to investigate techniques for automatic feature selection [22], such as those described in [9], where a novel machine learning-based feature selection method is used for predicting candidates features usefulness.

Acknowledgements. This work has received funding from the European Union's Horizon 2020 Research and Innovation programme under the BlueBRIDGE project (Grant agreement No: 675680).

References

1. MATLAB - MathWorks. https://www.mathworks.com/products/matlab.html
2. Antoniou, G., Van Harmelen, F.: A Semantic Web Primer. MIT press, Cambridge (2004)
3. Bischof, S., Martin, C., Polleres, A., Schneider, P.: Collecting, integrating, enriching and republishing open city data as linked data. In: Arenas, M., Corcho, O., Simperl, E., Strohmaier, M., d'Aquin, M., Srinivas, K., Groth, P., Dumontier, M., Heflin, J., Thirunarayan, K., Staab, S. (eds.) ISWC 2015. LNCS, vol. 9367, pp. 57–75. Springer, Cham (2015). doi:10.1007/978-3-319-25010-6_4
4. Bizer, C., Heath, T., Berners-Lee, T.: Linked data-the story so far. Semantic Services, Interoperability, Web Applications: Emerging Concepts, pp. 205–227 (2009)
5. Cheng, W., Kasneci, G., Graepel, T., Stern, D., Herbrich, R.: Automated feature generation from structured knowledge. In: CIKM, pp. 1395–1404. ACM (2011)
6. Ermilov, I., Lehmann, J., Martin, M., Auer, S.: LODStats: the data web census dataset. In: Groth, P., Simperl, E., Gray, A., Sabou, M., Krötzsch, M., Lecue, F., Flöck, F., Gil, Y. (eds.) ISWC 2016. LNCS, vol. 9982, pp. 38–46. Springer, Cham (2016). doi:10.1007/978-3-319-46547-0_5
7. Fafalios, P., Baritakis, M., Tzitzikas, Y.: Configuring named entity extraction through real-time exploitation of linked data. In: WIMS 2014, p. 10. ACM (2014)
8. Fafalios, P., Yannakis, T., Tzitzikas, Y.: Querying the web of data with SPARQL-LD. In: Fuhr, N., Kovács, L., Risse, T., Nejdl, W. (eds.) TPDL 2016. LNCS, vol. 9819, pp. 175–187. Springer, Cham (2016). doi:10.1007/978-3-319-43997-6_14
9. Katz, G., Shin, E.C.R., Song, D.: Explorekit: automatic feature generation and selection. In: ICDM 2016, pp. 979–984. IEEE (2016)
10. Lehmann, J., Isele, R., Jakob, M., et al.: Dbpedia-a large-scale, multilingual knowledge base extracted from wikipedia. Semant. Web **6**(2), 167–195 (2015)
11. Mendes, P.N., Jakob, M., García-Silva, A., Bizer, C.: Dbpedia spotlight: shedding light on the web of documents. In: I-SEMANTICS, pp. 1–8. ACM (2011)
12. Mountantonakis, M., Tzitzikas, Y.: On measuring the lattice of commonalities among several linked datasets. Proc. VLDB Endow. **9**(12), 1101–1112 (2016)
13. Mynarz, J., Svátek, V.: Towards a benchmark for LOD-enhanced knowledge discovery from structured data. In: KNOW@ LOD, pp. 41–48 (2013)
14. Narasimha, V., Kappara, P., Ichise, R., Vyas, O.: Liddm: a data mining system for linked data. In: Workshop on LDOW, vol. 813 (2011)
15. Paulheim, H., Fümkranz, J.: Unsupervised generation of data mining features from linked open data. In: Proceedings of WIMS 2012, p. 31. ACM (2012)
16. Pennock, M., Day, M.: Managing and preserving digital collections at the British library. Managing Digital Cultural Objects: Analysis, discovery and Retrieval, p. 111 (2016)
17. Hommeaux, E.P., Seaborne, A., et al.: Sparql query language for RDF. In: W3C Recommendation, 15 January 2008
18. Ristoski, P., Bizer, C., Paulheim, H.: Mining the web of linked data with rapidminer. Web Semant. Sci. Serv. Agents World Wide Web **35**, 142–151 (2015)
19. Ristoski, P., Vries, G.K.D., Paulheim, H.: A collection of benchmark datasets for systematic evaluations of machine learning on the semantic web. In: Groth, P., Simperl, E., Gray, A., Sabou, M., Krötzsch, M., Lecue, F., Flöck, F., Gil, Y. (eds.) ISWC 2016. LNCS, vol. 9982, pp. 186–194. Springer, Cham (2016). doi:10.1007/978-3-319-46547-0_20

20. Vrandečić, D., Krötzsch, M.: Wikidata: a free collaborative knowledgebase. Commun. ACM **57**(10), 78–85 (2014)
21. Witten, I.H., Frank, E., Hall, M.A., Pal, C.J., Mining, D.: Practical Machine Learning Tools and Techniques. Morgan Kaufmann, San Francisco (2016)
22. Yang, Y., Pedersen, J.O.: A comparative study on feature selection in text categorization. In: ICML, vol. 97, pp. 412–420 (1997)
23. Zibran, M.F.: Chi-squared test of independence. Department of Computer Science, University of Calgary, Alberta, Canada (2007)

Can Plausibility Help to Support High Quality Content in Digital Libraries?

José María González Pinto[(⊠)] and Wolf-Tilo Balke

Institut für Informationssysteme,
Mühlenpfordstrasse 23, 38106 Braunschweig, Germany
{pinto,balke}@ifis.cs.tu-bs.de

Abstract. Presented herein is a novel approach to support high quality content in Digital Libraries by introducing the notion of *Plausibility* of new scientific papers when contrasted with prior knowledge. In particular, our work proposes a novel assessment of scientific papers to support the workload of reviewers. The proposed approach focus on a core component of a scientific paper: its claim. Our methodology exploits state of the art neural embedding representation of text and topic modeling on a Digital Library of scientific papers crawled from PubMed. As a proof of concept of the potential usefulness of the notion of Plausibility, we study and report experiments on documents with claims expressed as statistical associations. This type of claims is very often found in medicine, chemistry, biology, nutrition, etc. where the consumption of a drug, substance, product, etc., has an effect on some other type of entity such as a disease, another drug, substance, etc.

Keywords: Plausibility · Information discovery · Quality assessment

1 Introduction

For years, digital libraries have been a valuable and trustworthy source of information due to the carefully *curated quality* of their content. Since collections are continuously growing with increasing publication numbers, the main challenge to preserve content quality lies in the inclusion of new articles in a collection. Today, peer review is the key to assess new articles and thus help digital libraries preserve high quality content. However, with increasing numbers of publications reviewers are facing the problem of workload scalability: there is less and less time to do this valuable and necessary task. This has also been recognized by the community [1] and while nobody has a perfect solution there are many approaches to at least aid the process, such as expertise profiling, matching submissions with possible reviewers, or resolving paper biddings. In this work, we aim at supporting peer review not at the process level, but with a clear focus on document level. We aim at assessing a new scientific paper's *Plausibility* in the light of prior knowledge represented by some digital library collection. With this novel assessment, the question of how many reviewers a new paper needs can be adjusted by its respective degree of Plausibility: the less plausible it is (i.e. the more its inclusion would hurt the collections consistency), the more reviewers might be needed to come to a clear decision.

© Springer International Publishing AG 2017
J. Kamps et al. (Eds.): TPDL 2017, LNCS 10450, pp. 169–180, 2017.
DOI: 10.1007/978-3-319-67008-9_14

The notion of Plausibility in our work is based on the *knowledge-fit theory* from cognitive sciences [11]. Basically, it states that human plausibility judgements consist of two steps: firstly, a mental representation of current knowledge is built and secondly, an assessment examines how a new piece of information fits all prior knowledge. Of course, this is very hard to decide in general settings. Thus, we will focus our work on a particular type of documents to provide first insights on the general feasibility of the idea: in particular, we focus on documents containing *empirical claims* in the sense of *statistical associations between entities*. Empirical claims thus are given by sentences that express some kind of association between two entities and in what way one affects the other. Actually, our research shows that this simple type of claims can be found in many scientific papers: consider for instance medicine, chemistry, biology, nutrition, etc. where the consumption of a drug, substance, product, etc., has an effect on some other type of entity such as a disease, another drug, substance, etc.

What makes exactly this type of claims so interesting are findings like those reported by nutritional researchers in [2]. Basically, for 50 common basic foods the researchers performed literature searches using PubMed to obtain articles investigating the association between each ingredient and the respective cancer risk. To their surprise, 80% of the ingredients were indeed related to cancer risk. But what was even more surprising: out of 264 single-study assessments 191 (72%) concluded that the tested food was associated with an increased (n = 103) or a decreased (n = 88) risk *at the same time* [2]. What does that say about the concept of Plausibility? How can we account for this type of situations and still provide a consistent instantiation of Plausibility over digital libraries? And how many of this type of claims are there anyway?

As opposed to the first two questions, the last one is easy to answer. To estimate this number, we used a similar linguistic query pattern as in [3]: *(help AND prevent) OR (lower AND risk) OR (increase OR increment AND risk) OR (decrease OR diminish AND risk) OR (factor AND risk) OR (associated AND risk).* Even with this simple filter there are currently almost 1 million articles in PubMed with empirical claims in the form of statistical associations. Figure 1 provides the cumulative number of articles per year. We can even observe a clear increase of the number of articles dealing with empirical claims every year.

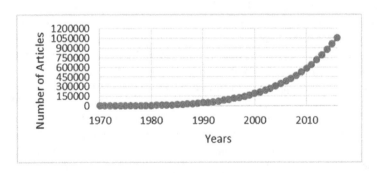

Fig. 1. Accumulated number of articles containing empirical claims in PubMed per year.

To tackle the challenge of the first two questions, in this paper we develop a data driven approach relying on a novel integration of state of the art neural embedding representation of text and generative topic models to operationalize the concept of Plausibility. Our goal is to provide a way to assess the consistency of each new document with respect to the current knowledge (i.e. the state of the art) so that we can answer questions such as: is a new document consistent with current knowledge? Do we have documents in our collection supporting or contradicting a new document? And can we represent our collection in a way such that we can derive a decision to reflect the consistency of new knowledge in the light of current knowledge?

To accomplish this, we first need to operationalize the concept of Plausibility. As a proof of concept, we then implement a new architecture integrating these ideas and providing first insights by analyzing empirical claims. In summary, our contributions are:

1. Firstly, a representation of document collections that combines topic modeling with a neural embedding to exploit two relevant metadata elements: conclusions and abstracts.
2. Secondly, a query facility to find semantically similar claims that may support or contradict a new document's claim.
3. And thirdly, a mechanism to finally assess the total Plausibility of a new document, e.g. to verify its consistency with respect to a collection's representation.

Our paper is organized as follows: in Sect. 2 we provide relevant related work. We then propose a general architecture with the formalization of Plausibility in Sect. 3. In Sect. 4 we present the experimental setting to evaluate our proposed solution with a discussion of our findings. Afterwards, we present concluding remarks and outline future work in Sect. 5.

2   Related Work

Many attempts to model arguments for different purposes exist in literature. Particularly relevant for our work is the body of research dealing with the semantic annotation of claims of scientific articles in the biomedical domain. For instance, in [3], a model is developed for the annotation of scientific hypotheses and claims in natural language using as a case study Alzheimer Disease. Nanopublications [4–6], promoted by the Concept Web Alliance, models core scientific statements with associated context and it is used for data integration across chemical and biological databases [7]. A more detailed model of scientific papers in the biomedical domain is Micropublications [8]. The model specified as an OWL 2 Vocabulary (the ontology language for the Semantic Web[1]) is developed around the idea that scientific claims are defeasible arguments [9, 10]. Thus, they support natural language statements, data, methods, materials specifications, discussion, challenge, and disagreement. In our work, we built on these ideas and represent one core component of scientific papers: claims. Moreover, we attempt to operationalize the notion of Plausibility from Cognitive Sciences. In particular,

---

[1] https://www.w3.org/TR/owl-overview/.

a Plausibility theory that has been empirically proven to be strongly correlated with human judgements [11].

3 Methodology

In this section, we formalize the concept of plausibility and the problem we aim to solve. Plausibility in this work builds on the knowledge-fit theory from Cognitive Sciences. The theory states that human judgements consist of two steps: firstly, a mental representation of prior knowledge that allow us to comprehend and make sense of the world; secondly, assess how new knowledge fits this prior knowledge. Therefore, to operationalize Plausibility we need to formally define (a) how to represent our current knowledge of a Digital Library and (b) how to determine Plausibility of a new document given (a).

Let's revisit the findings of [2] to better explain the rationale behind our proposed methodology. In [2] it was found that for some substances, there were papers that concluded that a given substance was a factor that increases the risk of cancer, while some other papers studying the same substance concluded the opposite. Our proposed decision process that accounts for this type of situations is as follows: if a new document agrees with our current view it is considered "plausible", otherwise "not plausible". Let's explain the difficult situation: if in our Digital Library, we have documents with claims that at the same time contradict as well as support the new claim, we decided to label the case as "controversial". At this point, we can stop and deliver a weighted claim measure based on the semantic similarity of the claims that support versus those that contradict the new claim. However, our hypothesis is that we can still do better: try to identify if the context of the documents in our Digital Library exhibit some characteristics that makes them belong to different groups e.g. "possible worlds". If we can find a possible world where the new document will fit and in this world the claims agree e.g. the world is consistent, then we can proceed to assess the plausibility of the new document as before. Otherwise, if the possible world is inconsistent, then we again have the "controversial" situation.

To operationalize this process, we hypothesized that we need to provide our Plausibility data driven approach with a representation capable of capturing two relevant and related components: (a) the semantics of the relationship of the entities in the collection and (b) how each claim context exhibit certain characteristics that makes it part of a possible world within the collection. Thus, towards this goal, we turn our attention to two aspects to assess the Plausibility of a new article:

1. *Claim of the paper*: the statement(s) of the contribution(s) of the paper. In a curated Digital Library, this is usually found in the conclusion metadata.
2. *Context of the claim of the paper:* the surrounding text of the claim that provides the "explanation" of how the authors of the paper reached the claim(s) stated in the paper.

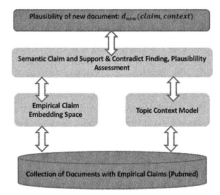

Fig. 2. Architecture of plausibility.

Formally, we can define our Plausibility problem as:

Definition 1. Document Plausibility Problem: Given a document collection with empirical claims $D = \{d_1 \ldots d_n\}$, and a new claim in document d_{new}, we aim at finding how *consistent* is the claim in d_{new} with respect to D.

We approach this problem by breaking it down into three tasks:

1. Representation of the collection of documents D.
2. Finding documents in the representation of D with semantic similar claims that support or contradict a new claim in document d_{new}.
3. Calculate the Plausibility of the claim of d_{new}.

In Fig. 2, we show the components of our proposed methodology that we further explain in this section.

3.1 Representation

Let's begin with how we model a document collection in our work. We consider a collection of documents $D = \{d_1 \ldots d_n\}$, where each document d_i in D is a tuple (*Claim, BagOfContext*).

Where:

- *Claim* is a sentence that represents an empirical claim. In other words, a sentence that contains an association between two entities and in particular, how one of them affects the other. In this work, we use the conclusion metadata of each paper to find such sentences.
- *BagOfContext* is a vector space model representation of the context of the claim. The context in this work is the abstract of the paper.

Empirical Claim Embedding Space. Because of the relevance in our work of the empirical claims, we use Neural Network language models to compute an embedding representation of them [12–15]. Embedding language models have shown interesting semantic properties to be able to find related concepts, related paragraphs, analogies, etc. [14, 16]. In this work, we rely on such representations to capture claim specific semantics. Moreover, we use this representation to find not only semantically similar claims but also to distinguish between claims that express supporting or contradicting positions with respect to the claim of the document we want to assess. In our experiments, we use the embedding implementation of [14]. It is relevant to mention that we decided to use the embedding space because it benefits our approach to find highly related claims. The idea is that entities used in similar contexts with respect to the effect on another entity are related and might help in the absence of explicit knowledge. In our experiments, we first train the word embedding in the entire collection of documents over the abstracts. Then, every claim is represented as a weighted point of embedded words. It is this representation that is used to query the embedding space over other claims to assess the Plausibility of a new document.

Topic Context Model. We use a generative probabilistic model to represent the *BagOfContext* of each $d_i \in D$. In particular, the Latent Dirichlet Allocation (LDA). This model is an instance of a general family of mixed membership models for decomposing a collection into multiple latent components (topics). In LDA it is assumed that words of each document arise from a mixture of topics, where each topic is a multinomial over a fixed vocabulary. The topics are shared by all documents in the collection, but the topic proportions vary stochastically across documents, as they are randomly drawn from a Dirichlet distribution [17, 18]. In this work, we employ this powerful representation to operationalize the idea of possible worlds to account for cases where we find claims that support and contradict each other in our collection. In particular, our hypothesis is that by the instantiation of the possible world idea, we can provide additional insights to understand what we have called "controversies". Thus, we operationalize the idea of possible worlds, with a representation of the context of a claim given by its latent mixture of topics.

3.2   Finding Semantic Similar Claims

Finding similar claims in our work is a crucial step given the embedding space representation. Moreover, given the claim of a document we would like to assess, we also need to distinguish between claims that support or contradict it. For this task, we proceed as follows: given the embedding space of claims, we first find similar claims by computing distance similarities as in [19, 20]. Because this distance is highly efficient in using the embedding space against some other alternatives [19], we rely on it in our first step. In Fig. 3 we show an illustrative example of the semantics captured in the claim embedding space as given by the distance computation named Word Mover's Distance (WMD) of [19]. Please observe how entities such as "tomato sauce" and "lycopene" end up close to each other in the embedding space because of the semantics captured by the WMD. Thus, instead of endless list of synonyms, we rely on this type of representation to find similar entities used in similar contexts. Next, we distinguish between supporting or contradicting claims. As a proof of concept, we focus on claims expressing "increase" vs "decrease" associations that express clearly contradictory positions. Thus, for the claims that we investigate in our experiments we distinguish these two positions. A simple textual pattern mechanism with synonyms to

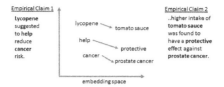

Fig. 3. Illustrative example of the power of the embedding space using WMD distance.

the words "increase" and "decrease" were used to distinguish between the two. Synonyms in this work are words related to "increase" and "decrease" as captured in the embedding space.

3.3   Computing Plausibility

In this section, we formally define how we determine the Plausibility of a new document. Plausibility in our work is the consistency of a new claim with current body of knowledge in terms of its agreement at the claim level or at the context level when a possible world consistency holds. Let d_{new} be a document that is currently not in our collection and we would like to know how plausible it is, given our current knowledge. Let $ClaimOf(d_{new})$ be the claim of document d_{new}. Let $DocSimClaim(d_{new})$ be the set of documents dealing with semantic similar claims to $ClaimOf(d_{new})$. Moreover, let $DocsContradict$ and $DocsSuppport$ be the documents that contradict and support respectively the $ClaimOf(d_{new})$. The distinction between these two sets of documents is given by finding first if the claim is in an "increase" or "decrease" association. Afterwards, we just map every claim found to one of the two groups. To be able to determine the Plausibility of the new document, we proceed as follows:

1. If $DocsContradict$ is empty and $DocsSuppport$ is not, then $ClaimOf(d_{new})$ is plausible.
2. If $DocsContradict$ is not empty and $DocsSuppport$ is empty, then $ClaimOf(d_{new})$ is not plausible.
3. If both $DocsContradict$ and $DocsSuppport$ are not empty, we initiate our quest for a possible world consistency:
   a. First, we need to find how document d_{new} would fit in our Topic Model representation of the collection. Let $topicOf(d_{new})$ be that topic. This is found by posterior inference and selecting the topic with the highest probability value.
   b. Second, if in $topicOf(d_{new})$ the claims agree, we call the world consistent and proceed to verify the consistency of $ClaimOf(d_{new})$. If the possible world is not consistent, we declare a "controversial" situation regarding the claim of the new document.
4. If $DocsContradict$ and $DocsSuppport$ are both empty, then $ClaimOf(d_{new})$ calls for a special assignment of resources to manually assess its value.

4 Experiments and Findings

To demonstrate and evaluate our proposed Plausibility measure, we performed experiments with two primary goals. Firstly, we wanted to gather valuable insight into the notion of finding similar semantic claims in our corpus that either support or contradict a new document's claim. Because we do not have a ground truth, we manually observed the claims and set a threshold that to the best of our understanding can lead to highly related claims that may or may not support each other with respect to a disease. After this experimentation, we set a threshold of 0.50 for the experiments that we report here. Secondly, as a proof of concept, we needed to compare our approach to experts work. In particular, we chose the results reported in [2] that we mentioned in the Introduction. Thus, we first retrieved all the documents related to two of the ingredients reported in [2] salt and lycopene. To retrieve the documents, we used the query pattern mentioned in our Introduction. We chose these two cases to acknowledge the scope of our tool since they represent different situations: salt was found to be one of the few exceptions of the analysis regarding its risk effect that was not subject of "controversy" due to contradicting findings. On the other hand, lycopene represent a situation that cannot be plausible from our perspective: the increase/decrease effect at the same time. Thus, the goal of studying these two cases was to see if we could find a suitable explanation. For every document, we extracted the empirical claims contained in the conclusions section when available. Unfortunately, not all the documents contained this valuable metadata. We just ignored them and put it aside the task of metadata generation that is necessary to improve the knowledge representation of our work.

Thus, the collection of documents used for these experiments consist of 87 k documents. We used this collection to train our embedding representation using default parameters as given by open source project Gensim [21]. For every experiment, we first selected one document at random and considered it as the new document and proceeded to assess its Plausibility.

The first case that we report here is the association of "salt" and "cancer". The document to evaluate was "*Salt intake and gastric cancer risk according to helicobacter pylori infection, smoking, tumour site and histological type*" [22]. The claim of the paper that we used to query our semantic embedding space is a very simple one: "*our results support the view that salt intake is an important dietary risk factor for gastric cancer, and confirms the evidence of no differences in risk according to h. pylori infection and virulence, smoking, tumour site and histological type.*" After querying our semantic embedding space, we retrieved some related claims in other papers. Some examples are:

- "Dietary salt intake was directly associated with risk of gastric cancer in prospective population studies, with progressively increasing risk across consumption levels."
- "Improved dietary habits, reducing salt consumption and eradication of h. pylori infection may provide protection against gastric cancer in Turkey."
- "These data suggest that high intake of salt and smoked and pickled food may be associated with a high risk of gastric cancer, and this association could be due to intragastric formation of nitrosamines"

In this particular case, the "new" claim finds support in our current knowledge and our approach states a "plausible" situation. This is an example of how our approach could help a reviewer to assess if the new document is consistent with current body of knowledge. Basically, it could allow the reviewer to find similar studies dealing with the specified entities probably in similar ways.

Next, we report on a second experiment. In this second experiment we take "lycopene" as one of the ingredients where there was evidence of being in a situation that we call "controversial". Remember that "controversial" means that "lycopene" was found in an increased and decreased risk association in different research papers. In this particular case, we found 197 documents related to the association within our collection. We selected a document with the following claim: *"this study does not support a role for lycopene in prostate cancer prevention"* [23].

We found in our collection claims that both support and contradict the new document's claims. This leads to a controversial situation as defined in our methodology section. However, we can take our second step to make a final decision. In this case, the new document fits better in a possible world that is not consistent. Thus, we conclude that this is a controversial situation in need of human experts to carefully look into the new document. One should notice the level of complexity of this case. For instance, in the community curated archive Wikipedia entrance of lycopene[2], one can find the following: The FDA (US Food and Drug Administration), in rejecting manufacturers' requests in 2005 to allow "qualified labeling" for lycopene and the reduction of various cancer risks, stated:

"…no studies provided information about whether lycopene intake may reduce the risk of *any of the specific forms* of cancer. Based on the above, FDA concludes that there is no credible evidence supporting a relationship between lycopene consumption, either as a food ingredient, a component of food, or as a dietary supplement, and any of these cancers." Furthermore, two more experts of medical panels cited in the entrance of the Wikipedia page also confirmed this situation.

To get a better assessment of the potential of our approach, we performed simulated experiments with a selection of the 80 most recent meta-analyses found in our collection with respect to three other diseases: hypertension, diabetes and asthma in addition to cancer. A "meta-analysis" is a systematic review that uses statistics analysis to be able to combine several research papers on a particular topic. One characteristic of meta-analyses is that it may never be possible to include all the papers that deal with a particular phenomenon. Usually, researchers query a digital library using keywords to get a candidate set of papers and after that, they manually decide which candidates can be included in the analysis. Depending on the methodology chosen by the researchers, the final number of articles vary. In this set of experiments, we proceeded as follows: we took out of our collection the meta-analysis, and then we queried our representation using the claim of the meta-analysis. If we could agree with the claim of the meta-analysis in at least one possible world, then we consider that as a positive outcome. After our experimentations, our best result was a kappa of 0.7746 with a 95% confidence interval (0.7875, 0.9549). Notice that because the criteria that the experts

---

[2] https://en.wikipedia.org/wiki/Lycopene.

use to include and/or exclude some papers in a meta-analysis are beyond our current text mining processing, we included all papers as given by our query pattern. However, one caveat of this type of experiments is the training time of the embedding and the LDA hyper parameters. In this particular setting, we trained the word embeddings with 100 dimensions and LDA with 8,000 iterations with a fixed 300 topics in a collection of 315 k documents.

To provide insights of our results, let's look at one of the cases where our approach failed. Consider the findings of [24], where the study of alcohol regarding prostate cancer was analyzed. As stated in the paper, a total of 340 studies were found in the exploratory search, but only 27 satisfied the inclusion criteria of the researchers (manual assessment). For this case, we found a "controversial" situation. In other words, our proposed approach did not agree with the meta-analysis in any possible world. More specifically, all the possible worlds were inconsistent and our tool stated a "controversial" situation. Moreover, the researchers reported "Our study finds, for the first time, a significant dose–response relationship between level of alcohol intake and risk of prostate cancer starting with low volume consumption" [24]. Of course, this is an expert assessment and our tool is not aiming at replacing a decision but instead helping to detect situations that may require a better administration of reviewers, especially in cases of "controversy" where clearly major care should be taken.

Discussion. Our results look promising and there are some issues that we noticed during our experiments. Firstly, the assessment of the degree of association between the claims is something that only domain experts can properly adjust. For instance, the idea that "tomato sauce" and "lycopene" can be considered similar enough to retrieve claims that associated both of them with "cancer" depends on what the experts would consider "related". Moreover, the idea of considering or not considering related types of a disease, such as "prostate cancer", "lung cancer", "gastric cancer", etc., in the retrieval of "related" claims is again questionable. In our experiments, we did notice a difference when we filtered the results to restrict the retrieval to the specified entities. Nevertheless, we envision an application where the reviewer can actually experiment with this feature of our approach. Secondly, the final decision of "controversial" with the idea of the possible world explanation did help to some extend but stayed below expectations in the experiments. One possible explanation is the criteria of inclusion/exclusion of articles in a meta-analysis and the methodology used to assess its conclusion. These two aspects are of course beyond our approach capacities and not in the scope of what we want to achieve. And third, our approach could accurately find controversial situations as confirmed by the meta-analysis experiments. However, this was only possible when we did not restrict the entities to exact matches but instead expand them to the most related ones as motivated in the work of [19].

5   Conclusions

We introduced a novel approach to assess the Plausibility of a new document to support peer review not at the process level, but with a clear focus at the document level. Our results look promising towards the goal of novel management of resources in peer review.

In particular, the question of how many reviewers a new paper needs can be adjusted by its respective degree of Plausibility. Of course, our experiments also reveal future work that is needed to crystalize our vision. For instance, assuming that "tomato sauce" and "lycopene" can be considered similar enough to retrieve papers that associate both of them with "cancer" depends on the goal of the analysis. And this is something that domain experts can properly adjust. Thus, in future experiments we will provide an online application to allow users of our system to personalize the degree of associations between the claims. Hopefully, we will learn some patterns to adapt to new users and new domains. The idea of "possible worlds" proved itself to be useful but in some domains one might consider a more restricted view and instead of proportions of topics as latent descriptors of documents, one might be interested in a hard clustering approach. We will also incorporate this notion into our system in the future.

Finally, the model of claims that we currently have must be extended to cope with other domains. To do that, we will need to account for more advanced model's representation of arguments in scientific papers. We are aware that the incipient field of Argumentation Mining in the last few years has shown tremendous potential to envision more powerful applications. We will also explore that line of research in future work.

References

1. Price, B.Y.S., Flach, P.A.: Computational support for academic peer review: a perspective from artificial intelligence. Commun. ACM **60**(3), 70–79 (2017)
2. Schoenfeld, J.D., Ioannidis, J.P.A.: Is everything we eat associated with cancer? A systematic cookbook review. Am. J. Clin. Nutr. **97**, 127–134 (2013)
3. Ciccarese, P., Wu, E., Wong, G., Ocana, M., Kinoshita, J., Ruttenberg, A., Clark, T.: The SWAN biomedical discourse ontology. J. Biomed. Inform. **41**, 739–751 (2008)
4. Groth, P., Gibson, A., Velterop, J.: The anatomy of a nanopublication. Inf. Serv. Use **30**, 51–56 (2010)
5. Velterop, J.: Nanopublications: the future of coping with information overload. LOGOS J. World B. Community **21**, 3–4 (2010)
6. Kuhn, T., Barbano, P.E., Nagy, M.L., Krauthammer, M.: Broadening the scope of nanopublications. In: Cimiano, P., Corcho, O., Presutti, V., Hollink, L., Rudolph, S. (eds.) ESWC 2013. LNCS, vol. 7882, pp. 487–501. Springer, Heidelberg (2013). doi:10.1007/978-3-642-38288-8_33
7. Groth, P., Loizou, A., Gray, A.J.G., Goble, C., Harland, L., Pettifer, S.: API-centric linked data integration: the open PHACTS discovery platform case study. J. Web Semant. **29**, 12–18 (2014)
8. Clark, T., Ciccarese, P.N., Goble, C.A.: Micropublications: a semantic model for claims, evidence, arguments and annotations in biomedical communications. J. Biomed. Semant. **5**, 28 (2014)
9. Toulmin, S.: The Uses of Argument. Ethics, 70, vi, 264 (1958)
10. Verheij, B.: The Toulmin argument model in artificial intelligence. In: Argumentation in Artificial Intelligence, pp. 219–238 (2009)
11. Connell, L., Keane, M.T.: A model of plausibility. Cogn. Sci. **30**, 95–120 (2006)
12. Bengio, Y., Ducharme, R., Vincent, P., Janvin, C.: A neural probabilistic language model. J. Mach. Learn. Res. **3**, 1137–1155 (2003)

13. Turian, J., Ratinov, L., Bengio, Y.: Word representations: a simple and general method for semi-supervised learning. In: Proceedings of 48th Annual Meeting of the Association for Computational Linguistics, pp. 384–394 (2010)
14. Mikolov, T., Chen, K., Corrado, G., Dean, J.: Distributed representations of words and phrases and their compositionality. Nips, pp. 1–9 (2013)
15. Mikolov, T., Yih, W., Zweig, G.: Linguistic regularities in continuous space word representations. In: Proceedings of NAACL-HLT, pp. 746–751 (2013)
16. Le, Q., Mikolov, T.: Distributed representations of sentences and documents. Int. Conf. Mach. Learn. ICML **2014**(32), 1188–1196 (2014)
17. Blei, D.M.: Probabilistic topic models. Commun. ACM **55**, 77 (2012)
18. Blei, D.M., Ng, A.Y., Jordan, M.I.: Latent Dirichlet allocation. J. Mach. Learn. Res. **3**, 993–1022 (2003)
19. Kusner, M.J., Sun, Y., Kolkin, N.I., Weinberger, K.Q.: From word embeddings to document distances. In: Proceedings of 32nd International Conference on Machine Learning, vol. 37, pp. 957–966 (2015)
20. Pele, O., Werman, M.: Fast and robust earth mover's distances. In: Proceedings of the IEEE International Conference on Computer Vision, pp. 460–467 (2009)
21. Sojka, P., ˇReh ˚uˇrek, R.: Software framework for topic modelling with large corpora. In: Proceedings of the LREC 2010 Workshop on New Challenges for NLP Frameworks, pp. 45–50. ELRA, Valletta, Malta (2010)
22. Peleteiro, B., Lopes, C., Figueiredo, C., Lunet, N.: Salt intake and gastric cancer risk according to Helicobacter pylori infection, smoking, tumour site and histological type. Br. J. Cancer **104**, 198–207 (2011)
23. Kristal, A.R., Till, C., Platz, E.A., Song, X., King, I.B., Neuhouser, M.L., Ambrosone, C.B., Thompson, I.M.: Serum Lycopene concentration and prostate cancer risk: results from the prostate cancer prevention trial. Cancer Epidemiol. Biomarkers Prev. **20**, 638–646 (2011)
24. Zhao, J., Stockwell, T., Roemer, A., Chikritzhs, T., Bostwick, D., et al.: Is alcohol consumption a risk factor for prostate cancer? A systematic review and meta–analysis. BMC Cancer **16**, 845 (2016)

Classifying Document Types to Enhance Search and Recommendations in Digital Libraries

Aristotelis Charalampous and Petr Knoth[(⊠)]

CORE, Knowledge Media Institute, The Open University, Milton Keynes, UK
{aristotelis.charalampous,petr.knoth}@open.ac.uk

Abstract. In this paper, we address the problem of classifying documents available from the global network of (open access) repositories according to their type. We show that the metadata provided by repositories enabling us to distinguish research papers, thesis and slides are missing in over 60% of cases. While these metadata describing document types are useful in a variety of scenarios ranging from research analytics to improving search and recommender (SR) systems, this problem has not yet been sufficiently addressed in the context of the repositories infrastructure. We have developed a new approach for classifying document types using supervised machine learning based exclusively on text specific features. We achieve 0.96 F1-score using the random forest and Adaboost classifiers, which are the best performing models on our data. By analysing the SR system logs of the CORE [1] digital library aggregator, we show that users are an order of magnitude more likely to click on research papers and thesis than on slides. This suggests that using document types as a feature for ranking/filtering SR results in digital libraries has the potential to improve user experience.

Keywords: Document classification · Academic search · Recommender systems for research · Text mining · Metadata quality · Document aggregation

1 Introduction

Over the last 15 years, there has been a significant growth in the number of institutional and subject repositories storing research content. However, each repository on its own is of limited use, as the key value of repositories comes from being able to search, recommend and analyse content across this distributed network. While these repositories have been established to store primarily research papers, they contain, in fact, a variety of document types, including theses and slides. Services operating on the content from across this repository network should be able to distinguish between document types based on the supplied metadata.

However, metadata inconsistencies are making this very difficult. As we show later in the study, ˜62% of documents in repositories do not have associated

© Springer International Publishing AG 2017
J. Kamps et al. (Eds.): TPDL 2017, LNCS 10450, pp. 181–192, 2017.
DOI: 10.1007/978-3-319-67008-9_15

metadata describing the document type. Moreover, when document type is specified, it is typically not done using an interoperable vocabulary.

Consequently, digital library aggregators like CORE [1], OpenAIRE [2] and BASE [3] face the challenge of offering seamless SR systems over poor quality metadata supplied by thousands of providers. We hypothesise that by understanding the document type, we can increase user engagement in these services, for example, by means of filtering or re-ranking SR systems results.

In this paper, we develop a novel and highly scalable system for automatic identification of research papers, slides and theses. By applying this identification system, we analyse the logs of CORE' SR systems to see if we can find evidence of users preferring specific document type(s) over others.

The contributions of the paper are:

- Presenting a lightweight, supervised classification approach for detecting *Research*, *Slides* and *Thesis*, based on a small yet highly predictive set of features extracted from textual descriptors of (scientific) articles, reaching an F1-score of 96.2% with the random forest classifier.
- A publicly exposed and annotated dataset [4] of approximately $11.5\,k$ of documents for the sake of comparison and reproducibility.
- Proposing a modified CTR metric, balanced QTCTR, to analyse historical SR systems' logs to evaluate user engagement with the proposed content types in digital library systems, showing our users' inclination towards research and theses over slides.

The rest of the paper is organised as follows. Firstly, we discuss related work, followed by the presentation of our current data state. Secondly, we outline our approach and present results of the classification approach and the analysis of current user engagement using our modified CTR metrics. Finally, we end with a discussion before concluding the paper.

2   Related Work

The library community holds traditionally metadata records as a key enabler for resource discovery. Systems, such as BASE and WorldCat[1], have been almost solely relying on metadata in their search services until today. But as such approach, as opposed to services indexing the content, cannot guarantee metadata validity, completeness and quality, nor can achieve acceptable recall [1], some have started to believe that aggregative digital libraries have failed due to the interoperability issues facing OAI-PMH data providers. In fact, [5] specifically argues that the fact that BASE and OpenAIRE do not (or cannot) distinguish between document types of the records they harvest makes them "not as effective as users might assume".

While automatic document categorisation using structural and content features has been previously widely studied [6–8], little work has been done on the

---

[1] https://www.worldcat.org/.

issue of document type categorisation in the context of digital libraries until the recent study Caragea et al. [9]. They experimented with (1) *bag-of-words*, (2) document *URL tokens* and (3) document *structural features* to classify academic documents into several types. Their set of 43 manually engineered *structural features* have shown significant performance gain over conventional *bag-of-words* models in these highly diverse data collections.

Unlike previous work in standard approaches to text categorisation, summarised in [10], we use a subset of file and text specific characteristics, selectively gathered from [9]. The reduced dimensionality, as a result of the subset's minimal size, allows for scalable integration in ingestion pipelines of SR systems. In addition to the previous work, our study is to our knowledge the first to understand whether the integration of these document type classification systems can lead to more effective user engagement in SR systems.

3   Data - Current State

CORE is a global service that provides access to millions of (open access) research articles aggregated from thousands of OA repositories and journals at a full text level. CORE offers several services including a search engine, a recommendation system, an API for text-miners and developers as well as some analytical services. As of April 2017, CORE provides access to over 70 million metadata records and 6 million full texts aggregated from $2,461$ data providers. From the available metadata descriptors, a directly available field to categorise records, at a certain extent, is the *dc:subjects* field. While mostly available, currently 92% of cases, only a small minority contain clear descriptions of the document type. More specifically, ~30.0 of records are marked as `article`, ~7.3% are marked as `thesis` and 0% as `slides`. This means that we do not have any type document type indication for ~62% of our data.

Table 1 lists the top re-occurring terms that are most indicative of the three document types we are interested in. This provides empirical evidence of the poor adoption of interoperable document type descriptors across data providers. Finally, from the ~6 million full text entries that CORE contains, 8.5 million

Table 1. Most popular terms found in the **dc:subjects** field with >1% occurrence

Term name	Term frequency
Article	0.1366
info:eu-repo/semantics/article	0.0866
Journal articles	0.0385
Thesis	0.0205
info:ulb-repo/semantics/openurl/article	0.0017
info:eu-repo/semantics/doctoralthesis	0.0106
info:eu-repo/semantics/bachelorthesis	0.0101

unique *dc:subjects* field terms are currently recorded (one record can contain multiple subjects fields).

4   Approach

While one approach to address the problem of poor or missing document type descriptors can be to create guidelines for data providers, we believe this approach is slow, unnecessarily complex and does not scale. Instead, we aim to develop an automated system that infers the document type from the full text.

The assumptions we make for this study follow several observations on the textual features of documents stored in CORE:

- **F1: Number of authors:** The more authors involved in a study, the more likely a document is a research paper as opposed to slides or thesis.
- **F2: Total words:** These were tokenised from the parsed text content using the *nltk*[11] package. Intuitively, the lengthier a document is, in terms of total written words and amount of pages, the more likely it is a thesis.
- **F3: Number of pages:** Research papers tend to have a fewer number of pages compared to theses and slides.
- **F4: Average words per page:** Calculated as $\frac{\#\text{total words}}{\#\text{total pages}}$. The fewer words written per page on average, the more likely the document type is *slides*.

We extract F2-F4 from their respective `pdf` files with pdfMiner [12]. F1 is extracted from the supplied metadata. We then apply one of the classifiers, described later in Sect. 5.2, to predict the document type given these features.

5   Experiments

5.1   Data Sample

Our first goal was to create a sufficiently large ground truth dataset. Data labelling took place with a rule-based method applied to the CORE dataset. More specifically, we used a set of regular expressions on the dc:subjects field and the document's title as follows:

- Subjects fields for which entries include the keyword "thesis" or "dissertation" were labelled as *Thesis*.
- Subjects fields for which entries do **not** include the keyword "thesis" or "dissertation" and their title does **not** include the keyword "slides" or "presentation" were labelled as *Research*.
- Subject fields for which entries do **not** include the keyword "thesis" or "dissertation" and their title includes the keyword "slides" or "presentation" were labelled as *Slides*.

While this rule-based labelling process produced a sufficiently large number of samples for the *Research* and *Thesis* classes, it has not yielded a satisfactory sample size for the *Slides* class. To address this issue, we have mined pdfs and metadata from SlideShare[2] using their openly accessible API.

We wanted the total size of the sample to satisfy two criteria, a confidence level of 95% at a confidence interval of 1%. The equation to calculate the necessary size of the data sample is:

$$n = \frac{Z^2 \hat{p}(1 - \hat{p})}{c^2} \tag{1}$$

where, Z is the Z score, $\hat{p}$ is the percentage probability of picking a sample and c is the desired confidence interval. Given a Z score of 1.96 for a 95% confidence level, a confidence interval of 0.01 and a sample proportion p of 0.5 (used as it is the most conservative and will give us the largest sample size calculation), this equation yields ~9.6 k samples.

We have gathered these 9.6k samples and additionally extended the dataset by 20% to form a validation set, resulting in 11.5k samples. To produce a sample with a representative balance of classes, we limited slides to take up to 10% of the final dataset, 55% for research and the remaining 35% for theses entries. We also ensured that all the pdfs in the data sample are parsable by pdfminer.

Finally, we addressed the issue of missing values for feature F1, which SlideShare did not provide in over 97% of cases, by applying multivariate imputations [13]. To improve our knowledge of the feature distributions prior to applying the imputations for the *Slides* class, we relied on extra data from Figshare[3].

To visualise the dimensionality and data variance in the resulting dataset, we have produced two and three dimensional projections of our data, using techniques introduced by [14]. On small datasets ($< 100k$ data points) these do not require much tuning of hyper-parameters and, out of manual inspection from a limited range of hyper-parameters, we decided to use perplexity of 30 and a theta of 0.5. As Fig. 1 suggests, there is sufficient evidence of data sparsity.

5.2 Feature Analysis and Model Selection

We have experimented with: Random Forest (RF), Gaussian Naive Bayes (GNB), k Nearest Neighbours (kNN), Adaboost with Decision trees (Adaboost) and linear kernel Support Vector Machines (SVM).

We followed a standard 10-fold cross-validation approach to evaluate the models with an extra 20% of the data left aside for model validation. The class balance discussed was preserved in each fold evaluation by applying stratified splits on both test and validation sets, simulating a representative distribution of categories in the CORE dataset. All features used were compared against their normalised and log-scaled counterparts to check for any possible performance improvements. We have also optimised for a small range of hyper-parameters

---

[2] https://www.slideshare.net/.
[3] https://figshare.com/.

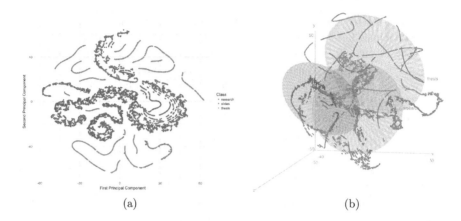

Fig. 1. Data variance visualisation using (a) two and (b) three dimensional projections on the corresponding principal components.

for each machine learning algorithm using parameter sweeps, recording the best achieved performance for each algorithm class. The evaluation results are presented in Table 3.

Two baseline models have been used to assess the improvement brought by the machine learning classifiers. The approaches used are:

- **Baseline 1:** Random class assignment with probability weights corresponding to the dataset's class balance.
- **Baseline 2:** A rule-based approach based on statistically drawn thresholds for each feature and class respectively, using the upper 0.975 and lower 0.025 quantiles.

An analysis was carried out on the assembled dataset to form Baseline 2, based on feature distributions' percentiles. Distributions from the sample dataset largely followed a right skewed normal distribution (Fig. 2), proving such a model should be a suitable candidate to evaluate against. To avoid overfitting, outliers were removed using Tukey's method [15], which was preferred due to its independence on the data distribution, omitting values outside of the range:

$$(Q1 - 1.5 * IQR) > Y > (Q3 + 1.5 * IQR) \tag{2}$$

where, Y is the set of acceptable data points, $Q1$ is the lower quartile, $Q3$ is the upper quartile and $IQR = Q3 - Q1$ is the interquartile range.

The acquired thresholds for Baseline 2 are listed in Table 2. To assign a particular example a document type t, all its features must fall within the boundaries specified. When this method fails, we assign the majority class (*Research*).

Table 2. Percentile thresholds (upper 0.975 and lower 0.025 quantiles) for Baseline 2, following outlier removal.

Feature	Research	Document type	
		Slides	Thesis
F1	$1 \leq x \leq 5$	$1 \leq x \leq 8$	$==1$
F2	$1227 \leq x \leq 19,151$	$94 \leq x \leq 7340$	$15,184 \leq x \leq 210,720$
F3	$3 \leq x \leq 41$	$1 \leq x \leq 75$	$47 \leq x \leq 478$
F4	$208 \leq x \leq 927$	$8 \leq x \leq 723$	$198 \leq x \leq 530$

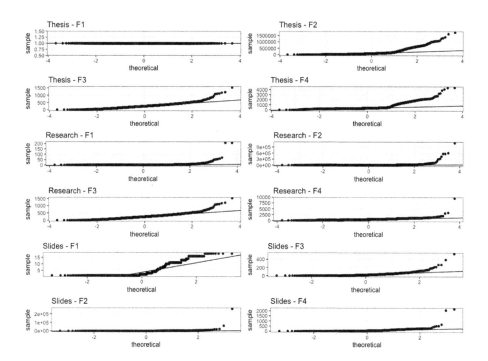

Fig. 2. Normal Q-Q Plots split by document type and feature.

5.3 Results

The evaluation results, presented in Table 3, show that all our models outperform the baselines by a large margin. However, baseline 2 demonstrates a perhaps surprisingly good performance on this task. Random forest and Adaboost are the top performers achieving about 0.96 in F1-score on both the test and validation sets. While we cannot distinguish which model is better at the 95% confidence level and 1% confidence interval, see Sect. 5.1, we decided to productionise random forest due to the model's simplicity.

Table 3. Test and validation set results on weighted evaluation metrics across all algorithms.

	Measure	Algorithm						
		RF	GNB	kNN	Adaboost	SVM	Baseline 1	Baseline 2
Test results	Precision	**0.962**	0.9431	0.949	0.9580	0.8968	0.4926	0.5688
	Recall	**0.9623**	0.9414	0.9497	0.9569	0.8933	0.3270	0.4762
	F1-score	**0.9623**	0.9416	0.9496	0.9573	0.8695	0.3270	0.5154
Validation results	Precision	0.9567	0.9356	0.9453	**0.9607**	0.8435	0.5572	0.6362
	Recall	0.9553	0.9338	0.9454	**0.9605**	0.8741	0.4570	0.6565
	F1-score	0.9558	0.9337	0.9453	**0.9606**	0.8311	0.4570	0.5945

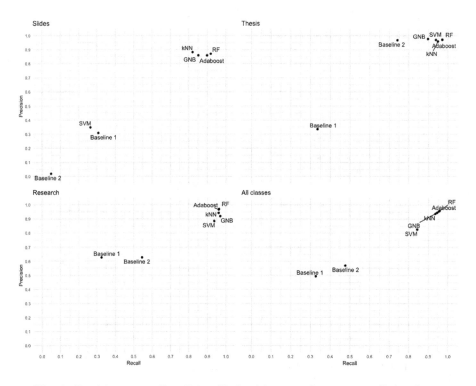

Fig. 3. Precision versus Recall for all algorithms on the test set split by class.

Figure 3 shows a breakdown of the final precision/recall performances according to the assigned document type. This indicates that a particularly significant improvement of the machine learning models over the baselines is achieved on the *Slides* class. However, as only about 10% of documents in the dataset are slides, the baselines are not so much penalised for these errors in the overall results.

To evaluate the importance of individual features, a *post-hoc* analysis was carried out. We fitted the models of our selected algorithms with a *single* feature

Table 4. Classifiers' performance with individual feature groups across all algorithms on the test set in descending order, based on their contribution.

Features	Average weighted F1-score				
	RF	GNB	kNN	Adaboost	SVM
Only: F2	0.8825	0.7661	0.8702	**0.8839**	0.1868
Only: F3	0.8436	0.8412	0.8414	0.8424	**0.8441**
Only: F1	**0.8007**	0.6819	0.8007	**0.8007**	0.6441
Only: F4	**0.7036**	0.4745	0.6919	0.7018	0.3506
All features (RF)	**0.9623**	0.9416	0.9496	0.9573	0.8695

group at a time. In this scenario, we have recorded high precision performances. Individual feature contributions do not vary widely, except in the case of F4 and the overall performance of the SVM classifier. F1-3 are the most predictive features. We list our findings in Table 4.

6   Can the Model Help Improve User Engagement in SR Systems?

We applied the random forest model to classify existing content in CORE. Joining the document type information with CORE's SR systems' user logs, enabled us to analyse document type user preferences in CORE's SR systems.[4] We followed the intuition that if we can find that users prefer clicking in SR results on one document type over another, this will provide the argument for using document type information in SR systems to better serve the needs of these users.

A traditional metric to measure the popularity of a link is the Click-Through Rate (CTR), measured as:

$$CTR_T = \frac{|Clicks|}{|Impressions|} \tag{3}$$

However, we cannot use CTR directly to assess whether people are more likely to click on certain document types than others in the SR system results. This is because we serve, on average, 66.7% *Research*, 27.2% *Thesis* and 6.1% *Slides* impressions across our SR engines. Consequently, the CTR metric would be biased towards the *Slides* class. This is due to the fact that when an action is made on an impression set, the class most represented in the set will benefit from this action on average the least. Put differently, this is accounted to the class imbalance.

To address this problem, we extend CTR to put *impression equality* into perspective with the following process. We group impressed items in sets Q,

---

[4] It should be noted that as CORE provides thumbnails on its SR results pages, users get an idea of the document type prior to accessing it.

Table 5. Modified click-through rate metrics performance on CORE's SR systems.

Metric	Engine	Impression set positions					
		Any position			Top position		
		Research	Slides	Thesis	Research	Slides	Thesis
QTCTR	Search	0.13685	0.01878	0.32358	**0.03818**	0.00389	0.01829
	Recommender	**0.00675**	0.00074	0.00361	**0.00482**	0.00046	0.00204
RQTCTR	Search	0.08186	0.00142	**0.10061**	**0.02284**	0.00029	0.00569
	Recommender	**0.00488**	0.00003	0.00079	**0.00348**	0.00002	0.00045

reflecting the documents served following a query submission (in case of the recommender, the query is a document with respect to which we recommend)[5]. We assign to each impression set a type q_t based on the types of document(s) clicked in the results list. In case multiple clicks to distinct document types are made in response to a query, we generate multiple impression sets derived from it, each assigned to one of them.

We then calculate the *Query Type Click-Through Rate* (*QTCTR*) as a fraction of the number of queries which resulted in a click to a given document type over the number of all queries:

$$QTCTR = \frac{|Q_T|}{|Q|} \tag{4}$$

QTCTR tells us the absolute proportion of queries that result in clicking on a particular document type. We can regularise/normalise *QTCTR* to reflect the imbalance of impression types, forming the *Regularised Query Type Click-Through Rate* (*RQTCTR*). We include impression sets with no interaction in this calculation.

$$RQTCTR = \frac{|Q_T|}{|Q|} * \frac{|Impressions_T|}{|Impressions|} \tag{5}$$

The *QTCTR* and *RQTCTR* values from the CORE's SR systems, for the three different document types, are presented in Table 5. The shows that there is noteworthy difference in preference for *Research* type documents and *Thesis* over *Slides* by an order of one magnitude. This is true for clicks generated on any document in an impression set and when the click was on top positioned document. The *QTCTR* results also reveal that many people in CORE are looking for theses. We believe this is due to the fact that CORE is one of the few systems (in not the only one) that aggregates theses from thousands of repositories at a full-text level.

---

[5] The number of impressions generated in response to a query can vary across queries. In our case, it can be from zero to ten for search and from zero to five for the recommender.

7   Scalability Analysis

There exists a linear relationship between the number of features (N) and prediction latency [16], expressed with the complexity of $O(N*M)$, where M are the number of instances. The low number of features and model complexity, with our deployed model having < 10 trees and < 5 maximum nodes for each, the latency amounts to slightly over 0.0001 seconds per prediction[6]. Due to CORE's continuously ongoing repository harvesting processes, the minimal feature extraction requirements will allow for new additions to be streamlined immediately after their processing, in comparison with the latency associated with the feature extraction process expected from [9]. This indicates the high scalability of our approach and applicability across millions of documents.

8   Future Work

In promoting the current solution within CORE's systems, and making it accessible to users worldwide, we aim to:

- Expose document type classification models as a service, with online model updating, through CORE's public API.
- Boost *Research* documents in our SR engines and negatively boost *Slides* to aid faster retrieval of preferred content.
- Evaluate the shift of user engagement as a direct effect of such changes in our services and adjusting our search/recommendation strategies accordingly.
- Enhance user engagement analysis by cross-validation of our observations here metrics such as the *dwell time*, a metric proven to be less unaffected by position, caption or other form of bias in SR results [17].
- Extend the model in further iterations to also discern between sub-types of the *Research* and *Slides* classes, such as theoretical, surveys, use case or seminal research papers as well as slides corresponding to conference papers and lecture/course slides respectively.

9   Conclusions

We have presented a new scalable method for detecting document types in digital libraries storing scholarly literature achieving 0.96 F1-score. We have integrated this classification system with the CORE digital library aggregator. This enabled us to analyse the SR system logs of to assess whether users prefer certain document types. Using a our Regularised Query Type Click-Through Rate (RQTCTR) metric, we have confirmed our hypothesis that the document type can contribute in finding a viable solution to improving user engagement.

Acknowledgements. This work has been partly funded by the EU OpenMinTeD project under the H2020-EINFRA-2014-2 call, Project ID: 654021. We would also like to acknowledge the support of Jisc for the CORE project.

---

[6] This excludes network overhead from the API call and the feature extraction process.

References

1. Knoth, P., Zdráhal, Z.: CORE: three access levels to underpin open access. D-Lib Mag. **18**(11/12) (2012)
2. Rettberg, N., Schmidt, B.: Openaire-building a collaborative open access infrastructure for european researchers. Liber Q. **22**(3) (2012)
3. Summann, F.: Bielefeld academic search engine: a scientific search service for institutional repositories. In: Open Scholarship 2006 Conference (2006)
4. Classifying document types to enhance search and recommendations in digital libraries - Dataset, https://figshare.com/articles/Classifying_document_types_to_enhance_search_and_recommendations_in_digital_libraries/4834229. Accessed 21 Apr 2017
5. Poynder, R.: Q&A with CNI's Clifford Lynch: Time to re-think the institutional repository? The Open Access Interviews (2016)
6. Sebastiani, F.: Machine learning in automated text categorization. ACM Comput. Surv. (CSUR) **34**(1), 1–47 (2002)
7. Qi, X., Davison, B.D.: Web page classification: features and algorithms. ACM Comput. Surv. (CSUR) **41**(2), 12 (2009)
8. Ghosh, S., Mitra, P.: Combining content and structure similarity for xml document classification using composite SVM kernels. In: 19th International Conference on Pattern Recognition, ICPR 2008, pp. 1–4. IEEE (2008)
9. Caragea, C., Wu, J., Gollapalli, S.D., Giles, C.L.: Document type classification in online digital libraries. In: AAAI, pp. 3997–4002 (2016)
10. Aphinyanaphongs, Y., Fu, L.D., Li, Z., Peskin, E.R., Efstathiadis, E., Aliferis, C.F., Statnikov, A.: A comprehensive empirical comparison of modern supervised classification and feature selection methods for text categorization. J. Assoc. Inf. Sci. Technol. **65**(10), 1964–1987 (2014)
11. Bird, S.: NLTK: the natural language toolkit. In: Proceedings of the COLING/ACL on Interactive Presentation Sessions, pp. 69–72. Association for Computational Linguistics (2006)
12. Shinyama, Y.: Pdfminer: Python PDF parser and analyzer (2015), http://www.unixuser.org/euske/python/pdfminer/. Accessed 08 Apr 2017
13. Buuren, S., Groothuis-Oudshoorn, K.: Mice: Multivariate imputation by chained equations in r. J. Stat. Softw. **45**(3) (2011)
14. van der Maaten, L., Hinton, G.: Visualizing data using t-SNE. J. Mach. Learn. Res. **9**, 2579–2605 (2008)
15. Tukey, J.W.: Comparing individual means in the analysis of variance. Biometrics, 99–114 (1949)
16. Pedregosa, F., Varoquaux, G., Gramfort, A., Michel, V., Thirion, B., Grisel, O., Blondel, M., Prettenhofer, P., Weiss, R., Dubourg, V., et al.: 7. computational performance - scikit-learn 0.18.1 documentation, http://scikit-learn.org/stable/modules/computational_performance.html. Accessed 08 Apr 2017
17. Kim, Y., Hassan, A., White, R.W., Zitouni, I.: Modeling dwell time to predict click-level satisfaction. In: Proceedings of the 7th ACM International Conference on Web Search and Data Mining, pp. 193–202. ACM (2014)

Understanding the Influence of Hyperparameters on Text Embeddings for Text Classification Tasks

Nils Witt[1(✉)] and Christin Seifert[2(✉)]

[1] ZBW-Leibniz Information Centre for Economics, Kiel, Germany
n.witt@zbw.eu
[2] University of Passau, Passau, Germany
christin.seifert@uni-passau.de

Abstract. Many applications in the natural language processing domain require the tuning of machine learning algorithms, which involves adaptation of hyperparameters. We perform experiments by systematically varying hyperparameter settings of text embedding algorithms to obtain insights about the influence and interrelation of hyperparameters on the model performance on a text classification task using text embedding features. For some parameters (e.g., size of the context window) we could not find an influence on the accuracy while others (e.g., dimensionality of the embeddings) strongly influence the results, but have a range where the results are nearly optimal. These insights are beneficial to researchers and practitioners in order to find sensible hyperparameter configurations for research projects based on text embeddings. This reduces the parameter search space and the amount of (manual and automatic) optimization time.

Keywords: Document embeddings · Hyperparameter optimization · Natural language processing

1 Introduction

Many applications in the natural language processing domain require the tuning of machine learning algorithms. Since there is no superior algorithm or model per se [19], machine learning models have to be carefully chosen and tuned. This tuning involves adaptation of learning parameters and model hyperparameters. A common approach to hyperparameter tuning is manual search: algorithm developers select sensible parameter choices from their experience and then repeatedly evaluate the models and adapt parameters. While manual search is the most time-consuming approach, it has the advantage that algorithm designers get some insights into parameter influences. But since humans are unable to reason in multi-dimensional spaces, they hardly achieve globally optimal results. A brute-force approach is grid search, where all (exponential many) parameter configurations from a pre-defined set are tested on a high-dimensional grid. For

© Springer International Publishing AG 2017
J. Kamps et al. (Eds.): TPDL 2017, LNCS 10450, pp. 193–204, 2017.
DOI: 10.1007/978-3-319-67008-9_16

cases where some parameters are less important than others random search [1] has been shown to outperform grid search. Both, random search and grid search are easy to implement, easily parallelizable and generally provide better results than manual search [1]. The Gaussian Process Optimizaton (GPO) approach aims to predict the next promising parameter configuration based on the previously observed performance w.r.t. a loss function and therefore searches the parameter space efficiently [18]. Despite its superior performance, the disadvantages of GPO is that another set of hyperparameters must be tuned (e.g., the choice of the acquisition function and covariance functions) and less insight about the parameter influence to a machine learning model is generated.

As in every machine learning problem, the performance of text classification applications strongly depends on the representation of the features. The most common representation is the bag of words representation in the vector-space model for texts [11]. More recently, distributed representations have been proposed, most prominently, word2vec [12] and GloVe [14] for representing terms and doc2vec for representing documents [9]. Text embedding-based techniques have outperformed the state-of-the-art in several tasks. Among these are document classification [5] and word sense disambiguation [4]. But again, the training of such models requires hyperparameter tuning, for instance, the size of the embedding vector, the learning rate and the number of negative samples.

In this paper, we show that the doc2vec model is very sensitive to settings of hyperparameters in classification tasks and analyze their influence. The goal of this paper is to provide insights into sensible parameter configurations for researchers and practitioners. More specifically, our research questions are the following:

RQ1-: What is the overall variance in accuracy for different hyperparameter settings? (Sect. 5.1)

RQ-2: Are optimal hyperparameters found on small datasets predictive for optimal hyperparameter settings on larger training sets? (Sect. 5.2)

RQ-3: How significant is the influence of the hyperparameters on the accuracy? (Sect. 5.3)

RQ-4: Are there interrelation between hyperparameters? That is, are there parameters that depend on each other and thus need to be considered jointly? (Sect. 5.4)

2   Related Work

As our work focuses on hyperparameter optimization of document embeddings we review work from these research areas.

Text Embeddings. The representation of words as dense vectors (word embeddings) has become popular for many natural language processing tasks. The popularity roots in the ability of these embeddings to encode semantic linearities such as `king - man + woman = queen` and the resulting state-of-the-art performance in several natural language processing tasks. The first approach to word embeddings is word2vec [12], which has been successfully used in applications like document classification [5] and zero-shot learning [16]. Word2vec can be trained using two different architectures, cbow and skip-gram. In cbow the vector of the input word is predicted by the vectors of the surrounding context words. The skip-gram architecture predicts the context word vectors by the input word vector. In both cases the number of context words is a hyperparameter of this model, such that vectors of words which appear in the same context will become similar whereas vectors from randomly sampled words (i.e., negative sampling) will become dissimilar [13]. Doc2vec is an extension of word2vec that learns embeddings for word sequences like paragraphs or entire documents [9]. In doc2vec, each document is treated as a word. Similar to word2vec, doc2vec can be trained using two different architectures dbow (distributed bag of words, similar to skip-gram) and dmpv (distributed memory model of paragraph vectors, similar to cbow). Dbow predicts the document words based on the document vector. Dmpv concatenates the document vector and the vectors of words in a window to predict a document word. Model hyperparameters of doc2vec are the choice of the architecture, the size of the context windows, the size of the embeddings, the number of negative samples and whether hierarchical sampling of words is used. We base our evaluation of hyperparamaters on document representations of doc2vec and embed the evaluation in a text classification task.

Hyperparameter Optimization. The majority of machine learning algorithms exposes hyperparameters which must be tuned carefully. Traditionally, the optimization is carried out by humans, which likely leads to suboptimal results because of inferior human intuition about multi-dimensional functions. In settings with sufficient computing resources, grid search approaches evaluate all combinations of hyperparameter values and are easily parallelizable. But, with an increasing number of parameters and values the computation becomes intractable. Bergstra and Bengio have found that randomly choosing values finds better models and requires less time than exhaustive grid-search [1] because less time is spent exploring parameters that have little influence. Bergstra et al. presented two hyperparameter optimzation algorithms and compared them to human experts and random search [2]. Their tree-structured Parzen approach showed superior performance over the Gaussian process (GP) [15] approach, while both outperformed manual and random search (in some cases with notable margins). Further advances were made by Snoek et al., who used Bayesian Optimization (BO) with GP priors to enhance the state of the art on the CIFAR-10 dataset by over 3% [18]. Despite its superior performance, the disadvantages of BO is that another set of hyperparameters have to be tuned (e.g., the choice of the kernel and the scopes for the hyperparameters) and – because of the missing manual parameter setting, test, evaluation cycle – less insight about the parameter influence to the machine learning model is generated. Thus, in order to understand hyperparameter influence, we apply grid search in our evaluation.

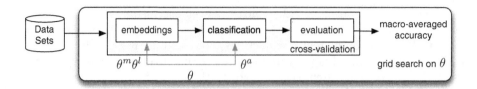

Fig. 1. Overview of the approach. Grid search is executed on the hyperparameter vector θ, the model is evaluated using cross-validation.

3   Approach

Our goal to investigate the influence of hyperparameters on the model performance, lead to questions about (i) overall performance variation, (ii) influence of single variables, the (iii) interrelation between variables and (iv) how the size of the training set influences optimal hyperparameters. To assess model performance, we choose the task of text classification, as text classification has been extensively studied [17] and comparative baselines are available. Also, this is a task for which doc2vec embeddings are especially suited, as doc2vec tends to build clusters of similar documents. The overall approach is depicted in Fig. 1. First, for each dataset feature representations are calculated using doc2vec. This step is governed by model hyperparameters of doc2vec θ^m (e.g., the size of the embeddings) and learning parameters θ^l (e.g., the learning rate). Then, a classification model is trained, whereas the classifier is governed by the hyperparameters of θ^a (such as k when using the k-Nearest neighbor classifier), composing the complete vector of hyperparameters θ. The grid search is employed on the hyperparameter vector θ. We use 5-fold cross-validation and report the results in terms of (macro-averaged) accuracy – the most common evaluation measure for text classification.

Because exhaustive search, even on a moderate number of parameters[1] requires considerable computational and memory resources, we use a two-stage strategy. In the first stage, we exhaustively search the hyperparameter space on a limited number of training samples. The initial parameters are derived from the literature [8,12]. In the second stage we train models on a larger training set but on a restricted grid using the best performing hyperparameter combinations. This allows us to compare the performance against state-of-the-art approaches using a similar amount of training data. Subsequently we also carry out a Bayesian optimization to find an optimal hyperparameter configuration and contrast the corresponding results to the results obtained using grid search.

4   Experimental Settings

We perform the experiments on well-known datasets for text classification and evaluate the overall accuracy, the influence of single parameters, their interrelation

---

[1] For example, 6 parameters with 3 values each results in $3^6 = 729$ combinations. Even worse, with 5-fold cross validation this results in $5 \cdot 3^6 = 3645$ models

Table 1. Overview of the hyperparameter space θ in the stages S1, S2.

θ	Description	Type	Values (S1)	Values (S2)
θ_{arch}	doc2vec architecture	nominal	dbow, dm	dbow
θ_{hs}	hierarchical sampling (off/on)	boolean	0, 1	1
θ_{ns}	number of negative samples	integer	1, 5, 20	5, 20
θ_d	embedding size	integer	2, 3, 8, 24, 64	24, 64, 256
θ_{win}	context window size	integer	5, 20, 50	5, 30, 100
θ_{ts}	number of training documents	integer	10^3, 10^4	10^4, 10^5
θ_α	learning rate (log scale)	real	0.001, 0.01, 0.1	0.01, 0.1, 1
θ_{epoch}	number of training iterations	integer	5, 30, 50	30, 100, 250
θ_k	number of nearest neighbors	integer	1, 5, 25	10, 50
Number of models			**4860**	**324**

and the stability of optimal parameters with varying sizes of the training set. The source code of the experiments as well as additional material is available online[2].

Parameter Configuration. Table 1 provides an overview of the parameter settings for the two stages of the experiments. Stage 1 trains more models than Stage 2 but with fewer training examples, thus satisfying memory and time constraints. The values for the first stage were chosen to cover a wide range from extremely low to extremely high values also including values used in related work. The values for the second stage are refinements over the first stage, while the most promising configurations were chosen and their range was narrowed. Some values were omitted (e.g. $\theta_{arch} = 1$) whereas others were added ($\theta_d = 256$ because larger embeddings sizes seem to be promising). Generally, the accuracy of this model increases with the size of the training data [3]. Hence, θ_{ts} is considered an external constraint rather than a value to be optimized. For the classifier we used k-nearest neighbor approach, thus the hyperparameters set for the classification model θ_a only consists of the number of neighbors.

Datasets. For the experiments, we chose two datasets from the domain of text classification, that exhibit different characteristics, i.e., the order of magnitude of contained documents and number of classes ,and are well-studied (e.g. [5–7]), The amazon[3] dataset, which was also used in [5] contains 12.8M user reviews for products assigned to four different categories (*Home and Kitchen, Electronics, Books* and *Movies and TV*). The dataset is strongly imbalanced with 8.9M *Books*, 1.69M *Electronics*, 1.7M *Movies and TV* and 0.55M *Home and Kitchen* reviews. This implies that a trivial categorizer can achieve 69% accuracy. The average length of the reviews is 796 tokens (i.e. sequences of characters

---

[2] http://doi.org/10.5281/zenodo.495086
[3] http://jmcauley.ucsd.edu/data/amazon/

surrounded by whitespaces). The 20newsgroups[4] dataset consists of 18.846 newsgroup articles categorized into 20 groups with an average length of 1902 tokens per article. The same preprocessing was carried out for both datasets. In order to reduce the noise in the text, short tokens (with less than 3 characters), quotation marks, punctuation marks, whitespaces (except for space characters) etc. were omitted[5]. Also, tokens that appeared less than three times where ignored. We did not use the available train/test splits but randomly generated the splits during the 5-fold cross validation runs. Further, we generated subsets with 1k, 10k (20newsgroups, amazon) and 100k documents (amazon) by random sampling.

5   Results

In this section we first describe the results of the experiments w.r.t. the overall performance, the influence of single parameters and the interrelation of parameters. We report these results on the training dataset with 10k documents. Further, we assess the stability of optimal parameters across varying sizes of the training dataset (1k, 10k and 100k documents).

5.1   Overall Performance

In terms of overall accuracy, results vary greatly across different parameter settings. Table 2 provides an overview of the results obtained on the 20newsgroup dataset with 10k training examples sorted by the rank of the model, where rank 1 is assigned to the best performing model. Table 2 also shows that the influence of single parameters on the model accuracy is not obvious, e.g., while worst performing models tend to have larger values for the number of negative samples θ_{ns}, some of the best performing models also have a value of $\theta_{ns} = 20$. The variation of accuracy across models is equally prominent on the amazon dataset (best performing model accuracy 0.9244, worst 0.1540) when trained on 10k documents, confirming that model hyperparameter settings are crucial for successful application of the learning algorithm. While the accuracy varies greatly across hyperparameter settings, the behavior depends on the dataset. As shown in Fig. 2 for some settings (e.g., 20newgroups, 10k training documents $\theta_d = 64$) accuracy decreases slowly from 73% and then drops rapidly (at approximately rank 230). This means that most of the models are quite similar in accuracy, i.e., most parameter configurations are "good", but some yield very low accuracy. In comparison, other settings (e.g., 20newsgroups, 1k training documents $\theta_d = 64$) show a steady decrease of accuracy, meaning that there are only some very good models, but many average- and bad-performing models.

We also observed that a well tuned doc2vec model achieves performance comparable to approaches using more complex features and the same classifier (k-NN). On the amazon dataset doc2vec achieves an accuracy of 0.924

---

[4] http://qwone.com/~jason/20Newsgroups/20news-bydate.tar.gz
[5] Details available in ipython notebooks https://doi.org/10.5281/zenodo.809860

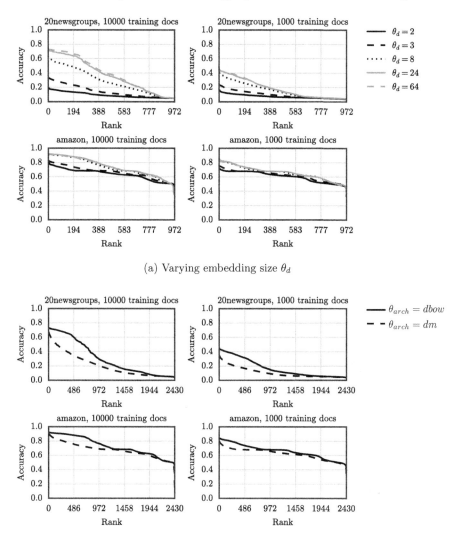

(a) Varying embedding size θ_d

(b) Varying model architecture θ_{arch}

Fig. 2. Classifier rankings for (a) different embedding sizes (θ_d) and (b) model architectures θ_{arch}. Accuracy obtained with 5-fold cross validation. Similar plots for other variable combinations are available via http://doi.org/10.5281/zenodo.495086

(with 10k training examples) comparable to 0.926 in [5]. On the 20newsgroups doc2vec achieves an accuracy of 0.73 with grid-search and 0.74 with Bayesian optimization, which is comparable to 0.73 in [5].

Table 2. Excerpt of the classifier results for the 20newsgroups dataset with 10k training examples ordered by model performance, showing accuracy a and standard deviation (averaged over cross-validation folds) of best and worst performing models for grid search (GS) and results from the Bayesian optimization (BO).

Meth.	Rank	θ_k	θ_α	θ_{arch}	θ_{hs}	θ_{epoch}	θ_{ns}	θ_d	θ_{win}	a (stdev)
GS	1	5	.10	dbow	1	30	20	64	5	**.7335** (.006)
GS	2	5	.01	dbow	1	50	20	64	5	**.7296** (.007)
GS	3	5	.01	dbow	1	50	5	64	20	**.7294** (.005)
GS	4	5	.01	dbow	1	50	5	64	50	**.7292** (.006)
GS	5	5	.10	dbow	1	30	20	64	20	**.7280** (.006)
GS	4856	25	.001	dbow	0	5	20	8	5	**.0449** (.004)
GS	4857	5	.001	dbow	0	5	5	24	20	**.0446** (.002)
GS	4858	25	.001	dbow	0	5	20	24	5	**.0445** (.006)
GS	4859	5	.001	dbow	0	5	5	8	20	**.0440** (.000)
GS	4860	25	.001	dm	0	5	5	8	20	**.0424** (.001)
BO	1	10	.0396	dbow	1	14	11	245	40	**.7415** (.003)
BO	2	8	.0893	dbow	1	15	15	58	48	**.7389** (.002)
BO	3	15	.0223	dbow	1	28	17	96	13	**.7341** (.004)

Table 3. Best classifications accuracy w.r.t the training set size using grid search. 18.8k is the size of the 20newsgroups dataset. Order of elements of optimal θ as in Table 2.

	10k	100k/18,8k
amazon	.9244	.9516
	$\theta^* = (25, 0.1, cbow, 1, 50, 5, 24, 20)$	$\theta^* = (10, 0.1, cbow, 1, 30, 5, 256, 5)$
20newsgroups	.7335	.8034
	$\theta^* = (5, 0.1, cbow, 1, 30, 20, 64, 5)$	$\theta^* = (10, 0.1, cbow, 1, 30, 5, 256, 5)$

5.2   Varying Training Data Set Size

The experiments confirm, that machine learning models benefit from more training data [3] (see Table 3). The accuracy gain of 7% on the 20newsgroups dataset when the training size is raised from 10k to 18.8k indicates that the accuracy could be enhanced even further, if more training data were available. The scaling on the amazon dataset is different: 10k training examples are sufficient to obtain good results. A tenfold increase in training samples only leads to a 2.7% accuracy gain. As the amount of training data increases, the ideal value for the learning rate also changes. Using 100k training documents, configurations using $\theta_\alpha = 0.01$ outplay those with $\theta_\alpha = 0.1$. But correspondingly, more iterations are necessary (100 or more compared to 30). Altogether, the learning rate is crucial; Fig. 3b (right) depicts that models with $\theta_\alpha = 1$ fall back to the performance of trivial classifiers. θ_d changes only slightly with respect to the training size, as shown by Fig. 3b (left) and Fig. 3a (left). With 100k training models using 64 dimensions perform marginally better than 24 dimensions, whereas the situation with 10k training examples is vice versa. A similar relation can be observed on the 20newsgroups dataset but with $\theta_d = 64$ and $\theta_d = 256$, respectively. In Fig. 3b (left) we also see that models using $\theta_d = 256$ overfit the data on the amazon dataset, which leads to a declining performance.

5.3 Parameter Influence

We plotted the model accuracy for different hyperparameter values as exemplified in Fig. 2 for the model architecture θ_{arch} and the embedding size θ_d. The plots were created by collecting the results of all models where one parameter was set to a specific value (e.g. the solid black line in Fig. 2a depicts the accuracy of all models that used two dimensional embeddings). These models were then ordered by accuracy. In terms of the model architecture, the distributed bag of words ($\theta_{arch} = dbow$) models outperformed the *distributed memory* ($\theta_{arch} = dm$) models in every scenario as depicted in Fig. 2b. Similar behaviour was observed for parameter θ_{hs}, models using hierarchical sampling which generally outperformed models not using hierarchical sampling. Very small embedding sizes ($\theta_d \in \{2, 3, 8\}$) have a strong negative impact on the accuracy. Larger embeddings sizes ($\theta_d \in \{24, 64\}$) yield more accurate classifiers as depicted in Figs. 2a and 3. Interestingly, the best results with comparatively small embedding sizes (e.g. 24 dimensions) are similar to those achieved with higher embedding sizes. Further, we found little to no effect of window size θ_{win}, the number of negative samples θ_{ns} and the number of nearest neighbors θ_k on the accuracy. Finally, we found a strong influence and interrelation on the accuracy when varying the learning rate θ_α and the number of epochs θ_{epoch}. Thus, when choosing these parameters they must be considered jointly as discussed subsequently.

5.4 Interrelation of Parameters

The experiments show a interrelation of the learning rate (θ_α) and the number of epochs (θ_{epoch}) (see Fig. 4 (left)). Good accuracy is achieved when a high learning rate is combined with few epochs. Likewise, a small learning rate in combination with many epochs gains similar results (see Fig. 4 at 50 epochs). It must be pointed out though that the training time mainly depends on the epochs, which makes a setting with a high learning rate and few epochs favorable when training time is crucial, since models with high learning rates are prone to overfitting. But in scenarios where accuracy is the top priority and the training time is negligible, a smaller learning rate with more epochs is favorable. Apart from the interrelation between the learning rate and the epochs our experiments found no additional interrelations (as exemplified by Fig. 4 (right)).

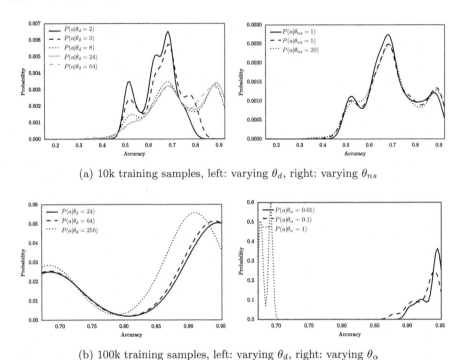

(a) 10k training samples, left: varying θ_d, right: varying θ_{ns}

(b) 100k training samples, left: varying θ_d, right: varying θ_α

Fig. 3. Approximation of classifier accuracy a on amazon dataset as probability density estimated using Gaussian Kernel Density Estimation. Similar plots for other parameters are available via http://doi.org/10.5281/zenodo.495086

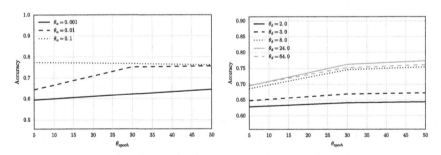

Fig. 4. Parameter interrelations on amazon dataset with 10k training samples. Left: To achieve optimal results θ_α and θ_{epoch} must be considered jointly, the parameters depend on each other. Right: To achieve optimal results θ_d is tuned without considering θ_{epoch}, the parameters are independent of each other.

6   Discussion

The experiments showed that, in general, hyperparameter settings have a huge impact on the accuracy (Sect. 5.1). Further, we found four categories of hyperparameters: those with no influence on the accuracy, those with a clear optimal value, those with many near-optimal values and those with a strong inter-relation among them. Hyperparameters of the first category are the windows size θ_{win} and the number of negative samples θ_{ns}. The missing effect of θ_{win} indicates that for this application a larger context is not predictive. Lau and Baldwin report $\theta_{ns} = 5$ being the best choice for the task of duplicate detection and determining semantic textual similarity [8], but performance for other values is not reported. The second category, hyperparameters with a clear optimal value, were found to be the type of architecture and whether hierarchical sampling was used. In our experiments the dbow architecture outperformed the dm architecture in every scenario (see Fig. 2), which accords with the literature [8–10]. Similarly, classifiers with hierarchical sampling performed better than those without. Thus, these parameters do not require much consideration, as they are binary and one configuration always outperforms the other. The size of the embeddings belongs to the third category, hyperparameters with many reasonable values. Very small embedding sizes ($\theta_d = \{2, 3, 8\}$) have a strong negative impact on the accuracy. But beyond that magnitude, there is a broad range of reasonable values (24 to 256) that can yield good classifiers (Sect. 5.1). The best results are achieved when θ_d is tuned according to the difficulty of the task at hand. On the relatively simple amazon dataset (four classes) an embedding size of 24 to 64 was optimal whereas the 20newsgroups dataset (20 classes) required 64 to 256 dimensions. All the learning hyperparameters fall in the fourth category. The individual parameters from the parameter subset θ_l (i.e., θ_{ts}, θ_α and θ_{epoch}) must be considered jointly and in connection with the desired scenario.

7   Summary

We presented a study on hyperparameters for document classification tasks using document embeddings, concretely doc2vec. Experiments on a text classification task showed that the window size and the number of negative samples have negligible influence, while the dbow and hierarchical sampling yield the best performance. For the size of the embeddings vectors there is a range of reasonable values (24–256). Model parameters and learning parameters showed no interrelation and can be tuned separately, while all learning parameters (θ_{ts}, θ_α, θ_{epoch}) must be considered jointly. Those insights can be used in further research or by practitioners to sensibly select initial hyperparameter configurations manually or restrict grid-search or Bayesian optimization approaches, which reduces optimization time substantially.

References

1. Bergstra, J., Bengio, Y.: Random search for hyper-parameter optimization. J. Mach. Learn. Res. **13**, 281–305 (2012)
2. Bergstra, J.S., Bardenet, R., Bengio, Y., Kégl, B.: Algorithms for hyper-parameter optimization. In: Advances in Neural Information Processing Systems, pp. 2546–2554 (2011)

3. Halevy, A., Norvig, P., Pereira, F.: The unreasonable effectiveness of data. IEEE Intell. Syst. **24**(2), 8–12 (2009)

4. Iacobacci, I., Pilehvar, M.T., Navigli, R.: Embeddings for word sense disambiguation: an evaluation study. In: Proceedings of Annual Meeting of the Association for Computational Linguistics, vol. 1, pp. 897–907 (2016)

5. Kusner, M.J., Sun, Y., Kolkin, N.I., Weinberger, K.Q., et al.: From word embeddings to document distances. ICML **15**, 957–966 (2015)

6. Lan, M., Tan, C.L., Low, H.B.: Proposing a new term weighting scheme for text categorization. AAAI **6**, 763–768 (2006)

7. Larochelle, H., Bengio, Y.: Classification using discriminative restricted Boltzmann machines. In: Proceedings of the 25th International Conference on Machine Learning, pp. 536–543. ACM (2008)

8. Lau, J.H., Baldwin, T.: An empirical evaluation of doc2vec with practical insights into document embedding generation. CoRR abs/1607.05368 (2016)

9. Le, Q.V., Mikolov, T.: Distributed representations of sentences and documents. In: Proceedings of International Conference on Machine Learning. JMLR Workshop and Conference Proceedings, vol. 32, pp. 1188–1196 (2014). JMLR.org

10. Liu, Y., Liu, Z., Chua, T.S., Sun, M.: Topical word embeddings. In: AAAI, pp. 2418–2424 (2015)

11. Manning, C.D., Raghavan, P., Schütze, H.: Introduction to Information Retrieval. Cambridge University Press, New York (2008)

12. Mikolov, T., Chen, K., Corrado, G., Dean, J.: Efficient estimation of word representations in vector space. CoRR abs/1301.3781 (2013)

13. Mikolov, T., Sutskever, I., Chen, K., Corrado, G.S., Dean, J.: Distributed representations of words and phrases and their compositionality. In: Advances in Neural Information Processing Systems, pp. 3111–3119 (2013)

14. Pennington, J., Socher, R., Manning, C.D.: Glove: global vectors for word representation. In: Proceedings of Empirical Methods in Natural Language Processing, pp. 1532–1543. EMNLP (2014)

15. Rasmussen, C.E., Williams, C.K.I.: Gaussian Processes for Machine Learning (Adaptive Computation and Machine Learning). The MIT Press, Cambridge (2005)

16. Sappadla, P.V., Nam, J., Loza Mencía, E., Fürnkranz, J.: Using semantic similarity for multi-label zero-shot classification of text documents. In: Proceedings of European Symposium on Artificial Neural Networks, Computational Intelligence and Machine Learning, vol. ESANN. d-side publications, Bruges, Belgium, April 2016

17. Sebastiani, F.: Machine learning in automated text categorization. ACM Comput. Surv. **34**(1), 1–47 (2002)

18. Snoek, J., Larochelle, H., Adams, R.P.: Practical Bayesian optimization of machine learning algorithms. In: Proceedings of International Conference on Neural Information Processing Systems, pp. 2951–2959. NIPS, USA (2012)

19. Wolpert, D.H.: The lack of a priori distinctions between learning algorithms. Neural Comput. **8**(7), 1341–1390 (1996)

Digital Humanities

Europeana: What Users Search for and Why

Paul Clough[1([⊠])], Timothy Hill[2], Monica Lestari Paramita[1], and Paula Goodale[1]

[1] University of Sheffield, Sheffield, UK
{p.d.clough,m.paramita,p.goodale}@sheffield.ac.uk
[2] Europeana, The Hague, Netherlands
timothy.hill@europeana.eu

Abstract. People use digital cultural heritage sites in different ways and for various purposes. In this paper we explore what information people search for and why when using Europeana, one of the world's largest aggregators of cultural heritage. We gathered a probability sample of 240 search requests from users via an online survey and used qualitative content analysis complemented with Shatford-Panofsky's mode/facet analysis for analysing requests to visual archives to investigate the following: (i) the broad type of search task; (ii) the subject content of searches; and (iii) motives for searching and uses of the information found. Results highlight the rich diversity of searches conducted using Europeana. Contributions include: collection and analysis of a comprehensive sample of Europeana search requests, a scheme for categorising information use, and deeper insights into the users and uses of Europeana.

Keywords: Search tasks · Digital cultural heritage · Europeana

1   Introduction

Users from diverse backgrounds are coming to cultural heritage websites and information services with increasingly varied goals, tasks and information needs [19]. There is a need to provide systems that go beyond keyword-based search and support wider information seeking behaviours, such as browsing and exploration [11,22]. Users' individual differences (e.g., age, gender, domain knowledge and learning style), search task and context (e.g., location and time), are known to affect the ways in which people search for information. Typically people search such sites not as an end in itself, but rather as part of their broader work (and leisure) tasks and informational goals: "Searching is merely a means to an end - a way to satisfy some underlying goal ... 'why are you performing that search?"' [16, p. 13]. Having a better understanding of users, their goals and tasks can therefore help with the design of more effective information systems.

Task-based information retrieval is a popular area of study. Byström and Hansen [4] characterise tasks at three levels: (i) information intensive *work tasks*; (ii) information *seeking tasks*; and (iii) *information search* (or information retrieval) tasks. The work task is often a trigger for someone's interaction with

© Springer International Publishing AG 2017
J. Kamps et al. (Eds.): TPDL 2017, LNCS 10450, pp. 207–219, 2017.
DOI: 10.1007/978-3-319-67008-9_17

a search system. However, focus has increasingly turned to non-work settings, e.g., casual-leisure, where other factors such as curiosity or free time can trigger people's information seeking activity [7]. For cultural heritage information services, such as Europeana, users carrying out both work and casual-leisure tasks will initiate their interactions and therefore should be studied [18]. In this paper we provide an in-depth study of tasks for Europeana, mainly at the level of the search task, which has previously not been conducted. In particular we consider the following research questions: [RQ1]: What information do users search for using Europeana? and [RQ2]: What do users use this information for? The paper is structured as follows: Sect. 2 describes related work; Sect. 3 describes the methodology used in this study; Sect. 4 describes the categorisation of search tasks; Sect. 5 provides the results and Sect. 6 concludes the paper and provides avenues for future work.

2   Related Work

2.1   Goals, Tasks and Activities

The terms 'task', 'goal' and 'activity' are often used interchangeably when describing users' seeking behaviours. More formally a *task* is "what someone does to achieve a goal" [9, p. 56]. In the context of information seeking, a task is defined as "the manifestation of an information seeker's problem and ... what drives information seeking actions" [14, p. 36]. Tasks are driven by underlying *goals* (the purpose or intent of the activity) and can be differentiated based on the specificity of the goal, the quantity of information to be searched and the expected time to complete the task. The lowest level of task (search task) involves a user searching using a specific application (e.g., search engine). White [21] defines a search task as an "atomic information need resulting in one or more queries." Ingwersen and Jarvelin [10, p. 20] define a search task as "... a means to obtaining information to fulfil a work task, and include information need generation, information interaction and search task solving." Tasks invoke *activities*, which can occur at multiple levels [4]. For example, at the level of information seeking this could include query formulation, results examination, etc. Studies have been carried out to investigate the broad range of information-related activities people undertake, particularly on the web. For example, Sellen et al. [17] describe six types of activity carried out on the web: finding, information gathering, browsing, transacting, communicating and housekeeping. Similarly, Kellar et al. [12] use the following categories (and sub-categories): (i) information seeking (fact-finding, information gathering, browsing); (ii) information exchange (transaction and communication); and (iii) information maintenance. In this paper we consider search task level as people search using Europeana.

2.2   Search Tasks and Goals in Cultural Heritage

Amin et al. [1] investigated the information seeking behaviours of cultural heritage experts as they carry out their daily search activities. This included identifying their search motivations, types, sources and tools, and categories of information task (based on [12]). For experts, a majority of search tasks involved complex information gathering (e.g., finding information to compare similarities and differences between objects). Contrasting with experts, Skov [18] carried out a study of online museum visitors in an everyday life information-seeking context. Based on results from a web-based questionnaire and follow-up interviews with 24 participants, the information needs of enthusiasts were identified and generally were found to be for well-defined known items and not for more exploratory information needs, e.g. "Seek information on King Christian the Tenth's hunting weapons (writing a journal article)." In this study we specifically consider types of information seeking and searching tasks from multiple user groups (e.g., professionals and casual users).

A number of prior studies have also been conducted to understand Europeana's users. For example, Europeana's 2014 survey [8] showed that the majority of users were in the 25–54 age group and many (27%) were first-time users with 72% visiting the site a few times a month or less. The most common reason for visiting the site was exploration within a topic (32%) with finding out more about Europeana a close second (30%). Most respondents came to Europeana through a link from another website. Results and further data collected from user studies and transactions logs has served to inform a series of Europeana personas: "each persona represents many users and a set of personas represents a spectrum of the target user groups" [15, p. 106]. In addition, user studies lead to the specification of two distinct types of Europeana user (see D3.1): (i) 'culture vultures' and (ii) 'culture snackers'. The former group are dedicated enthusiasts and professionals: they have domain expertise and likely lifelong enthusiasts of cultural heritage (likely to be returning users and mainly wanting to use Europeana to find resources to use in their own work, gain knowledge, expertise or inspiration). The latter group are more representative of the novice or general user who come with lower levels of technical/domain expertise and typically engage for general interest. Our work complements these existing studies.

3   Methodology

Various approaches have been employed to investigate search tasks, including diary studies and interviews [1], analysing samples from query logs [2,3] and pop-up web surveys [3]. In this study we made use of a web-based pop-up survey from which we could gather responses from actual users of Europeana as they carried out their searching activities. Such approaches are commonly employed in web surveys (e.g., usability testing and gathering feedback) and in the future we plan to aggregate this with other methods of data collection to provide richer insights of users' information searching behaviour.

3.1   Pop-up Web Survey

The pop-up web survey is a form of intercept survey where systematic sampling is used to intercept visitors of a website. Creating the pop-up survey involved

Table 1. Questions (first 6 out of 10) used in the pop-up survey.

No.	Question	Response
1	How often do you visit Europeana?	[Everyday, At least once a week, At least once a month, Less than once a month, This is my first visit]
2	How would you identify yourself	[Cultural heritage enthusiast, Student, Academic, Teacher, Cultural heritage professional, Other]
3	How did you get to Europeana today?	[Via a link from a search engine, Via a link from social media, I knew about the site already so came directly here, Via a link from teaching resources, Other]
4	What information are you looking for right now?	[Open]
5	Why are you looking for this information?	[Open]
6	After finding this information, you will:	[Look for more information on the same topic using Europeana, Look for more information using other resources, Browse Europeana (e.g., look for other interesting things), Have completed everything you need to do, Other]

multiple iterations (including pilot testing), particularly in relation to question design. We attempted to balance participant time and effort against the need to capture sufficient detail about users' current search activities in a fairly unconstrained manner. We therefore devised a set of 10 questions (the first 6 shown in Table 1, the remainder asking participants' level of subject knowledge and suggestions for further improvements to Europeana) that could be shown to users at any point during their interaction with Europeana.

The design of Q4 and Q5, the main focus of this paper, were modelled on Broder's pop-up survey [3] to investigate users' search goals. The wording of other questions was based on prior literature and surveys, including past Europeana studies. To aid users' interpretation of questions we provided additional text. For example, in the response options for Q2: "Cultural heritage enthusiast (e.g., hobbyist, genealogist, amateur historian)". Also, in Q4 and Q5 we provided example text to ensure sufficient input. For example, in Q4 we provided the following examples: "'I want to find an image of the Mona Lisa', 'I'm trying to explore what's available in Europeana on World War I', 'I am looking for photographs of Sheffield in the 1980s', 'I am looking for artwork by Leonardo Da Vinci', or 'Don't know/nothing specific' ".

The survey was administered using the Hotjar service Europeana routinely employs for user surveys. The survey was administered in English and was shown to 30% of users (later increased to 66% to increase response rates) who visited Europeana using desktop or tablet devices. The survey was triggered when users scrolled halfway down either a search results page, or a Europeana item page. Users who completed the survey were given the opportunity to enter a prize draw to win a €50 Amazon voucher. In addition to the questions posed in the pop-up survey, Hotjar also captured the date and time of submission and the respondent's country of origin. The study was approved by the University of Sheffield's Ethics Committee.

3.2   Data Analysis

The majority of analysis effort required for this paper related to the free-text responses for Q4 and Q5. These were used, along with responses from other questions, to investigate the following aspects of users' search tasks: (i) the broad *type* of search activity; (ii) the *subject content* of the search request; and (iii) the *motive* for conducting the search and *use* of the information found. The general approach adopted in analysing the data was *qualitative content analysis* based on Zhang and Wildemuth [24]. This was mainly an inductive approach, but informed by existing frameworks where applicable. For example, we utilised an approach for analysing requests to archives and libraries serving audiovisual content [2]. Following the development of various categorisation schemes (see Sect. 4) we involved a further researcher to validate the scheme (a sample of 50 responses, achieving around 76% agreement) after which we discussed differences and refined the scheme (and amended our coding) where necessary. For statistical analysis IBM's SPSS (version 22.0.01) was used.

3.3   Participants

The pop-up survey ran for 2 weeks (21 March – 4 April 2017) and elicited responses from 240 users of Europeana from 48 different countries (Spain 12.9%, US 8.9%, Italy 8.9%, France 7.1%, Germany 6.7%, UK 6.3%, Netherlands 4.2%, Sweden 3.3%, Hungary 3.3%, Brazil 2.9%). The majority of users were first time visitors to Europeana (27.1%); with 26.3% visiting at least once a month, 22.9% visiting less than a month, 20% visiting at least once a week and 3.8% visiting every day. Participants mainly came to Europeana having already known about the site (48.8%); with 34.2% arriving via a link from a search engine; 5.8% via a link from teaching resources; and 5% from a link via social media. The majority of respondents (30.4%) described themselves as academic. This group was followed by cultural heritage enthusiasts (24.6%); cultural heritage professionals (18.3%); students (13.3%); school teachers (4.6%); and others (8.8%).

4   Analysis of the Search Requests

One of the major challenges was analysing the rich data provided by the free-text responses describing users' search requests: Q4 (mean $= 10.1$ words, min $= 1$, max $= 49$) and Q5 (mean $= 8.3$ words, min $= 1$, max $= 72$). In the end we made use of the categorisation scheme by Armitage and Enser [2] for analysing the subject content of user requests for still and moving visual images. This approach has been applied in various previous studies [5,6] and proved to be readily applicable to Europeana's search requests (Q4), which commonly refer to audiovisual content. In this approach to subject analysis search requests are represented in a 2×2 matrix of unique/non-unique, refined/unrefined queries. Unique (or specific) subjects are "those concerned with named individuals, one-off events, singular objects or location" [2] (p. 288) - for example, 'images of Stuttgart', 'rare old images or texts about constantinople'. Non-unique (or general) subjects concern more generic subjects, kinds of people, events and places. For example: 'I am looking for images that convey the scope of humanitarian aid today' and 'I want to find informations about caricatures'.

In situations where the request contains both unique and non-unique aspects (e.g., 'Rio carnivals') then refinements can be used (e.g., a non-unique subject 'carnivals' refined by location 'Rio'). Although conceptually this offers a simple approach to analysing requests there are a number of difficulties faced when applying this in practice, especially in determining between the main subject of the request and its refiners [2]. For example, 'maps of Dublin' - is the request a general subject (maps) refined by location or vice-versa? Also, in the simple categorisation details of the subject content of the request are lost; therefore, Armitage and Enser [2] make use of Shatford-Panofsky's modes of image analysis in the form of mode/facet analysis. In this approach each subject element of the query is categorised as specific or general capturing aspects of 'who', 'what', 'where' and 'when' (see Sect. 4.2).

4.1   Categorisation of Search Task

Search tasks can be categorised in various ways, such as by goal or intent, complexity, search tactics and moves, timeframe and specificity [13,23]. Many of the prior schemes, however, are specific to web search and less suited to cultural heritage. In addition, there are a multitude of definitions[1] categorising information seeking and searching tasks. Toms [20] distils search tasks into two main categories: (i) specific item or information object (finding specific pieces of information, e.g. known-item, fact-finding, closed, transactional and navigational, name of person/organisation, etc.); and (ii) general topical search (finding information about a topic, e.g. informational, open, etc.). In this study we coded the search requests (mainly Q4) based mainly on the specificity of subjects expressed in the search request and search goal. The preliminary analysis of requests as unique or non-unique was useful in identifying whether people may be searching for specific subjects (unique) versus more general topic searches (non-unique). We used the following categories:

---

[1] For example, see the database of search tasks developed by Wildemuth et al.: https://ils.unc.edu/searchtasks/ (site visited: 20 June 2017).

Specific-item search: Search for specific item (i.e., known-item) typically expressed precisely (e.g., using title of book), e.g., "Boletín Oficial de Instrucción Pública", "I am looking for the 1919 film 'Les fetes de la victoire.' "

By named author: Search for information by a specific named author (or provider), e.g., "to look at paintings by Henriette Ronner", "I am searching for images of artifacts from the Regional Archaeological Museum Plovdiv." If referring to a known-item, however, we treat this as a specific-item.

Specific-subject search: Find information for specified (or named) subject (i.e., person, place, location, etc.) forming the main subject of the request, e.g., "I am looking for pictures of Stuttgart", "I'm looking for plans and images of Clermont-Ferrand."

General topical search: Find information for general subject, e.g., "Italian medieval illuminations", "Looking at examples of art made by women."

Browsing/Exploring: Used to identify searches where the user has no specific goal, e.g., "I am trying to explore the world through what is available in Europeana", "I'm just browsing your collections."

Ambiguous or unclear: Examples where the search request is unclear or difficult to determine category, e.g., "I'm an Opera lover", "book."

4.2   Categorisation Based on Mode/Facet Analysis

Analysis of the subject of the search request was based on the approach described in Armitage and Enser [2]. Components of the search request were categorised using the following codes:

- **General object/thing** (e.g., 'paintings', 'explorers accounts')
- **Specific object/thing** (e.g., 'Prelude, Op. 28, No. 7, by Frédéric Chopin')
- **General person/group** (e.g., 'working women', 'historical figures')
- **Specific person/group** (e.g., 'Saint Francis of Assisi')
- **General location** (e.g., 'public places', 'where my ancestors lived')
- **Specific location** (e.g., 'Spain', 'Norfolk')
- **General event/action** (e.g., 'working', 'privatization of school system')
- **Specific event/action** (e.g., 'Great War', 'black death')
- **General time** (e.g., 'medieval', 'today')
- **Specific time** (e.g., '1940', 'XIX century')

We also introduced additional codes we felt useful for analysing the search requests and adding further detail:

- **General subject** (e.g., 'art', 'history')
- **Creator** or **Provider** (e.g., "paintings by Van Gogh", "items from Vienna National Museum")
- **Nationality** (e.g. 'Icelandic art works')
- **Language** (e.g., 'books written in Italian')

- **Availability** (e.g., 'free open-source 3d models', 'public domain')
- **Response** (e.g., 'looking for a *nice* painting)

The categories were then used to identify the subject components of a search request. For example, "Great War photographs taken on exactly 100 years ago" would be coded as 'Specific event/action (Great War) + Specific time (100 years ago)'. The following example "I want to find information about old routes/-path in the South West of Spain" would be coded as 'General object/thing (old routes/path) + Specific location (Spain)'. During the coding, each type of subject category is applied just once (e.g., if multiple specific people are mentioned this is recorded as just one occurrence of 'Specific person/group'). In practice the requests are typically short enough that multiple occurrences of the same type do not occur. Finally, analysis is also performed to identify the **Medium** category, i.e. terms in the request where the user specifically refers to a media type (e.g., image, video, text, etc.).

4.3 Categorisation of Motives and Use

A final part of the analysis considered why people were searching for information during their current activity. This typically elicited from users a specific purpose for searching Europeana (e.g., work task or personal interest) and often the use to which the information gathered would be put. No prior suitable scheme could be found to categorise information use for our data, therefore we create a taxonomy for the various motives given by users for their search:

To create a new work: In this category, the purpose of the user is ultimately to create some new cultural artefact of some kind. The most common examples of this kind of task in the responses are monographs, articles, and visual art-pieces. This category can be subdivided in terms of:

- **task closure**: works can be considered *'open-ended'* if the user is the person who chiefly decides upon the form and content of the artefact produced (e.g., academic research). Works for which the form and subject are assigned by others can be considered *'closed'* (e.g., school/university assignment). In cases where this is not apparent, the task closure were coded as *'not specified'*.
- **modification**: this sub-category describes the extent to which the found content will be transformed by the user in production of the new work. At one end are *'remediated'* cases in which the user is looking for 'inspiration': here, the contribution of the found content to the end product may be completely unrecognisable to anyone except the artist who created it. At the other end (*'unmediated'*) are tasks in which the user is simply looking for an image to illustrate, e.g. a presentation or pamphlet, where the found content is essentially cut-and-pasted into position. While judgements of degree of remediation are necessarily to some extent subjective, where the user does not specify guidance can be found in the kind of output envisaged – monographs and articles will typically involve significant remediation; presentations, flyers and Tweets will normally demand less. The user's anticipated next steps are

also indicative: if the user considers that the task will be essentially complete once the content is found, they presumably envisage little modification being required.

- **type of output**: this sub-category defines the kind of output produced, e.g. textual.

For example, "a work of Edmund Husserl" (Q4) and "to write a paper" (Q5) would result in 'Create new work - Open-ended - Remediated - Textual'.

Professional activity: This category is intended to capture the activity of (chiefly) academics and cultural heritage professionals where the focus is purely research- or monitoring-oriented, and no precise output from the search is anticipated. For instance, a researcher may simply be attempting to keep abreast of current developments in their field, or a curator may be checking up on how their institution's content is displayed on the Europeana platform itself. Note that this category does not cover casual users who are simply 'checking out the site'; the search task must be specifically focused upon some job- or learning-oriented task.

Personal interest: The information will be used for personal or general interest. This interest may be of one of two types: *'transient'* or *'sustained'*. Transient interest is a focus that lasts for the length only of a single session: although the user's interest is piqued, they have little prior knowledge of or investment in the topic or object being searched for. Users who enter the site via social media links will often be of this type. Sustained interest lasts over the course of more than one session. Users will often speak of having a 'collection' of items related to their search, or describe antecedent searches that have led them to this point. Genealogical research can also be considered a sustained interest.

Teaching: The user is a person in a teaching role, and using the site to produce teaching resources - e.g., lesson plans and assignments.

Other: This category includes any other activities not included above.

5   Results and Discussion

5.1   RQ1: What Information Are Users Searching For?

Results show that the largest single search category (47.1%) of tasks is *general topical search*. This is followed by *specific-subject searches* (24.6%); *specific-item searches* (11.3%); searches by *named author* (7.1%) and *browse/explore* (7.1%). Broken down by group, we observe that the highest proportion of specific-item searches (63%) come from academics, while the highest proportion of browse/explore searches (29.4%) come from cultural heritage enthusiasts. We also note differences based upon referrer: the greatest proportion of general topical searches (51.3%) come from people who already knew about the site and so came directly to it; whereas the greatest proportion of specific-item searches (48.1%) come from people coming to Europeana via a search engine link.

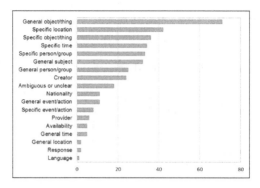

Fig. 1. Frequency of occurrence for each mode/facet in search requests.

The mode/facet analysis helps to provide insights into the subject content of search requests. First, we calculate the frequency of occurrence of each type of mode/facet (not including *Medium*). As shown in Fig. 1, the most frequent mode/facet is *general object/thing* (71 occurrences), followed by *specific location* (42 occurrences). Search requests comprise an average of 1.53 modes/facets (min = 1, max = 5). The most common combinations are "Creator + Specific object/thing", such as "I want to find some information about a painting of Willem van de Velde, 'Het kanonschot"' (9 occurrences), and "Creator + General object/thing" (8 occurrences), e.g., "I am looking for artworks by Leonardo da Vinci". We find that the *Medium* mode/facet is commonly used to refine the search (81 occurrences), e.g., 'images of Stuttgart' and 'I am looking for photographs of The Trachian tomb near to village of Mezek, Bulgaria.'

5.2   RQ2: Why Are Users Searching for the Information?

The results of analysing the search requests based on users' motives for conducting their search activities and the potential uses of the information once found also provide interesting insights into how users search Europeana. Table 2 shows the breakdown of search requests based on the analysis of motives and use carried out in Sect. 4.3 and cross-tabulated by search task.

The majority of users (37.1%) were searching Europeana with the intention of using the information found to *create a new work*, e.g. "to write a book", "to prepare an exhibition", "to use images for a presentation", and "to find additional material for my PhD-thesis." Inspecting this category more closely through the use of the sub-categories, we find that in 83.9% of cases the users were involved in 'open-ended' tasks (e.g., scholarly research), 14.9% in 'closed' tasks (e.g., school assignment), and 1.1% unspecified. Furthermore, in the modification sub-category, we found that 36.8% of users represent 'unmediated' cases, i.e. they would be making use of the information found (typically images) without modification (e.g., to illustrate an article or presentation), whilst 57.5% are 'remediated' cases. Our results also show that 64.4% of the newly created works would be textual in form (e.g., academic article); with 6.9% in a visual form; and 3.4% in audiovisual form.

Under the category of *personal interest* (27.5% of search tasks) we find that users are typically cultural heritage enthusiasts (e.g., family historians), with Europeana serving as one of their genealogical resources. We categorised 57.6% of cases of personal interest as 'sustained', i.e. the users are likely to have an interest in the topic beyond their current search activity on Europeana; 13.6% were categorised as 'transient'. Examples of responses for personal interest include "to enrich my personal archive" and "inspiration and general interest."

We categorised 20.8% of search tasks as *professional activities*, with examples including "it's my job", "fits in with my research project" and "to check whether the information was correct." Finally, 7.9% of search tasks were categorised under the *teaching* category, e.g., "to illustrate a university lecture." Breaking down motivation by search task (Table 2), in the case of specific-item searches information from 48.1% of searches is used to create a new work, commonly reflecting the greater search for specific-items by academics. In contrast, for specific-subject searches the majority of search tasks are split between personal interest (44.1%) and creating a new work (42.4%). The results highlight, again, the differences obtained based on the user's search task. (Dataset available for download from: http://bit.ly/europeanaSearchTasks)

Table 2. Cross-tabulation of users' motivation for searching vs. search task.

	Browse/ explore	By named author	General topical search	Specific-item search	Subject-specific search	Total
Create new work	17.6%	23.5%	37.2%	48.1%	42.4%	37.1%
Personal interest	35.3%	29.4%	22.1%	11.1%	44.1%	27.5%
Professional activity	11.8%	41.2%	26.5%	22.2%	5.1%	20.8%
Teaching	17.6%	5.9%	5.3%	14.8%	8.5%	7.9%
Other			0.9%			0.4%
Ambiguous/unclear	17.6%		8%	3.7%		6.3%
Total	100%	100%	100%	100%	100%	100%

6   Conclusions and Future Work

Digital cultural heritage sites, such as Europeana, are being used by increasingly diverse groups of users with varying needs and goals. In this paper we have investigated, through gathering a sample of users' search requests from a web-based survey, the types of searches conducted on Europeana, users' typical motives for searching and common uses of the information found. Existing methods for analysing the subject content of search requests to audiovisual archives were used to better understand the searches. A new scheme was designed for categorising users' search motives and subsequent uses of information found. As well as providing insights into search behaviour for Europeana, the results also help better understand search tasks more generally in cultural heritage across a wider range of users types than previously studied. We recognise there are limitations in our study (e.g., subjectivity in the coding, use of online survey only to elicit search requests) and therefore aim to pursue a number of avenues for further work. This includes validating and developing a more refined categorisation scheme, conducting deeper analysis of the current dataset, and combining the data from this study with data derived from other sources, such as search logs and diary studies, to gain deeper insights into aspects of users' search activity.

Acknowledgements. This study was funded by the European Commission under 'Europeana DSI-2'. We thank users of Europeana for participating in the online survey.

References

1. Amin, A., van Ossenbruggen, J., Hardman, L., van Nispen, A.: Understanding cultural heritage experts' information seeking needs. In: Proceedings of JCDL 2008, NY, USA, pp. 39–47. ACM, New York (2008)
2. Armitage, L.H., Enser, P.G.B.: Analysis of user need in image archives. J. Inf. Sci. **23**(4), 287–299 (1997)
3. Broder, A.: A taxonomy of web search. In: ACM Sigir Forum, vol. 36, pp. 3–10. ACM (2002)
4. Byström, K., Hansen, P.: Conceptual framework for tasks in information studies. J. Am. Soc. Inform. Sci. Technol. **56**(10), 1050–1061 (2005)
5. Chen, H.: An analysis of image queries in the field of art history. JASIST **52**(3), 260–273 (2001)
6. Choi, Y., Rasmussen, E.M.: Searching for images: the analysis of users' queries for image retrieval in American history. J. Am. Soc. Inform. Sci. Technol. **54**(6), 498–511 (2003)
7. Elsweiler, D., Wilson, M.L., Lunn, B.K.: Chapter 9 Understanding casual-leisure information behaviour. In: New Directions in Information Behaviour, chap. 9, pp. 211–241
8. Europeana: Results of the Europeana User Survey 2014 - Europeana Professional, June 2014, http://pro.europeana.eu/blogpost/results-of-the-europeana-user-survey-2014. Accessed 14 Mar 2016
9. Hackos, J.T., Redish, J.: User and Task Analysis for Interface Design (1998)
10. Ingwersen, P., Järvelin, K.: Information retrieval in context: Irix. In: ACM SIGIR Forum, vol. 39, pp. 31–39. ACM (2005)
11. Johnson, A.: Users, use and context: supporting interaction between users and digital archives. What Are Archives?: Cultural and Theoretical Perspectives: A Reader, pp. 145–164 (2008)
12. Kellar, M., Watters, C., Shepherd, M.: A field study characterizing web-based information-seeking tasks. JASIST **58**(7), 999–1018 (2007)
13. Li, Y.: Exploring the relationships between work task and search task in information search. J. Am. Soc. Inform. Sci. Technol. **60**(2), 275–291 (2009)

14. Marchionini, G.: Information Seeking in Electronic Environments, vol. 9. Cambridge University Press, Cambridge (1997)
15. Rasmussen, K., Petersen, G.: Personas. In: Dobreva, M., O'Dwyer, A., Feliciati, P. (eds.) User Studies for Digital Library Development, chap. 10, pp. 105–113. Facet (2012)
16. Rose, D.E., Levinson, D.: Understanding user goals in web search. In: Proceedings of WWW 2004, pp. 13–19. ACM (2004)
17. Sellen, A.J., Murphy, R., Shaw, K.L.: How knowledge workers use the web. In: Proceedings of the SIGCHI conference on Human factors in computing systems, pp. 227–234. ACM (2002)
18. Skov, M.: Hobby-related information-seeking behaviour of highly dedicated online museum visitors. Inf. Res. **18**(4) (2013)
19. Skov, M., Ingwersen, P.: Exploring information seeking behaviour in a digital museum context. In: Proceedings of IIIX 2008, NY, USA, pp. 110–115 (2008), http://doi.acm.org/10.1145/1414694.1414719
20. Toms, E.: Task-based information searching and retrieval. In: Ruthvan, I., Kelly, D. (eds.) Interactive Information Seeking, Behaviour and Retrieval, pp. 43–59. Facet Publishing (2011)
21. White, R.W.: Interactions with Search Systems. Cambridge University Press, Cambridge (2016)
22. Whitelaw, M.: Generous interfaces for digital cultural collections. Digital Humanities Quarterly **9**(1) (2015)
23. Wildemuth, B.M., Freund, L.: Search tasks and their role in studies of search behaviors. In: Proceedings of HCIR 2009 (2009)
24. Zhang, Y., Wildemuth, B.M.: Qualitative analysis of content. In: Wildemuth, B.M. (ed.) Applications of Social Research Methods to Questions in Information and Library Science, pp. 308–319. Libraries Unlimited, Westport (2009)

Metadata Aggregation: Assessing the Application of IIIF and Sitemaps Within Cultural Heritage

Nuno Freire[1]([✉]) [ID], Glen Robson[2] [ID], John B. Howard[3] [ID],
Hugo Manguinhas[4], and Antoine Isaac[4] [ID]

[1] INESC-ID, Lisbon, Portugal
nuno.freire@tecnico.ulisboa.pt
[2] National Library of Wales, Aberystwyth, UK
glen.robson@llgc.org.uk
[3] University College Dublin, Dublin, Ireland
john.b.howard@ucd.ie
[4] Europeana Foundation, The Hague, The Netherlands
{hugo.manguinhas,antoine.isaac}@europeana.eu

Abstract. In the World Wide Web, a very large number of resources is made available through digital libraries. The existence of many individual digital libraries, maintained by different organizations, brings challenges to the discoverability and usage of the resources. A widely-used approach is metadata aggregation, where centralized efforts like Europeana facilitate the discoverability and use of the resources by collecting their associated metadata. This paper focuses on metadata aggregation in the domain of cultural heritage, where OAI-PMH has been the adopted solution. However, the technological landscape around us has changed. With recent technological accomplishments, the motivation for adopting OAI-PMH is not as clear as it used to be. In this paper, we present the first results in attempting to rethink Europeana's technological approach for metadata aggregation, to make the operation of the aggregation network more efficient and lower the technical barriers for data providers. We (Europeana and data providers) report on case studies that trialled the application of some of the most promising technologies, exploring several solutions based on the International Image Interoperability Framework (IIIF) and Sitemaps. The solutions were trialled successfully and leveraged on existing technology and knowledge in cultural heritage, with low implementation barriers. The future challenges lie in choosing among the several possibilities and standardize solution(s). Europeana will proceed with recommendations for its network and is actively working within the IIIF community to achieve this goal.

Keywords: Metadata · Cultural heritage · Metadata aggregation · Web technology · Standards

© Springer International Publishing AG 2017
J. Kamps et al. (Eds.): TPDL 2017, LNCS 10450, pp. 220–232, 2017.
DOI: 10.1007/978-3-319-67008-9_18

1   Introduction

In the World Wide Web, a very large number of resources are made available through digital libraries. The existence of many individual digital libraries, maintained by different organizations, brings challenges to the discoverability and usage of the resources by potentially interested users.

An often-used approach is metadata aggregation, where a central organization takes the role of facilitating the discovery and use of the resources by collecting their associated metadata. Based on these aggregated datasets of metadata, the central organization (often called aggregator) is in a position to further promote the usage of the resources by means that cannot be efficiently undertaken by each digital library in isolation. This scenario is widely applied in the domain of cultural heritage (CH), where the number of organizations with their own digital libraries is very large. In Europe, Europeana has the role of facilitating the usage of CH resources from and about Europe, and although many European CH institutions do not yet have a presence in Europeana, it already holds metadata from over 3,500 providers[1].

In this paper, we present the first results of our work on rethinking Europeana's technological approach for metadata aggregation. Our goal is to make the operation of the aggregation network more efficient and lower the technical barriers for data providers to contribute to Europeana. Our approach was to undertake case studies with real CH collections and systems from data providers of the Europeana network. The case studies were based on promising technologies, which have been identified in our previous work. The results achieved make the following contribution to digital library research:

- A functional analysis for innovative use of state of the art technologies, based on a large network of data providers – the Europeana Network.
- A real-world application experience of open standards, thus contributing for their future improvement.

The paper will describe the technological approach to metadata aggregation most prevalent in CH in Sect. 2. Further details regarding metadata aggregation in CH are presented in Sect. 3. The series of case studies that were performed are presented in Sect. 4, along with their analysis. Section 5 summarizes the outcomes.

2   State of the Art: Metadata Aggregation in Cultural Heritage

In the CH domain, the technological approach to metadata aggregation has been mostly based on the OAI-PMH protocol, a technology initially designed in 1999 [1]. OAI-PMH was originally meant to address shortcomings in scholarly communication by providing a technical interoperability solution for discovery of e-prints, via metadata aggregation.

---

[1] Source: http://statistics.europeana.eu/europeana [consulted on 27th of April 2017].

The CH domain embraced OAI-PMH, since discovery of resources was only feasible if based on metadata instead of full-text [2]. In Europe, OAI-PMH was the technological solution adopted by Europeana since its start, to aggregate metadata from its network of data providers and intermediary aggregators.

However, the technological landscape around our domain has changed. Nowadays, with the technological improvements accomplished by network communications, computational capacity, and Internet search engines, the discovery of e-prints is largely based on full-text processing, thus the newer technical advances, such as ResourceSync [7], are less focused on metadata. Within the CH domain metadata-based discovery remains the most widely adopted approach since a lot of material is not available as full-text. The adoption of OAI-PMH for this purpose is not as clear as it used to be, however. OAI-PMH was designed before the key founding concepts of the Web of Data [3]. By being centered on the concept of repository, instead of focusing on resources, the protocol is often misunderstood and its implementations fail, or are deployed with flaws that undermine its reliability [2]. Another important factor is that OAI-PMH predates REST [4]. Thus, it does not follow the REST principles, further bringing resistance and difficulties in its comprehension and implementation by developers in CH institutions.

An additional aspect relevant for our work, is that CH institutions are increasingly applying technologies designed for wider interoperability on the World Wide Web. Particularly relevant are those related with Internet search engine optimization and the IIIF (International Image Interoperability Framework[2]). Regardless of the metadata aggregation process for Europeana, CH institutions are already interested in developing their systems' capabilities in these areas. By exploring these technologies, the participation in Europeana may become much less demanding for these institutions.

3 Characterization of Metadata Aggregation in Cultural Heritage

The CH domain has specific characteristics that influence how metadata aggregation is done. We consider the following to be the most influential:

- Several sub domains compose the CH domain: Libraries, Archives, and Museums.
- Each sub-domain applies its specific resource description practices and data models.
- All sub-domains embrace the adoption and definition of standards based solutions addressing description of resources, but to different extents. A long-time standardization tradition has existed in libraries, while it is more recent in archives and museums.
- Interoperability of systems and data is scarce across sub-domains, but it is common within each sub-domain, both at the national and the international level.
- Adopted standards tend to use XML-based data models, while models based on relational data are rare.

---

[2] International Image Interoperability Framework - http://iiif.io/#international-image-interoperability-framework.

- Organizations typically have limited budgets to devote to information and communication technologies, thus the speed and extent of innovation and adoption of new technologies is slow.

In this context, a common practice has been to aggregate metadata using an agreed data model that allows the data heterogeneity between organizations and countries to be dealt with in a sustainable way. The models typically seek to meet two main requirements: (a) retaining the semantics of the original data from the source providers; (b) supporting the information needs of the services provided by the aggregator.

These two requirements are typically addressed in a way that keeps the model complexity low, with the intention of simplifying the understanding of the model by all kinds of providers and to keep a relatively low barrier for both providers and aggregators to implement data conversion solutions,

Another relevant aspect of metadata aggregation is the sharing of the sets of metadata from the providing organizations to the aggregator. The metadata is transferred to the aggregator, but it continues to evolve at the data provider side, thus the aggregator needs to periodically update its copy of the data. In this case, the needs for data sharing can be described as a data synchronization problem across organizations.

In the CH domain, OAI-PMH is also the most well-established solution to address the data synchronization problem. Since OAI-PMH is not restrictive in terms of the data model to be used, it enables sharing of metadata for aggregation according to the data model adopted for each aggregation case. The only restriction imposed by OAI-PMH is that the data must be represented in XML.

In the case of Europeana, the technological solutions around the Europeana Data Model (EDM) (Europeana, 2016) have always been under continuous improvement. However, the solution for data synchronization based on OAI-PMH has not been reassessed since its adoption. In the case studies presented in the following sections we address mainly the data synchronization problem. Since the aggregation solution of Europeana is based on EDM, the data synchronization can be addressed with a wide variety of technologies because EDM follows the principles of the Web of Data, and can be serialized in XML and in RDF formats.

4   Case Studies

Our earlier work addressed the data synchronization problem by reviewing the state of the art and emerging Web technologies [5]. In the continuation of this work, we identified two key technologies, on which we based the case studies presented in this paper: IIIF (International Image Interoperability Framework) and Sitemaps[3].

IIIF is a family of specifications that were conceived to facilitate systematic reuse of image resources in digital repositories maintained by CH institutions. It specifies several HTTP based web services [6] covering access to images, the presentation and structure of complex digital objects composed of one or more images, and searching within their content. IIIF's strength resides in the presentation possibilities it provides for end-users.

---

[3] Sitemaps XML format: https://www.sitemaps.org/protocol.html.

From the perspective of data acquisition, however, none of the IIIF APIs was specifically designed to support metadata aggregation. Nevertheless, within the output given by the IIIF APIs, there may exist enough information to allow HTTP robots to crawl IIIF endpoints and harvest the links to the digital resources and associated metadata.

Sitemaps allow webmasters to inform search engines about pages on their sites that are available for crawling by search engine's robots. A Sitemap is an XML file that lists URLs of the pages within a website along with additional metadata about each URL (i.e. when it was last updated, how often it usually changes, and how important it is in comparison to other URLs within the same site) so that search engines can more efficiently crawl the site. Sitemaps is a widely-adopted technology, supported by all major search engines. Many content management systems support Sitemaps out-of-the-box and Sitemaps are simple enough to be manually built by webmasters when necessary. Moreover, there are Sitemaps extensions, like Google's Image Sitemaps[4] and Video Sitemaps[5], which have potential usage in metadata aggregation.

We have identified other promising Web technologies (for example, technologies related with the social web or the web of data [5]). We have chosen IIIF because it is getting increasingly traction in CH. Moreover, it is a community developed, open framework. Our requirements and suggestions for metadata aggregation may thus be incorporated into future versions. The choice for Sitemaps was motivated by its wide usage within the Europeana data providers. In addition, Sitemaps also provides a very simple technological solution, with a very low implementations barrier.

We have undertaken several case studies to investigate the feasibility of performing metadata aggregation via IIIF and/or Sitemaps. These studies were conducted in cooperation with data providers of the Europeana Network[6], which were actively deploying these two technologies within their own information systems.

IIIF played a double role in our work. It was used as the data source from where the source metadata was aggregated from providers, and as a technology that can be used with other suitable web-based technologies to facilitate aggregation processes. We have specifically studied how the functionality available in the IIIF Presentation API could be used to provide similar aggregation functionality as OAI-PMH and Sitemaps.

The following subsections will describe the case studies and how the two technologies were used for metadata aggregation.

4.1   Crawling IIIF Services from IIIF Inventories Using the Presentation API

The first case study was exploratory and targeted at IIIF in general. It was performed solely by Europeana with the objective of evaluating the functional capabilities of IIIF, the amount of data sources available, the maturity and compliance of the IIIF

---

[4] https://www.google.com/schemas/sitemap-image/1.1/.

[5] https://developers.google.com/webmasters/videosearch/sitemaps.

[6] The source code of the prototypes developed in the case studies is openly available at https://github.com/nfreire/Open-Data-Acquisition-Framework.

implementations and the quality of the available metadata. For this purpose, a IIIF aware Web Crawler was prototyped and the results of crawling several IIIF services were evaluated. All source code, collected samples and results of this case study may be consulted online[7].

To find available IIIF services to crawl, we have identified two crowdsourced listings of existing IIIF services, both provided in machine readable ways: IIIF Top Level Collections[8] and the iiif-universe[9]. From these listings, we have chosen 13 collections containing CH resources. The crawler was pointed to these IIIF collections and attempted to fully harvest them.

Alongside the full harvest, the crawler extracted a sample of IIIF Manifests[10] from each service, for later analysis. The extracted samples were manually inspected for the availability of (references to) descriptive metadata, the semantic granularity of the model in which they are available, and the availability of machine readable licensing information for re-use.

This early exploration revealed that IIIF contains all the necessary elements for automatic metadata aggregation. Some of these elements are, however, not mandatory for implementation, thus they will not be available in every IIIF service. The following optional elements of IIIF APIs must be provided by data providers, to enable metadata harvesting for Europeana:

- Structured metadata: the metadata available in the output of IIIF (manifests) is intended for end-user presentation, thus it cannot fulfill Europeana's ingestion requirements. But this can be overcome by using the optional links to structured metadata as specified in IIIF (using a seeAlso property[11]). When these are correctly populated (which our study confirmed to be not always the case), they enable crawlers to obtain structured metadata, such as EDM, Dublin Core, etc.
- IIIF Collection indicating the objects for Europeana: In IIIF, it is not required that the endpoint implements a mechanism to make publicly known all the digital objects that it makes available.

The implications for data providers in the provision of these two aspects for Europeana was addressed in the case studies that follow.

4.2   Crawling IIIF Services via IIIF Presentation API Collections Made Available by Europeana Providers

The IIIF Presentation API offers a Collection construct to represent groups of objects[12]. Although not all IIIF services make Collections available, they are often provided.

---

[7] https://github.com/nfreire/IIIF-Manifest-Metadata-Harvesting.

[8] https://docs.google.com/spreadsheets/d/1apQKFkfBV89BvycaBPN6v-LjeaKaVVMaMUsY6L4KRJo/edit#gid=0.

[9] https://github.com/ryanfb/iiif-universe.

[10] http://iiif.io/api/presentation/2.1/#manifest.

[11] http://iiif.io/api/presentation/2.1/#seealso.

[12] http://iiif.io/api/presentation/2.1/#collection.

Even when Collections are not available, the implementation effort would be very low for the provider. By making a IIIF collection known to Europeana, all the resources it references can be crawled and their metadata harvested by Europeana.

In this case study, Europeana worked with two data providers from the Europeana Network: The National Library of Wales and University College Dublin. At the starting point of the cases studies, both organizations had IIIF services available, but neither of them had EDM metadata available through IIIF and neither had IIIF Collections available for Europeana.

Implementation of a IIIF Collection was easily achieved in both cases.

Regarding the implementation of EDM metadata, it was a straightforward task for the National Library of Wales, since the library had recently deployed an EDM conversion of their MODS[13] metadata for other purposes. University College Dublin was also successful in implementing an EDM conversion and including it in their IIIF endpoint output. However, in the case of University College Dublin, the support of the Europeana ingestion team was required for implementing the EDM conversion and obtaining valid EDM metadata.

We identified an additional issue for metadata aggregation from IIIF services - IIIF collections do not provide the modification timestamp of resources. This aspect has an impact in the efficiency of the harvesting process. It becomes relevant in very large collections with hundreds of thousands of resources, where re-harvesting of resources that have not changed should be avoided. Given the importance of efficiency for the aggregation of large datasets, this problem still needs to be addressed by Europeana (and the IIIF community in general). To overcome it, other technologies may be used in conjunction with IIIF. Examples are Sitemaps and HTTP headers, which we have evaluated in the use cases described in the remainder of this section.

4.3   Crawling IIIF Services Referenced by Sitemaps

The issue of harvesting efficiency identified in the previous case study has been brought to the attention of the IIIF community. Discussions have been started for achieving a standard mechanism or recommendations to address it within the IIIF framework, in the context of a new IIIF Discovery Technical Specification Group[14]. The general opinion among the IIIF community was that IIIF Collections were designed for different purposes, i.e., to support use cases of end-users interacting with the IIIF viewers. The use of Collections for metadata harvesting purposes is therefore not an optimal solution from the point of view of the design of the IIIF framework.

In this context of these discussions, we conducted a supporting case study, where we experimented with Sitemaps-based solutions – both standard Sitemaps and Sitemaps with extensions.

---

[13] Metadata Object Descriptive Schema (MODS) is a schema for a bibliographic element set: http://www.loc.gov/standards/mods/.

[14] http://iiif.io/community/groups/discovery/.

4.3.1   Standard Sitemaps

The providers have created Sitemaps listing the specific resources that should be aggregated by Europeana. Europeana implemented a prototype for a IIIF harvester based on these Sitemaps, and the solutions were deployed and tested successfully with real datasets.

When using standard Sitemaps, the identifiers of the IIIF Manifests are present in the Sitemap as would any other URL pointing to a web page. That is, the Sitemap XML directly references the IIIF Manifest in the <loc> element, as shown in Fig. 1.

```
<url>
   <loc>https://data.ucd.ie/api/img/collection/ivrla:3573</loc>
   <lastmod>2014-08-24T04:09:09.716Z</lastmod>
</url>
```

Fig. 1. Example of URL data in a Sitemap from University College Dublin. The *loc* element references a IIIF Manifest.

This solution presents some shortcomings, however. The critical issue is that such Sitemap cannot be used for efficiently representing web pages and IIIF Manifests at the same time. The crawler would need to fetch the content of every URL in order to verify when it refers to a IIIF resource or to a webpage. We thus carried out further experiments with two alternative solutions based on extensions of Sitemaps: one with elements from the IIIF namespace and another with elements from the ResourceSync namespace [7].

4.3.2   Sitemaps Extended with Elements from the IIIF Namespace

Our goal is to extend Sitemaps to better contextualize and relate the IIIF resource with the end-user access webpages of the digital library.

In our first extension, we make explicit the availability of the resource via IIIF, in the Sitemap, and make it possible to relate the resource with IIIF Collections and end-user webpages. The example in Fig. 2 contains the end-user access location from the digital library in the <loc> element. The link to the IIIF Manifest is made via a

```
<url>
   <loc>http://newspapers.library.wales/view/3679651</loc>
   <iiif:Manifest
xmlns:iiif="http://iiif.io/api/presentation/2/">http://dams.llgc.org.uk/iiif/newspaper/issue/36796
51/manifest.json</iiif:Manifest>
   <dcterms:isPartOf>http://dams.llgc.org.uk/iiif/newspapers/3679650.json<dcterms:isPartOf>
   <lastmod>2014-11-08</lastmod>
</url>
```

Fig. 2. Example of URL data in a Sitemap from the National Library of Wales, with references to the webpage of the resource, the IIIF Manifest and its IIIF Collection.

<iiif:Manifest> element[15] and a link to a IIIF collection which the resource belongs to) is made via a <dcterms:isPartOf> element.

As shown by the ResourceSync related research [8], these extended Sitemaps remain compatible with Internet search engines.

4.3.3   Sitemaps Extended with Elements from the ResourceSync Namespace

Our last case study uses extensions from the ResourceSync namespace [7]. The end-user access location from the digital library in still referenced in the <loc> element. The link to the IIIF Manifest is made by a <rs:ln> element with the attribute 'rel' set to 'alternate', and a link to a IIIF collection made by a <rs:ln> element with the attribute 'rel' set to 'collection'. The values of the 'rel' attribute is defined in the ResourceSync specification. To explicitly state that these links lead to IIIF resources, the attribute 'conformsTo' from Dublin Core is also included in the <rs:ln> elements.

This extension provides the same expressive capabilities of the extension based on IIIF elements for relating all the relevant resources. Its main motivation is that by using elements from ResourceSync, we expect that the Sitemap can be better interpreted when used beyond the context of IIIF (Fig. 3).

```
<url>
   <loc>https://digital.ucd.ie/view/ucdlib:38491</loc>
   <rs:ln rel="alternate" href="https://data.ucd.ie/api/img/manifests/ucdlib:38491"
type="application/json" dcterms:conformsTo="http://iiif.io/api/presentation/2.1/"/>
   <rs:ln rel="collection href="https://digital.ucd.ie/view/ucdlib:38488"
type="application/json" dcterms:conformsTo="http://iiif.io/api/presentation/2.1/"/>
   <lastmod>2014-08-24T04:09:09.716Z</lastmod>
</url>
```

Fig. 3. Example of URL data in a Sitemap from University College Dublin, with references to the webpage of the resource, the IIIF Manifest and its IIIF Collection, and the indication of the IIIF API version in use.

4.4   Crawling IIIF Services via the IIIF Presentation API and HTTP Cache Headers

Another option to solve the issue of harvesting efficiency was experimented between Europeana and the National Library of Wales. The idea is to extend the solution based on IIIF Collections by using HTTP cache control [9]. Here, the IIIF service is required to have the implementation of some HTTP cache headers for the URLs that provide access to the IIIF resources.

In this solution, the Europeana IIIF crawler must include, in all the requests for IIIF manifests, the HTTP header *If-Modified-Since,* which will contain the timestamp of the last time the resource was harvested. The IIIF service then only needs to send the IIIF

---

[15] IIIF Presentation Ontology: http://iiif.io/api/presentation/2.

```
<url>.
    <loc>https://digital.ucd.ie/view/ucdlib:38509</loc>
    <rs:ln rel="describedby"         href="https://data.ucd.ie/api/edm/v1/ucdlib:38509"
dcterms:conformsTo="http://www.europeana.eu/schemas/edm/"/>
    <rs:ln rel="collection" href="https://data.ucd.ie/api/img/collection/ucdlib:38488"/>
    <video:video>
        <video:thumbnail_loc>https://digital.ucd.ie/get/ucdlib:38509/thumbnail
        </video:thumbnail_loc>
        <video:description>Irish poet Catherine Ann Cullen reads her poem 'Meeting at the
Chester Beatty' in UCD Library's Special Collections.</video:description>
        <video:player_loc allow_embed="yes">
https://player.vimeo.com/video/111413587</video:player_loc>
        <video:duration>00:02:51.04</video:duration>
        <video:family_friendly>yes</video:family_friendly>
        <video:live>no</video:live>
    </video:video>
    <lastmod>2015-09-10T17:14:26.523Z</lastmod>
</url>
```

Fig. 4. Example of URL data using the Sitemaps Video extension from University College Dublin. The Sitemap was extended to allow the association of EDM metadata.

manifest if an update has happened since that time in the manifest, the metadata or the resource itself. In case of deletion of the resource, the IIIF service returns a response with the HTTP Status code 404 *Not Found*. The assumption behind this solution is that the IIIF service can efficiently query the timestamp of the resources it serves and, in this way, save time and processing resources by not having to assemble and transmit the IIIF manifest back to the IIIF crawler.

We measured the performance of the crawling process for a collection of 500 resources in National Library of Wales's IIIF implementation. The outcome was a reduction around of 50% in the total time for crawling the 500 resources when they were not modified. The measurements were made at several points in time, different time of the week and the day, to prevent strong measurement bias due to variations in the user load of the IIIF service (our experiment was *in vivo* in the sense that it used the production service of the National Library of Wales and thus had to share bandwidth with "real" users).

4.5   Crawling (Non-IIIF) Resources Referenced by Sitemaps Extensions: Video and Image

An additional case study was performed with Sitemaps extensions used for better retrieval of image[16] and video[17] content within the Internet search engines that focus on

---

[16] https://www.google.com/schemas/sitemap-image/1.1/.

[17] https://developers.google.com/webmasters/videosearch/sitemaps.

Table 1. Summary of the main conclusions taken from the case studies

Method	Summary
Crawling IIIF Universe services	No contact or intervention required from the data provider. Only a very limited number of registered IIIF services provide the components of the IIIF API required for the metadata harvesting process
Crawling IIIF services based on IIIF Collections	Very simple to implement by data providers. No resource modification timestamp available in IIIF, thus it is applicable only to small and medium sized datasets. IIIF Collections are intended for usage by IIIF viewers, thus it's use for harvesting may appear to be a deviation from its purpose
Standard Sitemaps	Simple to implement for data providers. (and simpler if resource modification timestamps are not implemented). Reuse of Sitemaps originally created for search engines may be impractical, since there is no information available to distinguish Web pages from IIIF resources
Sitemaps extended with elements from the IIIF namespace	Simple to implement for data providers. Better contextualization of the IIIF resources, webpages, and their relations
Sitemaps extended with elements from the ResourceSync namespace	Simple to implement for data providers. Better contextualization of the IIIF resources, webpages and their relations. ResourceSync elements provide semantics associated with harvesting purposes
Crawling IIIF services based on IIIF Collections and HTTP cache headers	Enables a more efficient harvesting process, applicable to large datasets, but the IIIF specifications do not cover the use of HTTP headers. Implementation may not be possible for providers that are unable to modify their IIIF service
Crawling by Sitemaps Image and Video extensions	Enables reuse of the image and video Sitemaps made by data providers for Internet search engines. However, the metadata available may not fulfil the minimum requirements for making resources available in Europeana. For our case, an extension is required for linking to EDM metadata

these kinds of media. Just like search engines, metadata aggregators may also use the media specific metadata for their purposes.

Although we do not have much information about the usage of these types of Sitemaps in CH institutions, some cases are known to exist. One of them is the University College Dublin, which uses both extensions for images and videos.

Since images were being addressed in the context of IIIF metadata aggregation, we focused this case study on the analysis of the video extension.

From our metadata aggregation perspective, the main issue is that the metadata available through these extensions does not fulfil the minimum data requirements for making the resources available into Europeana. The solution adopted with University College Dublin was to further extend the Video Sitemaps with elements from ResourceSync that allow for the association of the EDM metadata, as shown in Fig. 4.

5   Future Work and Conclusion

In this paper, we presented the first results of our work on innovating Europeana's technological approach for CH metadata aggregation. Our primary goal is to find a solution that will make the continuous operation of the aggregation network more efficient and lower the technical barriers for data providers to contribute to Europeana. We conducted successful case studies with several technological options using deployed technologies and existing knowledge in CH institutions. A summary of the main conclusions from the case studies, is shown in Table 1.

Now, the challenge is to choose one of the several possibilities and work on establishing a best practice within the community. To achieve this, Europeana is working with the IIIF community in the context of the IIIF Discovery Technical Specification group and will proceed with recommendations targeted at its own partner network.

Acknowledgments. We would like to acknowledge the supporting work by Valentine Charles, from the Europeana Foundation, in the analysis of metadata samples collected during the case studies, and the IIIF community in general for the criticism and discussions of our work. This work was partially supported by Portuguese national funds through Fundação para a Ciência e a Tecnologia (FCT) with reference UID/CEC/50021/2013, and by the European Commission under the Connecting Europe Facility, telecommunications sector, grant agreement number CEF-TC-2015-1-01.

References

1. Lagoze, C., Van de Sompel, H., Nelson, M., Warner, S.: The Open Archives Initiative Protocol for Metadata Harvesting, Version 2.0 (2002), http://www.openarchives.org/OAI/2.0/openarchivesprotocol.htm
2. Van de Sompel, H., Nelson, M.L.: Reminiscing About 15 Years of Interoperability Efforts. D-Lib Mag. **21**(11/12) (2015). doi:10.1045/november2015-vandesompel, http://www.dlib.org/dlib/november15/vandesompel/11vandesompel.html
3. Berners-Lee, T.: Linked Data Design Issues. W3C-Internal Document (2006), http://www.w3.org/DesignIssues/LinkedData.html
4. Kenneth, L., Richardson, L., Ruby, S.: Restful Web Services. O'Reilly (2007)
5. Freire, N., Manguinhas, H., Isaac, A., Robson, G., Howard, J.B.: Web technologies: a survey of their applicability to metadata aggregation in cultural heritage. In: 21st International Conference on Electronic Publishing (2017)

6. Snydman, S., Sanderson, R., Cramer, T.: The International Image Interoperability Framework (IIIF): A community & technology approach for web-based images. In: Archiving 2015 (2015), http://purl.stanford.edu/df650pk4327

7. NISO. "ResourceSync Framework Specification", National Information Standards Organization (2014), http://www.niso.org/apps/group_public/download.php/12904/z39-99-2014_resourcesync.pdf

8. Klein, M., Van de Sompel, H.: Extending sitemaps for ResourceSync. In: Proceedings of the 13th ACM/IEEE-CS Joint Conference on Digital Libraries (JCDL 2013), pp. 277–280. ACM, New York (2013). Doi:http://dx.doi.org/10.1145/2467696.2467733

9. Fielding, R., Gettys, J., Modul, J., Frystyk, H., Masinter, L., Leach, P., Berners-Lee, T.: Hypertext Transfer Protocol—HTTP/1.1. Network Working Group. The Internet Society. Request for Comments: 2616 (1999) https://www.w3.org/Protocols/rfc2616/rfc2616.html

A Decade of Evaluating Europeana - Constructs, Contexts, Methods & Criteria

Vivien Petras[✉] and Juliane Stiller

Berlin School of Library and Information Science,
Humboldt-Universität zu Berlin, Dorotheenstr. 26, 10117 Berlin, Germany
{vivien.petras,juliane.stiller}@ibi.hu-berlin.de

Abstract. This meta-analysis of 41 evaluation studies of the Europeana Digital Library categorizes them by their constructs, contexts, criteria, and methodologies using Saracevic's digital library evaluation framework. The analysis shows that system-centered evaluations prevail over user-centered evaluations and evaluations from a societal or institutional perspective are missing. The study reveals, which Europeana components have received focused attention in the last decade (e.g. the metadata) and can serve as a reference for identifying gaps, selecting methodologies and re-using data for future evaluations.

Keywords: Digital library evaluation · Evaluation constructs · Europeana · Meta-analysis

1 Introduction

Almost a decade ago, Europeana, the European digital library, museum and archive was launched [43]. It was certainly not the first digital library (DL) in existence; by that time, the DL field had been well established[1]. Today, however, Europeana belongs to an elite group of DLs that has not only managed to go beyond the prototype stage of a research project, but has achieved exemplary status for other DLs. Europeana maintains a trailblazer role in metadata modelling, licensing, aggregating large and heterogeneous volumes of content and providing multilingual access to its collections - at least in its domain of cultural heritage information.

Europeana is an ecosystem of different stakeholders, collections, usage scenarios and services with the web-based portal as its primary access point to cultural heritage material. Having been part of its development from its first steps as the European Digital Library Network (EDLnet), we reflect on what progress has been achieved in this almost decade of development. Our particular lens of analysis in this paper is evaluation. Through the evaluation of a DL, we identify its important components, its strengths and its weaknesses. By applying

---

[1] Already twenty years ago, the first European Conference on Research and Advanced Technology for Digital Libraries (ECDL, now TPDL) was held in Pisa, Italy [39].

© Springer International Publishing AG 2017
J. Kamps et al. (Eds.): TPDL 2017, LNCS 10450, pp. 233–245, 2017.
DOI: 10.1007/978-3-319-67008-9_19

Tefko Saracevic's framework of DL evaluation [44], we use its structure-giving dimensions to inform a discussion on future developments in Europeana. Our study reviews 41 different evaluations of Europeana, from general surveys of user motivations and usability evaluations of the portal to specific evaluations of system components or the content.

The paper is structured as follows: Sect. 2 briefly reviews some of the frameworks for DL evaluation that were considered for this meta-analysis and then describes Saracevic's framework in more detail. Section 3 describes Europeana and some of the challenges for its development. Section 4 presents the analysis of Europeana evaluations with focus points that have been identified as neglected in previous evaluations. In Sect. 5, we conclude with more recommendations for future development.

2   Digital Library Evaluation Frameworks

From the mid-1990s to the early 2000s, the large NSF-funded DL initiatives dominated DL research in the US. Large evaluation initiatives were developed: the one described by Marchionini for Perseus is a quintessential example [33]. Researchers at Rutgers university, led by Saracevic [44], summarized and aggregated evaluation aspects in their evaluation frameworks [29,59,60,63]. The 5S research team at the University of Virginia [17,23] also developed models for the evaluation of DLs [24] and presented an automatic approach for the assessment of DL components [34].

Concurrently, the EU-funded DELOS Network of Excellence on Digital Libraries developed not only its reference model for DLs [6], but created an evaluation framework for it as well. A first categorization lists the three major parameters of DLs that need to be considered in evaluation: the data/collection, the technology and the users and uses [18]. The Interaction Triptych Evaluation Model developed by DELOS [19] refines and renames these components to content, system and users. It defines three axes of evaluation between these components: usability refers to the quality of interactions between the system and the users, usefulness to the relationship between content and users and performance to the relationship between system and content. It also describes criteria and methodologies to evaluate the relationships between the components of a DL. The Digital Library Evaluation Ontology DiLEO [57] formally models the strategic and procedural elements of evaluation efforts, integrating different frameworks.

The most recent evaluation framework was developed in the Multifaceted Evaluation of Digital Libraries (MEDaL) study [61], which reviewed 85 papers and 5 project websites and performed a two-round Delphi study to identify ten dimensions for DL evaluation, also describing evaluation objectives, criteria and measures.

Saracevic's Digital Library Evaluation Framework

Many evaluation frameworks, including DELOS and MEDaL, base their concepts on the dimensions provided by Tefko Saracevic's DL evaluation framework [44], which is also used in this study. In the framework, five elements, which are needed to describe a DL evaluation, are defined:

1. **Construct:** What is evaluated? Describes the aspect, which is the focus of the evaluation, for example the metadata or the search functionality.
2. **Context:** Which perspective is used for the evaluation? Saracevic distinguishes the user-centered perspective (with social, institutional or individual levels), the interface perspective and the system-centered perspective (with engineering, process and content levels).
3. **Criteria:** Which objective is evaluated? Saracevic names library criteria such as information accuracy, information retrieval criteria such as relevance, and HCI and interface criteria such as usability.
4. **Measures:** How are the criteria evaluated? Defines the operationalization of a criterion, e.g. precision for the evaluation of relevance.
5. **Methodology:** Which approach, instrument or tool is used for data collection and analysis for the evaluation?

Saracevic applied his framework to review 80 evaluation studies of DLs [45]. He finds that the system-centered approach is used more often than the human- or usability-centered approach as the context and that the most often evaluated criteria were usability, system performance and usage. Surveys were the methodology most often used in the evaluations, followed by structured interviews, focus groups, observations and task accomplishment. Saracevic does not discuss measures, which we will also not analyze here, because they are usually very specific to a particular evaluation.

Using their DiLEO ontology, Tsakonas et al. analyzed ca. 220 evaluation studies published between 2001–2011 in the JCDL and ECDL/TPDL conferences [56]. They found that system-centered contexts are employed most often with effectiveness, performance measurement and technical excellence as the main criteria. The most frequently used methodologies were laboratory experiments and surveys.

This paper is probably closest to these analyses, but focuses on evaluation studies of just one DL - Europeana. Zooming in on a particular DL should allow to compare evaluation results, but this is not as simple, as is discussed in the conclusion to this paper.

3   A Meta-Analysis of Europeana Evaluations

3.1   Europeana

Europeana[2] is the DL for accessing Europe's cultural and scientific heritage. Originally developed as an answer to the Google Books project [43], it has evolved into a DL, which aggregates and organizes European digital heritage in its many manifestations. Europeana's slogan "we transform the world with

---

[2] http://www.europeana.eu.

culture"[3] encompasses the vision of a network of all stakeholders in the cultural heritage sector.

The development of Europeana started in 2007 with the conceptualizing work of the European Digital Library network EDLnet, which culminated in the public launch of the Europeana portal in November of 2008 [43]. The Europeana portal now provides a single access point to over 53 million cultural heritage objects from over 3200 institutions across Europe. While Europeana also makes its data available via API access points, the portal is its most visible representation.

Europeana's challenges derive from the heterogeneity and multilinguality of its content, its providers and its audiences [27,42], making data quality, data openness and value creation its biggest priorities[3]. In almost a decade, Europeana has undergone several large-scale developmental steps - from changes in the layout and design of the portal to the modelling of the content and its functionalities. The following analysis describes the evaluation efforts accompanying these changes.

3.2   Methodology

To accumulate relevant studies, we started with a list of publications created by the Europeana Task Force for Enrichment and Evaluation that aggregates evaluations in the Europeana community [28, Appendix B]. Additionally, we searched for documents in Google Scholar and Web of Science[4], which focused solely on Europeana or used Europeana in comparison for an evaluation. The date range for the selected studies is between the launch in November 2008 until early 2017. The collected sample includes some deliverables from various Europeana satellite projects named in the Task Force document. However, we did not review all 50 projects listed on Europeana websites[5] systematically to capture evaluation efforts that were not published in a journal or conference venue, assuming that the most important ones were included in the Task Force list. Based on the abstracts of the result sets, we extracted 55 papers, which we then reviewed in detail. The criteria for extraction was that the paper focused on a evaluation of Europeana. We found three different types of evaluations:

- 38 evaluations with Europeana as the object,
- 3 evaluations using Europeana data, and
- 14 meta-studies, which named Europeana as a use case.

For the detailed analysis, we looked at 41 publications, dating from 2010–2016, which conducted an evaluation with Europeana itself or the data from Europeana. In the sample, there are several publications, which use data from the same study, but describe different aspects or different results. For methodological reasons, we counted these separately in order to reflect their different *Constructs* or *Contexts* of evaluation.

---

[3] http://strategy2020.europeana.eu/.

[4] Search terms: Europeana and (user* or evaluat* or study*) and variations.

[5] http://pro.europeana.eu/get-involved/projects/project-list.

For each of the 41 publications, we extracted information related to Saracevic's five elements of DL evaluation: *Construct, Context, Criteria, Measures, Methodology*. The extraction was done as close to the source as possible. Next, we followed a grounded theory approach [22] and discussed the extractions for each element to cluster the information into groups and determine categories. For example, different evaluation objects within Europeana such as metadata, enrichments or multilingual features were subsumed under the *Construct* category "Europeana component". For the *Context*, Saracevic's suggested perspectives were applied, dividing this element into a user-centered, system-centered and interface perspective and their subcategories (cf. Table 2).

4   Framing Europeana Evaluations

4.1   Constructs and Contexts of the Evaluations

We identified five different *Constructs* of Europeana that were evaluated (see Table 1). Most studies concentrate on the Europeana portal in general or one of its components (particularly various data quality aspects such as metadata completeness or the effectiveness of automatic vocabulary enrichments). Due to the previous Europeana funding structure based on satellite projects that deliver components, a number of studies describe services, which were planned to be integrated with Europeana, but were evaluated separately from the main portal. Because they base their assumptions and criteria on Europeana's objectives, they were also included. Similarly, information retrieval evaluations on Europeana data also utilized Europeana's assumptions on user objectives and their information needs.

Table 1. Constructs used in the evaluations

Construct	Description	Number
Europeana DL	Evaluations of the portal or services overall	17
Europeana component	Studies focusing on an aspect of Europeana, e.g. metadata quality	7
External service	Evaluations of services developed for Europeana, often as part of a project, e.g. PATHS[a]	9
Algorithms	Studies using Europeana data to conduct evaluations of (search) algorithms	3
Europeana in comparison	Evaluations on DL, which compared Europeana to similar services and DLs	5

[a]http://www.paths-project.eu/.

Figure 1 shows the number of studies and their evaluation perspective, i.e. *Context*. The Venn diagram shows which *Contexts* studies have in common: 17 studies are purely system-centered, 5 purely user-centered, whereas 7 encompass all *Contexts* in the same study.

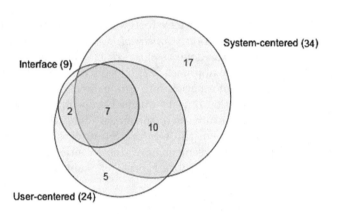

Fig. 1. Overlap of Contexts in different studies

Table 2 associates the five identified *Constructs* with their respective *Contexts*. The bold figures represent the overall number of studies for the particular *Construct* and *Context*, namely user-centered, interface or system-centered. The numbers in brackets indicate the reference. Note that the same publication can contain results that evaluate from a user-centered and a system-centered perspective, consequently, studies appear more than once in the matrix.

Equivalent to other comparative studies [45,56], system-centered evaluations are more often attempted than user-centered or interface-focused studies. It is logical that the interface is mainly evaluated when looking at the *Construct* of the portal as a whole, not when evaluating external services or components, which may not even be visible in the interface. The few algorithm studies are all system-centered, which is to be expected. Zooming in on the levels within the individual perspectives, a lack of user-centered evaluations on the social or institutional levels is apparent. Since many institutions both contribute and use Europeana, it is puzzling that we found no studies that evaluate the use of Europeana in an institutional setting. We hypothesize that the value of Europeana on a social, i.e. societal, level is "determined" through non-publication channels, for example the continued political and financial support by the EU. It is not surprising that many evaluations in our sample deal with data quality (categorized under the content *Context*) as this is an ongoing issue for Europeana. While processes are also in the focus of evaluations, we could not find any evaluations of engineering aspects. This is also not surprising as they are usually not published.

External services and Europeana components are evaluated much more from the system-centered perspective than a user-centered perspective. This may point to a lack in user-centered quality assessments for these *Constructs*. However, another explanation may be that components and services can only show their value to a user if they are integrated with other functionalities and thus cannot be evaluated individually from the user-centered perspective.

Table 2. Contexts used in the evaluations (categorized by construct)

	Europeana DL	External service	Europeana component	Algorithms	Europeana in comparison
# of Studies	17	9	7	3	5
User-centered:	**15**	**4**	**1**	**0**	**4**
Social	2 [58,62]	0	0	0	0
Institutional	0	0	0	0	0
Individual	14 [7–9,13,15, 16,20,21,35– 37,48,55,58]	4 [5,26,46,47]	1 [49]	0	4 [2,14,50,53]
Interface	**6** [13,15,16,35,48, 55]	**2** [26,47]	**1** [49]	**0**	**0**
System-centered:	**11**	**8**	**7**	**3**	**5**
Engineering	0	0	0	0	0
Process	7 [8,9,11,20,25, 36,37]	8 [3,4,10,26,28, 32,46,47]	5 [12,38,51, 52,54]	3 [1,40,41]	0
Content	10 [8,9,11,13,15, 20,35–37,48]	4 [26,28,32,46]	7 [12,31,38, 49,51,52,54]	0	5 [2,14,30,50,53]

4.2 Criteria and Methodologies of the Evaluations

We found that almost every reported evaluation defined its own different *Criteria* (objectives of analysis). Following, we list broad *Criteria* from each *Context* whereas for each criteria example studies are given. Usability and the effectiveness of the interaction design, user behavior and algorithm performance were used as objectives more often than others, but many *Criteria* are employed in Europeana evaluations.

User-Centered Context

- impact of Europeana on society and education [58,62]
- value of Europeana services wrt. mutuality[6], usability & reliability [35]
- value of multilingual services [20,21,49]
- usability & effectiveness of interaction design [5,13,15,26,46,47]
- effectiveness of search functionalities [7]
- usage patterns and criteria of Human Computer Interactions [8,9,20,36,37], such as task completion and time performance
- behaviour of particular user groups [16]
- value of engagement and access features [50,53].

---

[6] The study defines this term as a criteria to act as multiplier for member institutions.

System-Centered Context

- metadata quality [12,30,31]
- impact of semantic enrichments [38,52,54] and components of workflow [51]
- performance of item similarity algorithms [3,4,10,25,26]
- content characteristics compared to other DLs [2,14]
- usage of particular content [8,9,36,37]
- accessibility of content [2]
- information retrieval criteria, e.g. precision of search results [1,11,40,41,46]
- performance of enrichment tools [32].

Interface Context

- user foci in interface [13,55] and look and feel [48].

Table 3 categorizes the *Methodologies* used in the different evaluations. Most studies first defined their own *Criteria* and then assessed the quality of a component or service based on these *Criteria*. These are usually expert assessments that are performed on quantitative or qualitative data. A gold standard is commonly used for algorithm testing or the quality of a service where a particular outcome can be expected, for example, automatic enrichment of keywords in the metadata. Criteria-based, gold standard-based evaluations often assess the DL from the system-centered perspective. Logfile analyses are used for both system- and user-centered evaluations. User-centered methodologies such as usability studies usually take more effort and occur not as often.

Table 3. Methodologies used in the evaluations (multiple possible)

Method	Description	Number
Criteria-based	Certain criteria were determined to assess a service or algorithm	16
Gold standard-based	Use of a manually created gold standard to assess performance	9
Logfile analysis	Uses an automatically created logfile of user interactions	8
Usability study	Several methods to assess usability of a service, e.g. user studies, interviews, surveys	7
Impact study	Expert assessment of the overall value of a service	2

While it may be challenging to identify *Criteria* that have not been evaluated for Europeana, we argue that *Criteria* and *Methodologies* to rate the overall impact and added value of Europeana are still missing. Europeana has recognized this gap and is about to implement an Impact Framework.[7] Evaluations targeting aspects such as user satisfaction or performance of algorithms refer to established methods and criteria. Defining success criteria specifically for cultural heritage DLs seems to be a gap that should receive more attention in future evaluations.

---

[7] More information can be found here: https://impkt.tools/.

5   Conclusions

The meta-analysis showed that Europeana has been evaluated from many perspectives with a vast number of *Criteria* and many different *Methodologies*. The categorization of evaluation *Constructs* and *Contexts* showed that societal and institutional perspectives appear less often than system-centered quantitative approaches and that only a marginal number of studies tried to assess the impact of Europeana on different stakeholders. Concluding, we argue for more standardization with regard to evaluations of large-scale DLs and how the evaluation process can be improved.

Establish an Evaluation Archive

Our analysis showed that evaluations usually evolve individually and rarely refer to previous results or to similar efforts in the Europeana ecosystem. Logging data and other evaluation outcomes are rarely re-used, except within the same research groups. The lack of published documentation and coordination between evaluations calls for a more concerted effort. Europeana has been the target of over 50 satellite projects, which probably included more unpublished evaluation efforts. Europeana as an ecosystem needs an evaluation archive, which helps to build a common memory in the community and promotes learning from past results.

Track Improvements Over Time

We found that Europeana development has progressed due to evaluations, although implementations cannot be traced back to certain evaluation results. In general, more coordination and documentation is needed to learn from previous experiences and also track improvements over time. Europeana has recognized the importance of evaluation by integrating permanent activities into their ecosystem, for example by integrating a logging framework in the portal to understand user interactions better. The Europeana Statistics Dashboard[8] provides current interaction log- and content-based statistics for providers and users, while the Europeana Data Quality Committee[9] develops standards to continuously improve the data quality. System development should be traced alongside evaluation efforts.

Standardize Evaluations

The heterogeneity of methodologies and criteria used makes it hard to draw concrete solutions for Europeana development in general and aggravates the reuse of the data. Here, a more standardized format is desirable that allows to compare results over time. However, it is important to note that the evaluations, which created gold standards from Europeana data (for example [40]) or developed framework or experimental set-ups for evaluation (for example [32,52]) can be re-used in other evaluation efforts. This means that Europeana evaluations have pushed the envelope further and also contributed to DL evaluation research in general.

---

[8] http://statistics.europeana.eu/.
[9] http://pro.europeana.eu/page/data-quality-committee.

With evaluation as an integral part of development in Europeana, these efforts can hopefully be better organized so that new services and partners can learn from the large experience that has already been accumulated.

References

1. Akasereh, M., Malak, P., Pawłowski, A.: Evaluation of IR strategies for Polish. In: Przepiórkowski, A., Ogrodniczuk, M. (eds.) NLP 2014. LNCS, vol. 8686, pp. 384–391. Springer, Cham (2014). doi:10.1007/978-3-319-10888-9_38
2. van den Akker, C., van Nuland, A., van der Meij, L., van Erp, M., Legêne, S., Aroyo, L., Schreiber, G.: From information delivery to interpretation support: evaluating cultural heritage access on the web. In: WebSci 2013, pp. 431–440. ACM (2013)
3. Aletras, N., Stevenson, M.: Computing similarity between cultural heritage items using multimodal features. In: LaTeCH 2012, pp. 85–93. ACL (2012)
4. Aletras, N., Stevenson, M., Clough, P.: Computing similarity between items in a digital library of cultural heritage. J. Comp. Cult. Heri. 5(4), 16:1–16:19 (2012)
5. Aloia, N., Concordia, C., Gerwen, A.M., Hansen, P., Kuwahara, M., Ly, A.T., Meghini, C., Spyratos, N., Sugibuchi, T., Tanaka, Y., Yang, J., Zeni, N.: Design, implementation and evaluation of a user generated content service for Europeana. In: Gradmann, S., Borri, F., Meghini, C., Schuldt, H. (eds.) TPDL 2011. LNCS, vol. 6966, pp. 477–482. Springer, Heidelberg (2011). doi:10.1007/978-3-642-24469-8_54
6. Candela, L., Castelli, D., Ferro, et al.: The DELOS digital library reference model. Version 0.96. Technical report, DELOS Network of Excellence on Digital Libraries (2007)
7. Ceccarelli, D., Gordea, S., Lucchese, C., Nardini, F.M., Tolomei, G.: Improving Europeana search experience using query logs. In: Gradmann, S., Borri, F., Meghini, C., Schuldt, H. (eds.) TPDL 2011. LNCS, vol. 6966, pp. 384–395. Springer, Heidelberg (2011). doi:10.1007/978-3-642-24469-8_39
8. CIBER Research Ltd.: Europeana 2012–2013: usage and performance update. Technical report, CIBER Research Ltd. (2013). http://bit.ly/2rgWjrl
9. Clark, D., Nicholas, I., Rowlands, I.: D3.1.3 Report on best-practice and how users are using the Europeana service. Technical report, D 3.1.3, EuropeanaConnect (2011)
10. Clough, P., Otegi, A., Agirre, E., Hall, M.: Implementing recommendations in the PATHS system. In: Bolikowski, Ł., Casarosa, V., Goodale, P., Houssos, N., Manghi, P., Schirrwagen, J. (eds.) TPDL 2013. CCIS, vol. 416, pp. 169–173. Springer, Cham (2014). doi:10.1007/978-3-319-08425-1_17
11. Şencan, I.: The performance evaluation of the information retrieval system of the Europeana website. In: IMCW 2013, p. 59 (2013)
12. Dangerfield, M.C., Kalshoven, L.: Report and recommendations from the task force on metadata quality. Technical report, Europeana (2015). http://bit.ly/2rM113M
13. Dani, A., Chatzopoulou, C., Siatri, R., Mystakopoulos, F., Antonopoulou, S., Katrinaki, E., Garoufallou, E.: Digital libraries evaluation: measuring Europeana's usability. In: Garoufallou, E., Hartley, R.J., Gaitanou, P. (eds.) MTSR 2015. CCIS, vol. 544, pp. 225–236. Springer, Cham (2015). doi:10.1007/978-3-319-24129-6_20
14. Dickel, J.: Digitale Bibliotheken im Vergleich: Europeana & WDL. Perspektive Bibliothek 4(1), 45–67 (2015)
15. Dobreva, M., Chowdhury, S.: A user-centric evaluation of the Europeana digital library. In: Chowdhury, G., Koo, C., Hunter, J. (eds.) ICADL 2010. LNCS, vol. 6102, pp. 148–157. Springer, Heidelberg (2010). doi:10.1007/978-3-642-13654-2_19

16. Dobreva, M., McCulloch, E., Birrell, D., Ünal, Y., Feliciati, P.: Digital natives and specialised digital libraries: a study of Europeana users. In: Kurbanoğlu, S., Al, U., Lepon Erdoğan, P., Tonta, Y., Uçak, N. (eds.) IMCW 2010. CCIS, vol. 96, pp. 45–60. Springer, Heidelberg (2010). doi:10.1007/978-3-642-16032-5_5

17. Fox, E.A., Gonçalves, M.A., Shen, R.: Theoretical foundations for digital libraries: the 5s (societies, scenarios, spaces, structures, streams) approach. Synth. Lect. Inform. Concepts, Retrieval, Serv. 4(2), 1–180 (2012)

18. Fuhr, N., Hansen, P., Mabe, M., Micsik, A., Sølvberg, I.: Digital libraries: a generic classification and evaluation scheme. In: Constantopoulos, P., Sølvberg, I.T. (eds.) ECDL 2001. LNCS, vol. 2163, pp. 187–199. Springer, Heidelberg (2001). doi:10.1007/3-540-44796-2_17

19. Fuhr, N., Tsakonas, G., Aalberg, T., et al.: Evaluation of digital libraries. IJDL 8(1), 21–38 (2007)

20. Gäde, M.: Country and language level differences in multilingual digital libraries. Dissertation, Humboldt-Universität zu Berlin, April 2014

21. Gäde, M., Petras, V.: Multilingual interface preferences. In: IIiX 2014, pp. 231–234. ACM (2014)

22. Glaser, B.G., Strauss, A.: The discovery of ground theory (1967)

23. Gonçalves, M.A., Fox, E.A., Watson, L.T., Kipp, N.A.: Streams, structures, spaces, scenarios, societies (5s): a formal model for digital libraries. ACM TOIS 22(2), 270–312 (2004)

24. Gonçalves, M.A., Moreira, B.L., Fox, E.A., Watson, L.T.: "What is a good digital library?" - a quality model for digital libraries. IPM 43(5), 1416–1437 (2007)

25. Gonzalez-Agirre, A., Rigau, G., Agirre, E., Aletras, N., Stevenson, M.: Why are these similar? Investigating item similarity types in a large digital library. JASIST 67(7), 1624–1638 (2016)

26. Griffiths, J., et al.: D5.2 evaluation of the second PATHS prototype. Technical report (2014). http://bit.ly/2rh3hwJ

27. Hill, T., Charles, V., Stiller, J., Isaac, A.: Searching for inspiration: user needs and search architecture in Europeana collections. In: ASIS&T, pp. 1–7. Wiley (2016)

28. Isaac, A., Manguinhas, H., Stiller, J., Charles, V.: Final report on evaluation and enrichment. Technical report, Europeana (2015). http://bit.ly/2rcKuHr

29. Joo, S., Xie, I.: Evaluation constructs and criteria for digital libraries: a document analysis. In: Recent Developments in the Design, Construction, and Evaluation of Digital Libraries: Case Studies, pp. 126–140. IGI Global (2013)

30. Kapidakis, S.: Comparing metadata quality in the Europeana context. In: PETRA 2012, p. 25:1–25:8. ACM (2012)

31. Király, P.: A metadata quality assurance framework. Technical report, Europeana (2015). https://pkiraly.github.io/metadata-quality-project-plan.pdf

32. Manguinhas, H., Freire, N., Isaac, A., Stiller, J., Charles, V., Soroa, A., Simon, R., Alexiev, V.: Exploring comparative evaluation of semantic enrichment tools for cultural heritage metadata. In: Fuhr, N., Kovács, L., Risse, T., Nejdl, W. (eds.) TPDL 2016. LNCS, vol. 9819, pp. 266–278. Springer, Cham (2016). doi:10.1007/978-3-319-43997-6_21

33. Marchionini, G.: Evaluating digital libraries: a longitudinal and multifaceted view. Libr. Trends 49(2), 304 (2000)

34. Moreira, B.L., Gonçalves, M.A., Laender, A.H., Fox, E.A.: Automatic evaluation of digital libraries with 5squal. J. Informetrics 3(2), 102–123 (2009)

35. Navarrete, T.: Europeana as online cultural information service. Study Report (2016). http://bit.ly/2rcMsaT

36. Nicholas, D., Clark, D.: The second digital transition: to the mobile space - an analysis of Europeana. Learn. Publishing **26**(4), 240–252 (2013)
37. Nicholas, D., Clark, D., Rowlands, I., Jamali, H.R.: Information on the Go: a case study of Europeana mobile users. JASIST **64**(7), 1311–1322 (2013)
38. Olensky, M., Stiller, J., Dröge, E.: Poisonous India or the importance of a semantic and multilingual enrichment strategy. In: Dodero, J.M., Palomo-Duarte, M., Karampiperis, P. (eds.) MTSR 2012. CCIS, vol. 343, pp. 252–263. Springer, Heidelberg (2012). doi:10.1007/978-3-642-35233-1_25
39. Peters, C., Thanos, C. (eds.): ECDL 1997. LNCS, vol. 1324. Springer, Heidelberg (1997)
40. Petras, V., Bogers, T., Toms, E., Hall, M., Savoy, J., Malak, P., Pawłowski, A., Ferro, N., Masiero, I.: Cultural heritage in CLEF (CHiC) 2013. In: Forner, P., Müller, H., Paredes, R., Rosso, P., Stein, B. (eds.) CLEF 2013. LNCS, vol. 8138, pp. 192–211. Springer, Heidelberg (2013). doi:10.1007/978-3-642-40802-1_23
41. Petras, V., Ferro, N., Gäde, M., Isaac, A., Kleineberg, M., Masiero, I., Nicchio, M., Stiller, J.: Cultural Heritage in CLEF (CHiC) Overview 2012. In: CLEF (2012)
42. Petras, V., Hill, T., Stiller, J., Gäde, M.: Europeana - a search engine for digitised cultural heritage material. Datenbank-Spektrum **17**(1), 41–46 (2017)
43. Purday, J.: Think culture: Europeana.eu from concept to construction. Electron. Libr. **27**(6), 919–937 (2009)
44. Saracevic, T.: Digital library evaluation: toward an evolution of concepts. Libr. Trends **49**(2), 350–369 (2000)
45. Saracevic, T.: Evaluation of digital libraries: an overview. In: Notes of the DELOS WP7 Workshop on the Evaluation of Digital Libraries, Padua, Italy (2004)
46. Schindler, A., Gordea, S., Biessum, H.: The Europeana sounds music information retrieval pilot. In: Ioannides, M., Fink, E., Moropoulou, A., Hagedorn-Saupe, M., Fresa, A., Liestøl, G., Rajcic, V., Grussenmeyer, P. (eds.) EuroMed 2016. LNCS, vol. 10059, pp. 109–117. Springer, Cham (2016). doi:10.1007/978-3-319-48974-2_13
47. Schlötterer, J., Seifert, C., Wagner, L., Granitzer, M.: A game with a purpose to access europe's cultural treasure. ECIR **2015**, 13–18 (2015)
48. Schweibenz, W.: Eine erste evaluation der Europeana: Wie Benutzer das look & feel des prototypen der Europäischen digitalen bibliothek beurteilen. Inf. Wissenschaft und Praxis **61**(5), 277–284 (2010)
49. Stiller, J., Gäde, M., Petras, V.: Multilingual access to digital libraries: the Europeana use case. Inf.-Wissenschaft und Praxis **64**, 86–95 (2013)
50. Stiller, J.: From curation to collaboration. Dissertation, Humboldt-Universität zu Berlin, Philosophische Fakultät I (2014)
51. Stiller, J., Isaac, A., Petras, V.: EuropeanaTech task force on a multilingual and semantic enrichment strategy. Technical report, Europeana (2014). http://bit.ly/2soSHs7
52. Stiller, J., Olensky, M., Petras, V.: A framework for the evaluation of automatic metadata enrichments. In: Closs, S., Studer, R., Garoufallou, E., Sicilia, M.-A. (eds.) MTSR 2014. CCIS, vol. 478, pp. 238–249. Springer, Cham (2014). doi:10.1007/978-3-319-13674-5_23
53. Stiller, J., Petras, V.: A framework for classifying and comparing interactions in cultural heritage information systems. In: Cultural Heritage Information: Access and management. Facet Publishing (2015)

54. Stiller, J., Petras, V., Gäde, M., Isaac, A.: Automatic enrichments with controlled vocabularies in Europeana: challenges and consequences. In: Ioannides, M., Magnenat-Thalmann, N., Fink, E., Žarnić, R., Yen, A.-Y., Quak, E. (eds.) EuroMed 2014. LNCS, vol. 8740, pp. 238–247. Springer, Cham (2014). doi:10.1007/978-3-319-13695-0_23

55. Sykes, J., Dobreva, M., Birrell, D., McCulloch, E., Ruthven, I., Ünal, Y., Feliciati, P.: A new focus on end users: eye-tracking analysis for digital libraries. In: Lalmas, M., Jose, J., Rauber, A., Sebastiani, F., Frommholz, I. (eds.) ECDL 2010. LNCS, vol. 6273, pp. 510–513. Springer, Heidelberg (2010). doi:10.1007/978-3-642-15464-5_69

56. Tsakonas, G., Mitrelis, A., Papachristopoulos, L., Papatheodorou, C.: An exploration of the digital library evaluation literature based on an ontological representation. JASIST 64(9), 1914–1926 (2013)

57. Tsakonas, G., Papatheodorou, C.: An ontological representation of the digital library evaluation domain. JASIST 62(8), 1577–1593 (2011)

58. Valtysson, B.: Europeana - the digital construction of Europe's collective memory. Inf. Commun. Soc. 15(2), 151–170 (2012)

59. Xie, H.: Evaluation of digital libraries: criteria and problems from users' perspectives. Libr. Inf. Sci. Res. 28(3), 433–452 (2006)

60. Xie, H.I.: Users' evaluation of digital libraries (DLs): their uses, their criteria, and their assessment. IPM 44(3), 1346–1373 (2008)

61. Xie, I., Matusiak, K.K.: Chapter 10 - evaluation of digital libraries. In: Discover Digital Libraries, pp. 281–318. Elsevier, Oxford (2016)

62. Yankova, I., Velev, E., Nusheva, K., Sonia, S.: The European digital library-factor for long-life learning in arts and cultural studies. Qual. Quant. Methods Libr. 4(4), 965–971 (2015)

63. Zhang, Y.: Developing a holistic model for digital library evaluation. JASIST 61(1), 88–110 (2010)

On the Uses of Word Sense Change for Research in the Digital Humanities

Nina Tahmasebi[1]([✉]) and Thomas Risse[2]

[1] Språkbanken, University of Gothenburg, Gothenburg, Sweden
nina.tahmasebi@gu.se
[2] University Library J.C. Senckenberg, Frankfurt, Germany
t.risse@ub.uni-frankfurt.de

Abstract. With advances in technology and culture, our language changes. We invent new words, add or change meanings of existing words and change names of existing things. Unfortunately, our language does not carry a memory; words, expressions and meanings used in the past are forgotten over time. When searching and interpreting content from archives, language changes pose a great challenge. In this paper, we present results of automatic word sense change detection and show the utility for archive users as well as digital humanities' research. Our method is able to capture changes that relate to the usage and culture of a word that cannot easily be found using dictionaries or other resources.

1 Introduction

When interpreting the content of historical documents, knowledge of changed word senses play an important role. Without knowing that the meaning of a word has changed we might falsely place a more current meaning on the word and thus interpret the text wrongly. As an example, the phrase *an awesome concert* should be interpreted as a positive phrase today. However, *an awesome leader* in a text written some hundred years ago, should be interpreted as a negative phrase, i.e., one to fear. The interpretation depends on the time of writing and not on the context terms and is thus not a pure disambiguation problem. Instead, we consider this as manifestation of **word sense change**.

The emergence of large digital and historical archives gives us a chance to learn these changes and to utilize them for research, both in linguistic research and in the digital humanities. It also gives us the possibility to feed our results back to the archives for better search and interpretation of results, thus opening them up for the public. Researchers can follow a word over time, query for specific kinds of change or mine for events that co-occur with language changes.

In this paper, we present and discuss results of automatic word sense change detection utilizing induced word senses. In Tahmasebi et al. [20] the induced word senses were evaluated on historical data and shown to provide good quality sense approximation. In Tahmasebi [21] we present the details of the word sense change detection algorithm. In this paper, we focus on analyzing and interpreting the results of word sense change.

© Springer International Publishing AG 2017
J. Kamps et al. (Eds.): TPDL 2017, LNCS 10450, pp. 246–257, 2017.
DOI: 10.1007/978-3-319-67008-9_20

We measure the time between an expected change in word sense and the corresponding found change to investigate not only *if* but *when* changes can be found and with which time delay. The delay aspect is of particular interest for linguists and concept historians. Why is there a time delay and how does it differ between regions, media and time? There is evidence that our language changes quicker in social media [8], can we see this also in modern traditional media? We believe that by capturing cultural changes in addition to sense changes, our results can be of importance for the digital humanities and social sciences.

2   State of the Art

The first methods for automatic word sense change detection were based on context vectors; they investigated semantic density (Sagi et al. [19]) and utilized mutual information scores (Gulordava and Baroni [7]) to identify semantic change over time. Both methods detect signals of change but neither aligns senses over time or determines what changed.

Topic-based models (where topics are interpreted as senses) have been used to detect novel senses in one collection compared to another by identifying new topics in the later corpus (Cook et al. [2]; Lau et al. [12]), or to cluster topics over time (Wijaya and Yeniterzi [25]). A dynamic topic model that builds topics with respect to information from the previous time point is proposed by Frermann and Lapata [6] and again sense novelty is evaluated. Topics are not a 1-1 correspondence to word senses (Wang and McCallum [24]) and hence new induction methods aim at inferring sense and topic information jointly (Wang et al. [23]). With the exception of Wijaya et al. that partition topics, no alignment is made between topics to allow following diachronic progression of a sense.

Graph-based models are utilized by Mitra et al. [14,15] and Tahmasebi [21] and aim to reveal complex relations between a word's senses by (a) modeling senses per se using WSI; and (b) aligning senses over time. The models allow us to identify individual senses at different periods in time and Tahmasebi also groups senses into linguistically related concepts.

The largest body of work is done using word embeddings of different kind in the last years (Basile et al. [1]; Kim et al. [10]; Zhang et al. [26]). Embeddings are trained on different time-sliced corpora and compared over time. Kulkarni et al. [11] project words onto their frequency, POS and word embeddings and propose a model for detecting statistically significant changes between time periods on those projections. Hamilton et al. [9] investigate both similarity between a priori known pairs of words, and between a word's own vectors over time to detect change. [1,9,11] all propose different methods for projecting vectors from different time periods onto the same space to allow comparison.

Methods for detecting change based on word embeddings do not allow us to recover the senses that have changed and therefore, no way of detecting *what* changed. Most methods show the most similar terms to the changing word as a method to illustrate what happens. However, the most similar terms will only represent the dominant sense and not reflect changes among the other senses or capture stable parts of a word. The advantage of word embeddings over *e.g.* graph-based models is the inherent semantic similarity measure where otherwise often resources like WordNet [13] are used. In addition, compositionality methods can be used to find labels to help users better understand the results.

Due to a lack of proper evaluation methods and datasets, all presented papers have performed different, non-comparable evaluations. Most previous work have opted to pre-determine a set of words for further evaluation, both positive and negative examples of word sense change, rather than to evaluate the top terms outputted by the system thus needing evaluation for each new set of parameters.

3   Methodology

As a basis for our analysis we consider automatically induced word sense clusters. Each cluster represents a distinct time period and consists of a set of nouns and noun phrases of length two, i.e., **terms**. These clusters are approximations of word senses and to some extent capture also contexts. Throughout the paper we use **word senses** and **clusters** interchangeably. A **concept** consists of senses that are related (i.e., polysemous) following Cooper [3].

To model word sense change, we should allow each sense to change individually; worst case, this results in a graph where, for each time period $t \in T$ and a maximum number of sense S, we have in the order of S^T edges representing sense similarity. Even for a small number of time periods, this graph becomes infeasible to evaluate and investigate. Therefore, we reduce this complexity by first considering coherent senses over time (units) and then following the units over time. Units that are related are placed in a *path*. A unit can contain an arbitrary number of clusters, so to get a good representation of a unit, we create a centroid called a *unit representative*. We measure **similarity between units** as similarity between the unit representatives.

Individual *senses* s_w for a word w at one point in time are captured by *clusters*. A unit $u(w)$ captures a coherent sense s_w over a period of time and allows some change within s_w, e.g., broadening and narrowing. A path corresponds to a concept by grouping all units that are related (polysemous).

Our methodology consists of three steps. Firstly, deriving word sense clusters. Secondly, finding coherent senses by merging clusters into units and representing these with their unit representatives. Thirdly, relating units into paths by comparing unit representatives.

We find the word senses using an unsupervised word sense induction algorithm called *curvature clustering* (Dorow et al. [5]). The algorithm calculates clustering coefficient in a co-occurrence graph built with nouns and noun phrases that appear in the text separated with *and, or* and *commas*. Nodes with low clustering coefficient are removed and the graph falls apart into clusters that represent word senses. These clusters were shown to have 85% precision [20]. To the best of our knowledge, the curvature clustering method is the only induction method that has been properly evaluated on historical texts.

An Example For the details of the algorithm, we refer to Tahmasebi [21] and instead give an example to illustrate the workings. We start with three time points t_1, t_2, t_3 and unit sets $U_{t_1}(w) = \{u_1\}$, $U_{t_2}(w) = \{u_2, u_3\}$ and $U_{t_3}(w) = \{u_4, u_5\}$ for the target word *tape*. In this first iteration, each unit represents one cluster.

$u_1 = \{stereo, cassette, tape, record, radio\}$,
$u_2 = \{pin, thread, tape, silk, chair, cotton\}$,
$u_3 = \{video, cassette, tape, record\}$,
$u_4 = \{tape, sparkplug cable, wire, clip\}$,
$u_5 = \{television, record, tape, video, book, film, magazine, video industry\}$.

In the first step, similarity between pairs (u_1, u_2) and (u_1, u_3) is measured. Pairs are ranked according to the highest similarity and the pair with the highest similarity is merged. In this case, u_1 and u_3 are merged into $u' = \{u_1, u_3\}$ because u_3 is an almost subset of u_1. The unit representative consists of the terms $\{cassette, tape, record\}$. The pair (u_1, u_2) is removed because u_1 is already merged with one unit from $U_{t_2}(w)$.

The resulting merged set is $U_{[t_1, t_2]}(w) = \{\{u_1, u_3\} = u', u_2\}$. At time t_3, unit u_4 and u_5 are compared to the two units in $U_{[t_1, t_2]}(w)$. u_5 is merged with u' resulting in $u'' = \{u_1, u_3, u_5\}$. u_4 remains a single unit and is placed in $U_{[t_1, t_3]}(w)$ without being merged. When we merge two units, we add up all their clusters and build a new representative. When unit u_5 is merged with $u' = \{u_1, u_3\}$ we consider this to be a broadening because the single unit u_5 has a broader sense than the merged unit u'. The resulting unit set consists of $U_{[t_1, t_3]}(w) = \{\{u_1, u_3, u_5\} = u'', u_2, u_4\}$.

As a final step, to create paths, we measure similarity between the pairs (u'', u_2) and (u'', u_4). In this example, no units are related into paths which tells us that there are three different concepts for *tape*, one regarding *sewing tape*, one regarding *scotch tape* and one regarding *musical tape* which later includes also the *video tape*, matching well the main senses of tape but also capturing *sewing tape*, a sense less common today (OED [17]).

4   Experiments

The aim of our experiments is to find the quality and degree (i.e., recall) to which word sense change can be found using our word sense change detection and to investigate the utility of the results for research communities outside of linguistics. There exist no standard datasets or automatic evaluation metrics for word sense change. In addition, evaluation is a hard task because the outcome is specific to the collection and inherent location in mind; *when was a term used for the first time in the collection with the correct corresponding sense?* Therefore, in our experiments, we opt for a simplified, manual evaluation. We evaluate the found change for each term against the main changes of the term according to a set of knowledge sources and do not take completeness into account, *e.g.* by ignoring fine-grained sense differentiations.

We use *The Times Archive*, a large sample of modern English spanning 1785–1985. The collection is OCRed and corrected for OCR errors using the OCR Key method (Tahmasebi et al. [20]). We append the *New York Times Annotated Corpus*, a modern collection spanning 1987–2007, and disregard the annotations to treat both corpora the same. In total, the corpora span 222 years.

4.1   Testset

As a testset, we manually chose a set of 23 terms which we know have experienced word sense change during the past centuries. The main changes for each term were found using Wikipedia, dictionary.com and the Oxford English Dictionary, see extract in Table 1, and the automatically found changes were compared against the manually found counterpart. In addition, we considered the words *automobile, bitch, camera, car, cinema, computer, internet, mail, memory, phone, racism, record, train, travel*. We consider major changes in usage as well as changes to sense. In cases where multiple (fine-grained) senses were available, we opted to accept the widest sense. *E.g.* for the term *rock* we consider a *music* sense without any distinction between different types of rock music, because our dataset is unlikely to have fine-grained sense differentiations. If a clear time point cannot be pinpointed, we choose the earliest possible. For comparison purposes we also chose a set of 11 terms (*deer, export, mirror, symptom, horse, ship, paper, newspaper, bank, founder, music*) that have experienced minimal change during the investigated period, i.e., **stable terms**. The full testset can be found in [16].

We consider individual senses and their changes as separate events, *e.g.* an added sense and later a changed sense are two separate events. We have **35 change events** and **26 non-change events**. The change category consist of evolved senses (*e.g.* broadening and narrowing) and novel senses (related, i.e., polysemous senses and unrelated, i.e., homonymic sense).

The existing senses are also split into two categories, *existing -stable* (senses that belong to words that do not change over the entire dataset) and *existing -evo* (stable senses of words that have changes to their other senses).

4.2   Evaluation

For each experiment, we measure **recall** as the proportion of expected change events that were found; and **average time delay** as the difference in time between the *expected*, according to our ground truth, and the *found* events.

Recall is straightforward and measures the portion of expected change found, according to our ground truth. The expected time of change is trickier; true expected time of change for a given term is the first time that it was used in the collection with the correct corresponding sense. We do not know this time and therefore we approximate it using two different time points. The first expected time point is the *time of definition* or *time of invention* of a term w, $t_{DI}(w)$, in a given dictionary or knowledge resource. However, that an invention has been made does not necessarily correspond to newspapers reporting on it frequently. *E.g.* the *computer* was invented in its modern form in the 1940s, but was not mentioned in newspapers often in the early 40's, most likely due to WWII.

Table 1. Description of change for some terms used in the evaluation. WWI occurred during 1914–1918, WWII occurred during 1939–1945.

Term	Year	Description
tape	1960–1965	Common household use
aeroplane	1908	First modern aircraft design
aeroplane	WWI	First test as weapon
aeroplane	WWII	Large scale war weapon
rock	1950–1960	Birth of rock-and-roll music
gay	1985–1990	Recommended instead of *homosexual*
tank	1916	First tank in battle
cool	1964	Slang used for self-control
flight	WWI-WWII	First commercial flights non-war related
flight	after WWI	Commercial aviation grows rapidly
mouse	1965	The computer mouse was introduced
mouse	1980–1985	Common usage with computers like Macintosh 128K
telephone	1839	First commercial use in Great Western Railway
telephone	1893	28k subscribers in Sweden, highest density in the world.[a]
telephone	1914	USA twice the phone density than any other country

[a] This can be found in [18] and corresponds to usage change rather than lexical change.

Therefore, as a second expected time point, we consider the *first cluster evidence*, $t_{CE}(w)$, indicating the first time the term appears in a cluster and hence can be used for tracking. If the term is present with the corresponding sense in the collection before the time of the first cluster evidence, it means that it has either been mentioned very few times, or that the clustering algorithm could not find it. This time point represents the first possible time point for the tracking, given the curvature clustering algorithm for extracting word sense clusters. The true expected time lies in the interval $[t_{DI}, t_{CE}]$. Finally, we have the time point when our method detects the change event, $t_{found}(w)$.

The time delay is $T_{DI}(w) = t_{found}(w) - t_{DI}(w)$ and $T_{CE}(w) = t_{found}(w) - t_{CE}(w)$. The average time delay is summed over all words, $AT_{DI} = \frac{\sum_{\forall w} T_{DI}(w)}{|w|}$ and $AT_{CE} = \frac{\sum_{\forall w} T_{CE}(w)}{|w|}$.

Experimental Set-Up. We differentiate between change events, stable senses of changing words and stable senses of stable words. We provide an upper limit to our change detection (Upper) by considering only if the change event is present in our units, disregarding the relation to other senses. This provides a measure of how much can be found in our clusters and implicitly measures the quality of the induction algorithm for change detection. For our change detection, we expect the evolved senses to appear inside a unit, the polysemic senses should be found within an existing path to illustrate the relatedness to other senses, and the homonym senses should be in their own path to show the lack of relatedness.

5   Experimental Results

We will present the experimental results on recall followed by average time delay.

5.1   Recall

Table 2 shows the recall of our experiment. Our upper bound shows that we are able to find 95% of all changes and stable senses among our clusters, giving us an upper bound on the recall of 95%. The only senses that are not found are the first senses for *Internet* and *computer*, and *bitch* in its offensive sense, most likely because of few mentions in the dataset.

Table 2. Recall and time delay for all terms in the testset, where BC is the best case and *All* is the all class experiments. The *value* in **bold** represents delay time from first cluster evidence AT_{CE} and the second represents time of definition AT_{DI}.

	Recall		Avg. time delay	
	Upper	All	Upper	All
Evolved sense	0.91	0.71	**4.9**–17.4	**12.0**–21.2
Existing – evo	1.00	1.00	**11.7**–59.0	**11.7**–59.0
Existing – stable	1.00	1.00	**2.7**–20.5	**2.7**–20.5
Average excl. stable	0.94	0.80	**7.1**–30.7	**11.9**–35.4
Total average	0.95	0.84	**6.3**–28.7	**9.9**–32.1

For the change events, we are able to find 71% of them in the way we expect in relation to the other senses. The ones that are missing are the polysemous novel senses. By looking at examples from this class, it is obvious that the linguistic definition is very hard to detect automatically. *E.g.* the term *memory* in a digital sense is related to *human memory*, but rarely used in similar context. *mouse* used in a *computer mouse* sense has no words in common with the *animal* sense, *train* as a *mechanic train* with a locomotive differs largely from a *train of people or vehicles* (*e.g.* funeral train) and the *musical tape* is related to the *sewing tape* because of the shape but share no common words. Therefore, our method cannot place them in the correct path but chooses to place them in their own path. Excluding the polysemic senses, our recall is 92% for the change events.

Table 3 shows units for the term *rock* corresponding to three paths. The first unit represents the stone senses and the last unit the Rock, paper, scissors game both in their own path. The remaining three units are placed in a path for the *music* sense, $u_2 \rightarrow u_3 \rightarrow u_4$. A future direction of investigation is to find why the first music sense appears first in 1979.

Table 3. Extract of units for *rock*. Units display some internal clusters and terms.

Year	Cluster terms
	Unit u_1: 1951–2003 (Stone)
1951	rock, sand, mud, clay, rain, ward, stone
1987	gravel, rock, sand, asphalt
1998	gravel, rock, sand
2003	dirt, calcined, clay, rock, stone, sand, gravel, moy sand
	Unit u_2: 1979–2006 (First music cluster)
1979	rock, jazz, marriage, advice bureau
1987	classical, soul, drug, rockabilly, sex, folk, funk, gospel
1995	jazz, reggae, rock, funk, rap, hard rock, punk
2006	chamber music, bluegrass, soul, blue, funk
	Unit u_3: 1987–2003 (Modern music)
1987	rap, opera, calypso, drug, sex, drama
1995	grunge, punk, alternative rock, hiphop, blue, rock
2003	irish music, mexican, mixing rock, appalachian song, rock, hiphop
	Unit u_4: 1988–2007 (Rock & Roll lifestyle)
1988	rock, roll, sex, african, drug
2001	fantasy of sex, sex, rock, drug, roll, capture
2006	guitarist, songwriter, freeassociates about religion, rock,
	Unit u_5: 2000–20075 (Game)
2000	rock, paper, scissors

False Positives. Precision is not well understood w.r.t. word sense change detection when units can consist of 70–80 clusters and paths can contain hundreds of units. Instead, we analyze false positives by looking at the average number of change events per word. On average there are 3 paths per word and 5.3 units per path for change words and 13.3 for stable words. Among the changing words, we have an average of 2.2 change events and thus we would expect around 2 false positives (5.3 units mean 4 change events on average out of which we expect 2 to be correct). Among the stable words, all change events and thus different units are per definition wrong, that means on average 13.3 false positives. However, there are some words that stand out, *horse*, *bank* and *music* are very common words and have, in average, 47.5, 21.4 and 24.9 units per path when we would expect only one. For these we observe very long spanning units with 206, 197 and 204 years. Excluding these words, the average number of unit per path drops to 6.6 and represents 5 change events.

Though this is an approximation of the false positive rate, it does tell us that the number of elements to manually filter is limited and thus the results can be of great use for researchers and digital archive users. The true utility of the paths will be determined in future work with researchers from the digital humanities as well as normal users of digital archives.

5.2   Average Time Delay

Table 2 shows the average time delays for our experiments. Values marked in bold are delay times with respect to first cluster evidence, AT_{CE} and the second values are with respect to time of definition AT_{DI}. At best, we can find evidence in our units 7.1 years after the time the changes appear in our clusters and 30.7 years after being invented or defined in a dictionary. We consider the true time delay to be between 7.1–30.7 years. To appear in the paths as we expect, the time delay is slightly longer, between 11.9–35.4 years. If we split the time delays into the change categories, we have 16.1–20.9 year for the evolved senses, 5.8–27.8 for the polysemous senses and 1.6–19.8 years for the homonymic senses.

For **existing senses** we see something interesting; the existing senses for words that later have a change event have significantly longer average time delays compared to existing senses of stable terms, 11.7 compared to 2.7. One possible explanation is that words are less likely to change their meanings, if they are commonly used. The long time delays compared to definition is likely due to the choice of words in the stable category. The papers might not often discuss the *bitch* as a female dog, *train* as a train of people or the *car* as a wheeled, usually horse-drawn conveyance and hence we cannot detect these senses with our induction method, thus the longer time delays for stable senses of evolving words. On average, we find that excluding the existing senses of stable terms we have an average time delay of 7.1–30.7 years for any evidence to appear in a unit, 11.9–35.4 for our method to find the change in its expected form. Including existing senses, delay times decrease to 6.3–28.7 and 9.9–32.1 respectively.

6   Discussions

Our experiments show that we are able to find much of the expected word sense changes as well as the stable senses. We depend on automatically induced word senses that are grouped into *units* to capture individual senses over time. Units are then grouped into *paths* that capture concepts for a term.

The utility of using a method that differentiates between senses of a word are plentiful. For example, the word *rock* has a stable sense of *stone* in our dataset and then, in the 20th century, adds a sense of *music style*. The *music* sense evolves with different kinds of music and adds a *rock-and-roll lifestyle* sense in the same path as the *music* sense, clearly showing that these senses are related.

Also among words that are considered the same meaning over time, we can find changes that reflect usage and culture. For example, the *telephone* was firstly mentioned in contexts that related to the entire community or to houses in general, *1882, hydraulic lift, electric light, telephone, lift*. Then, slowly, it became something that belonged inside each apartment, *1977, television set, freezer, telephone, refrigerator, cooker, washing machine* and then a tool for (mass) communication *1997, telephone, television, radio, newspaper.* The word *aeroplane* is firstly defined as a flying machine, *1908 airship, aeroplane, balloon, aeroplane construction,* then as a means of transportation *1914 plane, aeroplane, motor bicycle, motor lorry, car* and finally as a weapon of war *1917 piping, gun, aeroplane, shafting, tank, infantry.* The word *travel* had only senses related to a literature genre *1803 literature, science, art, travel, voyage* before we could see evidence in the early 20th century of actual travel *1906 full board, travel, best hotel.* It is important to note that our datasets represent different dialects, British (The Times) and American (New York Times) which could lead to changes that are due to dialectal differences rather than sense changes. Among our test set, we have only three words (*gay, phone* and *telephone*) where the expected change lies in the period up to 1985 (The Times) and the found changes is in the period after and hence bridges this dialectal gap. In addition, for the *All* experiment, the *computer* sense of *mouse* was found in 1995, the expected was in the 1960s and the first cluster evidence in 1985. For the remainder of the words, the expected and found changes lie in the same dataset and hence they do not suffer from risk of dialectal interference.

The results of word sense change can be used to help users of a digital archive to understand the content in the archive when the language has changed over time. Senses that have changed can be marked and examples can be presented to help interpret the older sense. Language changes will be an increasing problem as we store more social media content in our archives [8]. The advantages of automatically detecting sense changes from the archive directly rather than relying on an outside reference, *e.g.* a dictionary are also obvious; dictionaries are meant as references and do not model how people use the language. But the results of word sense change detection can also be useful for exploring an archive and the culture represented there; *E.g.* what was the updake of the *telegraph*?[1]. They can also be used for language teaching and learning [4].

There is a need for temporal sentiment analysis which can only be made reliably after having detected word sense change, to be able to differentiate between *awesome leaders* of different times but also to answer research questions like what the attitude towards *rhetoric* has been over time [22].

7   Conclusions and Future Work

In this paper we presented results for a word sense change detection method that relies on induced word senses as a basis for detecting word sense change. We present analysis of the results and show that these can have an impact for research also in the digital humanities, where the *when, how* and *why* of language change are important. We show that our method, in addition to finding word sense change, also finds cultural and usage change. Our method detects change in the correct form 11.9 years after the first cluster evidence and is the first work to report such time analysis. Given the 222 year timespan, we consider this delay

---

[1] https://sweclarin.se/sites/sweclarin.se/files/videos/invigning_2016/
Johan-Jarlbrink.mp4.

to be a good starting point for future work and for analysis regarding differences between data sources, place of publication and time periods.

It remains future work to find the best way to preserve and utilize found change. Temporal indexing structures, information retrieval and presentation techniques as well as scalability issues are future directions for research in the field of automatic detection of word sense change. Preferably, digital archives should be stored with existing concurrent dictionaries and resources, and be word senses disambiguated to ensure long-term semantic access.

Acknowledgments. This work has been funded in parts by the project "Towards a knowledge-based culturomics" supported by a framework grant from the Swedish Research Council (2012–2016; dnr 2012-5738). This work is also in parts funded by the European Research Council under Alexandria (ERC 339233) and the European Community's H2020 Program under SoBigData (RIA 654024). We would like to thank Times Newspapers Limited for providing the archive of The Times for our research.

References

1. Basile, P., Caputo, A., Luisi, R., Semeraro, G.: Diachronic analysis of the italian language exploiting google Ngram. In: Proceedings of Third Italian Conference on Computational Linguistics (CLiC-it 2016) (2016)
2. Cook, P., Lau, J.H., McCarthy, D., Baldwin, T.: Novel word-sense identification. In: Proceedings of COLING 2014, Dublin, Ireland, pp. 1624–1635, August 2014. http://www.aclweb.org/anthology/C14-1154
3. Cooper, M.C.: A mathematical model of historical semantics and the grouping of word meanings into concepts. Comput. Linguist. **32**(2), 227–248 (2005)
4. Dejica, D., Hansen, G., Sandrini, P., Para, I.: Language in the Digital Era. Challenges and Perspectives. De Gruyter, Berlin (2016)
5. Dorow, B., Eckmann, J.P., Sergi, D.: Using curvature and markov clustering in graphs for lexical acquisition and word sense discrimination. In: Proceedings of the Workshop MEANING-2005 (2005)
6. Frermann, L., Lapata, M.: A bayesian model of diachronic meaning change. TACL **4**, 31–45 (2016)
7. Gulordava, K., Baroni, M.: A distributional similarity approach to the detection of semantic change in the Google Books Ngram corpus. In: Proceedings of the GEMS 2011 Workshop on GEometrical Models of Natural Language Semantics, GEMS 2011, pp. 67–71. Association for Computational Linguistics (2011)
8. Hamilton, W.L., Leskovec, J., Jurafsky, D.: Cultural shift or linguistic drift? comparing two computational measures of semantic change. In: Proceedings of the Conference on Empirical Methods in Natural Language Processing (2016)
9. Hamilton, W.L., Leskovec, J., Jurafsky, D.: Diachronic word embeddings reveal statistical laws of semantic change. CoRR abs/1605.09096 (2016)
10. Kim, Y., Chiu, Y.I., Hanaki, K., Hegde, D., Petrov, S.: Temporal analysis of language through neural language models. In: Workshop on Language Technologies and Computational Social Science (2014)
11. Kulkarni, V., Al-Rfou, R., Perozzi, B., Skiena, S.: Statistically significant detection of linguistic change. In: Proceedings of the 24th International Conference on World Wide Web, pp. 625–635. ACM (2015)

12. Lau, J.H., Cook, P., McCarthy, D., Newman, D., Baldwin, T.: Word sense induction for novel sense detection. In: EACL 2012, 13th Conference of the European Chapter of the Association for Computational Linguistics, pp. 591–601 (2012). http://aclweb.org/anthology-new/E/E12/E12-1060.pdf
13. Miller, G.A.: WordNet: a lexical database for english. Commun. ACM **38**, 39–41 (1995)
14. Mitra, S., Mitra, R., Maity, S.K., Riedl, M., Biemann, C., Goyal, P., Mukherjee, A.: An automatic approach to identify word sense changes in text media across timescales. Nat. Lang. Eng. **21**(05), 773–798 (2015)
15. Mitra, S., Mitra, R., Riedl, M., Biemann, C., Mukherjee, A., Goyal, P.: That's sick dude!: automatic identification of word sense change across different timescales. In: Proceedings of the 52nd Annual Meeting of the Association for Computational Linguistics, ACL 2014 USA, pp. 1020–1029 (2014). http://aclweb.org/anthology/P/P14/P14-1096.pdf
16. Tahmasebi, N., Risse, T.: Word Sense Change Test Set (2017). https://doi.org/10.5281/zenodo.495572
17. OED, O.E.D. (2017). http://www.oed.com/view/Entry/197656?rskey=8IY6gT$&$result=1$&$isAdvanced=false#eid. Accessed 02 May 2016
18. Roslin Bennett, A.: The Telephone Systems of the Continent of Europe. Longmans Green and CO., London (1895). http://archive.org/stream/telephonesystems00bennrich#page/332/
19. Sagi, E., Kaufmann, S., Clark, B.: Semantic density analysis: comparing word meaning across time and phonetic space. In: Proceedings of the Workshop on Geometrical Models of Natural Language Semantics, GEMS 2009, pp. 104–111. ACL (2009). http://dl.acm.org/citation.cfm?id=1705415.1705429
20. Tahmasebi, N., Niklas, K., Zenz, G., Risse, T.: On the applicability of word sense discrimination on 201 years of modern english. Int. J. Dig. Libr. **13**(3–4), 135–153 (2013). doi:10.1007/s00799-013-0105-8
21. Tahmasebi, N.N.: Models and algorithms for automatic detection of language evolution. Ph.D. thesis, Gottfried Wilhelm Leibniz Universitt Hannover (2013). http://edok01.tib.uni-hannover.de/edoks/e01dh13/771705034.pdf
22. Viklund, J., Borin, L.: How can big data help us study rhetorical history? In: Clarin Annual Conference (2016)
23. Wang, J., Bansal, M., Gimpel, K., Ziebart, B.D., Clement, T.Y.: A sense-topic model for word sense induction with unsupervised data enrichment. TACL **3**, 59–71 (2015)
24. Wang, X., McCallum, A.: Topics over time: a non-markov continuous-time model of topical trends. In: Proceedings of the 12th ACM SIGKDD International Conference on Knowledge Discovery and Data Mining, KDD 2006, USA, pp. 424–433. ACM (2006)
25. Wijaya, D.T., Yeniterzi, R.: Understanding semantic change of words over centuries. In: Proceedings of the 2011 International Workshop on DETecting and Exploiting Cultural diversiTy on the Social Web, DETECT 2011, pp. 35–40. ACM, New York (2011)
26. Zhang, Y., Jatowt, A., Tanaka, K.: Detecting evolution of concepts based on cause-effect relationships in online reviews. In: Proceedings of the 25th International Conference on World Wide Web, pp. 649–660. ACM (2016)

Entities

Multi-aspect Entity-Centric Analysis of Big Social Media Archives

Pavlos Fafalios[1], Vasileios Iosifidis[1(✉)], Kostas Stefanidis[2], and Eirini Ntoutsi[1]

[1] L3S Research Center, University of Hannover, Hanover, Germany
{fafalios,iosifidis,ntoutsi}@l3s.de
[2] Faculty of Natural Sciences, University of Tampere, Tampere, Finland
kostas.stefanidis@uta.fi

Abstract. Social media archives serve as important historical information sources, and thus meaningful analysis and exploration methods are of immense value for historians, sociologists and other interested parties. In this paper, we propose an *entity-centric* approach to analyze social media archives and we define measures that allow studying how entities are reflected in social media in different time periods and under different aspects (like popularity, attitude, controversiality, and connectedness with other entities). A case study using a large Twitter archive of 4 years illustrates the insights that can be gained by such an entity-centric multi-aspect analysis.

1 Introduction

Social networking services have now emerged as central media to discuss and comment on breaking news and noteworthy events that are happening around the world. In Twitter, for example, every second around 6,000 tweets are posted, which corresponds to over 350,000 tweets per minute, 500 million tweets per day and around 200 billion tweets per year[1].

Such user-generated content can be seen as a comprehensive documentation of society and is therefore of immense historical value for future generations [7]. Although there are initiatives and works that aim to collect and preserve social media archives (e.g., the Twitter Archive at the Library of Congress [25]), the absence of meaningful access and analysis methods still remains a major hurdle in the way of turning such archives into useful sources of information for historians, journalists and other interested parties [7]. Besides, when exploring archived data, analysts are not interested in the documents per se, but instead they want to see, compare, and understand the behavior of (and trends about) entities, like companies, products, politicians, music bands, songs and movies, thus calling for entity-level analytics over the archived data [22].

In this paper, we propose an *entity-centric multi-aspect* approach to analyze social media archives. Our approach allows tracking of how entities are reflected

---

[1] http://www.internetlivestats.com/twitter-statistics/ (June 21, 2017).

© Springer International Publishing AG 2017
J. Kamps et al. (Eds.): TPDL 2017, LNCS 10450, pp. 261–273, 2017.
DOI: 10.1007/978-3-319-67008-9_21

in a collection of user-generated content (e.g., tweets) and how such information evolves over time and also with respect to other entities. Specifically, we define measures for the temporal analysis of an entity in terms of its: *popularity*, *attitude* (predominant sentiment), *sentimentality* (magnitude of sentiment), *controversiality*, and *connectedness* to other entities. A distinctive characteristic of our approach is that it does not rely on service-specific labels (like #hashtags and @mentions), but it exploits *entity linking* and thus can be applied over any type of time-annotated texts.

We examine the insights gained by the proposed measures by analyzing a large collection of billions of tweets spanning a period of 4 years. Such analytics enable to answer questions like:

– *How did the popularity of Greek Prime Minister, Alexis Tsipras, evolve in 2015? Were there any "outlier" periods, i.e., periods of extremely high or low popularity? What were the entities discussed in social media together with Alexis Tsipras during these periods?*
– *How did the predominant sentiment about Donald Trump and Hillary Clinton vary during 2016? Were there any controversial time periods related to these two politicians, i.e., time periods in which there were many positive and negative tweets? How did the "connectedness" of Trump with the entity 'Abortion' evolve during 2016?*

In a nutshell, we make the following contributions:

– We introduce a multi-aspect entity modeling and propose a set of measures for capturing important entity features in a given time period. A sequence of such captures comprises a multi-variate time series in which each point is a multi-aspect description of the entity at a certain time period. We demonstrate the usefulness of our approach through illustrative examples.
– We provide an open source distributed library for computing the proposed measures efficiently.
– We analyze a large Twitter archive (spanning 4 years and containing billions of tweets) and make publicly available the entity- and sentiment- annotations of this archive. This dataset can foster further research in related topics (like event detection, topic evolution, entity recommendation, concept drift).

The rest of this paper is organized as follows: Sect. 2 provides some background and related works. Section 3 details the multi-aspect entity description and the proposed measures. Section 4 presents a library for the distributed computation of the measures. Section 5 presents the results of a case study. Finally, Sect. 6 concludes the paper and identifies interesting directions for future research.

2   Background and Related Work

We first discuss the required background and then we describe related works and how they differ from our approach.

2.1   Entity Linking and Sentiment Analysis

Our analysis is based on two different types of annotations applied in the short texts of a social media archive (like a Twitter archive): *entity linking* and *semantic analysis*.

Entity Linking. In our problem, an *entity* is anything with a distinct, separate and meaningful existence that also has a "web identity" expressed through a unique URI (e.g., a Wikipedia/DBpedia URI). This does not only include persons, locations, organizations, etc., but also events (e.g., *US 2016 presidential election*) and concepts (e.g., *Democracy*). Each entity is associated with a unique URI, while several labels/names can be used to refer to this entity. For example, for the entity *Barack Obama* (https://en.wikipedia.org/wiki/Barack_Obama), possible names are "Barack Obama", "Obama" and "former President Obama". There is a plethora of tools that automatically extract entities from plain text and link them to knowledge bases like Wikipedia/DBpedia [5,10,14] (for a survey on entity linking and resolution, see [9]). In our experiments, we use Yahoo FEL [5] which has been specially designed for linking entities from short texts to Wikipedia.

Sentiment Analysis. Sentiment analysis refers to the problem of assigning a sentiment label (e.g., positive, negative) or sentiment score to a document [15]. We opt for the latest and we use SentiStrength, a robust tool for sentiment strength detection on social web data [21]. SentiStrength assigns both a positive and a negative score (since both types of sentiment can occur simultaneously). The score of a positive sentiment strength score ranges from $+1$ (not positive) to $+5$ (extremely positive). Similarly, negative sentiment strength scores range from -1 (not negative) to -5 (extremely negative).

2.2   Related Work

The availability of web-based application programming interfaces (APIs) provided by social media services (like Twitter and Facebook) has led to an "explosion" of techniques, tools and platforms for social media analytics. The work in [4] surveys analytics tools for social media as well as tools for scraping, data cleaning and sentiment analysis on social media data. There is also a plethora of works on exploiting social media for a variety of tasks, like opinion summarization [13], event and rumor detection [3,16], topic popularity and summarization [2,23], information diffusion [11], popularity prediction [18], and reputation monitoring [1]. Below, we discuss works related to temporal analysis of topics and entities in social media.

[20] proposes a query-answering framework to allow entity search in social networks by exploiting the underlying social graph and temporal information. [24] studies how to incorporate social attention in the generation of timeline summaries. It proposes capturing social attention for a given topic by learning users' collective interests in the form of word distributions from Twitter. A more recent work on the same topic focuses on how to select a small set of

representative tweets to generate a meaningful timeline, which provides enough coverage for a given topical query [23]. [2] performs a spatiotemporal analysis of tweets, investigating the time-evolving properties of the subgraphs formed by the users discussing each topic. The focus is on the network topology formed by follower-following links on Twitter and the geospatial location of the users. [6] introduces a catalogue of metrics for analyzing hashtag-based communication on Twitter, while [18] tackles the problem of predicting entity popularity on Twitter based on the news cycle. [8] investigates whether semantic relationships between entities can be learned by analyzing microblog posts published on Twitter. The evaluation results showed that co-occurrence based strategies allow for high precision and perform particularly well for relations between persons and events. Our entity-to-entity connectedness scores are also based on entity co-occurrences (more in Sect. 3).

To our knowledge, our work is the first that models *multi-aspect entity-centric analytics* for social media archives. The proposed measures capture the multi-aspect behavior of an entity in different time periods and can be exploited in a variety of tasks, like entity evolution, event detection, and entity recommendation.

3   Multi-aspect Entity Measures

We propose a multi-aspect description of an entity in terms of its: *popularity* (how much discussion it generates), *attitude* (predominant sentiment), *sentimentality* (magnitude of sentiment), *controversiality* (whether there is a consensus about the sentiment of the entity), *connectedness* to another entity, and *network* (strongly connected entities). All these measures are computed for a given time period (e.g., July 2014, 10–20 June 2013, June-August 2015). Below, we formally introduce these measures by classifying them into: *single-entity measures* and *entity-relation measures*.

First, let C be a collection of short texts (e.g., tweets) covering the time period $T = [t_s, t_e]$ (where t_s, t_e are two different time points with $t_s < t_e$), and let U be the total set of users who posted these texts. Let also E denote a finite set of entities, e.g., all Wikipedia entities.

3.1   Single-Entity Measures

Popularity. Let $e \in E$ be a given entity and $T_i \subseteq T$ a given time period. Let also $C_i \subseteq C$ be the collection of short texts posted during T_i. The popularity of e during T_i equals to the percentage of *texts* mentioning e during that period. Formally:

$$popularity_c(e, T_i) = \frac{|C_{e,i}|}{|C_i|} \tag{1}$$

where $C_{e,i} \subseteq C_i$ denotes the set of texts mentioning e during T_i.

Using the above measure, an entity can be very popular even if it is discussed by a few users but in a large number of texts. A more fine-grained indication of popularity is given by the number of different users discussing the entity.

In that case, if $u_c \in U$ denotes the user who posted the text c, the popularity of an entity $e \in E$ during T_i can be defined as the percentage of different *users* discussing e during that period, i.e.:

$$popularity_u(e, T_i) = \frac{|\cup_{c \in C_{e,i}} u_c|}{|\cup_{c \in C_i} u_c|} \quad (2)$$

We can now combine both aspects (percentage of *texts* and *users*) in one popularity score using the following formula:

$$popularity_{c,u}(e, T_i) = popularity_c(e, T_i) \cdot popularity_u(e, T_i) \quad (3)$$

An entity has now a high popularity score if it is discussed in many tweets and by many different users.

Attitude and Sentimentality. We use two measures (proposed in [12] for the case of questions and answers) for capturing a text's *attitude* (predominant sentiment) and *sentimentality* (magnitude of sentiment). First, for a text $c \in C$, let $s_c^+ \in [1, 5]$ be the text's positive sentiment score and $s_c^- \in [-5, -1]$ be the text's negative sentiment score (according to SentiStrength, c.f. Sect. 2.1). The attitude of a text c is given by $\phi_c = s_c^+ + s_c^-$ (i.e., $\phi_c \in [-4, 4]$) and its sentimentality by $\psi_c = s_c^+ - s_c^- - 2$ (i.e., $\psi_c \in [0, 8]$).

We now define the *attitude* of an entity e in a time period T_i as the average attitude of texts mentioning e during T_i. Formally:

$$attitude(e, T_i) = \frac{\sum_{c \in C_{e,i}} \phi_c}{|C_{e,i}|} \quad (4)$$

Likewise, the *sentimentality* of an entity e in a time period T_i is defined as the average sentimentality of texts mentioning e during T_i:

$$sentimentality(e, T_i) = \frac{\sum_{c \in C_{e,i}} \psi_c}{|C_{e,i}|} \quad (5)$$

Controversiality. An entity e can be considered controversial in a time period T_i if it is mentioned in both many positive and many negative texts. First, let $C_{e,i}^+$ be the set of texts mentioning e during T_i with strong positive attitude, i.e., $C_{e,i}^+ = \{c \in C_{e,i} \mid \phi_c \geq \delta\}$, where $\delta \in [0, 4]$ is a strong attitude threshold (e.g., $\delta = 2.0$). Likewise, let $C_{e,i}^-$ be those with strong negative attitude, i.e., $C_{e,i}^- = \{c \in C_{e,i} \mid \phi_c \leq -\delta\}$. We now consider the following formula for entity *controversiality*:

$$controversiality(e, T_i) = \frac{|C_{e,i}^+| + |C_{e,i}^-|}{|C_{e,i}|} \cdot \frac{min(|C_{e,i}^+|, |C_{e,i}^-|)}{max(|C_{e,i}^+|, |C_{e,i}^-|)} \quad (6)$$

Intuitively, a value close to 1 means that the probability of the entity being "controversial" is high since there is a big percentage of texts with strong attitude (first part of the formula) and also there are both many texts with strong positive attitude and many texts with strong negative attitude (second part of the formula).

3.2   Entity-Relation Measures

Entity-to-Entity Connectedness. We define a *direct-connectedness* score between an entity $e \in E$ and another entity $e' \in E$ in a time period T_i, as the number of texts in which e and e' co-occur within T_i. Formally:

$$direct\text{-}connectedness(e, e', T_i) = \frac{|C_{e,i} \cap C_{e',i}|}{|C_{e,i}|} \qquad (7)$$

Notice that the relation is not symmetric. We consider that if an entity e_1 is strongly connected with an entity e_2, this does not mean that e_2 is also strongly connected with e_1. For example, consider that e_1 is mentioned in only 100 texts, e_2 in 1M texts, while 90 texts mention both entities. We notice that e_2 seems to be very important for e_1, since it exists in 90/100 of e_1's texts. On the contrary, e_1 seems not to be important for e_2, since it exists in only 90/1M of its texts.

Two entities may not co-occur in texts, but they may share many common co-occurred entities. For example, both *Barack Obama* and *Donald Trump* may co-occur with entities like *White House*, *US Election* and *Hillary Clinton*. For an input entity $e \in E$ and another entity $e' \in E$, we define an *indirect-connectedness* score which considers the number of *common entities* with which e and e' co-occur in a time period T_i:

$$indirect\text{-}connectedness(e, e', T_i) = \frac{|(\cup_{c \in C_{e,i}} E_c) \cap (\cup_{c \in C_{e',i}} E_c)|}{|(\cup_{c \in C_{e,i}} E_c)|} \qquad (8)$$

where $E_c \subseteq E$ is the entities mentioned in text c. Also in this case, the relation between the two entities is not symmetric.

Entity k-Network. This measure targets at finding a list of entities strongly connected to the query entity in a given time period T_i. First, we define a connectedness score between an entity $e \in E$ and a set of entities $E' \subseteq E$ within T_i, as the average direct-connectedness score of the entities in E'. Formally:

$$connectedness(e, E', T_i) = \frac{\sum_{e' \in E'} direct\text{-}connectedness(e, e', T_i)}{|E'|} \qquad (9)$$

The k-Network of an entity e during T_i is the set of k entities $E' \subseteq E$ with the highest average connectedness score. Namely:

$$k\text{-}Network(e, T_i) = \underset{E' \subseteq E, \; |E'|=k}{\operatorname{argmax}} connectedness(e, E', T_i) \qquad (10)$$

In simple terms, the k-Network of an entity e consists of the k entities with the highest *direct-connectedness* scores.

3.3   Discussion

The above presented measures capture the multi-aspect behavior of a given entity at a certain time period. In the long run, a multi-variate time series is formed where each point represents the multi-aspect description of the entity at a certain period in time.

An important characteristic of our approach is that we can support both entity-specific queries referring to a single entity and cross-entity queries involving more than one entities (e.g., a category of entities). This is achieved through the *entity linking* process in which entities are extracted from the texts and are linked to knowledge bases like Wikipedia/DBpedia. In that way, we can collect a variety of properties for the entities extracted from our archive. This enables us to aggregate information and capture the behavior of sets of entities. For example, by accessing DBpedia, we can collect a list of German politicians, derive their popularity and then compare it with that of another set of entities.

Although the proposed analysis approach is generic and can be applied over different types of social media archives, it is clear that the quality of the generated data depends on the quality of the input data. Twitter, for example, provides 1% random sample, which though is subject to bias, fake news and possibly other adversarial attacks. In our case study (detailed in Sect. 5), although we remove spam, we do not take similar actions to deal with bias and other data peculiarities. This also means that high profile entities might occupy a big volume in the archive, whereas long-tail entities might be underrepresented or not represented at all. Except for the quality of the original data, the different preprocessing steps (spam removal, entity linking, sentiment analysis) are also prone to errors. This means that, especially for small archives, the data produced by the proposed measures are also prone to errors. For instance, regarding the entity linking task, selecting a very low threshold for the confidence score of the extracted entities can result in many false annotations, which in turn can affect the quality and reliability of the produced time-series.

4   Library for Computing the Measures

For computing the measures, we provide an Apache Spark library. Apache Spark[2] is a cluster-computing framework for large-scale data processing. The library contains functions for computing the proposed measures for a given entity and over a specific time period. It operates over an annotated (with entities and sentiments) dataset split per year-month (the dataset should be in a simple CSV format). The library is available as open source[3].

The time for computing the measures highly depends on the dataset volume, the used computing infrastructure as well as the available resources and the load of the cluster at the analysis time. The Hadoop cluster used in our experiments for analyzing a large Twitter archive of more than 1 billion tweets consisted of 25 computer nodes with a total of 268 CPU cores and 2,688 GB RAM (more about the dataset in the next section). Indicatively, the time for computing each of the measures was on average less than a minute (without using any index, apart from the monthly-wise split of the dataset).

---

[2] http://spark.apache.org/.
[3] https://github.com/iosifidisvasileios/Large-Scale-Entity-Analysis.

5   Case Study: Entity Analytics on a Twitter Archive

In this section, we first describe the results of the analysis and annotation of a large Twitter archive. Then, we present examples of case studies illustrating the insights gained from the proposed measures.

5.1   Annotating a Large Twitter Archive

We analyzed a large Twitter archive spanning 4 years (January 2014 - January 2017) and containing more than 6 billion tweets. The tweets were collected through the Twitter streaming API. Our analysis comprised the following steps: (i) filtering (filtering out re-tweets, keeping only English tweets), (ii) spam removal, (iii) entity linking, and (iv) sentiment analysis. The filtering step reduced the number of tweets to about 1.5 billion tweets (specifically, to 1,486,473,038 tweets). For removing the spam tweets, we trained a Multinomial Naive Bayes (MNB) classifier over the HSpam dataset [19]. This removed about 150 million tweets. The final dataset consists of 1,335,324,321 tweets from 110,548,539 users. Figure 1 shows the number of tweets per month on the final dataset.

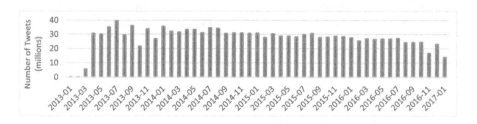

Fig. 1. Number of tweets per month.

For the *entity linking* task, we used Yahoo FEL [5] with a confidence threshold score of -3. Totally, 1,390,286 distinct entities were extracted from the tweets collection. On average, each tweet contains about 1 entity (specifically, 0.95), while FEL returned no entity for about 550 million tweets. For each extracted entit y, we also store the confidence score provided by FEL. Thereby, data consumers can select suitable confidence ranges to consider, depending on the specific requirements with respect to precision and recall. For *sentiment analysis*, we used SentiStrength [21]. The average sentimentality of all tweets is 0.92, the average attitude 0.2, while 622,230,607 tweets have no sentiment (-1 negative sentiment and 1 positive sentiment). Table 1 shows the number of tweets per attitude value.

Table 1. Number of tweets per attitude value.

Attitude:	-4	-3	-2	-1	0
Number of tweets:	2,234,887	34,666,708	68,812,370	104,628,022	670,484,267
Attitude:	1	2	3	4	
Number of tweets:	301,635,430	138,197,637	13,610,492	1,054,508	

The annotated dataset is publicly available in CSV format[4]. We make the dataset available so anyone interested can use it together with the library (described in Sect. 4) to extract the measures for any entity at the desired level of temporal granularity.

5.2   Case Studies

Entity Popularity. Figure 2 (left) shows the popularity of *Alexis Tsipras* (Greek prime minister) within 2015. We notice that his popularity highly increased in July. Indeed, in July 2015 the Greek bailout referendum was held following the bank holiday and capital controls of June 2015. This event highly increased the popularity of the Greek prime minister. Moreover, by comparing the trend of the two different popularity scores (Formulas 1 and 2), we notice that, during June and July 2015, the percentage of different users discussing about *Alexis Tsipras* increased in bigger degree compared to the percentage of tweets, implying that more people were engaged in the discussion.

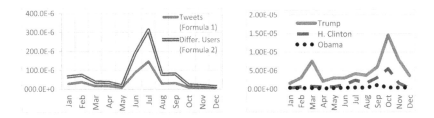

Fig. 2. Popularity of *"Alexis Tsipras"* in 2015 (left); Popularity of *"Donald Trump"*, *"Hillary Clinton"* and *"Barack Obama"* in 2016 (right).

Likewise, we can compare the popularity of multiple entities within the same time period. For example, Fig. 2 (right) shows the popularity of *Donald Trump*, *Hillary Clinton* and *Barack Obama* within 2016 (according to Formula 3). We notice that *Donald Trump* is much more popular in all months. We also notice that, in October 2016 the popularity of *Donald Trump* and *Hillary Clinton* highly

---

[4] http://l3s.de/~iosifidis/tpdl2017/. For each tweet the dataset includes the following information: ID, user (encrypted), post date, extracted entities, positive and negative sentiment values. The text of the tweets is not provided for copyright purposes.

increased compared to the other months. This is an indicator of possible important events related to these two entities in October 2016 (indeed, two presidential
general election debates took place in that period).

Entity Attitude and Sentimentality. Figure 3 (left and middle) depicts the
attitude and sentimentality of *Donald Trump* and *Hillary Clinton* within 2016.
We notice that both entities had constantly a negative attitude, however that of
Hillary Clinton was worse in almost all months. Moreover, we notice that *Hillary
Clinton*'s attitude highly decreased in May 2016 (possibly, for example, due to
a report issued by the State Department related to Clinton's use of private
email). Regarding sentimentality, we notice that for the majority of months
the tweets mentioning *Donald Trump* are a bit more sentimental than those
mentioning *Hillary Clinton*. In general, we notice that the values of both attitude
and sentimentality are relatively small and close to zero. This is due to the very
big number of tweets with no sentiment (almost half of the tweets).

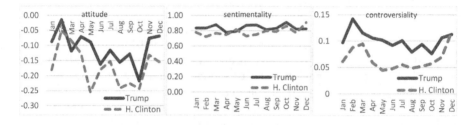

Fig. 3. Attitude (left), sentimentality (middle) and controversiality (right) of *"Donald
Trump"* and *"Hillary Clinton"* in 2016.

Entity Controversiality. Figure 3 (right) shows the controversiality of *Donald
Trump* and *Hillary Clinton* within 2016 (using $\delta = 2.0$). We notice that *Donald
Trump* induces more controversial discussions in Twitter than *Hillary Clinton*,
while February was his most "controversial" month, probably because of his
references to some debatable topics (like abortion) during his campaign trail.
It is interesting also that *Hillary Clinton*'s controversiality has an exponential
increment from September to December 2016.

Entity-to-Entity Connectedness. Figure 4(a) depicts the connectedness of
Alexis Tsipras with the concept *Greek withdrawal from the eurozone* within 2015.
We notice that these two entities are highly connected in June and July, while
after August, their connectedness is very close to zero. Indeed, important events
related to Greece's debt crisis took place in June and July 2015, including the
bank holiday, the capital controls and the Greek bailout referendum. Likewise,
Fig. 4(b) shows the connectedness of both *Donald Trump* and *Hillary Clinton*
with the concept *Abortion* in 2016. Here we notice that the connectedness is
almost constant for *Hillary Clinton*, while for *Donald Trump*, there is a very
large increment in March and April.

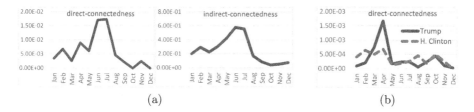

(a)   (b)

Fig. 4. (a) Connectedness of *"Alexis Tsipras"* with *"Greek withdrawal from the euro-zone"* in 2015 (Formulas 7 and 8); (b) Connectedness of *"Donald Trump"* and *"Hillary Clinton"* with *"Abortion"* in 2016 (Formula 7).

Entity k-Network. Figure 5 shows the 10-Network of *Alexis Tsipras* in three different time periods (April, July and October, 2015). We notice that there are three general entities that exist in all time periods (*Greece, Athens, Reuters*). For April and July, we notice that the 10-Network contains 4 common entities (*Syriza, Referendum, Greek withdrawal from the eurozone*, and *Yanis Varoufakis*), while for July and October, *Austerity* is the only common entity (probably related to the approval of strict measures required by the creditors). For April, the 10-Network contains three entities related to Russia (due to Tsipra's visit in Moscow to meet Russian president Vladimir Putin), while for October, it contains two entities related to European migrant crisis (probably due to Tsipra's visit in Lesvos island).

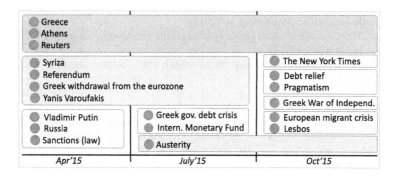

Fig. 5. 10-Network of *Alexis Tsipras* in April, July and October 2015.

6   Conclusion

We have proposed an entity-centric and multi-aspect approach to analyze social media archives, and we defined measures that allow studying how entities are reflected in social media and how entity-related information evolves over time. We believe that the proposed analysis approach is the first step towards more advanced and meaningful exploration of social media archives, while it can facilitate research in a variety of fields, such as information extraction, sociology, and digital humanities.

As part of our future work, we plan to exploit the rich amount of generated data for *prediction* of entity-related features. In particular, given an entity, our focus will be on how we can predict future values of the proposed measures (e.g., popularity or attitude in a given horizon). We also intend to study approaches on *understanding* and *representing* the dynamics of such evolving entity-related information, for instance, as done in [17] for the case of RDF datasets.

Acknowledgements. The work was partially funded by the European Commission for the ERC Advanced Grant ALEXANDRIA under grant No. 339233.

References

1. Amigó, E., Carrillo-de-Albornoz, J., Chugur, I., Corujo, A., Gonzalo, J., Meij, E., Rijke, M., Spina, D.: Overview of RepLab 2014: author profiling and reputation dimensions for online reputation management. In: Kanoulas, E., Lupu, M., Clough, P., Sanderson, M., Hall, M., Hanbury, A., Toms, E. (eds.) CLEF 2014. LNCS, vol. 8685, pp. 307–322. Springer, Cham (2014). doi:10.1007/978-3-319-11382-1_24
2. Ardon, S., Bagchi, A., Mahanti, A., Ruhela, A., Seth, A., Tripathy, R.M., Triukose, S.: Spatio-temporal analysis of topic popularity in Twitter. arXiv preprint arXiv:1111.2904 (2011)
3. Atefeh, F., Khreich, W.: A survey of techniques for event detection in twitter. Computat. Intell. **31**(1) (2015)
4. Batrinca, B., Treleaven, P.C.: Social media analytics: a survey of techniques, tools and platforms. AI & Society **30**(1) (2015)
5. Blanco, R., Ottaviano, G., Meij, E.: Fast and space-efficient entity linking for queries. In: WSDM (2015)
6. Bruns, A., Stieglitz, S.: Towards more systematic Twitter analysis: metrics for tweeting activities. Internat. J. Soc. Res. Method. **16**(2) (2013)
7. Bruns, A., Weller, K.: Twitter as a first draft of the present: and the challenges of preserving it for the future. In: 8th ACM Conference on Web Science (2016)
8. Celik, I., Abel, F., Houben, G.-J.: Learning semantic relationships between entities in Twitter. In: Auer, S., Díaz, O., Papadopoulos, G.A. (eds.) ICWE 2011. LNCS, vol. 6757, pp. 167–181. Springer, Heidelberg (2011). doi:10.1007/978-3-642-22233-7_12
9. Christophides, V., Efthymiou, V., Stefanidis, K.: Entity Resolution in the Web of Data. Synthesis Lectures on the SemanticWeb: Theory and Technology. Morgan & Claypool Publishers, San Rafael (2015)
10. Ferragina, P., Scaiella, U.: Tagme: on-the-fly annotation of short text fragments (by Wikipedia entities). In: CIKM (2010)
11. Guille, A., Hacid, H., Favre, C., Zighed, D.A.: Information diffusion in online social networks: a survey. ACM SIGMOD Record **42**(2) (2013)
12. Kucuktunc, O., Cambazoglu, B.B., Weber, I., Ferhatosmanoglu, H.: A large-scale sentiment analysis for Yahoo! answers. In: WSDM (2012)
13. Meng, X., Wei, F., Liu, X., Zhou, M., Li, S., Wang, H.: Entity-centric topic-oriented opinion summarization in Twitter. In: Proceedings of the 18th ACM SIGKDD International Conference on Knowledge Discovery and Data Mining. ACM (2012)

14. Moro, A., Raganato, A., Navigli, R.: Entity linking meets word sense disambiguation: a unified approach. Trans. Assoc. Computat. Linguist. **2** (2014)
15. Pang, B., Lee, L., et al.: Opinion mining and sentiment analysis. Found. Trends® Inf. Retrieval **2**(1–2) (2008)
16. Qazvinian, V., Rosengren, E., Radev, D.R., Mei, Q.: Rumor has it: Identifying misinformation in microblogs. In: Proceedings of the Conference on Empirical Methods in Natural Language Processing (2011)
17. Roussakis, Y., Chrysakis, I., Stefanidis, K., Flouris, G., Stavrakas, Y.: A flexible framework for understanding the dynamics of evolving RDF datasets. In: Arenas, M., Corcho, O., Simperl, E., Strohmaier, M., d'Aquin, M., Srinivas, K., Groth, P., Dumontier, M., Heflin, J., Thirunarayan, K., Staab, S. (eds.) ISWC 2015. LNCS, vol. 9366, pp. 495–512. Springer, Cham (2015). doi:10.1007/978-3-319-25007-6_29
18. Saleiro, P., Soares, C.: Learning from the news: predicting entity popularity on Twitter. In: Boström, H., Knobbe, A., Soares, C., Papapetrou, P. (eds.) IDA 2016. LNCS, vol. 9897, pp. 171–182. Springer, Cham (2016). doi:10.1007/978-3-319-46349-0_15
19. Sedhai, S., Sun, A.: Hspam14: A collection of 14 million tweets for hashtag-oriented spam research. In: Proceedings of the 38th International ACM SIGIR Conference on Research and Development in Information Retrieval (2015)
20. Stefanidis, K., Koloniari, G.: Enabling social search in time through graphs. In: Web-KR@CIKM (2014)
21. Thelwall, M., Buckley, K., Paltoglou, G.: Sentiment strength detection for the social web. J. Am. Soc. Inform. Sci. Technol. **63**(1), 163–173 (2012)
22. Weikum, G., Spaniol, M., Ntarmos, N., Triantafillou, P., Benczúr, A., Kirkpatrick, S., Rigaux, P., Williamson, M.: Longitudinal analytics on web archive data: it's about time! In: CIDR (2011)
23. Yao, J.-G., Fan, F., Zhao, W.X., Wan, X., Chang, E., Xiao, J.: Tweet timeline generation with determinantal point processes. In: Proceedings of the Thirtieth AAAI Conference on Artificial Intelligence. AAAI Press (2016)
24. Zhao, X.W., Guo, Y., Yan, R., He, Y., Li, X.: Timeline generation with social attention. In: Proceedings of the 36th International ACM SIGIR Conference on Research and Development in Information Retrieval. ACM (2013)
25. Zimmer, M.: The Twitter Archive at the Library of Congress: Challenges for information practice and information policy. First Monday **20**(7) (2015)

A Comparative Study of Language Modeling to Instance-Based Methods, and Feature Combinations for Authorship Attribution

Olga Fourkioti, Symeon Symeonidis[✉], and Avi Arampatzis

Database and Information Retrieval Research Unit,
Department of Electrical and Computer Engineering,
Democritus University of Thrace, 67100 Xanthi, Greece
olgafour1@gmail.com, {ssymeoni,avi}@ee.duth.gr
http://www.nonrelevant.net

Abstract. We present a comparative study of language modeling to traditional instance-based methods for authorship attribution, using several different basic units as features, such as characters, words, and other simple lexical measurements, as well as we propose the use of part-of-speech (POS) tags as features for language modeling. In contrast to many other studies which focus on small sets of documents written by major writers regarding several topics, we consider a relatively large corpus with documents edited by non-professional writers regarding the same topic. We find that language models based on either characters or POS tags are the most effective, while the latter provide additional efficiency benefits and robustness against data sparsity. Moreover, we experiment with linearly combining several language models, as well as employing unions of several different feature types in instance-based methods. We find that both such combinations constitute viable strategies which generally improve effectiveness. By linearly combining three language models, based respectively on character, word, and POS trigrams, we achieve the best generalization accuracy of 96%.

Keywords: Authorship attribution · Text mining · Language models · Computational linguistics · Text categorization · Text classification · Machine learning

1 Introduction

Authorship attribution can be used in a broad range of applications. Apart from literary research where a document of disputed authorship is assigned to one candidate author, the plethora of anonymous electronic texts (e.g. emails, blogs, electronic messages, forums, source code, etc.) has rendered authorship attribution analysis indispensable to diverse areas dealing with real-world texts. These areas include civil law (e.g. in cases of disputed copyrights), criminal law (e.g. in order to identify the author of a suicidal note or a terrorist's proclamation),

© Springer International Publishing AG 2017
J. Kamps et al. (Eds.): TPDL 2017, LNCS 10450, pp. 274–286, 2017.
DOI: 10.1007/978-3-319-67008-9_22

forensics (e.g. in order to determine the author of source code of malignant software), and others.

The general approach to authorship attribution is based on the extraction of features that have a high discriminatory potential between the candidate authors, the so-called style markers, which is followed by a feature selection method and the training of a classifier [13]. Specifically, there are three main types of style markers reflecting a document's representation: lexical, character, and syntactic [18]. Current authorship attribution studies are dominated by lexical and character feature types meaning that a document is considered as a sequence of words or characters respectively.

There have been proposed many measures in an attempt to quantify the diversity of an author's vocabulary. Type token ratio (i.e. vocabulary size to the total number of tokens), the hapax legomena (i.e. words occurring once in a document), the hapax dislegomena (i.e. words occurring twice in a document) are some typical examples. The frequency of occurrence of context-free words (i.e. stop words) and the word or character n-gram approaches are some of the most important and effective methods in the field of authorship attribution.

In this paper, we conduct a comparative study of language modeling to traditional instance-based approaches. Our general approach is that we build a probabilistic language model for each author or train a classifier using as basic units of representation either characters, words, or part-of-speech (POS) tags. While POS tags have been used before in instance-based methods, e.g. [12], we first introduce them in this study as features in statistical language modeling for authorship attribution. We also experiment with linear combinations of several language models, as well as feeding unions of different feature types into instance-based methods. We are not aware of another study employing such combinations in the problem of authorship attribution.

Our experiments are conducted in a corpus consisting of 62,000 movie reviews written by 62 users. This corpus is selected because of its two interesting characteristics. First, it was edited by non-professional authors, rendering the task of authorship attribution more challenging and closer to the contemporary problem of authorship attribution on the Internet. Second, the numbers of documents and authors are both considered relatively large in comparison to traditional authorship attribution studies. Consequently, it would be interesting to see whether past results extend also to such large collections which are homogeneous in the sense that all authors treat the same topic.

2   Related Work

Previous work in authorship attribution studies focused mainly on the extraction of lexical features. Two are considered the state-of-the-art methodologies regarding the lexical approach: the multivariate vocabulary richness analysis and the frequency of occurrences of individual words [17].

Syntactic analysis has been less studied because of the limitations imposed by the language and the availability of a parser, a tool able to perform syntactic analysis of texts. However, in recent years there have been attempts to exploit syntactic information from texts. The idea behind this is that authors tend to unconsciously use specific syntactic patterns and particular sentence structures, which can be a reliable authorial fingerprint and facilitate authorship inference [18].

Baayen et al. [3] were the first to investigate the discriminatory potential of syntactic features for authorship attribution purposes. Based on a syntactically annotated corpus, which comprised around 20,000 words from two English books, they extracted rewrite rule frequencies. Their method outperformed the traditional word-based ones.

Stamatatos et al. [17] used Sentence and Chunk Boundaries Detector (SCBD), a robust and accurate NLP tool, to detect sentence and chunk boundaries in unrestricted Greek text. The text analysis was divided into three stylometric levels: token-level, phrase-level, and analysis-level. The two first levels consisted of measurements on sentence and phrase level respectively, such as length of noun phrases, length of verb phrases, noun phrase counts, and so on. The analysis level captured information omitted in the previous two levels regarding the way in which the input text was analyzed by the SCBD tool. Their method achieved 80% accuracy on 300 Greek texts written by 10 authors, containing a total of 333,744 words.

Recently, Pokou et al. [12] suggested the use of part-of-speech (POS) skip-grams in authorship attribution studies. Skip-grams are constructed like n-grams but allow a distance gap between adjacent POS tags. First, a set of training texts is pre-processed to become a sequence of POS tags, and a unique signature representing each author's style is extracted by using the most frequent part-of speech skip-grams in the training texts. Then, these signatures are used as a criterion to classify the test documents. In their experimentation, they used a collection of 30 texts, consisting of 2,615,856 words, written by 10 authors. Their method led to high classification accuracy.

Sidorov et al. in their recent work [16] introduced syntactic n-grams. Syntactic n-grams differentiate from classic n-grams because they take into account the position in which the elements are presented in the syntactic trees, not in the original text. Thus, they manage to capture syntactic relations between words. In the experiments conducted in a corpus of 39 documents written by 3 authors, Sidorov's method provided better results than the common n-gram approach for various n-gram lengths and types.

In this paper, we further investigate the use of syntactic information by building a separate language model for each author using part-of-speech tags as features to train the models. Our proposed method uses as features all the part-of-speech tags included in the training texts. No feature selection process to select the optimal number of features is required due to the fact that language models employ all features, and moreover, our set of features is already small consisting only of a handful of part-of-speech tags. For comparison, we also include in our experiments word and character level language models which were initially introduced by Fuchun Peng [11].

3   A Part-of-Speech Language Modeling Method

Our proposed method is based on the similarity-based paradigm. This approach includes the concatenation of all documents written by a certain author in a single profile, which is used for the extraction of the style-markers. For the evaluation process an attribution model is implemented to estimate the differences between every profile and an unseen text and the most likely author is chosen [18]. Specifically, we present a method for computer-assisted authorship attribution based on language models. This approach is composed of three phases described in the following subsections.

3.1   Preprocessing Phase

In this phase all the texts which are written by a certain author in the training corpus are concatenated in large files to create author profiles; thus, an author's profile is a union of all his considered known, or training, documents.

Using the Stanford NLP tagger [19], each word in author profiles is replaced by its corresponding part-of-speech tag from the Penn Treebank tagset [9]. Also, punctuation considered a useful literary style marker is preserved, because reviews are edited by non-professional authors who made wide use of punctuation. For example, consider the following text excerpt:

> *I bought Earthly Possessions because they filmed some scenes in New Jersey's Ocean Grove which is doubled as Perth, South Carolina. I really liked the chemistry between Sarandon and Dorff to surprise me.*

After the pre-processing step, it is transformed to a sequence of part-of-speech tags and punctuation marks:

PRP VBD NNP NNP IN PRP VBD DT NNS IN NNP NNP POS NNP NNP WDT VBZ VBN IN NNP, NNP NNP. PRP RB VBD DT NN IN NNP CC NNP TO VB PRP.

In this step, every document of the test set is also pre-processed to obtain this form.

3.2   Language Modeling Phase

In this phase, every author profile created in the prepossessing phase is used to build a separate language model for each author. Next, we present the basic mathematical principles related to a statistical model of language.

Let us denote as $w_1^N = (w_1, w_2, ..., w_N)$ a sequence of N words. The probability of observing this text fragment under a language model can be computed as the conditional probability of every word in the fragment given the previous ones, i.e.

$$P(w_1^N) = \prod_{i=1}^{N} P(w_i|w_1^{i-1}),$$
(1)

where $w_i^j = (w_i, w_{i+1}, .., w_{j-1}, w_j)$ is the sub-sequence from the i-th word to the j-th word, and w_i is the i-th word.

The above representation leads to a complex model because with a vocabulary size of V words, there are V^N possible sequences of the form $(w_1, w_2, ..., w_N)$. The above conditional probabilities constitute the free parameters of the model, which are learned from a training set. Obviously, even with reasonable magnitudes of V and N, we will never have enough training data to estimate V^N probabilities.

The need for a more simplified and compact model leads to the n-gram approach. In the n-gram model, without loss of generality, the probability of observing a new word is computed by taking into account only the previous $n-1$ words [1]. This approximation implies that the joint probability of the entire fragment can be calculated as

$$P(w_1^N) \approx \prod_{i=1}^{N} P(w_i | w_{i-(n-1)}^{i-1}), \tag{2}$$

where $w_{i-(n-1)}^{i-1} = (w_{i-(n-1)}, \ldots, w_{i-1})$, and n the selected n-gram. Using this approximation, the number of the potential free parameters of a model with vocabulary size V are limited to V^n.

Let us define as $\text{count}(w_i^j)$ the number of times the sub-sequence $w_i^j = (w_i, w_{i+1}, \ldots, w_j)$ appears in the training corpus. Then, the conditional probabilities of Eq. 2 can be estimated as

$$P(w_i | w_{i-(n-1)}^{i-1}) = \frac{\text{count}(w_{i-(n-1)}^{i})}{\text{count}(w_{i-(n-1)}^{i-1})}. \tag{3}$$

In practice, however, the probabilities in an n-gram model do not derive directly from the frequency counts, because it is likely for novel n-grams that were never explicitly witnessed in the training set to occur in the test set. Hence, a non-zero probability should be assigned to these unseen n-grams. There are many smoothing techniques used to confront this problem, including Good Turing discounting and back-off models. In this work we use the Witten-Bell discounting technique [6] because the size of the vocabulary is small containing 36 POS tags and 12 punctuation marks rendering smoothing not essential.

In this phase, every author profile created in the prepossessing phase, which consists of a sequence of tokens, is used to create a separate language model for each author by computing the aforementioned conditional probabilities (Eq. 3).

3.3   Authorship Attribution Phase

After the learning of models' parameters on the training corpus, we can classify unknown texts by how well each model predicts a text. For this purpose, we employ the Perplexity measure.

Given a test document $D = t_1^M = (t_1, t_2, ..., t_M)$, and considering an n-gram model, the intrinsic perplexity of the model on the test document is defined as:

$$\text{Perplexity}(D) = \sqrt[M]{\prod_{i=1}^{M} \frac{1}{P(t_i|t_{i-(n-1)}^{i-1})}}. \qquad (4)$$

The lower a model's perplexity, the more likely the model is to predict the document. Thus, in the last phase, every unknown document of the test set is supplied to each language model, the perplexity of each model on the document is estimated, and the most likely author is selected.

4   Experimental Evaluation

In this section, we present the experimental results of comparing our proposed part-of-speech language model to traditional approaches. First, we describe the corpus and evaluation measures used.

4.1   Corpus and Evaluation Measures

We have experimented with the IMDB62 dataset which consists of 62,000 movie reviews written by 62 users, with exactly 1,000 reviews per user. The data were crawled from www.imdb.com by downloading all the reviews by prolific reviewers who submitted more than 500 reviews each. All downloaded texts belong to the period of May 2009 in order to minimize the risk of change of authorial style over time. We have chosen to use the IMDB62 dataset for the following three reasons:

- The data collected are homogeneous, because all texts deal with the same topic, making it more challenging to distinguish the stylistic idiosyncrasies of each author.
- The texts are written by regular people, not professional authors. In the traditional authorship attribution approaches, the training instances belong mostly to professional writers, whose style and language have been cultivated through the years, and as a consequence are distinguishable among the authors [4]. However, this dataset is edited by non-professional writers and the question that arises is whether authors with similar training and background are able to imprint their texts with their own unique authorial fingerprint.
- The number of candidate authors is relatively large compared to the majority of traditional authorship attribution studies.

Thus, this dataset is more challenging that others typically used in authorship attribution studies. While it was used at least once before for authorship attribution, e.g. [15], there is no extensive evaluation of different classifiers, feature types, and language models on this corpus, before our study. Others, e.g. [14], used this dataset for other tasks, such as sentiment classification.

For our evaluation, the corpus was divided randomly into a training and a test set. We selected 70% of the initial data set as a training set and the remaining 30% as a test set. We employed two evaluation metrics commonly used in text classification tasks: generalization Accuracy and macro-averaged F_1 measure.

4.2 Runs and Baselines

In order to set a baseline for the evaluation of the proposed method, we consider four types of features previously employed successfully in authorship attribution studies [5], namely: simple lexical (SL), simple character trigrams (CTG), content words (CW), and part-of-speech trigrams (POST).

The structure unit of the simple lexical features (SL) are the words. Based on the words, there have been extracted simple measurements that can be useful. The features extracted at token level include: the average and standard deviation of the words per sentence in each text, the total number of sentences in each text, the relative frequency of the tokens that are alphabetic units (calculated by dividing the total number of alphabetic elements of a text by the total number of tokens of every text), the relative frequency of the words longer that 15 letters, and the relative frequency of words shorter than 4 letters. Also, the features that represent the diversity of the vocabulary belong in the same set of features. In order to quantify the vocabulary's diversity of each author, the type-token ratio V/N (vocabulary size to the total number of words) was used. Two vocabulary functions count the so-called hapax legomena and hapax dislgomena, i.e. the number of the words of the text that appear only once or twice, respectively. We also implemented Yule's metric [21], Honore's metric [2], and the entropy function.

All the aforementioned measurements constitute the simple lexical features (SL) group. The second class of features consists of character trigrams (CTG). Previous research has shown that trigrams perform better than other character n-grams in an English corpus [8]. The third class of features includes words with the highest appearance frequency, the so-called content words (CW). Apart from features that rely on the words or n-grams of texts, the extraction of syntactic information is a reliable literary footprint [18]; thus, the fourth class of features includes part-of-speech trigrams (POST).

While SL features are arithmetic and small in number, CTG, CW, and POST, are tens of thousands or more. In order to reduce the computational load, we resort to feature selection. For the selection of features that carry significant information for classification, we implemented a feature selection method known to be among the best for text classification tasks [20], namely, the chi-square (χ^2) metric. Based on this metric, we selected the top-500 most informative features among the 10,000 most common features in the training corpus, per CTG, CW, and POST. The choice of the 500 cutoff comes from our preliminary experiments and is also supported by previous research, see e.g. [5,7]. Using more features did not seem to improve effectiveness, at an extra computational cost.

Each document of the training set is processed to produce a feature vector, a numerical vector consisting of the frequency of each feature of the feature set occurring in the document. Feature vectors are then used to train classifiers, via several algorithms, which are then applied to the test set to calculate generalization accuracy and macro-averaged F_1.

We paired each of the above four feature classes to four classification algorithms commonly and successfully used in authorship attribution studies [22]: Multinomial Naive Bayes (MNB), a Support Vector Machine (SVM), k Nearest Neighbour (kNN), and Random Forest (RF). For each of those classifiers parametrization is needed, however, parameter optimization is beyond the scope of our work. Hence, we used the default settings of scikit-learn[1], i.e. the machine learning Python library we employed for our experiments. As a baseline, we will select the best performing feature-class/classifier combination per evaluation measure; this constitutes a rather strong baseline.

Regarding language modeling, we experimented with using as features characters (C/LM), words (W/LM), and our proposed part-of-speech tags (POS/LM). For building and applying statistical language models, we employed the SRILM[2] toolkit comprised of a set of C++ classes. Again, while SRILM has some parameters, parameter optimization is beyond the scope of our work, so we used the default values (i.e. the default setting of SRILM for n-grams is 3). Note that, in contrast to instance-based methods where each text is represented as feature vector, as aforementioned, in language models there is no feature selection step; all features are used.

Table 1. Feature types, machine learning methods, and language models

SL	Lexical features based on simple measurements
CTG	500 most informative char 3-grams among the 10^4 most common 3-grams
CW	500 most informative words among the 10,000 most common words
POST	500 most informative POS 3-grams among the 10^4 most common 3-grams
MNB	Multinomial Naive Bayes with default settings (scikitlearn)
SVM	Support Vector Machines with linear kernel and default settings (scikitlearn)
kNN	k nearest neighbour with default settings (scikitlearn)
RF	Random Forest with default settings (scikitlearn)
C/LM	Language model with default settings (SRILM) and characters as features
UW/LM	Language model with context length 1 and words as features (SRILM)
W/LM	Language model with default settings (SRILM) and words as features
POS/LM	Language model with default settings (SRILM) and POS-tags as features

In summary, the feature types, machine learning methods, and language models we experimented with, are given in Table 1. Due to the settings described above, CTG, CW, and POST, are directly comparable to C/LM, UW/LM, and POS/LM, respectively, since they are using the same feature sets. These are character 3-grams, word unigrams, and POS 3-grams, respectively. W/LM is an

---

[1] www.scikit-learn.org.
[2] www.speech.sri.com/projects/srilm/.

extra run using word 3-grams. The language model based on unigrams (UW/LM) is not expected to perform since it is trivial.

Furthermore, we have tried combining feature types by (a) feeding unions of them into the classifiers, and (b) taking linear combinations of several language models with equal weights by simply adding their perplexities. A MinMax normalization process preceded the linear combination of language models because the ranges of perplexities produced by using different units seemed incomparable. This achieved better effectiveness in preliminary experiments (not reported here).

4.3   Results

In this section we present a set of experiments we ran in order to assess the performance of language modeling in comparison to the aforementioned baseline methods.

Table 2. Accuracy and F_1 on the test set of the IMDB62 corpus, for a variety of feature types and learning algorithms. Best results per feature type and per measure are in bold typeface; worst are with italics

Features/learner	Accuracy %				F_1 %			
	MNB	SVM	kNN	RF	MNB	SVM	kNN	RF
SL	14.3	**37.8**	*12.4*	31.5	12.8	**37.5**	*11.7*	30.8
CTG	82.7	**85.6**	57.6	*57.6*	82.9	**86.0**	58.2	*56.8*
CW	86.1	**88.6**	60.5	*58.6*	86.0	**89.0**	60.6	*57.8*
POST	53.5	**58.7**	34.0	*25.5*	52.4	**58.1**	33.4	*24.5*

Table 2 shows the generalization accuracy and macro-average F_1-measure of each combination of features and learning algorithms for the IMDB62 corpus. As it can be seen, the k-nearest neighbour (kNN) and Random Forest (RF) classifiers perform poorly on all feature sets for both evaluation metrics. Multinomial Naive Bayes (MNB) proves to be an effective learning method for almost all feature types except SL features, but Support Vector Machines (SVM) are superior to all other learning algorithms for all feature types.

Regarding the feature sets, simple lexical measures (SL) perform very poorly in all classifiers, so they do not seem to provide information relevant to the recognition of an author. While part-of-speech trigrams (POST) are better, character trigrams (CTG) and content words (CW) constitute more effective and reasonable choices of feature sets because they perform far better than the former two. The use of the syntax frequency tags (POST) fails to adequately describe the broader syntax structures and gather all the information about the syntax profile of each author. In summary, the best-performing feature-class/classifier combination is CW/SVM, in both evaluation measures, with CTG/SVM being very competitive. We will use both these runs as baselines.

Table 3. Accuracy and F_1 on the test set of the IMDB62 corpus, for a variety of combinations of feature types and learning algorithms. Best results per feature type and per measure are in bold typeface; worst are with italics

Features/learner	Accuracy %				F_1 %			
	MNB	SVM	kNN	RF	MNB	SVM	kNN	RF
CTG+CW	86.6	**91.2**	*61.5*	61.6	86.5	**91.2**	62.0	*60.9*
CTG+POST	84.4	**87.1**	59.3	*55.2*	84.2	**87.0**	59.8	*55.9*
CW+POST	85.3	**87.8**	59.3	*55.5*	85.2	**87.9**	59.8	*54.6*
CTG+CW+POST	87.4	**91.7**	63.0	*61.3*	87.3	**91.8**	63.5	*60.5*

Table 3 shows the results for all feature unions, except SL which were proven very weak above. Concerning the learning algorithms, we reach similar conclusions as above, i.e. SVM performs best, MNB following, and kNN, RF are the worst. Regarding feature combinations, we see that taking unions of features is generally beneficial to effectiveness: all combinations show improved performance than all the individual feature-types combined, except when CW combined with POST. This means that POST provide additional useful information in most cases. The union of all feature types (CTG+CW+POST) is the best run so far, closely followed by CTG+CW, both when fed into SVM. We will also use both these runs as baselines in order to compare the language models based on different units.

Table 4. Accuracy and F_1 on the test set of the IMDB62 corpus, for a variety of combinations of feature types and learning algorithms or language models. Best results per feature/learner class and per measure are in bold typeface

Features/learner	Accuracy %	F_1 %
CW/SVM	**88.6**	**89.0**
CTG/SVM	85.6	86.0
CTG+CW+POST/SVM	**91.7**	**91.8**
CTG+CW/SVM	91.2	91.2
C/LM	**92.3**	**92.7**
UW/LM	13.6	18.7
W/LM	84.4	85.2
POS/LM	89.5	89.8
C/LM+W/LM	93.6	93.8
C/LM+POS/LM	94.9	95.0
W/LM+POS/LM	94.1	94.3
C/LM+W/LM+POS/LM	**95.9**	**96.0**

Table 4 shows the language model runs (3rd batch of results) in comparison to the previously chosen baselines (1st and 2nd batches), as well several language model combinations. The trivial language model on unigrams (UW/LM) fails, as expected. From the rest, the pretty standard language model on word 3-grams (W/LM) is the weakest one, which performs slightly worse than the weakest of two baselines (CTG/SVM). The proposed language model on part-of-speech 3-grams (POS/LM) comes slightly above (rather insignificantly) the strongest baseline (CW/SVM), however, it has stronger efficiency benefits. The language model based on characters (C/LM) achieves a much higher performance.

Regarding the linear combinations of language models, all of them achieve better performance than the single-feature as well as the combined-feature baselines. Again here, we leave out of the combinations the very weak UW/LM. The combination of all the rest three language models achieves the best accuracy and F_1 of around 96%.

5   Conclusions

Traditional methods for automated authorship attribution employ several feature types and learning algorithms for building author profiles. Most previous research has dealt with small heterogeneous collections where each professional author may have been strongly associated with a topic. Furthermore, the style and language used by professional authors have been cultivated throughout the years, consequently becoming distinguishable, making authorship attribution relatively an easier task. We considered larger collections, with many non-professional authors, writing on a specific topic (homogeneous collection) such as movies. Our contributions are the following.

First, we evaluated the performance of four different feature classes commonly used in past literature, paired with four commonly used classifiers for the task. We found that Support Vector Machines paired with words or character trigrams as features are the most effective. This result is in-line with previous research, e.g. [8], so past results with instance-based methods seem to extend to larger homogeneous collections.

Second, we proposed a language model based on part-of-speech units and evaluated its performance against the former methods and other language models based on standard units such as characters or words. Here, in contrast to past literature, e.g. [10], where the word-level language model provides the best results, our experiment demonstrates that character or POS level language models achieve better classification results.

Third, we investigated combinations of features in learning algorithms by simply taking unions, as well as combining language models based on different units by taking a linear combination of their individual perplexity scores. Both combination methods seem to work well, achieving better results than the individual feature classes or language models they combine.

While our proposed POS/LM method provides only a slight effectiveness benefit over the best-performing standard methods, it has important efficiency benefits:

(a) building a language model on a handful of POS tags is fast, much faster than using characters or words as units, and (b) feature selection is not required in language models. Also, the attribution method of POS/LM avoids data sparsity problems, making smoothing non-essential. The vocabulary used for this model consists of 36 syntactic labels and 12 commonly used punctuation marks, eliminating the possibility for an unseen trigram of syntactic labels to arise in the test phase. There are no limitations imposed on the vocabulary, and every word in the English vocabulary, as well as novel word n-grams that were never witnessed in the training set, can appear in the test set.

Regarding the combination methods, while feeding unions of features into some classifier has no extra parameters, taking linear combinations of language models introduces some extra parameters: the coefficients of the linear combination. We have so far simply assumed equal weights by adding MinMax-normalized perplexities, nevertheless, this still achieved the best results in this paper. In this respect, optimizing in the future these coefficients could lead to even better effectiveness.

Acknowledgement. We thank Nektarios Mitakidis, master's student at our department, for his valuable guidance during the early stages of this work.

References

1. Allamanis, M., Sutton, C.: Mining source code repositories at massive scale using language modeling. In: Proceedings of the 10th Working Conference on Mining Software Repositories, pp. 207–216. MSR 2013. IEEE Press, Piscataway (2013)
2. Antony, H.: Some simple measures of richness of vocabulary. Assoc. Literary Linguist. Comput. Bull. **7**(2), 172–177 (1979)
3. Baayen, H., van Halteren, H., Tweedie, F.: Outside the cave of shadows: using syntactic annotation to enhance authorship attribution. Literary Linguist. Comput. **11**(3), 121–132 (1996)
4. Baayen, H., Halteren, H.V., Neijt, A., Tweedie, F.: An experiment in authorship attribution. In: 6th JADT I(January), pp. 69–75 (2002)
5. Grieve, J.: Quantitative authorship attribution: an evaluation of techniques. Literary Linguist. Comput. **22**(3), 251–270 (2007)
6. Ismail, R.: Comparison of modified kneser-ney and witten-bell smoothing techniques in statistical language model of bahasa Indonesia. In: 2nd International Conference on Information and Communication Technology (ICoICT), pp. 409–412, May 2014
7. Koppel, M., Schler, J.: Exploiting stylistic idiosyncrasies for authorship attribution. In: IJCAI 2003 Workshop on Computational Approaches to Style Analysis and Synthesis, pp. 69–72 (2003)
8. Koppel, M., Schler, J., Argamon, S.: Computational methods in authorship attribution. J. Am. Soc. Inf. Sci. Technol. **60**(1), 9–26 (2009)
9. Marcus, M., Kim, G., Marcinkiewicz, M.A., MacIntyre, R., Bies, A., Ferguson, M., Katz, K., Schasberger, B.: The penn treebank: annotating predicate argument structure. In: Proceedings of the Workshop on Human Language Technology, HLT 1994, pp. 114–119 (1994)

10. Peng, F., Schuurmans, D., Wang, S.: Augmenting Naive Bayes classifiers with statistical language models. Inf. Retrieval **7**(3), 317–345 (2004)
11. Peng, F., Schuurmans, D., Wang, S., Keselj, V.: Language independent authorship attribution using character level language models. In: Proceedings of the Tenth Conference on European Chapter of the Association for Computational Linguistics, EACL 2003, vol. 1, pp. 267–274. Association for Computational Linguistics, Stroudsburg (2003)
12. Pokou, Y.J.M., Fournier-Viger, P., Moghrabi, C.: Authorship attribution using variable length part-of-speech patterns. In: Proceedings of the 8th International Conference on Agents and Artificial Intelligence, pp. 354–361 (2016)
13. Raghavan, S., Kovashka, A., Mooney, R.: Authorship attribution using probabilistic context-free grammars. In: Proceedings of the ACL 2010 Conference Short Papers, ACLShort 2010, pp. 38–42 (2010)
14. Seroussi, Y., Zukerman, I., Bohnert, F.: Collaborative inference of sentiments from texts. In: Bra, P., Kobsa, A., Chin, D. (eds.) UMAP 2010. LNCS, vol. 6075, pp. 195–206. Springer, Heidelberg (2010). doi:10.1007/978-3-642-13470-8_19
15. Seroussi, Y., Zukerman, I., Bohnert, F.: Authorship attribution with latent Dirichlet allocation. In: Proceedings of the Fifteenth Conference on Computational Natural Language Learning, pp. 181–189, CoNLL 2011. Association for Computational Linguistics, Stroudsburg (2011)
16. Sidorov, G., Velasquez, F., Stamatatos, E., Gelbukh, A., Chanona-Hernández, L.: Syntactic dependency-based n-grams as classification features. In: Batyrshin, I., Mendoza, M.G. (eds.) MICAI 2012. LNCS (LNAI), vol. 7630, pp. 1–11. Springer, Heidelberg (2013). doi:10.1007/978-3-642-37798-3_1
17. Stamatatos, E., Fakotakis, N., Kokkinakis, G.: Computer-based authorship attribution without lexical measures. Comput. Humanit. **35**(2), 193–214 (2001)
18. Stamatatos, E.: A survey of modern authorship attribution methods. J. Am. Soc. Inf. Sci. Technol. **60**(3), 538–556 (2009)
19. Toutanova, K., Manning, C.D.: Enriching the knowledge sources used in a maximum entropy part-of-speech tagger. In: Proceedings of the 2000 Joint SIGDAT Conference on Empirical Methods in Natural Language Processing and Very Large Corpora: Held in Conjunction with the 38th Annual Meeting of the Association for Computational Linguistics, EMNLP 2000, vol. 13, pp. 63–70 (2000)
20. Yang, Y., Pedersen, J.O.: A comparative study on feature selection in text categorization. In: Proceedings of the Fourteenth International Conference on Machine Learning, ICML 1997, pp. 412–420. Morgan Kaufmann Publishers Inc., San Francisco (1997)
21. Yule, G.U.: The Statistical Study of Literary Vocabulary. Cambridge University Press, Cambridge (1944)
22. Zhao, Y., Zobel, J.: Effective and scalable authorship attribution using function words. In: Lee, G.G., Yamada, A., Meng, H., Myaeng, S.H. (eds.) AIRS 2005. LNCS, vol. 3689, pp. 174–189. Springer, Heidelberg (2005). doi:10.1007/11562382_14

What Others Say About This Work? Scalable Extraction of Citation Contexts from Research Papers

Petr Knoth[1]([✉]), Phil Gooch[2], and Kris Jack[2]

[1] Knowledge Media Institute, The Open University,
Walton Hall, Milton Keynes MK7 6AA, UK
petr.knoth@open.ac.uk
[2] Mendeley Ltd., Elsevier B.V., 14-18 Finsbury Square, London EC2A, UK

Abstract. This work presents a new, scalable solution to the problem of extracting citation contexts: the textual fragments surrounding citation references. These citation contexts can be used to navigate digital libraries of research papers to help users in deciding what to read. We have developed a prototype system which can retrieve, on-demand, citation contexts from the full text of over 15 million research articles in the Mendeley catalog for a given reference research paper. The evaluation results show that our citation extraction system provides additional functionality over existing tools, has two orders of magnitude faster runtime performance, while providing a 9% improvement in F-measure over the current state-of-the-art.

Keywords: Information extraction · Citation extraction · Text-mining · Digital libraries

1 Introduction

There are already over 114 million academic papers on the Web [1]. With over 1 million papers published each year [2] and an estimated 10% year on year increase in the annual number of these outputs [3], researchers need tools to help them decide what to read. While recommendation systems for academic papers, such as those provided by Google Scholar, Mendeley Suggest [4] or CORE [5,6] have been created to address the problem of discovering relevant literature, more can be done to help users to effectively navigate through the network of scientific papers. One traditional yet effective way of discovering new and relevant content is by following the edges of the citation graph in the opposite direction, i.e. from the cited to the citing articles. Unfortunately this activity is, even in the most popular scholarly communication systems, not adequately supported. Although users can discover articles that cite a particular work, Google Scholar and similar services do not enable the user to quickly understand how important and relevant to their interest that citation link is, prior to accessing that document.

© Springer International Publishing AG 2017
J. Kamps et al. (Eds.): TPDL 2017, LNCS 10450, pp. 287–299, 2017.
DOI: 10.1007/978-3-319-67008-9_23

Similarly, users of academic digital libraries need to make choices about what to read. When presented with a particular article landing page, they need to decide if investing time in reading the article is worthwhile. Such decision is typically based on (a) the perceived relevance of the article to the current researcher's interest and (b) the importance or trust in the work.

While the former is typically assessed by scanning the abstract and title, the latter is today often evidenced using the paper's citation count, typically displayed on the article details page, the journal impact factor or other similar metric. However, all approaches relying on an aggregate function of citation counts to evidence the importance of an article face problems caused by the variety of situations in which people cite work [7]. As described by Eugene Garfield [8], the motivations for citing prior work include: paying homage to pioneers, giving credit for related work (homage to peers), identifying methodology, equipment, and the like, providing background reading, correcting one's own work, correcting the work of others, criticising previous work, substantiating claims, alerting researchers to forthcoming work, authenticating data and classes of fact (such as physical constants), identifying original publications in which an idea or concept was discussed, identifying the original publications describing an eponymic concept or terms, arguing against the work or ideas of others and disputing the claims of others to have been first with their work.

As a result, we believe researchers can benefit from leveraging citations in a qualitative rather than just quantitative way. Citation contexts, i.e. the text surrounding a citation, explain how the cited paper is used in this particular work. By extracting all these mentions from the full text of articles citing a document of interest, we can help researchers to quickly explore the ways in which a given paper was useful in other peoples' work, hence we can help them decide whether the work might be useful in their own work. Our assumption is that by enabling researchers to quickly interrogate the contexts in which a given paper is used, we can assist them in making a more informed choice about whether or not to read it.

Consequently, we address the problem of automatically retrieving and extracting citation contexts for a given research paper. The presented work brings the following contributions:

- We present a new, scalable tool which uses machine learning techniques to recognise and parse references from unstructured text and extracts the textual content surrounding their mentions.
- We report on the results of an end-to-end evaluation of this tool and discuss its advantages over existing solutions.

In addition to the above mentioned use cases, we believe this work could also be applied in other situations, in particular, (a) to improve browsing in digital libraries by enabling more focused navigation across resources (e.g. it would be possible to (hyper-)link to a particular fragment rather than just to a document), (b) as a tool to assist researchers/funders in understanding which claims from a work have been discussed and/or built upon in further work and

(c) as a supporting tool to understand the contribution of a researcher, a research group, organisation, etc., at a finer granularity.

2   Citation Contexts Extraction Method

The citation contexts extraction process consists of four stages depicted in Fig. 1. We first clean and pre-process the input text. We then process the text line by line classifying each as either a reference line or not (Sect. 2.1). The lines classified as a reference are then passed to a probabilistic parser based on Conditional Random Fields (CRFs) [9] that splits each reference into its constituent fields (Sect. 2.2), such as authors, title, year and venue. We subsequently use a set of regular expressions to link each reference to all its citations in the processed document extracting all the citation contexts (Sect. 2.3). Finally, we try to link each of the citation reference strings to a unique ID of the cited document (Sect. 2.4).

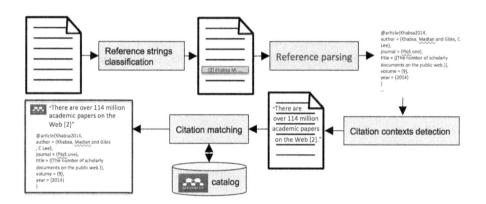

Fig. 1. The four stages of citation context extraction.

2.1   Reference Classification

The following features were used to train the classifier to distinguish the text of references from non-references at the line level:

- **(F1) Line length** (float): Character line length as a ratio to the mean line length in the document.
- **(F2) Is within reference or numbered block** (Bool) Follows a heading that signifies the start of a references block, or within a block of at least n consecutively numbered lines.
- **(F3) Date presence** (Bool)
- **(F4) Contains URL or DOI** (Bool)

– **(F5) Contains cue words** (Bool): Contains words that likely appear in a reference, such as "ibid", "et al." or "in press"
– **(F6) Contains publication word stem** (Bool): Contains a word from a list ("journal, adv, bull, proc, stud, biochem, etc")
– **(F7) Contains a page range** (Bool)
– **(F8) Contains volume information** (Bool) Contains words, such as "vol", or common "vol/issue/pages" stereotype pat- terns.
– **(F9) Contains editor information** (Bool): Words, such as (eds, ed. etc.)
– **(F10) Punctuation ratio** (float): Ratio of punctuation characters to the total line characters.
– **(F11) Capital letters ratio** (float): Ratio of capital letters to total line characters.
– **(F12) Camel case bigrams ratio** (float): Ratio of camel case bigrams to the line bigrams.
– **(F13) Starts with a numeric label** (Bool)
– **(F14) Surname/initials pair** (Bool): Contains a surname or initials pairs.

Using the above features, we have trained SVM, Random Forest, Logistic Regression, Decision Tree and CRF models. The first four models were created using the WEKA [10] software workbench. We used the CRF++ toolkit [11] implementation for the CRF model. This model uses, in addition to the above mentioned features, sequence information consisting of feature values of each line and the preceding and following 3 lines.

2.2   Reference Parsing

Given the plain text of a reference in any bibliographic format, the goal of reference parsing is to fill in a template consisting of fields, such as author, title, journal and year. To solve this problem we apply the AnyStyle parser[1], which is an open-source tool that uses CRFs to split a reference string into its constituent fields and output the result as BibTeX. The parser ships with a default model and training data that consist of 657 records of annotated data. To increase the accuracy of the parsing, we have retrained the CRF model using additional training data from Mendeley (see Sect. 4.2).

2.3   Context Extraction

Citation context extraction addresses the problem of locating all the links in the body of a paper to each reference and extracting the text surrounding them. There are three main approaches of connecting a citation to its reference we support:

– The reference is preceded with a number which is used as citation marker in the body of the document.

---

[1] http://anystyle.io/.

– The reference is preceded with an abbreviation created, for example, as a name & year, which is used as a citation marker in the body of the document.
– The citation is linked to a reference using a footnote.

We approach the problem of linking the citation to a reference by using a set of regular expressions. These regular expressions were manually curated and fine-tuned on a test set. Using a naive baseline method, the citation context snippet is then formed by a context window of 300 characters to the left and right of the position of the citation.

2.4   Citation Matching

The final step is to link each of the references cited in a given research paper to a unique document ID of the cited document. We use the catalog search functionality of the Mendeley API for this purpose. For each parsed reference, we compose a query using the following logic:

– If the reference string contains an identifier, such as a DOI, we use this identifier to look up the record.
– If we manage to extract the title from the reference string, the query contains the title plus year and author information, provided this is available.
– If the reference does not contain (or we don't manage to extract) an identifier nor a title, such as in "J. A. Maruhn and W. Greiner, Z. Phys. 251, 431 (1972).", we fall-back to a fuzzy lookup on author, source and year.

If the look up is successful for a reference, we record the document ID and normalise the parsed metadata based on the information in the Mendeley catalog.

3   Collecting Information About What Others Say About This Work

As our goal is to retrieve all the citation contexts for a given reference article, we need a fast way of determining the set of articles citing the reference article. We have considered two approaches of addressing this problem. In the first approach, we would apply the citation extraction tool described in Sect. 2 to create a catalogue of research articles and their citations. We would start by deduplicating research articles and adding all of them into the catalogue with their metadata. We would then process the full texts of all these papers. For each paper, we would extract and parse its references and would try to match each reference to this catalogue using a learnt similarity threshold for a metric, such as Jaccard coefficient. We would then generate a pair of citing catalogue ID and cited catalogue ID for each successfully matched tuple.

As this process contains many non-trivial steps where errors can occur, we have decided to opt for an alternative approach which relies on already

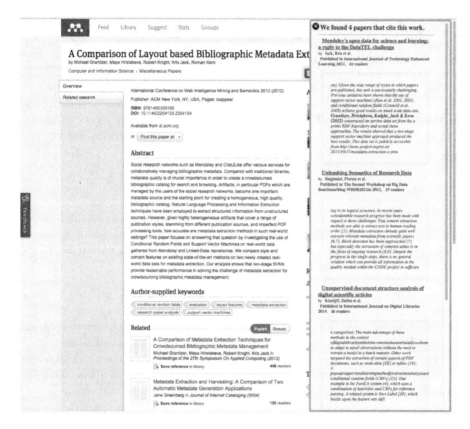

Fig. 2. The "What cites this work" browser bookmarklet allows the user to see how a given research paper is cited in other peoples work. Instead of having to search for the mentions of the paper, the citation context is automatically extracted and displayed to the user in the form of a snippet.

existing databases. More specifically, we use the Mendeley Catalogue as an already deduplicated database of research papers. The Mendeley Catalogue contains over 70 million unique research articles crowd-sourced from Mendeley users. As we host the full text documents uploaded by users on our servers, we can process these articles using the citation context extraction tool. In order to identify the articles citing a given document from the Mendeley Catalogue, we rely on information from Scopus. Scopus is one of the largest citation databases of peer-review literature. As the citation information in Scopus is automatically extracted and then manually curated, it is of high quality. The dump of the Scopus citation dataset we used in our experiments consisted of over 934 million citation pairs.

To enable a fast retrieval of document IDs citing a given paper, we first match Scopus citing-cited article pairs to the Mendeley Catalogue. We then group these pairs by the cited document aggregating all citing document IDs on one line. Finally, we index this dataset using Elasticsearch.

To retrieve the citation contexts for a given research paper, all we need to do now is to:

1. Query the citations index to retrieve Mendeley IDs of all documents citing a given reference document.
2. Download the full texts of the citing documents.
3. Process the documents using the citation context extraction tool retrieving only the citation contexts referring to the reference document.
4. Retrieve canonical metadata for each of the citing references from the Mendeley Catalogue to accompany the citation context with information about the source it comes from.
5. Rank the citations according to some criterion and reorder.

We have built a demonstrator in the form of a "What cites this work" browser bookmarklet that implements this method for articles in the Mendeley Catalogue. As the user launches the bookmarklet, the title and the DOI of the currently visited article is retrieved from the HTML page of the Mendeley Catalogue. The system then follows steps 1–5, where steps 2–3 are done in parallel. Our demonstrator ranks the retrieved citations according to the popularity of the citing article, i.e. based on the number of Mendeley readers who have that article in their library. An example of the result is shown in Fig. 2. While it has recently become the de facto standard in scientific databases and search engines to provide a link to a list of articles citing a given article, these systems typically do not show the context in which the article is cited. For example, the ACM Digital Library displays only the list of article titles that cite a given document and Google Scholar displays the list of titles with their abstracts but not with the citation contexts. We believe that using citation contexts as snippets is more informative and useful to the user.

4   Evaluation

There is a number of components that form our pipeline. In order to get a good understanding of the system's performance, we have to evaluate all of them.

4.1   Reference Classification

To train the reference classification models described in Sect. 2.1, we have created a training set of 1,000 randomly selected PDFs from the Mendeley Catalogue for which we have canonical citation data from Scopus. These PDFs were converted to a single text file consisting of about 300 k lines and all actual citations were labeled 1 and non-citations 0. The validation set consisted of 1,000 PDFs (365 k lines) randomly selected from PubMed labeled in the same way as the training data.

The results (Table 1) show that SVM, CRF and J48 were the top performers. However, the observed run-time performance of the CRF was much faster than the other two methods, which is why we decided to use this model within our tool.

Table 1. Evaluation of line-level citation classifiers.

Model	Precision	Recall	F-measure
SVM	0.998	0.998	0.998
CRF	0.996	0.997	0.997
J48	0.997	0.997	0.997
Random Forest	0.993	0.993	0.993
Log Regression	0.990	0.990	0.990

Table 2. Parsing error rate for evaluation data set of 26,000 citation records.

Model	Parse error rate
Baseline	22%
Retrained model	5%

4.2 Reference Parsing

One of the difficulties with parsing a citation string is dealing with noisy data that might be extracted from, for example, PDF files. There may be inconsistencies in text encoding, punctuation and use of white space. In addition, many different citation referencing styles need to be handled. We created training data representative of a wide range of citation styles and included noisy data exactly as extracted from the source document. The training set consisted of 600 manually structured citations from open access papers in the Mendeley catalogue in addition to the 657 training records supplied with the Anystyle parser[2].

In order to see whether the CRF model trained on the additional data performs better than the default model shipped with AnyStyle, we created an evaluation set of 26,000 structured citation records randomly selected from PubMed research papers. The size of the data sample gives us at 99.9% significance level a confidence interval of just below 0.1. The evaluation references were then compared with the system generated references to calculate a raw error rate based on character-level differences.

A comparison of the error rate between the system-generated references and those from PubMed are shown in Table 2. By retraining this parser on a more representative data sample that included an additional 600 records, we have reduced the error rate of the baseline citation parser from 22% to 5% - more than a four-fold error reduction. Our intuition behind the significant error reduction is that the default AnyStyle model was trained on too little and too clean data.

4.3 End-to-end Citation Extraction

We ran two end-to-end evaluations against 26,000 research article PDFs randomly selected from the Mendeley Catalogue and for which canonical citation data were available from the manually curated Scopus database. The gold data comprises the raw string value and structured citation (author, title, year, source), and the expected catalogue identifier for each reference cited.

This evaluation faced significant challenges, such as that the extracted citations might be in a different citation format than in the canonical record as well as that the PDF may be locked by the creator or may be a scan.

---

[2] http://anystyle.io/.

For the first evaluation (Fig. 3), each extracted citation was parsed with the CRF model described in Sect. 2.2 and the authors, title, year, and source and DOI fields extracted were used in a query to the Mendeley Catalogue lookup API. Results of this evaluation are shown in Table 3.

For the second evaluation, we attempted to match extracted citation strings against the canonical citation strings for each article. We also ran the same end-to-end evaluation with the state-of-the-art CERMINE [12] software on the same hardware to compare performance both in terms of accuracy and processing speed. Results of this evaluation are shown in Table 4.

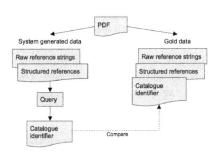
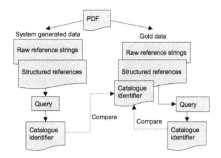

Fig. 3. Evaluation pipeline comparing retrieved identifiers with expected identifiers in the gold data set.

Fig. 4. Evaluation pipeline with optimal baseline data that uses the gold data to generate queries.

Table 3. End-to-end citation extraction and linking via parsing and querying.

Canonical citations	Extracted references	Matched references	False matches	Precision	Recall	F-measure
731,119	731,119	453,750	127,469	0.776	0.621	0.690

Table 4. End-to-end citation extraction and matching via hashing, P = precision, R = recall, F = F-measure, t = time in seconds.

Canonical citations	Extracted references	Matched references	False matches	Precision	Recall	F-measure	time (seconds)
Our system	887,191	657,277	238,696	0.734	**0.741**	**0.737**	$5.5x10^3$
CERMINE	887,191	524,248	187,331	**0.789**	0.591	0.676	$3.46x10^5$

Table 5. Optimal catalogue matching performance given a query string from the gold set.

Measure	Score
Precision	0.78
Recall	0.72
F-measure	0.76

Table 6. System catalogue matching performance given an automatically generated query string.

Measure	Score
Precision	0.78
Recall	0.63
F-measure	0.69

While the individual results for citation classification and citation parsing show high performance, the end-to-end results for citation network indicates that there is space for improvement. This may be because the matching of extracted citations against their canonical form needs to be more sophisticated or that the catalogue lookup is too imprecise. However, the end-to-end results still represent an overall 9% F-measure improvement on the previous state-of-the-art using the same evaluation set and using the same metrics, and our approach also runs two orders of magnitude faster (10^3 vs 10^5 seconds to complete) on the same hardware.

In order to distinguish catalogue lookup errors from errors in the citation extraction pipeline, we decided to compare the system performance against an optimal baseline. This optimal baseline helps us to answer the question of what would be the maximum achievable performance for locating the correct catalogue entry if perfect, structured citations could be extracted automatically from each evaluation article. To do this, for each structured citation in the gold set we generated a catalogue query (in the same way as was done for the system generated structured citation) and compared the identifier of the returned result with the expected identifier in the gold set (Fig. 4).

A comparison of the end-to-end system evaluation with the optimal baseline is shown in Tables 5 and 6. Table 5 shows the results of the optimal baseline evaluation when comparing the retrieved catalogue identifier given a "perfect" query string generated from the gold data, with the expected identifier in the gold set, for each document in the evaluation set. The idea here is that a successful lookup is not guaranteed even when the query is generated from the gold dataset. We want to see how this optimal baseline performs to be able to compare it with the citation extraction system performance.

Table 6 shows the end-to-end system results evaluated by comparing the retrieved catalogue identifier given a query string generated from automated extraction and parsing, with the expected identifier in the gold set, for each document in the evaluation set.

The results show that the system performs at $\frac{0.69}{0.76} = 91\%$ of the optimum that could be expected if a perfect query string could be generated for each document in the evaluation set.

5 Related Work and Discussion

Previous approaches such as ParsCit [13] use heuristics to identify the block of references at the end of the article and regular expressions to split these into individual references strings. In this work, we use CRFs to identify individual reference strings anywhere within the document, which allows references in, for example, footnotes to be extracted. Similar to ParsCit, our approach requires only plain text as input.

Other approaches, such as SectLabel [14] and pdfextract use rich-text features such as font size, position, and indentation to identify reference sections in order to improve extraction performance. Although our approach leads on our evaluation set excellent results without requiring such features, it may well be improved with the addition of them. This assumption is consistent with the findings of Kern & Kampfl [15] who enriched ParsCit with features, such as font information, reporting a slight improvement in parsing performance. While we have not yet performed a direct comparison with these approaches, one needs to consider the trade-off between our lightweight approach that allows reference extraction, parsing, context extraction and linking to be performed in real time, and more complex approaches that may not allow such real-time processing.

Our work addresses the following limitations of some existing tools:
Poor runtime performance (CERMINE [14], CrossRef pdfextract [1]). We ran CERMINE and pdfextract on our evaluation data set and hardware. CERMINE took 4 days while pdfextract failed to complete. The evaluation shows our tool runs two orders of magnitude faster than the state-of-the-art CERMINE tool.
Unrealistic reference formatting requirements. They require as input exactly one reference string per line (e.g. ParsCit [4]). Extracting candidate citations with exactly one citation per line is challenging, as many tools that extract text from formats such as PDF either preserve hard line breaks, or attempt to wrap the text, which works well for running paragraphs, but tends to glom multiple citations together (e.g. pdftotext). In contrast, our tool can deal with situations where a reference string is split across a number of consecutive lines which are prior to reference parsing reconnected.
Reference position requirements. They require citations to appear in a block towards the end of the document under a heading such as "References" or "Bibliography" and/or require the references to be formatted with hanging indents (e.g. ParsCit, pdfextract). In contrast, our tool assumes that references can appear anywhere in the document body, such as in footnotes. The lower reliance on document structure makes the tool also applicable to non-academic documents.

6 Future Work

There is a number of ways in which we can improve and apply the tool in the future. First we would like to implement more sophisticated logic for determining the citation context boundaries. At the moment this is only based on a fixed size character window. One option is to detect semantically coherent segments by applying the Text Tiling algorithm [16], using only the segment in which the citation occurs as the citation context. However, as this might significantly increase the runtime, we might want to opt for a more lightweight solution.

The second area of interest is the automatic classification of reasons for citation, for example, to the categories specified by Garfield [8] as listed at the beginning of the paper or Teufel [17]. Such work would also be closely related to the identification of influential citations [18]. This has the potential to improve the browsing capabilities of digital libraries and to be used as a feature in the development of new research evaluation metrics/scientometrics. Another strand of work constitutes the application of the citation context extraction tool to effectively construct a sentence/paragraph level co-citation matrix. As demonstrated in [19], such co-citation information could be used as a valuable feature in recommender systems.

7   Conclusions

We have successfully applied CRFs to address two problems: real-time extraction of bibliographic reference strings, from anywhere within the text of an article, and splitting those strings into structured queries to a large digital library of research papers. This approach is article-format and domain agnostic and can potentially be modified for any digital library. We have applied our method to an existing citation network to extract citation contexts and links, so that researchers can read more easily what others say about a given article.

References

1. Khabsa, M., Giles, C.L.: The number of scholarly documents on the public web. PLOS ONE **9**(5), 1–6 (2014)
2. Björk, B., Roos, A., Lauri, M.: Scientific journal publishing: yearly volume and open access availability. Inf. Res. **14**(1) (2009)
3. Bornmann, L., Mutz, R.: Growth rates of modern science: A bibliometric analysis. CoRR abs/1402.4578 (2014)
4. Hristakeva, M., Kershaw, D., Rossetti, M., Knoth, P., Pettit, B., Vargas, S., Jack, K.: Building recommender systems for scholarly information. In: 1st International Workshop on Scholarly Web Mining (SWM), International Conference on Web Search and Data Mining (2017)
5. Knoth, P., Zdráhal, Z.: CORE: three access levels to underpin open access. D-Lib Mag. **18**(11/12) (2012)
6. Knoth, P., Anastasiou, L., Charalampous, A., Cancellieri, M., Pearce, S., Pontika, N., Bayer, V.: Towards effective research recommender systems for repositories. In: Proceedings of Open Repositories 2017, Open Repositories (2017)
7. Pride, D., Knoth, P.: Incidental or influential? - challenges in automatically detecting citation importance using publication full texts. In: 21st International Conference on Theory and Practice of Digital Libraries (TPDL) (2017)
8. Garfield, E., et al.: Citation analysis as a tool in journal evaluation. American Association for the Advancement of Science (1972)

9. Lafferty, J., McCallum, A., Pereira, F.: Conditional random fields: probabilistic models for segmenting and labeling sequence data. In: Proceedings of the 18th International Conference on Machine Learning, pp. 282–289. Morgan Kaufmann, San Francisco (2001)

10. Hall, M., Frank, E., Holmes, G., Pfahringer, B., Reutemann, P., Witten, I.H.: The WEKA data mining software: an update. SIGKDD Explor. Newsl. **11**(1), 10–18 (2009)

11. Kudo, T.: CRF++: Yet another CRF toolkit. Software (2005), https://taku910. github.io/crfpp/

12. Tkaczyk, D., Szostek, P., Dendek, P.J., Fedoryszak, M., Bolikowski, L.: CERMINE - automatic extraction of metadata and references from scientific literature. In: 11th IAPR International Workshop on Document Analysis Systems, DAS 2014, Tours, France, 7–10 April 2014, pp. 217–221 (2014)

13. Councill, I.G., Giles, C.L., Kan, M.: Parscit: an open-source CRF reference string parsing package. In: Proceedings of the International Conference on Language Resources and Evaluation, LREC 2008, 26 May–1 June. European Language Resources Association, Marrakech, June 2008

14. Luong, M.T., Nguyen, T.D., Kan, M.Y.: Logical structure recovery in scholarly articles with rich document features. IJDLS **1**(4), 1–23 (2010)

15. Kern, R., Klampfl, S.: Extraction of references using layout and formatting information from scientific articles. D-Lib Mag. **19**(9/10) (2013)

16. Hearst, M.A.: Texttiling: segmenting text into multi-paragraph subtopic passages. Computat. Linguist. **23**(1), 33–64 (1997)

17. Teufel, S., Siddharthan, A., Tidhar, D.: Automatic classification of citation function. In: Proceedings of the 2006 Conference on Empirical Methods in Natural Language Processing, EMNLP 2006, pp. 103–110. Association for Computational Linguistics, Stroudsburg (2006)

18. Valenzuela, M., Ha, V., Etzioni, O.: Identifying meaningful citations. In: AAAI Workshops (2015)

19. Gipp, B., Beel, J.: Citation Proximity Analysis (CPA) - a new approach for identifying related work based on co-citation analysis. In: Larsen, B., Leta, J. (eds.) Proceedings of the 12th International Conference on Scientometrics and Informetrics (ISSI 2009), vol. 2. International Society for Scientometrics and Informetrics, Rio de Janeiro, July 2009. ISSN 2175-1935

Semantic Author Name Disambiguation with Word Embeddings

Mark-Christoph Müller$^{(\boxtimes)}$ (iD)

Heidelberg Institute for Theoretical Studies, Heidelberg, Germany
mark-christoph.mueller@h-its.org

Abstract. We present a supervised machine learning AND system which tackles semantic similarity between publication titles by means of word embeddings. Word embeddings are integrated as external components, which keeps the model small and efficient, while allowing for easy extensibility and domain adaptation. Initial experiments show that word embeddings can improve the Recall and F score of the binary classification sub-task of AND. Results for the clustering sub-task are less clear, but also promising and overall show the feasibility of the approach.

Keywords: Author name disambiguation · Semantic similarity · Word embeddings · Classification · Clustering · Machine learning · Deep learning

1   Introduction

Author name ambiguity can be observed in collections of (scientific) publications when several authors bear, or publish under, the same name. It is caused by the natural limitation of available person names, and by the fact that some names are much more frequent than others. It is further aggravated by the common publishing practice of initializing authors' first names. Author name *disambiguation* (**AND**) is the task of deciding, for a given pair of publications with the same author name, whether that name refers to the same author individual [4,18]. AND is a multi-facetted task, which comprises (1) content similarity, (2) co-author similarity, and (3) publication meta data similarity. In this paper, we present a supervised machine learning system which handles these three facets in a unified and extensible way. In particular, our system uses **word embeddings** (**WEs**) to deal with the content similarity facet of AND. WEs are employed to detect *semantic* content similarity or relatedness between pairs of publication titles, beyond surface-based string matching. To give just one illustrative example which exhibits neither title string nor co-author overlap, consider the following pair of publications from our data set.

A. Verma, **A. Kumar** (2004): <u>Articulatory</u> class based spectral envelope representation for <u>voice</u> fonts.
A. Karmakar, **A. Kumar**, R. K. Patney (2006): A Multiresolution Model of <u>Auditory</u> Excitation Pattern and Its Application to Objective Evaluation of Perceived <u>Speech</u> Quality.

© Springer International Publishing AG 2017
J. Kamps et al. (Eds.): TPDL 2017, LNCS 10450, pp. 300–311, 2017.
DOI: 10.1007/978-3-319-67008-9_24

Here, the author name *A. Kumar* does refer to the same person, but the only hint is in the semantic relatedness of the underlined tokens.

WEs have recently become popular in Deep Learning approaches to natural language processing (NLP). Full-blown Deep Learning models can take long to train and are technically demanding, often requiring specialized hardware. These requirements limit their practical applicability for digital libraries or online bibliographies. Our approach, in contrast, avoids these problems by (1) using only a simple machine learning model, which is fast and easy to train, and (2) keeping the WEs separate from the model. This way, the WEs are trained in a one-off effort, and they can easily be re-used and combined, even as the model architecture gets more complex. The rest of this paper is structured as follows: Sect. 2 provides a brief definition of AND and outlines how we cast AND as binary classification followed by clustering. Section 3 introduces the concept of WEs for the computation of semantic similarity and then provides a detailed description of our system. Section 4 describes and discusses our experiments, and Sect. 5 briefly reviews some related work. The paper concludes in Sect. 6.

2   Definition of AND

AND deals with *authorship records* [3], which consist of an author name and some representation of the publication content, co-authors, and other meta data. Publications with n authors yield as many authorship records, and for every author, the $n-1$ other records provide important information about the publication co-authors. Content and co-author similarity are two interrelated facets of AND, none of which is sufficient in isolation. *Content* similarity between two authorship records with the same author name is not necessary to establish author identity: The same author can produce publications on different topics, or even in completely distinct fields, which will not be very similar. High *co-author* similarity is normally a strong indicator for author identity [15,17], and is the sole information source for some AND approaches. On the other hand, the absence of common co-authors does not indicate non-identical authors, as one author can collaborate with distinct groups of colleagues. Additional meta data like publication year distance can have a mediating function here, as it can capture changes in an author's interests over time [6].

2.1   AND as Binary Classification Plus Clustering

We follow [6,14], and others in separating the AND task into binary classification of pairs of authorship records and subsequent clustering. Commonly, the input for an AND system is a list of so-called *blocks*, i.e. a list of sets of authorship records with a shared identical, or highly similar, name, and the output is a partitioning of each block into sub sets for the individual authors. In the binary classification paradigm, a single data instance represents two authorship records (from the same block, but from different publications) and a binary label which is 1 if the publications are authored by the same person, and 0 otherwise. In our approach, each instance represents the following information for each of the two authorship records: **Content information** subsumes various textual information. We assume that minimally the publication title is available, but the same representation is easily applied to other textual artifacts, like abstracts or full texts. Title words of each of the two authorship records are lowercased, cleaned of stop words, and represented as one list of complete and one list of stemmed tokens (created using the PorterStemmer). Having these two lists at our disposal allows us to use pre-trained WE resources that expect either format. Each title is also split into one list each of character 3-, 4-, and 5-grams. In addition, we apply a two-word window on the stemmed token list for each publication to obtain a list of word bi-grams. **Co-author information** for each of the two authorship records is represented as a list of normalized co-author names, *excluding* the shared author name. Normalization includes initialization of the first name and lowercasing of the entire string. A list of three co-authors could thus look like "f.harary m.lim d.wunsch". **Meta data** includes relational, first-order attributes of the pair of authorship records: *Publication year distance* is the absolute difference between both publications' year attributes. If the year is missing for one or both publications, the value is −1. *Publication venue match* is 1 if the publication venues of the two publications match, 0 if they are available but do not match, and −1 if one or both is unavailable.[1] After obtaining binary classification results for a given block, the individual decisions have to be combined into clusters. This is commonly done as a graph partitioning task (cf. Sect. 5), where binary classification confidence scores are used as edge weights in an undirected graph, and the author partitions are obtained by some graph algorithm. Alternatively, we create the graph in such a way that its *connected components* can directly be interpreted as clusters. We do this by employing different minimum positive confidence thresholds during graph creation (cf. Sect. 4.2).

3   A Deep Learning Model for AND

3.1   Word Embeddings for Semantic Similarity

The basic idea behind WEs is that distributional (i.e. co-occurrence) information derived from a large text corpus is represented in low-dimensional vector space in such a way that proximity in this vector space can be interpreted as similarity or relatedness. The vector representation for a single word is commonly given as a list of n real-valued numbers, where n is the dimensionality of the embedding. Two popular algorithms for learning WEs from texts are GloVe [13] and word2vec [10]. Apart from the desired dimensionality, both algorithms accept, among others, one parameter for the window size, and one for the minimum vocabulary count. The first parameter controls the maximum distance between words that are considered as co-occurrent, and the second parameter controls how often a word has to occur in the corpus in order to be considered at

---

[1] We only consider venue *identity* rather than *similarity* because our data set only contains abstract, uninterpretable venue identifiers.

all. Depending on the choice of parameters and the size of the corpus, training WEs can be computationally expensive. However, since they are supposed to capture universal, task-independent semantic relations, they can be utilized in diverse settings without the need for re-training. Several studies have focussed on the evaluation of WEs. [16] perform extensive experiments with diverse WEs, and evaluate how well they reproduce *human* semantic relatedness judgements, and how much they contribute to tasks like e.g. sentiment classification. [5], in a similar vein, evaluate several WEs on what they call *NLP* (=extrinsic) and *linguistic* (=intrinsic) tasks. While the level of granularity of WEs is the individual word, computing the semantic similarity of arbitrarily long word sequences (e.g. sentences or publication titles) requires that those sequences are reduced to single vectors that somehow capture the semantics of the whole sequence. A common way to do this is to average over the embeddings for the individual words: Given a collection of n-dimensional WE vectors and a sequence of i words, we retrieve the j vectors for those words that are covered in the collection (with $j <= i$), sum over the j values for each of the n dimensions, and divide each of the values by j. This yields one n-dimensional embedding vector for each word sequence, which can be compared to similar representations of other sequences, e.g. by means of computing the cosine similarity. This simple and efficient method has been shown to work surprisingly well, and is often used as a baseline in more complex, training-intensive systems, e.g. [9][2]. We prefer this simple heuristic over more powerful Deep Learning devices (like e.g. RNNs or LSTM networks, which maintain a notion of *ordering* in the reduced sequence) because (1) our preliminary experiments showed that they dramatically increase the technical complexity and training time for our system, rendering it difficult to use in a practical setting, and (2) we think that, for publication titles, the subtle differences conveyed by word order are negligible.

3.2 System Architecture and Components

Figure 1 shows the architecture of the binary classifier employed in our system, which is implemented as one multi-layer neural network with Keras [2] on top of Theano [19], and which is trained using the Adam optimizer. Input to the network is provided by the two authorship records (depicted as documents). The network consists of three auxiliary models (horizontal boxes), each of which focusses on a particular facet of the classification problem. Only the meta data attributes, due to their simplicity, do not have their own model. Each auxiliary model is a multi-layer neural network (cf. below for details) with sigmoid activation and a final softmax layer, which outputs the positive and negative class probability for each instance. The **simple co-author model** contains only two features, the cosine and the Jaccard similarity of the normalized co-author names (without the shared name). The model consists of one two-node hidden layer only. The simplicity of this model, which treats each name as one atomic

---

[2] [7] report similar baseline results with *summing* instead of *averaging* over the embeddings of a sequence.

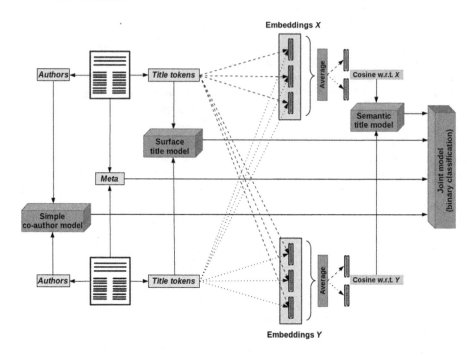

Fig. 1. System architecture: Binary Classifier

token, is a result of the author name structure found in the KISTI data set (cf.
Sect. 4.1), which, like most standard AND data sets [12], does not require any
sophisticated string similarity measures due to the absence of name variability.
As the focus of this paper lies on the semantic title model, the co-author model
is intended as a high-precision baseline only. The **surface title model** covers
the string-matching aspect of content similarity in terms of cosine and Jaccard
similarity of stemmed tokens, character 3-, 4-, and 5-grams, and word bi-grams,
resulting in a total of ten features. The model consists of one hidden layer with
ten nodes. Here, stemmed tokens are used in order to increase the coverage. The
features used in this model are more or less standard features found in many NLP
and IR systems. Despite their simplicity, surface-based features can handle a lot
of cases, and display a very reasonable performance (cf. the baseline results in
Sect. 4.2). This can make it difficult to demonstrate the contribution of semantic
features, because their effect is easily covered up by the effect of the surface-based
features. The **semantic title model** (including its preprocessing) is the most
interesting component of the binary classifier. The system can be supplied with c
collections of WEs. For each collection, the preprocessing component computes
the cosine similarity of the (stemmed or unstemmed) title tokens from the two
authorship records, which are then passed to the semantic title model as input
features. The model consists of three hidden layers with one node for each of the
c input features. For reasons of clarity, Fig. 1 shows only two collections of WEs

($c = 2$), but the system can accept arbitrarily many, including none, in which case the semantic title model remains inactive. One important feature of the semantic title model is that tokens that appear in the publication titles of *both* authorship records are completely removed from its input prior to generating the averaged WEs. This is motivated by the idea that perfect string identity can and should be handled in the surface title model, and that the semantic title model should be allowed to ignore these cases in favor of cases that involve actual vector space *similarity* rather than *identity*. Finally, the **joint model** (vertical box) is also a multi-layer neural network with sigmoid activation and a final softmax layer. It simply integrates the outputs of the softmax layers of the auxiliary models, as well as the meta data attributes, and produces the final classification. The joint model consists of two hidden layers with eight (with semantic title model) or six (without semantic title model) nodes. At test time, it outputs for each instance the positive and negative class probability, and the higher of the two probabilities determines the final classification for the instance. During training, each auxiliary model is presented with its respective sub set of features, computed from the instance representation described in Sect. 2.1 above, and with the binary label. Likewise, the joint model is presented with the outputs of the auxiliary models and the same binary label. Being part of a single network, all four models are trained simultaneously. However, we decouple the training of the models by computing a training error and corresponding loss for each model individually. This allows the different parts of the system to train at different speeds: the simple co-author model and the surface title model, e.g., converge quickly, while the semantic title model and the joint model, depending on the parameters and the number of WE collections, converge more slowly.

In order to obtain author clusters, we then employ NetworkX[3] to create an undirected graph containing one node for every authorship record in the block, and to add edges between all node pairs that were classified with a minimum positive confidence above a given (variable) threshold.

4 Data, Experiments, and Results

4.1 Data Set and Word Embeddings

Our task-specific data set is KISTI [8], which is derived from dblp data and consists of 41.674 authorship records from 37.613 publications, and 6.921 different authors. The data is pre-structured into blocks which are identified by a first name initial and a full last name. Each block contains authorship records from 1 to max. 71 different authors. DEV-TRAIN, DEV-TEST, and EVAL data was generated from the KISTI data set as follows: We randomly distributed the individual authors in each block (i.e. *y.chen_1, y.chen_2, ... y.chen_n*) into three sets of roughly the same size, ignoring blocks with less than six authors. Then, for each of the three sets, we paired all authorship records in the same block with each other, and created either a positive or a negative instance in the format

---

[3] https://networkx.github.io/.

described above in Sect. 2.1. Note that, of the various possible methods of creating data instances from the KISTI data set, this method makes it rather difficult for the system, because, at test time, all authors in DEV-TEST and EVAL are unseen. This yielded 190.009 DEV-TRAIN instances (41.8% pos, 58.2% neg.), 226.546 DEV-TEST (47.66% pos., 52.34% neg.), and 163.000 EVAL (42.95% pos., 57.05% neg.). Word level semantics is integrated into our system by means of pre-trained GloVe embeddings [13] and custom-built WEs trained on various text corpora. The GloVe embeddings[4] (**GloVe**) were trained on huge corpora (between 6 and 840 billion tokens) covering Wikipedia pages and other web data, as well as news wire texts. A second set of WEs (**dblp**) was trained on a text corpus of 3.5 million publication titles derived from a dblp XML dump[5]. We made sure to remove from the corpus the titles of all publications that are also contained in the KISTI data set. A third set of WEs (**MSAc**) was trained on a corpus of 6.5 million publication titles extracted from the Microsoft Academic Search dataset. Prior to training, for **dblp** and **MSAc**, special characters were removed, and the publication titles were lowercased, cleaned from stopwords, and stemmed with the PorterStemmer. A fourth set of WEs (**dblp+MSAc**) was trained on the concatenation of the dblp and the MSAc text corpora. All WEs were trained with the gensim[6] implementation of word2vec, using the CBOW variant. For all corpora, we employed several values for the parameters *dimensionality* ($d = 50, 100, 200, 300$), *minimum token count* ($mc = 5, 20$), and *window size* ($w = 5, 10$). There is one embedding file for each combination, resulting in 16 separate files per text corpus. Table 1 gives some statistics. Note that the total number of text tokens used for calculating the coverage of the embeddings is higher for the GloVe embeddings (10.734) because they contain unstemmed tokens, while all other embeddings were trained on stemmed tokens. Also, the glove.840B.300d embeddings are case-sensitive, which is why their coverage on our lowercased data set is smaller.

Table 1. Word embedding statistics

Name	# Tokens	Coverage
glove.6B.{100,200,300}d	400.000	9.151/10.734 (85%)
glove.42B.300d	1.917.494	9.944/10.734 (92%)
glove.840B.300d	2.196.017	9.511/10.734 (88%)
dblp.cbow.{50,100,200,300}d.mc5.{w5,w10}	56.081	6.522/7.263 (89%)
dblp.cbow.{50,100,200,300}d.mc20.{w5,w10}	23.738	6.081/7.263 (83%)
msac.cbow.{50,100,200,300}d.mc5.{w5,w10}	198.383	6.411/7.263 (88%)
msac.cbow.{50,100,200,300}d.mc20.{w5,w10}	74.255	5.924/7.263 (81%)
dblp+msac.cbow.{50,100,200,300}d.mc5.{w5,w10}	224.599	6.680/7.263 (91%)
dblp+msac.cbow.{50,100,200,300}d.mc20.{w5,w10}	84.104	6.383/7.263 (87%)

---

[4] https://nlp.stanford.edu/projects/glove/.

[5] http://dblp.uni-trier.de/xml/.

[6] https://radimrehurek.com/gensim/.

4.2 Experiments and Discussion

As mentioned in Sect. 2.1, our system consists of two parts, which we evaluate individually. The *binary classifier* is evaluated in terms of P, R, and F for retrieving positive instances. We train one binary classifier with each set of WEs individually, and with all sets of WEs (i.e. GloVe, dblp, and MSAc) at once. Note that we use *all* five (GloVe) resp. 16 (dblp, MSAc) embedding files per set *simultaneously*. Thus, the semantic title model has as many as 37 features when using all WEs at once. We do this in order to exploit potential complementarity of the WEs produced with different training parameters, and because our initial experiments gave no clear indication as to which parameters are optimal. The results for the binary classifier are given in Table 2, with the maximum P, R, and F values in bold. Each classifier is trained for 40 epochs, and we report the performance with the maximum F on DEV-TEST, along with the number of epochs that were required to reach that result (E).

Table 2. Binary classification results (max. F score reached after *E* epochs)

ID	WEs	Files	P	R	F	E
0	-	-	**82.48**	45.82	58.91	9
1	GloVe	5	81.21	47.22	59.72	9
2	dblp	16	76.96	67.18	71.74	20
3	MSAc	16	78.29	53.82	63.79	23
4	dblp+MSAc	16	76.38	65.29	70.40	23
5	GloVe, dblp, MSAc	37	75.67	**69.01**	**72.19**	13

As a baseline, we trained and tested the classifier with ID **0** without supplying any WEs, so that the semantic title model remains inactive. As expected, this classifier shows the worst performance (58.91 F), but not, however, by a large margin. The baseline binary classifier is clearly biased towards P, with the highest P and the lowest R of all binary classifiers. We see this as the result of the classifier's limitation to simple string matching, which prevents it from retrieving instances which are semantically related, but where this relatedness is not obvious on the surface. The worst non-baseline classifier is **1**, only slightly above baseline with 59.72 F. The other non-baseline classifiers are considerably better, with F scores of 63.79 (**3**), 70.40 (**4**), and 71.74 (**2**). Note that classifier **2** uses the WEs trained on the dblp corpus, which is in-domain in the sense that it bears the most similarity - although no publication overlap - to the KISTI data set. It yields by far the best binary classification performance of all WEs trained on individual corpora. Note also that classifier **4**, which is trained on the concatenation of the dblp and MSAc corpora, fails to improve, or even reach, the F score of the classifier trained on the dblp corpus alone (**2**). Thus, simply merging both corpora prior to WE training did not yield an improvement. As mentioned earlier, another way of integrating different sources of word level semantics into our system is by using *several* sets of WEs *simultaneously*, as in classifier **5**. And indeed, this configuration yields the best performance of all binary classifiers: although P drops to the lowest value of all, the associated gain in R is sufficient to also result in the best overall F of 72.19.

In the next step, we applied each classifier to the *clustering* task. Table 3 reports B^3 [1] results on DEV-TEST calculated by the CONLL scorer[7]. For each binary classifier from Table 2, we report one set of results for different minimum positive confidence threshold values (mpc = 0.5, 0.75, 0.9, 0.95).[8] The maximum P, R, and F values for each of the four sets are given in bold. In addition, the best F value for each classifier (i.e. for each *row* in Table 3) is underlined. The intuition behind the mpc thresholds is to increase the precision (P) of the clustering by only allowing high-confidence binary classifications to cause the clustering together of two authorship records. At the same time, the recall (R) will decrease, as fewer clusters are created, but we expect this tradeoff to be less severe for 'better', more discriminative binary classifiers.

Table 3. Clustering Results

ID	mpc = 0.5			mpc = 0.75			mpc = 0.9			mpc = 0.95		
	P	R	F	P	R	F	P	R	F	P	R	F
0	57.23	97.05	72.00	75.74	90.77	82.57	87.95	82.95	85.37	93.61	70.29	80.29
1	56.48	97.36	71.49	71.76	92.43	80.80	85.97	85.24	85.60	93.65	70.20	80.25
2	53.50	98.99	69.46	65.19	95.50	77.49	83.98	86.94	85.44	95.31	68.45	79.68
3	53.86	98.46	69.63	69.87	92.92	79.76	85.42	85.40	85.41	94.49	69.89	80.35
4	52.67	99.35	68.85	65.28	94.97	77.37	85.00	86.25	85.62	94.50	69.81	80.30
5	52.70	99.37	68.87	64.67	95.80	77.21	85.41	85.68	85.66	95.06	69.63	80.38

It can be seen that increasing the mpc threshold has the intended effect, as the P values for all classifiers increase from left to right without exception. Also as expected, the R values decrease from left to right. Up to and including mpc = 0.9, this decrease is compensated for by the associated increase in P, such that the F values do also increase. For mpc = 0.95, however, there is a sharp decline in R for all classifiers, which also causes the F values to drop. Thus, in our experiments, all binary classifiers reach their best F values for mpc = 0.9 (underlined). However, for all classifiers, including the baseline, these F values are extremely similar at around 85. Although the best binary classifier (**5**) performs also best in clustering with an F of 85.66, it does so by a negligible margin only. So, the observed differences in binary classification do not translate to similar differences in clustering, which is somewhat surprising. We also see, however,

---

[7] http://conll.github.io/reference-coreference-scorers/.

[8] mpc = 0.5 corresponds to *no threshold*, as 0.5 is the minimum confidence in a binary classification.

that all non-baseline classifiers (i.e. those with an active semantic title model) have a better R (at least 85.24 (**1**)) than the baseline classifier (82.95), which is the intended effect of the semantic title model.

5 Related Work

As a computational task, AND has a long history in both computer science and digital library science, and has been tackled with symbolic and heuristic approaches, as well as with supervised and unsupervised machine learning [4]. To our knowledge, WEs have not previously been used for the task. However, a similar approach for inventor disambiguation in patent data bases is described in [11]. [20] is the only work so far in which Deep Learning methods have been applied to AND. Their model consists of an ensemble of (an unreported number of) N multi-layer perceptrons (MLPs) with seven layers of 50 hidden units each. As data, [20] use 30.537 binary labeled pairs of authorship records (12.93% positive pairs, 87.07% negative) featuring a matching, or highly similar, Vietnamese author name. In total, the data set includes names and name variants of ten authors. Each data instance contains numerical scores representing the similarity of the two records' author names, co-author names, affiliations, paper keywords, and author interest keywords, respectively. Note that paper titles are apparently not used. The scores are calculated using the Jaccard, Levenshtein, Jaro, Jaro-Winkler, Smith-Waterman, and Mogne-Elkan measures. The model is trained by iteratively providing each of the N MLPs with a randomly selected sample from the training data set. In contrast to our work, every MLP is exposed to the full set of features of the selected instances during training, so that there is no 'division of labour' between the MLPs with respect to the individual facets of the AND task. At test time, a classification is obtained by simply averaging over the predictions of the individual MLPs, while we use a dedicated network (the **joint model**) to integrate the outputs of the three individual models. [20] report a binary classification accuracy of 99.31 on their 20% hold-out data set, which the authors claim significantly outperforms their earlier systems based on conventional machine learning. Given the strong negative bias in their data set (almost 90%), we argue that it would have been more appropriate to evaluate the binary classifier according to Precision, Recall, and F-measure for retrieving positive instances (like in the present work). No clustering of the binary decisions is performed, so a full comparison of [20] and our work is not possible. Even more importantly, the system comprises (in our terminology) only the simple co-author model and the surface title model, and does not address semantic similarity beyond the string level.

   The NC system described in [15] marks the current state of the art on the KISTI data set. It relies on manually encoded, domain-specific expert heuristics, which operate on automatically extracted string-similarity scores (including cosine similarity and tf*idf). The weights of these scores are manually tuned, rendering the approach completely unsupervised. Supervised training can optionally be employed to further optimize parameters. The system uses (co-)author names as well as publication and venue titles but, like [20], does not go beyond the string level. The output of the system are clusters, which are evaluated with the K score, which is roughly equivalent to F. On the KISTI data set [15] report a total K score of 94.00, which is obtained in a supervised setup by doing ten runs of two-fold (50%–50%) cross-validation per block, averaging the results per block, and again averaging the results for all blocks. Although the data set and the evaluation measure are the same or similar, the results of [15] cannot directly be compared with our results. One reason is that, in the block-wise cross-validation of [15], authors in the test split have probably also been in the training split, while in our approach, authors in test are always unseen.

6   Conclusion and Future Work

We presented the first AND system which tackles semantic similarity between publication titles by means of WEs. The system, although using some Deep Learning technology, aims at being practically usable and efficient. We found that adding WE-based semantic similarity can make a significant contribution to the binary classification part of the AND task, which is the most important result of this paper, and that WEs trained on in-domain corpora perform better than those trained on other, less similar corpora. Pre-trained general-purpose WEs (GloVe), although of high quality, were not helpful. We also found that complementarity of different WEs can best be exploited by using many independent WEs simultaneously, while training single WEs on concatenated corpora was not successful. Improvements observed in binary classification, however, did not clearly translate to improvements in clustering. In future work, therefore, we will improve the way our system exploits the individual binary, weighted classifications, e.g. by employing more powerful, state-of-the-art graph clustering algorithms. In order to create a competitive AND system, we also plan to extend the simple co-author model to be a more powerful component. All extensions and improvements will be able, if required, to make use of state-of-the-art methods in the Deep Learning 'ecosystem', which our system is already part of.

Acknowledgments. The research described in this paper was conducted in the project *SCAD – Scalable Author Name Disambiguation*, funded in part by the Leibniz Association (grant SAW-2015-LZI-2), and in part by the Klaus Tschira Foundation. We thank Florian Reitz (dblp) for data preparation and the anonymous TPDL reviewers for their useful suggestions.

References

1. Bagga, A., Baldwin, B.: Algorithms for scoring coreference chains. In: Proceedings of the 1st International Conference on Language Resources and Evaluation, Granada, Spain, 28–30 May 1998, pp. 563–566 (1998)
2. Chollet, F.: Keras (2015). https://github.com/fchollet/keras

3. Cota, R.G., Ferreira, A.A., Nascimento, C., Gonçalves, M.A., Laender, A.H.F.: An unsupervised heuristic-based hierarchical method for name disambiguation in bibliographic citations. J. Am. Soc. Inf. Sci. Technol. **61**(9), 1853–1870 (2010)

4. Ferreira, A.A., Gonçalves, M.A., Laender, A.H.: A brief survey of automatic methods for author name disambiguation. SIGMOD Rec. **41**(2), 15–26 (2012)

5. Ghannay, S., Favre, B., Estève, Y., Camelin, N.: Word embedding evaluation and combination. In: Proceedings of LREC 2016, Portorož, Slovenia, 23–28 May 2016 (2016)

6. Gurney, T., Horlings, E., van den Besselaar, P.: Author disambiguation using multi-aspect similarity indicators. Scientometrics **91**(2), 435–449 (2012)

7. Hu, B., Lu, Z., Li, H., Chen, Q.: Convolutional neural network architectures for matching natural language sentences. In: Advances in Neural Information Processing Systems 27: Annual Conference on Neural Information Processing Systems 2014, Montreal, Quebec, Canada, 8–13 December 2014, pp. 2042–2050 (2014)

8. Kang, I.-S., Kim, P., Lee, S., Jung, H., You, B.-J.: Construction of a large-scale test set for author disambiguation. Inf. Process. Manage. **47**(3), 452–465 (2011)

9. Kenter, T., de Rijke, M.: Short text similarity with word embeddings. In: Proceedings of CIKM 2015, New York, NY, USA, pp. 1411–1420 (2015)

10. Mikolov, T., Chen, K., Corrado, G., Dean, J.: Efficient estimation of word representations in vector space. In: Proceedings of the International Conference on Learning Representations (2013)

11. Monath, N., McCallum, A.: Discriminative hierarchical coreference for inventor disambiguation. Presentation at PatentsView Inventor Disambiguation Technical Workshop, September 2015

12. Müller, M.-C., Reitz, F., Roy, N.: Data sets for author name disambiguation: an empirical analysis and a new resource. Scientometrics **111**(3), 1467–1500 (2017)

13. Pennington, J., Socher, R., Manning, C.D.: GloVe: global vectors for word representation. In: Proceedings of the 2014 Conference on Empirical Methods in Natural Language Processing, Doha, Qatar, 25–29 October 2014, pp. 1532–1543 (2014)

14. Qian, Y., Zheng, Q., Sakai, T., Ye, J., Liu, J.: Dynamic author name disambiguation for growing digital libraries. Inf. Retrieval J. **18**(5), 379–412 (2015)

15. Santana, A.F., Gonçalves, M.A., Laender, A.H.F., Ferreira, A.A.: On the combination of domain-specific heuristics for author name disambiguation: the nearest cluster method. Int. J. Digit. Libr. **16**(3–4), 229–246 (2015)

16. Schnabel, T., Labutov, I., Mimno, D.M., Joachims, T.: Evaluation methods for unsupervised word embeddings. In: Proceedings of EMNLP 2015, Lisbon, Portugal, 17–21 September 2015, pp. 298–307 (2015)

17. Shin, D., Kim, T., Choi, J., Kim, J.: Author name disambiguation using a graph model with node splitting and merging based on bibliographic information. Scientometrics **100**(1), 15–50 (2014)

18. Smalheiser, N.R., Torvik, V.I.: Author name disambiguation. ARIST **43**(1), 1–43 (2009)

19. Theano Development Team: Theano: A Python framework for fast computation of mathematical expressions. arXiv e-prints, abs/1605.02688, May 2016

20. Tran, H.N., Huynh, T., Do, T.: Author name disambiguation by using deep neural network. In: Nguyen, N.T., Attachoo, B., Trawiński, B., Somboonviwat, K. (eds.) ACIIDS 2014. LNCS (LNAI), vol. 8397, pp. 123–132. Springer, Cham (2014). doi:10.1007/978-3-319-05476-6_13

Scholarly Communication

Towards a Knowledge Graph Representing Research Findings by Semantifying Survey Articles

Said Fathalla[1,3]([✉]), Sahar Vahdati[1], Sören Auer[4,5], and Christoph Lange[1,2]

[1] Smart Data Analytics (SDA), University of Bonn, Bonn, Germany
{fathalla,vahdati,langec}@cs.uni-bonn.de
[2] Fraunhofer IAIS, Sankt Augustin, Germany
[3] Faculty of Science, Alexandria University, Alexandria, Egypt
[4] Computer Science, Leibniz University of Hannover, Hanover, Germany
[5] TIB Leibniz Information Center for Science and Technology, Hannover, Germany
soeren.auer@tib.eu

Abstract. Despite significant advances in technology, the way how research is done and especially communicated has not changed much. We have the vision that ultimately researchers will work on a common knowledge base comprising comprehensive descriptions of their research, thus making research contributions transparent and comparable. The current approach for structuring, systematizing and comparing research results is via survey or review articles. In this article, we describe how surveys for research fields can be represented in a semantic way, resulting in a knowledge graph that describes the individual research problems, approaches, implementations and evaluations in a structured and comparable way. We present a comprehensive ontology for capturing the content of survey articles. We discuss possible applications and present an evaluation of our approach with the retrospective, exemplary semantification of a survey. We demonstrate the utility of the resulting knowledge graph by using it to answer queries about the different research contributions covered by the survey and evaluate how well the query answers serve readers' information needs, in comparison to having them extract the same information from reading a survey paper.

Keywords: Semantic metadata enrichment · Quality assessment · Recommendation services · Scholarly communication · Semantic publishing

1 Introduction

Despite significant advances in technology in the last decades, the way how research is done and especially communicated has not changed much. Researchers still encode their findings in sequential text accompanied by illustrations and wrap these into articles, which are mostly published in printed form or as semi-structured PDF documents online. We have the vision that ultimately researchers will rather work on a common knowledge base comprising

© Springer International Publishing AG 2017
J. Kamps et al. (Eds.): TPDL 2017, LNCS 10450, pp. 315–327, 2017.
DOI: 10.1007/978-3-319-67008-9_25

comprehensive descriptions of their research, thus making research contributions transparent and directly comparable. The current approach for structuring, systematizing and comparing research results is via survey or review articles. Such articles usually select a number of articles describing comparable research and (a) develop a common organization scheme with feature classifications, (b) provide a conceptualization of the research domain with mappings to the terminologies used in the individual articles, (c) compare and possibly benchmark the research approaches, implementations and evaluations described in the articles and (d) identify directions for future research. As a result, survey and review articles significantly contribute to structuring a research domain and make its progress more transparent and accessible. However, such articles still share the same deficiencies as their original research counterparts – the content is not represented according to a formal knowledge representation and not machine comprehensible, which prevents systematic identification of conceptualization problems as well as the building of intelligent search, exploration and browsing applications on top.

In this article, we describe how surveys for research fields can be represented in a semantic way resulting in a knowledge graph that describes the individual research problems, approaches, implementations and evaluations in a structured and comparable way. We present a comprehensive ontology for capturing the content of survey articles. The ontology is structured around four core concepts:

- *research problem* – describing a challenge in a particular field, possibly hierarchically decomposed into sub-problems,
- *approach* – describing attributes and features of particular research approaches,
- *implementation* – describing the implementation of an approach in a concrete technical environment,
- *evaluation* – describing the benchmarking of an implementation in a certain formally defined evaluation scenario.

As a result of structuring and representing research advances according to such a semantic scheme, they will become more comparable and accessible. For example, research addressing a certain problem can be automatically retrieved, approaches can be compared according to their features or w.r.t. evaluation results in a certain defined setting. In particular, we discuss possible applications and present an evaluation of our approach with the retrospective, exemplary semantification of a survey resulting in a knowledge graph comprising a comprehensive description of the respective research.

The ultimate aim of this work is to enable the provision of better and more intelligent services for the discovery of scientific work.

We illustrate our methodology with the example of the following three survey articles:

- Bringing Relational Databases into the Semantic Web: A Survey [12].
- A Survey of Current Link Discovery Frameworks [6].
- Querying over Federated SPARQL Endpoints —A State of the Art Survey [9].

The remainder of the article is structured as follows: We present an overview on related work in Sect. 2. The conceptualization of a knowledge graph of research advances is described in Sect. 3. We present a methodology for semantifying survey articles in Sect. 4. An evaluation describing typical usage scenarios and queries for exploring the knowledge graph in these scenarios is presented in Sect. 5. We conclude with an outlook on future work in Sect. 6.

2   Related Work

In the last decade, there has been a mass growth in scholarly communications due to the impact of the ubiquitous availability of the Internet, email, and web-based services on scholarly communication. The preparation of manuscripts as well as the organization of conferences, from submission to peer review to publication, have become considerably easier and efficient. Research is based on digital assets, such as datasets, services, and produces its output in digital form.

Capadisli's "linked research" approach starts with HTML and enriches it semantically, encapsulating publication meta-data and content [3]. Researchers are encouraged and enabled to announce their research so that they will be both authors and publishers. Research Articles in Simplified HTML (or RASH) is another Web-first format for writing HTML-based scholarly papers [8]. RASH enables a formal representation of the structure an article, which is linked to semantically related articles, thus supporting integration of data between papers. Both approaches scale up to a semantic representation of the full details of a research investigation, but hardly any author has made this effort manually.

Several efforts on developing ontologies as well as publishing reusable, machine-comprehensible (meta)data (i.e. *linked open data*) related to scholarly entities such as publications, scientific events, authors etc., have been carried out so far [10]. For example, the Springer LOD dataset[1] contains metadata about conference proceedings from the Lecture Notes in Computer Science series and aims at answering trust-related questions of different stakeholders. Bryl et al. mention questions such as "Should I submit a paper to this conference?", and point out that the data that is required for answering such questions is not easily available but, e.g., hidden in conference management systems [2]. The Semantic Web Dog Food (SWDF) dataset[2] and its successor *ScholarlyData*[3] are among the pioneers of datasets of comprehensive scholarly communication metadata. All these works support scholarly communication by giving end users easy access just to metadata about research-related entities, not to the research findings. However, none of them provides services to ease the process of gaining an overview of a field, which is what we introduce in this work.

---

[1] http://lod.springer.com/.
[2] http://data.semanticweb.org/.
[3] http://www.scholarlydata.org/dumps/.

3   Conceptualization

In different research disciplines there is, depending on the culture of that domain, a need for studying the literature on a specific topic to write a survey or review article, which facilitates comprehension of the topic. Experts in the field commonly create such reviews ready for the community. The readers of such review articles are often peer-researchers in the field, in particular also young researchers aiming to get an overview. Due to the representation of review articles as unstructured text, it is impossible to automatically extract and analyze information from them. In the remainder of this section, we introduce the concepts, terms and vocabularies that we defined for representing the content of review articles.

3.1   SemSur Ontology

SemSur, the Semantic Survey Ontology, is a core ontology for describing individual research problems, approaches, implementations and evaluations in a structured, comparable way. We describe its structure and contents, which captures detailed terminological knowledge about survey articles, e.g., evaluation method, hypothesis, benchmark, and experiment. SemSur is represented in the Web Ontology Language (OWL) and developed using Protégé 5.2.0 [5]. We defined new vocabularies in the OpenResearch namespace[4]. Table 1 shows the ontology statistics.

Table 1. Overview of SemSur ontology statistics.

Metrics	Count	Metrics	Count
Classes	197	Object properties	149
Data properties	78	Instances	220
Subclass relationships	234	Transitive properties	2
Inverse properties	14	Symmetric properties	2

3.2   Reuse of Ontological Knowledge

Technologies for efficient and effective reuse of ontological knowledge are one of the key success factors for developing ontology-based systems [11]. Therefore, the first step in building our knowledge graph is reusing vocabularies from related existing ontologies on the Web, since reuse increases the value of semantic data. We have selected the most closely related ontologies listed in the Linked Open Vocabularies (LOV) directory[5]. Existing related vocabularies are shown in Table 2.

---

[4] http://openresearch.org; prefix (or).
[5] Linked Open Vocabularies: http://lov.okfn.org/dataset/lov/vocabs.

Table 2. Prefixes and namespace URIs of reused vocabularies.

Prefix	Vocabulary	URI
dcterms	Dublin Core Metadata Initiative (DCMI)	http://purl.org/dc/terms/
swrc	Semantic Web for Research Communities	http://swrc.ontoware.org/ontology#
foaf	Friend of a Friend ontology	http://xmlns.com/foaf/0.1/
mls	Machine Learning Schema	https://www.w3.org/ns/mls#
deo	The Discourse Elements Ontology	http://purl.org/spar/deo/
lsc	Linked Science Core Vocabulary	http://linkedscience.org/lsc/ns#
doap	Description of a Project	http://usefulinc.com/ns/doap#

For modeling the top level metadata of a scientific article as a whole, we reuse the DC, SWRC and FOAF ontologies. The Dublin Core Metadata Initiative (DCMI)[6] provides a standard vocabulary for describing resources. *SWRC* (Semantic Web for Research Communities) describes research communities and relevant related concepts such as persons, organizations, bibliographic metadata and relationships between them. The FOAF ontology describes persons and their activities. For modeling the inner structure of a scientific article independently of the field of research we use DEO (Discourse Elements Ontology) and LSC (Linked Science Core). *DEO* is an ontology for describing the major elements within journal articles such as Abstract, Introduction, Reference List and Figures. *LSC* is designed for describing scientific resources including Publication, Researcher, Method, Hypothesis, and Conclusion. To model concepts of specific fields of research we use MLS and DOAP and may in future use additional ontologies. MLS is a standard schema published by the W3C Machine Learning Schema community group for machine learning algorithms, data mining, datasets, and experiments. DOAP (Description of a Project) is a vocabulary that describes software projects and related concepts.

Figure 1 gives an overview of a SemSur knowledge graph describing individual research problems, approaches, implementations and evaluations. For better readability of the visualization some classes are omitted. Namespace prefixes are used according to prefix.cc[7].

3.3   SemSur Classes

The SemSur ontology imports classes from the ontologies introduced in Sect. 3.2 in addition to its owns classes. Some of these classes need more specialization so we created respective subclasses. For instance, we added three subclasses Mathematical Model, ArchitecturalModel and PipelineModel for the Model class inherited from the MLS ontology. Another concern is the integration of imported ontologies. In other words, classes imported from an ontology

---

[6] Dublin Core Metadata Initiative: http://dublincore.org/.
[7] Namespace look-up tool for RDF developers: http://prefix.cc/.

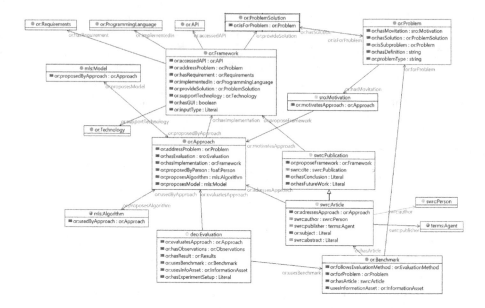

Fig. 1. Overview of a SemSur knowledge graph

should have a suitable relation with related classes found in other ontologies. For instance, the `Article` class (from SWRC) should have a relation with the `Conclusion` class (from LSC) with the relation *produces* (from LSC). Some of the reused classes are shown in Table 3.

3.4 SemSur Relations

SemSur provides a taxonomic class hierarchy. For instance, `Article` is a subclass of `Publication` as shown in Fig. 1. There are some transitive relations such `usesFramework` and `isSubproblem`. For instance, if X `isSubproblem` of Y and Y `isSubproblem` of Z then it could be inferred that X `isSubproblem` of Z. Also, there are some symmetric relations such as `hasRelatedProblem` and some

Table 3. Classes and relations reused by SemSur.

Ontology	Reused classes	Reused relations
SWRC	Article	year
foaf	Person	name
mls	Model, Algorithm, Information Entity	hasInput, hasOutput
lsc	Conclusion	Produces, timeAccepted
deo	FutureWork, Evaluation, Motivation	

inverse relations such as proposesModel and proposedByApproach. In addition, we borrow some relations from different ontologies as shown in Table 3.

3.5 SemSur Instances

Creating instances of classes is the last step of common knowledge engineering methodologies [1]. The required steps for creating a knowledge graph are: (1) identify the classes, (2) create instances of these classes, and (3) add values for the associated properties [7]. For example, creating the instance *ANAPSID-framework*, which is a specific adaptive query processing engine for SPARQL endpoints, requires (1) identify the Framework class, (2) create the instance, and (3) add values of properties such as hasGUI, platform and implementedIn. The complete instance is shown in Fig. 2.

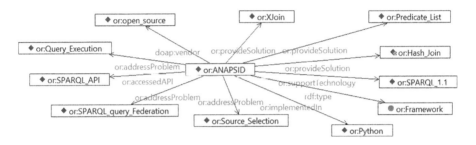

Fig. 2. SemSur ontology instance describing the ANAPSID framework.

SemSur contains a total of 220 instances for 14 classes. Our vision is that researchers who own a piece of research result or know about it create such instances as wiki pages where other researchers contribute to complete it. Providing semantic forms enable researchers of other domains easily create instances of research results from their community. Overall, we have 29 instances for classes, 95 instances for person, and 29 frameworks. 13 problems are instantiates with solutions and 7 without.

4   Methodology

The methodology of populating the SemSur knowledge graph is divided into two main phases: (1) select a narrow research field with many comparable approaches, problem and implementations, e.g. question answering, link discovery, SPARQL query federation or relationship extraction (2) build the knowledge graph comprising comprehensive descriptions of a specific research field and instantiating individual research articles in that field. The overall workflow of this study (see Fig. 3) comprises four steps: (1) Article selection, (2a) Formalization, (2b) Ontology development, and (3) querying the ontology to demonstrate its potential usage.

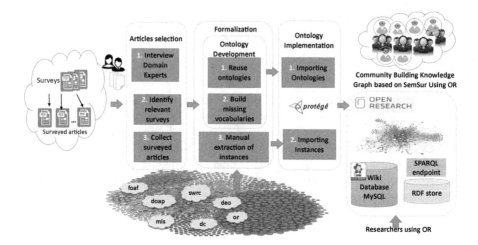

Fig. 3. Overview of the workflow for the proposed research knowledge graph population methodology.

Article selection: To demonstrate the feasibility of our approach, three survey articles for three different topics (mentioned in Sect. 1) have been selected by domain experts. These articles are used as main references by researchers in the domain to obtain an overview about frameworks, models, evaluation methods and research methodologies. From modeling these three survey articles, we collected all the 29 individual research articles covered by these surveys. These articles are addressed in the survey papers as references and we used them to create instances of the SemSur ontology.

Formalization: This step contains ontology development and manual extraction of the instances. As mentioned in Sect. 3.2, after conceptualization of the domain and interviewing experts, the SemSur ontology is instantiated. In the development of SemSur, we reused already existing and relevant ontologies with their proper suit and for the missing terms we developed our own. In parallel, we studied the articles and interviewed domain experts to extract instances describing the content of the 29 articles (done by the two first authors) based on ontology classes. This helped us to develop the ontology and also ease instance creation in the knowledge management system.

Ontology implementation: SemSur ontology was written in OWL using the Protégé, the open source ontology editor and knowledge management system. The extracted instances have been imported into Protégé, an example is shown in Fig. 2.

Querying the ontology: Querying SemSur is performed using the *Snap-SPARQL* query framework [4]. A list of 30 pre-defined queries (10 for each of the three surveys) has been created for evaluating the approach, which will be presented in Sect. 5. The questions were developed with the help of domain experts;

these are common questions or requirements on which new researchers in the domain often need to spend months of research to obtain such an overview.

To provide collaboratively editable and queryable version of instances, we use the semantic wiki-based platform [13] *OpenResearch.org (OR)*. We now enhanced its ontology by importing SemSur and asking the community to cover metadata about research results, e.g., developed tools, frameworks, methodologies, etc. Using the Semantic MediaWiki extension, we are able to represent the SemSur knowledge graph, to provide an environment for its curation and for creating overviews of the respective research domains (e.g., using the evaluation queries). A sample wiki page[8] of an instance is added to OpenResearch. In the right hand side, the information box is shown in which semantic representation of the instance and its properties based on SemSur is presented.

5   Evaluation

In this section we describe the method and the results of evaluating our approach.

5.1   Evaluation Method

We first succinctly introduce the evaluation setup and then discuss the result. The evaluation started with the phase of letting researchers first read the given overview questions and letting them try in their own way to find the respective answer. We followed these steps:

- A set of 10 predefined natural language queries has been prepared for evaluation Table 4. Then, asking participants to try to answer these queries using their own tools and services. The queries were chosen in increasing order of complexity.
- We implemented SPARQL queries corresponding to each of these queries to enable non-expert participants, not familiar with SPARQL, to query the knowledge graph.
- We asked researchers to review the answers of the pre-defined queries that we formulated based on the SemSur ontology. We asked them to tell us whether they consider the provided answers and the way queries are formulated comprehensive and reasonable.
- We finally asked the same researchers to fill in a satisfaction questionnaire with 18 questions[9].

As an example, the SPARQL implementation of Q5 is listed below. Figure 4 shows the results of this query using OR.

---

[8] http://openresearch.org/ANAPSID:_An_Adaptive_Query_Processing_Engine_for_SPARQL_Endpoints.
[9] https://goo.gl/eZC4UL.

```
SELECT DISTINCT ?Framework ?Problem ?subProblem ?solution ?platform ?hasGUI ?t
WHERE {
   ?Framework    or:addressProblem     ?subProblem .
   ?subProblem   or:isSubproblem       ?Problem .
   ?Framework    or:provideSolution    ?solution .
   ?solution     or:hasSolution        ?subProblem .
   ?Framework    or:hasGUI             ?hasGUI .
   ?Framework    or:supportTechnology  ?t .
   ?t            foaf:name             "SPARQL⌴1.1"
OPTIONAL {
   ?Framework    doap:platform         ?platform
} }
```

Table 4 shows the 10 sample SemSur knowledge graph evaluation queries for the three surveys. Note, that these are prototypical queries, which can be easily adapted to obtain similar information in other fields (we did the same for the other two surveys).

Acronym	Name	GUI	OS	Platform	Supports Technology	Implemented in	Addresses Problem
ANAPSID	An Adaptive Query Processing Engine for SPARQL Endpoints	True	Linux	ANAPSID	SPARQL 1.1	Python	SPARQL Query Federation Query_Execution Source_Selection
FedX	Optimization Techniques for Federated Query Processing on Linked Data	True	Windows	Sesame	SPARQL 1.0	Java	SPARQL Query Federation Query_Execution Source_Selection

Fig. 4. Sample overview query run on OR to show list of frameworks in SPARQL Federated Query

In the end, we asked participants to fill a questionnaire with 18 questions. The result of this evaluation is discussed in the following section.

5.2   Evaluation Results

To obtain answers of queries, 5 out of the 9 researchers immediately started with well-known standard Web search engines to explore the given topic. They tried to use several variations of keywords from the questions, e.g., "Federated Query Engines", "SPARQL Federation", etc. They also used digital libraries and scientific metadata services, e.g., ACM DL or Microsoft Academic Search, following the same approach and sometimes using advanced search options and filters. However, the retrieved results were either out of scope for the question but more related to the search keywords. All subjects unanimously agreed that the current way would not help them unless they explored more and read some survey articles on topic.

Table 4. 10 SemSur knowledge graph evaluation queries

Query #	Text
Q1	What are the possible strategies of "query execution" for DQE?
Q2	What are the programming languages used for implementing FQE over SPARQL endpoints?
Q3	Which evaluation metrics, information assets, results and benchmarks are used to evaluate LD frameworks?
Q4	What are the research problems related to database-ontology mapping?
Q5	What frameworks support SPARQL 1.1 or SPARQL 1.0 federation extension along with the platform, addressed problems, OS in which they can run, programming language used and have a GUI or not?
Q6	What are the frameworks that address the SPARQL query federation problem along with the articles where they are described, the publication year and authors names?
Q7	Which are the frameworks that solve the problem of query execution over a federation of SPARQL endpoints and support SPARQL 1.0?
Q8	What are the scientific articles that tackle the problem of generating RDF data from existing large quantities of data residing in relational databases?
Q9	What experiment setups should be considered for evaluating a DQE against SPARQL endpoints?
Q10	What are the motivations, the approaches and frameworks for current LD frameworks?

Overall, 8 researchers found it difficult to collect information and reach a conclusive overview of the research topics or related work using current methods. Six of the participants pointed out that for some of the overview questions, search engines were as good as the proposed system particularly when the framework name is part of the search keyword. They all agreed that for complicated questions our SemSur approach outperformed any existing approach/tool. Seven participants agreed that our system would be helpful for both new and experienced researchers. Two-thirds of them strongly agreed that the time and effort they spent to find such information using our system in comparison to other traditional ways is relatively low. Finally, 100% of the participants would like to use SemSur approach in their further research for studying the literature of a research topic or writing a survey article. Since the results of queries were shown to the participants in table view, the main feedback from all participants about possible improvements was to provide a better way of data representation.

6   Conclusions

In this article we presented SemSur, a Semantic Survey Ontology, and an approach for creating a comprehensive knowledge graph representing research findings.

We see this work as an initial step of a long-term research agenda to create a paradigm shift from document-based to knowledge-based scholarly communication. Our vision is to have this work deployed in an extended version of the existing OpenResearch.org platform.

We have created instances of three selected surveys on different fields of research using the SemSur ontology. We evaluated our approach involving nine researchers. As we see in the evaluation results, SemSur enables successful retrieval of relevant and accurate results without users having to spend much time and effort compared to traditional ways. This ontology can have a significant influence on the scientific community especially for researchers who want to create a survey article or write literature on a certain topic. The results of our evaluation show that researchers agree that the traditional way of gathering an overview on a particular research topic is cumbersome and time-consuming. Much effort is needed and important information might be easily overlooked. Collaborative integration of research metadata provided by the community supports researchers in this regard. Interviewed domain experts mentioned that it might be necessary to read and understand 30 to 100 scientific articles to get a proper level of understanding or an overview of a topic or sub-topics. A collaboration of researchers as owners of each particular research work to provide a structured and semantic representation of their research achievements, can have a huge impact in making their research more accessible. A similar effort is spent on preparing survey and overview articles.

Integrating our methodology with the procedure of publishing survey articles can help to create a paradigm shift. We plan to further extend the ontology to cover other research methodologies and fields. For a more robust implementation of the proposed approach, we are planning to use and significantly expand the OpenResearch.org platform and a user-friendly SPARQL auto-generation services for accessing metadata analysis for non-expert users. More comprehensive evaluation of the services will be done after the implementation of the curation, exploration and discovery services. In addition, our intention is to develop and foster a living community around OpenResearch.org and SemSur, to extend the ontology and to ingest metadata to cover other research fields.

Acknowledgments. This work has been supported by the H2020 project no. 645833 (OpenBudgets.eu). The authors would like to thank Prof. Maria-Esther Vidal and Afshin Sadeghi for their support. We also appreciate the help of all participants of the evaluation. This work was conducted using the Protégé resource, which is supported by grant GM10331601 from the National Institute of General Medical Sciences of the United States National Institutes of Health.

References

1. Antoniou, G., Van Harmelen, F.: A Semantic Web Primer. MIT Press, Cambridge (2004)
2. Bryl, V., et al.: What's in the proceedings? combining publisher's and researcher's perspectives. In: Proceedings of the 4th Workshop on Semantic Publishing (SePublica) (2014)
3. Capadisli, S., Riedl, R., Auer, S.: Enabling accessible knowledge. In: Conference for E-Democracy and Open Governement, p. 257 (2015)
4. Horridge, M., Musen, M.: Snap-SPARQL: a java framework for working with SPARQL and OWL. In: Tamma, V., Dragoni, M., Gonçalves, R., Ławrynowicz, A. (eds.) OWLED 2015. LNCS, vol. 9557, pp. 154–165. Springer, Cham (2016). doi:10.1007/978-3-319-33245-1_16
5. Musen, M.A.: The Protég'e project: a look back and a look forward. In: AI Matters. Association of Computing Machinery Specific Interest Group in Artificial Intelligence, vol. 1(4) (2015)
6. Nentwig, M., et al.: A survey of current link discovery frameworks. Semant. Web 8(3), 419–436 (2017)
7. Noy, N.F., McGuinness, D.L., et al.: Ontology development 101: A guide to creating your first ontology (2001)
8. Peroni, S., et al.: Research articles in simplified HTML: a web-first format for HTML-based scholarly articles. Technical report, PeerJ Preprints (2016)
9. Rakhmawati, N.A., et al.: Querying over federated SPARQL endpoints - a state of the art survey. In: CoRR abs/1306.1723 (2013)
10. Shotton, D.: Semantic publishing: the coming revolution in scientific journal publishing. Learned Publishing 22(2), 85–94 (2009)
11. Simperl, E.: Reusing ontologies on the semantic web: a feasibility study. Data Knowl. Eng. 68(10), 905–925 (2009)
12. Spanos, D.-E., Stavrou, P., Mitrou, N.: Bringing relational databases into the semantic web: a survey. Semant. Web 3(2), 169–209 (2012)
13. Vahdati, S., Arndt, N., Auer, S., Lange, C.: OpenResearch: collaborative management of scholarly communication metadata. In: Blomqvist, E., Ciancarini, P., Poggi, F., Vitali, F. (eds.) EKAW 2016. LNCS (LNAI), vol. 10024, pp. 778–793. Springer, Cham (2016). doi:10.1007/978-3-319-49004-5_50

Integration of Scholarly Communication Metadata Using Knowledge Graphs

Afshin Sadeghi[1]([envelope]), Christoph Lange[1,2], Maria-Esther Vidal[2],
and Sören Auer[3,4]

[1] Smart Data Analytics (SDA), University of Bonn, Bonn, Germany
i@afshn.com, langec@cs.uni-bonn.de
[2] Fraunhofer Institute for Intelligent Analysis and Information Systems (IAIS),
Sankt Augustin, Germany
vidal@cs.uni-bonn.de
[3] Computer Science, Leibniz University of Hannover, Hannover, Germany
[4] TIB Leibniz Information Center for Science and Technology, Hannover, Germany
soeren.auer@tib.eu

Abstract. Important questions about the scientific community, e.g., what authors are the experts in a certain field, or are actively engaged in international collaborations, can be answered using publicly available datasets. However, data required to answer such questions is often scattered over multiple isolated datasets. Recently, the Knowledge Graph (KG) concept has been identified as a means for interweaving heterogeneous datasets and enhancing answer completeness and soundness. We present a pipeline for creating high quality knowledge graphs that comprise data collected from multiple isolated structured datasets. As proof of concept, we illustrate the different steps in the construction of a knowledge graph in the domain of scholarly communication metadata (SCM-KG). Particularly, we demonstrate the benefits of exploiting semantic web technology to reconcile data about authors, papers, and conferences. We conducted an experimental study on an SCM-KG that merges scientific research metadata from the DBLP bibliographic source and the Microsoft Academic Graph. The observed results provide evidence that queries are processed more effectively on top of the SCM-KG than over the isolated datasets, while execution time is not negatively affected.

1 Introduction

Yearly thousands of research articles are published in journals and conference proceedings around the world. To conduct research and take advantage of the latest knowledge in an area, it is imperative for researchers to follow the work of other scientists. Therefore, metadata describing articles, authors, journals, calls and conferences can enable effective and efficient research communication. A data source can be rich in one aspect and insubstantial in other aspects. For example, the *DBLP* computer science bibliography database gathers ample information about publications in specific conferences but has sparse data about their keywords and no data about citations. Furthermore it lacks metadata on

© Springer International Publishing AG 2017
J. Kamps et al. (Eds.): TPDL 2017, LNCS 10450, pp. 328–341, 2017.
DOI: 10.1007/978-3-319-67008-9_26

publications in different fields of research. The *Microsoft Academic Graph* fills these gaps but is less complete in every scientific field. We claim that collecting research communication metadata from heterogeneous sources and integrating them in a queryable environment not only leads to a more robust knowledge base but also, thanks to increased completeness, enables more effective data analysis.

From the 2012 blog post in which Google used the term 'Knowledge Graph' for the first time [11], knowledge graphs have been an important subject of research, but still there does not exist a single, widely accepted definition of this term. Many authors refer to 'knowledge graphs' as a structured base of human knowledge in the form of a graph, with an emphasis on comprehensiveness and large scale [4,9]. Examples of famous knowledge graphs include *DBpedia*, *YAGO*, and *Freebase*.

In this work, we created an integrated graph of scientific knowledge from DBLP and the Microsoft Academic Graph and describe the challenges in matching, linking and integrating the datasets and our approach to addressing these challenges as a methodology that can be reused to build similar knowledge graphs. We present the application of semantic structure based similarity measures in instance matching and show that traditional linking frameworks such as *Silk* are capable of linking with high relative precision and recall, when they consider data semantics during the linking process.

The remainder of this paper is as follows: Sect. 2 describes DBLP and the Microsoft Academic Graph, and motivates the need for knowledge graph integration with concrete examples. Section 3 defines our concept of a knowledge graph for scholarly communication metadata (SCM-KG). Section 4 shows how the integrated knowledge graph is built. Section 5 reviews related work, and Sect. 6 reports on the evaluation of our approach. Finally, Sect. 7 concludes and provides an outlook to future work.

2   Motivating Example

In this example, we target the problem of data accuracy in DBLP and the Microsoft Academic Graph and show how creating a high-quality integrated knowledge graph from these heterogeneous sources helps to solve ambiguity problems.

DBLP[1] is an up-to-date dataset of publications, authors and conferences in the area of computer science. Information about an article includes the title and the year of publication; information about authors includes their most recent affiliation. DBLP rarely includes keywords of its publications and misses valuable information such as abstracts and information on the citation of articles. DBLP can be browsed online and is available for download as an XML dump; third parties also provide RDF dumps.

The **Microsoft Academic Graph** (henceforth called "MAG")[2] covers publications, authors, and conferences in all scientific areas. It is neither updated as

---

[1] http://dblp.l3s.de/dblp++.php, accessed on 10 April 2017.
[2] https://academicgraphwe.blob.core.windows.net/graph-2016-02-05 accessed on 10 April 2017.

regularly nor as complete as DBLP in the computer science area, but it includes abstracts, keywords, and citation relations. Further, for each publication, it covers the author affiliations at the time of publication. MAG is available as a relational database dump in CSV format.

In the latest DBLP version of April 2017, there are four authors named "Christoph Lange", indexed 0001 to 0004. When one of these four persons publishes a new article, the maintainers of DBLP face the challenge of linking the article to the right person using his affiliation but DBLP keeps only the current affiliation. By matching authors' publications and recent affiliations, we can link DBLP authors to MAG authors. Now, an old, unindexed publication by a researcher named "Christoph Lange" can be matched against the author and affiliation information in the unified knowledge graph and linked to the correct person entity – at least when no two different persons published at the same institution at different times. This example shows how combining multiple available data sources can solve an ambiguity problem.

3   SCM Knowledge Graph Concept

In this section, we first define basic principles of knowledge graphs and then our notion of a scholarly communication metadata knowledge graph (SCM-KG).

Identification. A key prerequisite for a knowledge graph is to uniquely identify things. All entities of interest should be uniquely identified by Universal/International Resource Identifiers (URI/IRI).

Representation. We need to ensure that information about these things can be easily understood by different parties. The *W3C Resource Description Framework* (RDF) has meanwhile evolved into the *lingua franca* of data integration.

Integration. For data exchange in a digitized domain to scale, organizations and involved people need to develop a common understanding of the data. Vocabularies define common concepts (classes) and their attributes (properties) and assign unique identifiers to them.

Coherence. Scholarly meta-data frameworks use a large number of data models and data exchange and serialization techniques including relational databases, XML, and JSON. Meanwhile transformation techniques for the RDF data model have been standardized by the W3C.

Access. Depending on the usage scenario, there are different requirements and possibilities for data access, such as push vs. pull or individual vs. bulk access. To support these scenarios, knowledge graphs should provide various methods to access data.

Coverage. Knowledge graphs should cover a sizeable, extensible area of knowledge stretching across several domains. Even though the field of scholarly publications is well defined with high-quality reference datasets, their incompleteness justifies the need for an integrated knowledge graph.

Knowledge Graph. Based on the principles introduced previously, a knowledge graph is a fabric of concept, class, property, relationships, and entity descriptions. It uses a knowledge representation formalism, typically RDF, RDF Schema, or OWL. It aims at a holistic representation of knowledge covering multiple sources, multiple domains, and different granularity. It can be *open* (e.g., DBpedia), *private* or *closed*. It includes schema data as well as instance data. Publishing our knowledge graph as LOD allows clients to easily consume it directly or by performing queries over an SPARQL endpoint. Additionally, it can be integrated with other data quite easily. Third parties who want to perform further integration would not have to install our pipeline but could also follow alternative approaches.

Applying these principles to the domain of scholarly communication requires:

- **Identification** is provided by a scholarly schema such as ORCID for authors, DOI for articles and books, or ISBN for books.
- Besides RDF-based representations, the XML schema of DBLP serves as a well-known **representation**.
- Common RDF-based vocabularies for knowledge **integration** include those from the SPAR family of ontologies[3].
- Regarding **coherence**, it is necessary to map data from a variety of sources, e.g., DBLP from XML and MAG from CSV.

There is not currently an integrated knowledge graph that satisfies all criteria of the definition given above, but besides DBLP and MAG and non-free data sources such as those of Google Scholar or ResearchGate, there are other open datasets, and their schemas could serve as sources for a more comprehensive integration. For example, *Scholarly Data* is a well-engineered RDF dataset on papers of Semantic Web conferences[4] and *OpenCitations*[5] is an open repository of scholarly citation data.

4 Building a Knowledge Graph

In this section, we step by step explore our general approach to build high quality knowledge graphs. We use the scientific communication domain as an example, although the methodology is domain-independent. Figure 1 shows the architecture of the overall system, called SCM-KG-PIP (SCM-KG creation Pipeline).

As input of SCM-KG-PIP, heterogeneous data arrives in different formats, such as CSV, RDF, web pages, or data returned by calling Web APIs. Our approach results in a high-quality, queryable semantic knowledge graph, using a unified schema.

The following subsections present the components of the SCM-KG-PIP architecture in detail and describe how we applied them to scholarly communication

---

[3] http://www.sparontologies.net/.
[4] http://www.scholarlydata.org.
[5] http://opencitations.net/.

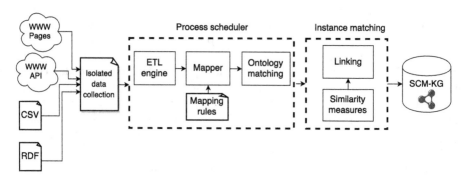

Fig. 1. The pipeline to create a knowledge graph from heterogeneous resources.

metadata to build a knowledge graph for that domain. The pipeline steps in order of execution are: (1) data acquisition, (2) ontology engineering, (3) mapping data to the ontology, (4) calculating similarity and instance matching, and (5) producing the KG and querying it. Steps (1)–(2) are carried out manually, steps (3)–(5) are executed automatically.

4.1 Data Acquisition

Data available in heterogeneous sources can be obtained in different ways. When they are available as structured dumps, e.g., as CSV, SQL or RDF, their structure may not match the target ontology. For example, the DBLP and OpenAIRE[6] datasets are available as RDF, and MAG is available as CSV. Data from Web APIs, another source of structured data, can be collected by gradual querying. Usually, the number of API calls in a specific time window is limited; therefore, throttling has to be applied to requests.

When structured data is not provided through open interfaces, one may be forced to resort to scraping data from web pages. Currently, Google Scholar and ResearchGate, two highly relevant sources of data about authors' current affiliations and recent publications, do not provide ways to access metadata other than by web scraping. Web scraping requires finding relevant pages, parsing them, and extracting the desired metadata from their content. In the concrete case of ResearchGate, we experimented with such a parser for author and publication metadata, implemented using the *Scrapy* Python framework[7], but found it hard to maintain, as, after just half a year, the content structure of the ResearchGate pages had changed significantly.

4.2 Ontology Engineering

Different structured data sources may use different schemas, e.g., DBLP and MAG model the same concepts (e.g., affiliation) differently. Creating an

---

[6] http://lod.openaire.eu.
[7] https://scrapy.org, accessed on 5 April 2017.

integrated knowledge graph requires a mapping step to accommodate these differences, e.g., that can model both an author's current affiliation and earlier ones.

In the SCM-KG pipeline, we reused subsets of existing vocabularies including the *SWRC ontology, Dublin Core* and *FOAF*[8] to create a core vocabulary. We created and matched classes for resources, i.e., nodes, in the source datasets to the core vocabulary modeled initially and instantiated it with one of the data sources.

When the initial vocabulary is missing a definition for a concept from a joining data source we created a new class for it. Thereupon, we linked this new concept to the existing classes by defining a new relation type in our ontology. As a concrete example, DBLP is missing a notion of fields of study but MAG has a distinct index of fields of study and also article keywords, and relates keywords of articles to fields of study. We related the articles integrated from DBLP in the SCM-KG to fields of study with a new RDF property given that we knew their relation to fields of study by integrating MAG.

Challenges in integrating occur with structured datasets whose schemas model the same concept in a way different from the ontology of the knowledge graph existing so far. Nguyen [7] has classified these challenges. As Nguyen describes, a conflict on the concept level occurs when classes with same name have different structures in two merged ontologies. We encountered this issue when mapping the affiliation property. We addressed it by keeping the more descriptive vocabulary in our ontology model and pruning the other, conflicting vocabulary from the model. The notion of an author's affiliation has a temporal dimension that *swrc:affiliation* used by DBLP does not cover, as it merely models the *current* affiliation, not the affiliation at the time a certain article was published. We simplified a temporal modeling approach proposed by Nuzzolese et al. [8] by following the reification pattern of MAG's *paperAuthorAffiliations* table, i.e., turning each ternary relation of a publication, its authors and their affiliations at the time of publication into a resource. A conflict on the instance level occurs when descriptions of identical instances in different ontologies are different. To resolve it we either could choose only one instance by fact checking their materialized instances against the real word or if possible extend the class of the instance such that it holds both conflicting descriptions for later check. For example, publication dates of some articles are different in MAG and DBLP and we had to find the correct year manually, e.g., via the homepages of their authors.

4.3   Mapping Data to an Ontology

Data acquired from different sources can follow a variety of data models (e.g., graph, relational, tree) or even be unstructured. Thus, having acquired the data, and having modeled a common integration ontology, the next step of constructing a knowledge graph is to convert all data into a common model. RDF is well suited as a target data model for integration and thanks to the wide availability

---

[8] SWRC: http://ontoware.org/swrc, FOAF: http://xmlns.com/foaf/spec/.

of mapping languages and tools for it, mapping data from different sources to RDF is practically feasible (cf. Sect. 3).

In our concrete situation, an RDF version of DBLP is already available and the CSV sources of MAG can be mapped to RDF. We developed a process scheduler with a command line interface to execute this step of the pipeline in general. For CSV sources such as MAG, the Sparqlify-CSV tool [5] maps the source ontology to the integration ontology. To use Sparqlify-CSV we expressed mapping rules in its intuitive Sparqlification Mapping Language [14].

In some cases direct mapping of CSV files is not possible. Therefore we implemented an ETL component to shape the data in the format required for the mapping by applying string manipulations. Using a process scheduler, we stream results of the ETL component into Sparqlify-CSV. To improve the performance of mapping, we run multiple parallel instances of the process scheduler. Each row in a CSV file and each set of triples that the Sparqlify-CSV mapping engine creates from it is semantically independent from the other rows. Based on this understanding, the scheduler executes the conversion in parallel processes. After breaking the big input files, e.g., from a size of 9 GB into 20 KB in-memory-processable chunks it creates queues that convert and map the data chunks in parallel and finally merges the respective mapping results. Section 6.3 presents a performance evaluation of this module.

4.4   Calculating Similarity and Instance Matching

In Sect. 4.2 we addressed how we mapped semi relational data to a common ontology but did not cover the level of mapping *instances* where multiple instances refer to the same real world thing. We therefore added a data linking step to our pipeline. First of all, we keep data integrated from different sources in separate URI namespaces to avoid clashes in case different sources use same identifiers. We then created "same as" links between different URIs referring to the same thing by instance matching. *Articles* can be matched by common title, publication year and, if provided, the name of the conference or journal. To increase linking coverage, we considered the incidence of variations of title strings in punctuation and letter cases that occurs in different datasets, and compared them using the Jaccard similarity measure. We implemented these conversions and comparisons and the linking of the articles using the Silk workbench [18]. A high-quality instance-level linking of *persons* is a challenge for the Silk Workbench. A mere triple based matching, as applied in Silk, fails to distinguish different persons with similar or even same names.

We tackled this problem using the semantic *relations* of the persons with their articles. In our data sources, persons only occur in the role of authors of publications; additionally, we can rely on links between papers as identified in the previous step. We leverage this semantics by embedding it into the author molecules[9]. First we create a hash for each article. Provided that instance matching of articles is performed in the last step and they are stored in the SCM-KG,

---

[9] Here, a "molecule" refers to a set of one node in the knowledge graph and the immediate links to its neighbors.

Fig. 2. Matching instances of an author in two datasets

we find those articles of a person that have been matched to an article in other datasets. We concatenate the IDs of these articles to a comma-separated string. We then associate this string immediately to the author via a new property of type *authorLinkedPaperIds*. We store these new links in the SCM-KG to use them in Silk subsequently.

By applying a substring similarity metric defined by Stoilos et al. [15] on the concatenated list of unique IDs of articles, we can discover if two instances of Person have common publications. The more common publications, the higher the value of this metric. Figure 2 depicts an example of this step of instance matching in action.

4.5   Producing and Querying a KG

Our objective in the final pipeline step is to store all the data in a form that is accessible via SPARQL queries. We employed the high-performance Apache Jena TDB as our RDF store. After importing our data into TDB we configured Apache Jena Fuseki 2 to make the data queryable using SPARQL 1.1, both from the command line and, via HTTP, from a SPARQL endpoint. The latter SPARQL endpoint enables the integration of Silk in the linking step and the resulting links are added to the KG in the end. Fuseki also supported the evaluation (cf. Sect. 6) by enabling us to query the dataset conveniently via a web frontend. To further improve performance, we employed the Cassandra big data database to cache query results of Fuseki. However, we did not consider Cassandra in the evaluation of the query execution time to have a fair comparison.

5   Related Work

Recent approaches toward constructing knowledge graphs, e.g., NOUS [2], Knowledge Vault [3] or NELL [1] focus on materializing a knowledge graph by inferring relations in the existing data. In comparison, our focus was to integrate data from heterogeneous resources and to increase the quality of the integrated knowledge graph. For that we evaluated the steps of the knowledge graph construction pipeline and optimized our pipeline based on that.

In a similar work, Szekely et al. [16] created a knowledge graph of human trafficking data; text and images from the Web were parsed and unstructured data was mapped to a vocabulary. In contrast, we resolved the challenge of structure variations of the data being integrated. As explained in Subsect. 4.3, we mapped semi-structured metadata into triples using Sparqlify-CSV. This step distinguishes our pipeline from the research of Szekely et al. They integrated data by building up a new ontology model while we modified the existing ontology model of the manually maintained DBLP and aggregated other vocabularies to it. The vocabulary used in DBLP has already a combination of the common vocabularies in describing the scientific metadata. Therefore, we accumulated other terms and vocabularies or modified the current model when the vocabulary of DBLP was not sufficiently describing the integrating data.

Another difference of the two works is the ETL component. From a technical perspective, Szekely et al. used the Karma framework [6] for data mapping. Their approach is limited as they apply ETL the Karma component used for mapping. ETL rules in Karma are in Python, while we implemented an efficient ETL component in C++. Furthermore, Szekely et al. enhanced their linking with image similarly measures, whereas we used semantics of the incoming data to increase the quality of instance matching.

Traverso et al. [17] suggested applying semantics in relation discovery in existing knowledge graphs. Similarly, we apply the concept of semantic molecular similarity, but we use the semantic relations in the network toward the linking of instances during the creation of a knowledge graph.

In a recent research, Danh Le-Phuoc et al. [10] integrated data from variety of resources including sensors, the Twitter social network and RSS resources of famous news websites to create a knowledge graph of things. Their pipeline similarly needs to process a holistic amount of data in batch and makes them queryable via a SPARQL endpoint. In contrast to our work, they process streaming data coming from resources that are much more loosely coupled in comparison to the resources in our pipeline.

In our experiment one of the data sources is in CSV format, i.e., semi-structured relational data. Many approaches have been investigated to map relational data to RDF, e.g., heuristic and rule-based methods, graph analysis, probabilistic approaches, reasoning, machine learning, etc. We chose a manual rule-based mapping method. This allows for vocabulary reuse but requires users to be familiar with popular Semantic Web vocabularies to choose the most suitable terms [13].

6   Evaluation and Results

We conducted an empirical evaluation to study the effectiveness of the proposed pipeline in creating a knowledge graph from different data sources in the domain of scholarly communication metadata (SCM-KG). We assessed the following research questions:

RQ1) Can relative answer completeness be enhanced when queries are executed against an SCM-KG instead of the original sources? Is the query execution time affected when queries are executed against an SCM-KG? **RQ2)** How accurate is the linking of the integrated dataset in terms of precision and relative recall? **RQ3)** How much data can be processed per second in the mapping and linking steps of the pipeline?

Datasets: For the evaluation, we chose a subset of authors and their papers from both DBLP and MAG [12]. This subsection involves all the metadata relevant to the WWW conference series in both datasets[10]. WWW has a long history, and this fraction of data covers all the vocabulary and structure used in the whole dataset. MAG was last updated on 5 February 2016, and we acquired the DBLP dataset on 10 November 2016 from the DBLP++ website[11]. We chose Apache Jena Fuseki as our triple store.

We executed each query 15 times, each time instantiated with a different author. We selected these 15 authors among the most publishing authors in WWW as found by another SPARQL query over the SCM-KG.

Queries: In the next two experiments, we defined queries and compared their results over the integrated knowledge graph with their evaluation on the isolated source datasets.

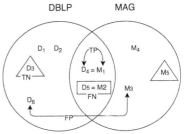

Fig. 3. Articles belonging to an author in DBLP and MAG. Arrows represent the matched instances.

Metrics: We evaluated how much the integration enhanced the accuracy and completeness of the query results. Some authors do not have a Google Scholar profile or any other "complete" publication list available, therefore the dataset completeness is calculated in a relative way. In the second experiment, we tested the quality of the linking in terms of relative precision and recall.[12] The D_4–M_1 connection in Fig. 3 is an example of a true positive link. When the equivalence of items is not discovered we consider that a false negative (FN). For example, the lack of a D_5–M_2 connection is a FN. When two articles are linked that are

---

[10] The integrated WWW dataset has 346,480 triples including the "same as" links between matched instances.
[11] http://dblp.l3s.de/d2r/sparql.
[12] In the process of linking articles by an author, true positives (TP) are articles whose metadata exist in both DBLP and MAG and their instances are correctly linked in the matching step.

not really equivalent we assume it as a false positive, such as the arrow connecting D_6 and M_3. When the instance matching step correctly does not relate two different articles, we consider this a true negative, depicted as a triangle in the diagram.

We also evaluated the data integration process by comparing the execution times of the queries provided above over the different datasets.

Implementation: Experiments 1 and 2 were run on a test platform with an Intel i7-4710HQ 2.5 GHz CPU and 16 GB 1333 MHz DDR3 RAM; the operating system was Mac OS 10.12. The test queries were executed on Jena Fuseki. In Experiment 3, we used a machine with 32 GB RAM and an Intel(R) Xeon(R) 3.00 GHz CPU with 16 cores; the operating system was openSUSE Linux. We implemented the process scheduler in C++ with a shell script frontend. SPARQL queries were executed to create triples for the semantic based similarity measurement. The process manager, Sparqlify-CSV mapping rules, ETL source code, and the test datasets evaluated are publicly available.[13]

6.1   Experiment One: Relative Completeness

Publications and the number of hits in the different datasets were collected. Queries were executed for each of the 15 selected authors over the three datasets and compared them in terms of relative completeness of the result sets. Comparing the number of WWW publications in MAG, DBLP, and SCM-KG, we observed that although DBLP contains more articles for the selected authors, there exist articles that are only included in MAG. The mapping and linking process allows for identifying common articles in both datasets; thus, the resulting dataset includes more articles for these authors.

Query response time for WWW publications in MAG, DBLP, and SCM-KG indicated that these queries had an average response time of 8.8 ms on DBLP, while equivalent queries on MAG had an average response time of 11.66 ms, and 12.8 ms was the average response time of their equivalent on the integrated graph. These values suggest that the integration did not affect query response time significantly.

6.2   Experiment Two: Linking Accuracy and Relative Coverage

In this survey we ran a SPARQL query over MAG and SCM-KG and evaluated how much the process of linking affected the integrity of the author entities in MAG.

We first defined a query that finds an author entity and his/her articles. It searches instances of authors by name. We observed that for cases like Ravi Kumar the query yields several different author entities instead of one. Likewise, his/her published articles were scattered between different author entities in MAG.

---

[13] http://afshn.com/re/scmkg.html, accessed on 5 April 2017.

By running the same query over SCM-KG, we observed that the instant matching of author entities in MAG and DBLP had brought these pieces of information together. To survey the indirect merging of authors in MAG, we considered the scattering of an author's articles into each extra instance of an author as a false negative, i.e., author instances in MAG that were equal but not found by the linking process; true positives correspond to merged instances of authors.

This query was executed for 15 selected authors. The comparison of indirect integrated duplicate author entries in MAG, due to instance matching between MAG and DBLP, indicates a correct linking (TP) with a precision of 1 in all cases, and an average recall value of 0.986. Secondly, we tested if, per author, the linked articles belonging to each author are linked to correct equivalent items between datasets. The linking performed in this experiment had a precision of 1 and an average recall of 0.982; these results show the positive effect of using semantic molecular relations in linking.

6.3   Performance Evaluation of the Mapping Process Scheduler and Linking

In the mapping step, the process scheduler generated 10 parallel processes that occupied approx. 99.5 percent of the available 16 CPU cores and 3.6 GB RAM. By the SCM-KG pipeline, we converted 96.88 GB of MAG and generated approx. 2.9 B triples from MAG and integrated them with 150 M triples from DBLP. The process scheduler could generate approx. 250,000 triples per second, that thanks to parallelization, is significantly faster than the original Sparqlify RDB2RDF transformation engine [5]. The instance matching process could find approximately 500 matches per second when tested on the Mac OS system mentioned in the introduction of Sect. 6.

7   Conclusions and Future Work

In this paper, we presented the concept of Scholarly Communication Metadata Knowledge Graph (SCM-KG), which integrates heterogeneous, distributed schemas, data and metadata from a variety of scholarly communication data sources. As a proof-of-concept, we developed an SCM-KG pipeline to create a knowledge graph by integrating data collected from heterogeneous data sources. We showed the capability of parallelization in rule-based data mappings, and we also presented how semantic similarity measures are applied to determine the relatedness of concepts in two resources in terms of the relatedness of their RDF interlinking structure. Results of the empirical evaluation suggest that the integration approach pursued by the SCM-KG pipeline is able to effectively integrate pieces of information spread across different data sources. The experiments suggest that the rule based mapping together with semantic structure based instance matching technique implemented in the SCM-KG pipeline integrates data in a knowledge graph with high accuracy. Although our initial use case addresses the scientific metadata domain, we generated billions of triples with high accuracy in mapping and linking, and we regard it capable at an industrial scale and in use cases demanding high precision. In the context of the OSCOSS project on Opening Scholarly Communication in the Social Sciences[14], the SCM-KG approach will be used for providing authors with precise and complete lists of references during the article writing process.

Acknowledgments. This work has been partially funded by the European Commission under grant agreements 643410 (OpenAIRE2020) and 644564 (BigDataEurope), and the DFG under grant agreement AU 340/9-1 (OSCOSS).

References

1. Carlson, A., Betteridge, J., Kisiel, B., Settles, B., Hruschka Jr., E.R., Mitchell, T.M.: Toward an architecture for never-ending language learning. In: Proceedings of the 24th AAAI (2010)
2. Choudhury, S., Agarwal, K., Purohit, S., Zhang, B., Pirrung, M., Smith, W., Thomas, M.: NOUS: construction and querying of dynamic knowledge graphs. In: ICDE (2017)
3. Dong, X., Gabrilovich, E., Heitz, G., Horn, W., Lao, N., Murphy, K., Strohmann, T., Sun, S., Zhang, W.: Knowledge vault: a web-scale approach to probabilistic knowledge fusion. In: SIGKDD (2014)
4. Ehrlinger, L., Wöß, W.: Towards a definition of knowledge graphs. In: SEMANTiCS (2016)
5. Ermilov, I., Auer, S., Stadler, C.: User-driven semantic mapping of tabular data. In: 9th International Conference on Semantic Systems, ISEM, pp. 105–112 (2013)
6. Knoblock, C.A., Szekely, P., Ambite, J.L., Goel, A., Gupta, S., Lerman, K., Muslea, M., Taheriyan, M., Mallick, P.: Semi-automatically mapping structured sources into the semantic web. In: Simperl, E., Cimiano, P., Polleres, A., Corcho, O., Presutti, V. (eds.) ESWC 2012. LNCS, vol. 7295, pp. 375–390. Springer, Heidelberg (2012). doi:10.1007/978-3-642-30284-8_32
7. Nguyen, N.T.: A method for ontology conflict resolution and integration on relation level. Cybern. Syst. **38**(8), 781–797 (2007)
8. Nuzzolese, A.G., Gentile, A.L., Presutti, V., Gangemi, A.: Semantic Web Conference Ontology - A Refactoring Solution. In: Sack, H., Rizzo, G., Steinmetz, N., Mladenić, D., Auer, S., Lange, C. (eds.) ESWC 2016. LNCS, vol. 9989, pp. 84–87. Springer, Cham (2016). doi:10.1007/978-3-319-47602-5_18
9. Paulheim, H.: Knowledge graph refinement: A survey of approaches and evaluation methods. Semantic Web **8**(3), 489–508 (2017)
10. Phuoc, D.L., Quoc, H.N.M., Quoc, H.N., Nhat, T.T., Hauswirth, M.: The graph of things: A step towards the live knowledge graph of connected things. J. Web Sem., 37–38 (2016)
11. Singal, A.: Introducing the knowledge graph: Things, not strings (2012)
12. Sinha, A., Shen, Z., Song, Y., Ma, H., Eide, D., Hsu, B.P., Wang, K.: An overview of microsoft academic service (MAS) and applications. In: WWW Companion (2015)
13. Spanos, D., Stavrou, P., Mitrou, N.: Bringing relational databases into the semantic web: A Survey. Semantic Web **3**(2), 169–209 (2012)

---

[14] http://afshn.com/re/oscoss.html.

14. Stadler, C., Unbehauen, J., Westphal, P., Sherif, M.A., Lehmann, J.: Simplified RDB2RDF mapping. In: Proceedings of the Workshop on Linked Data on the Web, LDOW 2015 (2015)
15. Stoilos, G., Stamou, G., Kollias, S.: A string metric for ontology alignment. In: Gil, Y., Motta, E., Benjamins, V.R., Musen, M.A. (eds.) ISWC 2005. LNCS, vol. 3729, pp. 624–637. Springer, Heidelberg (2005). doi:10.1007/11574620_45
16. Szekely, P., et al.: Building and using a knowledge graph to combat human trafficking. In: Arenas, M., Corcho, O., Simperl, E., Strohmaier, M., d'Aquin, M., Srinivas, K., Groth, P., Dumontier, M., Heflin, J., Thirunarayan, K., Staab, S. (eds.) ISWC 2015. LNCS, vol. 9367, pp. 205–221. Springer, Cham (2015). doi:10.1007/978-3-319-25010-6_12
17. Traverso-Ribón, I., Palma, G., Flores, A., Vidal, M.-E.: Considering semantics on the discovery of relations in knowledge graphs. In: Blomqvist, E., Ciancarini, P., Poggi, F., Vitali, F. (eds.) EKAW 2016. LNCS (LNAI), vol. 10024, pp. 666–680. Springer, Cham (2016). doi:10.1007/978-3-319-49004-5_43
18. Volz, J., Bizer, C., Gaedke, M., Kobilarov, G.: Silk - a link discovery framework for the web of data. In: Proceedings of the 2nd Linked Data on the Web Workshop, pp. 1–6 (2009)

Analysing Scholarly Communication Metadata of Computer Science Events

Said Fathalla[1,3]([✉]), Sahar Vahdati[1], Christoph Lange[1,2], and Sören Auer[4,5]

[1] Smart Data Analytics (SDA), University of Bonn, Bonn, Germany
{fathalla,vahdati,langec}@cs.uni-bonn.de
[2] Fraunhofer IAIS, Sankt Augustin, Germany
[3] Faculty of Science, University of Alexandria, Alexandria, Egypt
[4] Computer Science, Leibniz University of Hannover, Hannover, Germany
[5] TIB Leibniz Information Center for Science and Technology, Hannover, Germany
soeren.auer@tib.eu

Abstract. Over the past 30 years we have observed the impact of the ubiquitous availability of the Internet, email, and web-based services on scholarly communication. The preparation of manuscripts as well as the organisation of conferences, from submission to peer review to publication, have become considerably easier and efficient. A key question now is what were the measurable effects on scholarly communication in computer science? Of particular interest are the following questions: Did the number of submissions to conferences increase? How did the selection processes change? Is there a proliferation of publications? We shed light on some of these questions by analysing comprehensive scholarly communication metadata from a large number of computer science conferences of the last 30 years. Our transferable analysis methodology is based on descriptive statistics analysis as well as exploratory data analysis and uses crowd-sourced, semantically represented scholarly communication metadata from OpenResearch.org.

Keywords: Scientific events · Scholarly communication · Semantic publishing · Metadata analysis

1 Introduction

The mega-trend of digitisation affects all areas of society, including business and science. Digitisation is accelerated by ubiquitous access to the Internet, the global, distributed information network. Data exchange and services are becoming increasingly interconnected, semantics-aware and personalised. Further trends are crowd-sourcing and collaboration, open data as well as big data analytics. These developments have profound effects on scholarly communication in all areas of science. We particularly focus on computer science, where conferences and workshops are of paramount importance and a major means of scholarly communication. Online platforms and services such as *EasyChair*[1] or

---

[1] http://easychair.org.

© Springer International Publishing AG 2017
J. Kamps et al. (Eds.): TPDL 2017, LNCS 10450, pp. 342–354, 2017.
DOI: 10.1007/978-3-319-67008-9_27

CEUR-WS.org[2] automate and optimise scholarly communication workflows. A key question now is: What were the measurable effects of digitisation on scholarly communication in computer science? Of particular interest are the following questions: (*a*) Did the number of submissions increase? (*b*) Is there a proliferation of publications? (*c*) Can we observe popularity drifts? (*d*) Which events are more geographically diverse than others? We shed light on some of these questions by analysing comprehensive scholarly communication metadata from computer science conferences of the last 30 years. Large collections of such data are nowadays publicly available on the Web. Research has recently been conducted to browse and query such data [6,7], with a focus on authors, publications and research topics [4].

We analysed the evolution of key characteristics of scientific events over time, including frequency, geographic distribution, and submission and acceptance numbers. We analysed 40 conference series in computer science with regard to these indicators over a period of 30 years. Our analysis methodology is based on descriptive statistics analysis, exploratory data analysis and confirmatory data analysis. This article is organised as follows: Sect. 2 gives an overview on related work. Section 3 presents the methodology we used. Section 5 discusses the results of our evaluation. Section 6 concludes and outlines future work.

2   Related Work

Conference metadata and bibliography services. A lot of research has been performed to reveal information about scholarly communication from bibliographic metadata. *DBLP* and *DBWorld*[3], the most widely known bibliographic databases in computer science, provide information mainly about publications and events but also consider related entities such as authors, editors, conference proceedings and journals. *WikiCFP*[4] is a popular service for publishing calls for papers (CfPs). *Springer LOD* and *ScholarlyData*[5] publish as Linked Open Data metadata of conference related to computer science collected from Springer's traditional publishing process.

Conference series analysis. For various conference series, analyses similar to ours were performed by steering committee members or other members of the community. They often include the analysis of bibliographic data of each edition and rarely comprise comparisons with other events or editions of the same event series. A comprehensive analysis of the Principles of Database Systems (PODS) conference series includes detailed author analyses such as the distribution of the number of papers per author, which, for example, shows that two thirds of the authors are only involved in a single PODS publication (e.g., PhD students) but 10% are involved in 5 or more (e.g. active supervisors) [1]. It includes a relatively

---

[2] http://ceur-ws.org.
[3] http://dblp.uni-trier.de/, https://research.cs.wisc.edu/dbworld/.
[4] http://www.wikicfp.com/.
[5] http://lod.springer.com/, http://www.scholarlydata.org/dumps/.

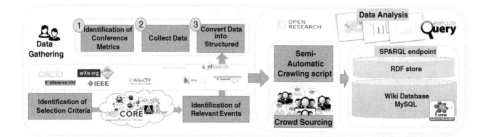

Fig. 1. Overall workflow of this study

short analysis of submission and acceptance rates for 10 years (2002–2011) that shows an increasing number of submissions in the beginning of the period, while they reduced in the last four years.

Literature Overview. Se and Lee proposed a list of alternative measures for ranking events [9]. The *goodness* of events (conferences and journals) is defined as the goodness of the articles published in these events. Biryukov and Dong addressed collaboration patterns among a research community using information of authors, publications and conferences [3]. Similarly, Aumüller and Rahm analysed affiliations of database publications using author information from DBLP [2]. A pilot study with a different focus analysed submissions to top technical conferences in computer science [5], while our analyses are about the quality of events considering different metrics than only metadata about publication and authors.

3   Method

The overall workflow of this study (see Fig. 1) comprises four steps: (1) identification of relevant events, (2) data gathering, (3) ingestion into the OpenResearch.org semantic scholarly communication data curation platform, and (4) data analysis.

Identification of Relevant Events. To identify a subset of high qualified events to which we can apply our evaluation, we collected all the metrics which are used by most of the well-known services. The analysis focuses only on conferences because of the high impact to the research community. However, all these metrics can be applied for more types of events. Depending on availability and re-usability of the metrics, the following set of criteria is finalized to be used in this study (see Table 1):

h-index Google Scholar Metrics (GSM)[6] provides ranked lists of conferences and journals by scientific field based on a 5-year impact analysis over the Google

---

[6] https://scholar.google.com/intl/en/scholar/metrics.html.

Table 1. Identification of relevant event criteria

Ranking metric	TPDL	WWW	PERCOM	COLT	EuroCrypt	CHI	CAV	PLDI
CORE rank	A*	A*	A*	A*	A*	A*	A*	A*
H5 index	74	66	28	22	50	83	39	45
Qualis	B1	A1	A2	A2	A1	A1	A1	A1

Scholar citation data. The ranking is based on the two metrics h5-index and h5-median. *Qualis*[7] uses the h-index as performance measure for conferences. Based on the h-index percentiles, the conferences are grouped into performance classes that range from A1 (best), A2, B1, ..., B5 (worst).

Mix of indicators. The *Computing Research and Education Association of Australasia* (CORE)[8] provides a ranking method for major conferences in computing. The ranking is determined by a mix of indicators including citation rates, paper submission, acceptance rates and the visibility and research track record of the key people hosting and managing the conference. Based on these metrics an event can be categorised into six classes A*, A, B, C, Australian, and unranked. The portal shows international event series in the first four categories.

Data Gathering. Data gathering is the process of collecting data from a variety of online sources in an objective and unbiased manner. We collected metadata about 40 conference series in different computer science sub-fields from different sources of metadata including title, series, sub-field, start date, end date, homepage, country and Twitter account. This information is available as Linked Data in the case of DBLP, and other structured forms, or semi-structured and unstructured in the case of WikiCFP, the ACM digital library[9], or conference.city[10]. The *OpenResearch.org* wiki[11] serves us both as an additional source of semantically structured data, and as a tool to support data analysis. At the time of writing, OpenResearch contains crowd-sourced metadata about more than 5000 conferences, 900 workshops and 350 event series. OpenResearch supports researchers in collecting, organising, sharing and disseminating information about scientific events, tools, projects, people and organisations in a structured way [8].

Data Preprocessing. In this step, we carried out several preprocessing tasks including:

Data Integration/Transformation: This step starts with identifying inadequate, incorrect, inaccurate or irrelevant data and then filling in missing data, deleting

---

[7] http://www.conferenceranks.com/.
[8] http://www.core.edu.au/.
[9] http://dl.acm.org/.
[10] http://www.conference.city/.
[11] http://openresearch.org.

the dirty data, and resolving inconsistencies. In the data integration process, we combine data from multiple sources into meaningful and valuable information. Transformation is the conversion of cleaned data values from unstructured formats into a structured format.

Conference Name Unification: Looking into the collected data we found that some events have changed their names once or more since they had been established. This led us to perform a unification process before beginning to analyse the data. The unification process integrates all events of a series with multiple names under its most recent name because it is important for the researchers

Table 2. Conference title and acronym evolution for Some Conferences.

Unified acronym	Acronym	Full conference title	Time span
IEEE VR	IEEE VR	IEEE Virtual Reality	1999–2017
	VRAIS	Virtual Reality Annual International Symposium	1993–1998
ASE	ASE	Automated Software Engineering	1997–2017
	KBSE	Knowledge-Based Software Engineering Conference	1990–1996
ISWC	ISWC	International Semantic Web Conference	2002–2017
	SWWS	Semantic Web Working Symposium	2001
FOCS	FOCS	Annual Symposium on Foundations of Computer Science	1975–2017
	SWAT	Annual Symposium on Switching and Automata Theory	1966–1974
	SWCT	Annual Symposium on Switching Circuit Theory and Logical Design	1960–1965
ISMAR	ISMAR	International Symposium on Mixed and Augmented Reality	2002–2017
	ISAR	International Symposium on Augmented Reality	2000–2001
	IWAR	International Workshop on Augmented Reality	1999
ISSAC	ISSAC	International Symposium on Symbolic and Algebraic Computation	1988–2017
	SYMSAC	Symposium on Symbolic and Algebraic Manipulation	1966,1971,1976,1981 and 1986
	EUROSAM	International Symposium on Symbolic and Algebraic Computation	1974, 1979, 1982 and 1984
SPLASH	SPLASH	Systems, Programming, Languages and Applications: Software for Humanity	2010–2017
	OOPSLA	Conference on Object-Oriented Programming, Systems, Languages, and Applications	1986–2009

who want to submit their work to know the recent name rather than the name that had been in use for the longest time, as shown in Table 2. For example, SPLASH is the unified name of a Conference on Object-Oriented Programming, Systems, Languages, and Applications which was named SPLASH from 2010 to 2017 and previously OOPSLA from 1986 to 2009, i.e., for 24 years.

Ingestion into OpenResearch. The collected data can be ingested into Open-Research.org in several ways using either single or bulk import. For single import, one should use semantic forms. The required steps for bulk import are: (a) Create a spreadsheet with the important information, (b) Export the spreadsheet to CSV, (c) import CSV file using OpenResearch's *ImportCSV* service.

4   Data Analysis

The heart of our work is an explorative analysis of the metadata of selected computer science conferences over the past 30 years.

Metrics and Analysis Tools. We first defined metrics, then chose suitable tools for computing them and evaluating the results of the computation.

We defined statistical metrics over numeric values, as well as metrics having other complex datatypes, focusing on conferences because of their high impact on research communities. We chose spreadsheets as the main tool to compute statistical metrics over numeric values; the evaluation of the results is supported by charts. OpenResearch provides further components for visual analytics, in particular for displaying non-numeric results (e.g., the conferences with the highest number of submissions). Even though spreadsheets are, in principle, based on the relational data model, they practically lack support for joins across sheets. Joins may be required for connecting information about events to information about related entities, such as persons participating in events. The SPARQL query language for RDF, which is supported by OpenResearch, facilitates such join computations. However, while SPARQL also supports basic statistical analysis via aggregate functions, this type of analysis is better supported by spreadsheets.

Statistical Analysis. Acceptance rate is defined as the ratio between submitted and accepted articles.

Continuity refers to how continuously a conference has been held over its history. We propose a formula $C = \min\{100\%, (E * R)/A\}$ to calculate the percentage of continuity for a specific conference where C stands for continuity, E for the number of editions of the event, R for the regularity of the event editions (1 for 'every year', 2 for 'every two years'), and A for the age, counting the number of years since the first time the event was established. Year is the granularity for this metric.

Geographical Distribution: Every event is held in a geographical *Location*. We consider it as a triple of City, Country, Continent. From the extension of this metric to event series, one can derive the number of distinct locations visited by an event. We map every distinct location to the number of times the event has taken place there (by city, country or continent). We can thus classify event series by their most frequent location, e.g., as a "German" or "European" series. Geographical Distribution of an event series increases the awareness of researchers about the existence of the event and its covered topics.

Time Distribution: Every event is held in a certain period of time each year. It is important for a researcher interested in a particular conference to know when this conference will be held in the year to know when to prepared and present their work.

Sub-field Popularity: In the sub-field popularity metric, we divided conferences into five groups, each of which is labelled with the sub-field of computer science they belong to. We considered two time intervals: three 10-years periods for accepted papers but three 5-years periods for submitted papers due to the difficulty to obtain information about the number of submitted papers for many conferences. Table 3 shows research communities and corresponding conferences investigated.

Table 3. CS Sub-fields and top conferences

Acronym	CS sub-field	Conferences
GRA	Computer Graphics	ACMMM, EuroGraphics, IEEE VR, SIGGRAPH
SEC	Computer Security	CCS, CRYPTO, EuroCRYPT, ASIACRYPT
PROG	ProgrammingLanguages	ICFP, PLDI, POPL, SPLASH
SE	Software Engineering	ICSE, FSE, ASE, FASE
DB	Database Systems	PODS, SIGMOD, ICDT, VLDB

Field Productivity: Field Productivity reveals how much interest there is in a computer science sub-field in a given year within the past 30 years. The Field Productivity (FP) for a sub-field (f) in a year (y), where $C_{i,y}^f$ is the number of publications for a conference i in year y and n is the number of conferences belonging to sub-field f, and m is the number of years in the time span of the study, is defined as $FP_y^f = \dfrac{\sum\limits_{i=1}^{n} C_{i,y}^f}{\sum\limits_{k=1}^{m}\sum\limits_{i=1}^{n} C_{i,y_k}^f}$.

Entity-Centric Visual Analytics. In contrast to spreadsheets and their charting facilities, OpenResearch makes it easy to generate visualisations that focus on entities rather than numbers. Besides geographical maps and ranked tables or lists, timelines are a prominent example of entity-centric visualisations. The input for a timeline is provided by a query in the MediaWiki expression language. The following code, for example, defines a timeline of events with upcoming submission deadlines:

```
{{#ask: [[Category:Event]]
        [[submission deadline::>{{CURRENTYEAR}}...]]
        [[Category:{{#urlget:field}}]]
 | ?title = Name          | ?abstract deadline
 | ?submission deadline | ?notification
 | ?Category:Conference = Conference
 | ?Category:Workshop   = Workshop
 | format=timeline       | sort=submission deadline}}
```

Similar types of queries that we have implemented in OpenResearch include:
– event series in a given field and their average acceptance rates, – countries
with a high number of events in a given field, – fields with decreasing numbers of accepted papers over years, In addition to querying the data inside the
OpenResearch wiki, queries to external SPARQL endpoints can be embedded
into wiki pages using the *LinkedWiki*[12] extension for MediaWiki.

Joins Across Entity Types. OpenResearch currently focuses on semantic
representation of CfPs as one wiki page per event, but including semantic relations to related entities, e.g., to document the role that a person had in the
organisation of an event. A concrete use case for querying this data is supporting the research community in taking decisions on what conference to submit
one's results to, or whether to accept invitations for assuming certain roles in
the organisation of a certain conference. Such queries often require joins across
multiple entity types. Simple queries of this kind can be implemented in the
MediaWiki expression language introduced in Sect. 4, more complex one require
SPARQL. The output of both kinds of queries can be a table, list, map, timeline,
etc.

Consider, for example, finding all roles that a person has ever had in events;
this requires joins between person and event entities:

```
SELECT ?event ?person ?hasRole WHERE {
    ?e       rdfs:label        ?event .
    ?e       ?hasRole          ?person .
    ?hasRole rdfs:subPropertyOf property:Has_person .
    ?person  rdfs:label         "PERSON NAME" .}
```

Geographical distribution and affiliation changes of persons in the role of
general chairs of events related to a certain field over last 10 years can be shown
on a map or graph by embedding a SPARQL query as follows into the wiki page
representing a certain field (i.e., in MediaWiki, a *category* page):

```
{{#sparql: SELECT ?event ?country ?person WHERE {
    ?e  a        category:Semantic_Web .
    ?p  property:Has_location_country ?country .
    ?p  property:Has_affiliation ?organization .
    [...]
    MINUS{ ?e  property:Has_general_chair  :person . }
    FILTER (?startDate >= "2007-01-01"^^xsd:date && ?endDate   <  "2017-01-01"^^xsd:date )
    } LIMIT 10  | format=maps}}
```

---

[12] https://www.mediawiki.org/wiki/Extension:LinkedWiki.

5   Observations

In this section we report detailed analysis results for 40 conference series over a period of 30 years according to six statistical analysis dimensions. The complete raw data is available at https://goo.gl/vnsXRe.

5.1   Statistical Analysis

Acceptance Rate. We have selected Fig. 3(a) shows the average acceptance rate for a sample of 10 conferences from different CS sub-fields in five consecutive 5-year periods from 1992 to 2016.

In all three periods, the average acceptance rate for all series falls into the range 17% to 26% in the time window of 25 years. The greatest acceptance rate ever was the one of COLT in the second period (45%), but decreased it to 36% in 2016. The average acceptance rate of EuroCrypt had increased to 33% by 2011 before decreasing to final 23%. The average acceptance rate of CCS dramatically decreased. The number of submissions to this series increased over time; however, the acceptance rate remained approximately the same. Only the average acceptance rate of EuroCrypt significantly increased to 33% in 2007–2011 and then decreased again to 20% in 2012–2016. A reason for decreasing acceptance rate can be increasing submissions, with the number of presentation slots at a conference being more or less constant over time.

Continuity. The continuity of conferences is calculated using the proposed formula in Sect. 4. For example, the continuity of CCS (ACM Conference on Computer and Communications Security) is 92% where it was held every year from 1993 except for two years in 1995 and 2003. Moreover, the continuity TPDL (The International Conference on Theory and Practice of Digital Libraries) is 100% where it is occurring every year since the first year of establishment. For illustration, the continuity of five conferences are shown in Table 4; for the others, the continuity is 100%. Overall we observed a very high continuity among the renowned conferences.

Table 4. Continuity of five conference series

Conference	Age	Editions	Regularity	Continuity (C)
ACMMM	23	22	1	96%
CCS	24	22	1	92%
CHI	35	34	1	97%
FOGA	27	13	2	96%
TPDL	21	21	1	100%

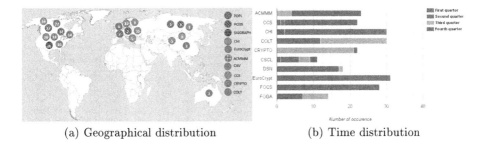

(a) Geographical distribution   (b) Time distribution

Fig. 2. Geographical and time distribution of events.

Geographical Distribution. The EUROCRYPT conference series has been held in a different country every year since 1987 but always in Europe. This is mostly related to the organisation committee in this series since it is a European committee. For the same reason, the SIGGRAPH series has been held every year since 1974 in different North American countries (mostly in the US).

The FOCS series has been held 26 times in the US, every year since 1989 in North America, and in Europe only for one edition in 2004. On the contrary, ISSAC has been moving between different countries of different continents such as Japan, Canada, Germany, etc., since its first edition. Figure 2(a) shows the Geographical Distribution of a sample of ten conference series randomly selected. The most geographically diverse conference series are EUROCRYPT (diversity by country in Europe). The most static conference is FOCS series that has been held 26 times only in the US for the past 25 years.

Time Distribution. Most editions of top conference series are held around the same month of each year; see Fig. 2(b). Namely, the PERCOM conference (IEEE International Conference on Pervasive Computing and Communications) has been held every year since 2003 in March and POPL (ACM SIGACT Symposium on Principles of Programming Languages) has been held every year since 1994 in January. Furthermore, almost all conferences in the study have been established around the same month. For example, EuroCrypt is always held in April or May and SIGGRAPH always held in July or August.

Sub-field Popularity. There are five groups labelled: Computer Security (SEC), Computer Graphics (GRA), Database Systems (DB), Programming languages (PROG) and Software Engineering(SE) each of which contains four top conferences belonging to this sub-field. Table 5 compares five CS communities in terms of the number of accepted and submitted papers. GRA communities made the largest number of submission in the whole time span, even though GRA submissions began to decrease since 2005 until they reached their minimum value in the last period. The average number of accepted papers (Fig. 3(c)) in GRA doubled in the first time frame and increased to almost 150% in the past 10 years, similarly in SE. The average number of accepted papers in DB slightly increased

Table 5. Accepted and submitted papers measures for five CS sub-fields over three 10-years and three 5-years intervals respectively

		Computer Graphics			Computer Security			Programming Languages			Software Engineering			Database Systems		
		Avg.	Min	Max	Avg.	Min	Max	Avg.	Min	Max	Avg.	Min	Max	Avg.	Min	Max
Accepted Papers	1987–1996	78.9	33	172	106.2	68	135	103.8	90	119	65.1	36	101	135.7	117	166
	1997–2006	198.4	113	261	130.1	86	157	110.6	102	125	116.1	105	130	161.9	121	206
	2007–2016	302.2	219	593	235.2	144	337	170.3	128	199	189.5	144	256	240.7	151	347
Submitted Papers	2000–2004	927	535	1,182	633	513	849	522	481	576	709	585	879	905	718	1,207
	2005–2009	1,188	1,090	1,454	855	607	988	594	568	635	904	803	1,038	1,250	1,166	1,348
	2010–2014	1,304	1,017	1,786	1,122	936	1,264	754	676	827	992	837	1,170	973	548	1,109

in the first period and then again increased in the last 10 years by 50%. Over all three periods, the GRA community has attracted most, and PROG has attracted the least submissions. Overall, there is an increasing number of submissions for all CS sub-fields we considered (Fig. 3(d)

Field Productivity. We calculated the Field Productivity for the five sub-fields in the study. The results are shown in Fig. 3(b).

We found that PROG and DB remained at the same FP with some ups and downs from 1987 to 2010 and then saw a slight increase. At the end of the 1980s and the early 1990s, GRA had the lowest FP with less than 1% until it began to increase to around 3% by 1993 and continued increasing to around 10% before decreasing to only 4% by the end of the period.

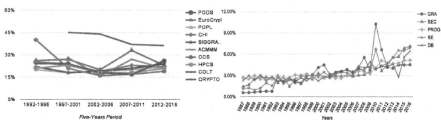

(a) Avg. Acceptance Rate for Ten Top Con- (b) Field Productivity for five CS sub-fields
ference Series

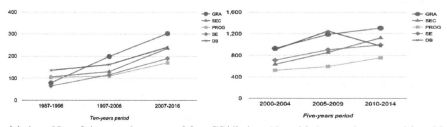

(c) Avg. No. of Accepted papers of five CS (d) Avg. No. of Submitted papers of five CS
sub-fields   sub-fields

Fig. 3. Visualisation of observations

Moreover, all fields had an FP around 3% from 1987 till 2006. For instance, FP of SE varied between 1.13% and 2.97%. In addition, GRA reached the maximum FP in 2010 with 10% and DB reaches the maximum FP with 6.45% in 2016. Overall, GRA has the highest FP with 5,795 publications over the other fields; the PROG community has the lowest FP with 3,707 publications. The DB community ranks second with 5,383 publications, followed by SEC with 4,715 publications and then PROG with 3,847 publications.

6 Conclusions

We presented a method for analysing scholarly communication metadata of scientific events. We combined descriptive and exploratory analysis with regard to a broad set of metrics, supported by spreadsheets, charts and queries in the OpenResearch semantic wiki. Up to our knowledge for the first time, we were able to empirically validate the often raised concern of a proliferation of submissions to major conferences. Also, we were able to calculate and demonstrate with our method a number of other indicators, such as a new way to calculate conference continuity, the popularity of different sub-fields, a new way to calculate field productivity or the geographic distribution of conferences. In addition to efficiency gains, the digitisation of scholarly communication also has negative impacts, most significantly the proliferation of submissions, which significantly increases the reviewing workload with an already noticeable knock-on effect on reviewing quality (one of the core features of peer-review). We plan to systematically investigate review quality in future.

In summary, we made the following observations:

- With the number of submissions to the top conferences having tripled on average in the last three decades, acceptance rates are going down slightly.
- Most of those conferences that are A- or A*-rated today have a long continuity.
- Geographical distribution is not generally relevant; some good conferences take place in the same location; others cycle between continents.
- Good conferences always take place around the same time of the year. This might mean that the community got used to them being important events.
- Some topics have attracted increasing interest recently e.g., database topics thanks to the 'big data' trend. This might be confirmed by further investigations into more recent, *emerging* events in such fields.

In further research, we aim to expand the analysis to other fields of science and to smaller events. Also, it is interesting to assess the impact of digitisation with regard to further scholarly communication means, such as journals (which are more important in fields other than computer science), workshops, funding calls and proposal applications as well as awards. Although large parts of our analysis methodology are already automated, we plan to further optimise the process so that analysis can be almost instantly generated from the OpenResearch data basis.

Acknowledgement. This work has been supported by the European Union for the Horizon 2020 project OpenBudgets.eu (GA no. 645833).

References

1. Ameloot, T.J., et al.: 30 Years of PODS in facts and figures. SIGMOD Rec. **40**(3), 54–60 (2011)
2. Aumüller, D., Rahm, E.: Affiliation analysis of database publications. SIGMOD Rec. **40**(1), 26–31 (2011)
3. Biryukov, M., Dong, C.: Analysis of computer science communities based on DBLP. In: Lalmas, M., Jose, J., Rauber, A., Sebastiani, F., Frommholz, I. (eds.) ECDL 2010. LNCS, vol. 6273, pp. 228–235. Springer, Heidelberg (2010). doi:10.1007/978-3-642-15464-5_24
4. Bryl, V., et al.: What's in the proceedings? Combining publisher's and researcher's perspectives. In: SePublica, vol. 1155 (2014). CEUR-WS.org
5. Committee on Global Approaches to Advanced Computing, Board on Global Science and Technology, Policy and Global Affairs Division. Appendix F: Pilot Study of Papers at Top Technical Conferences in Advanced Computing. In: The New Global Ecosystem in Advanced Computing: Implications for U.S. Competitiveness and National Security. National Academies Press (2012)
6. Diederich, J., Balke, W.-T., Thaden, U.: Demonstrating the semantic growbag: automatically creating topic facets for faceteddblp. In: ACM (2007)
7. Osborne, F., Motta, E., Mulholland, P.: Exploring scholarly data with rexplore. In: Alani, H., Kagal, L., Fokoue, A., Groth, P., Biemann, C., Parreira, J.X., Aroyo, L., Noy, N., Welty, C., Janowicz, K. (eds.) ISWC 2013. LNCS, vol. 8218, pp. 460–477. Springer, Heidelberg (2013). doi:10.1007/978-3-642-41335-3_29
8. Vahdati, S., Arndt, N., Auer, S., Lange, C.: OpenResearch: collaborative management of scholarly communication metadata. In: Blomqvist, E., Ciancarini, P., Poggi, F., Vitali, F. (eds.) EKAW 2016. LNCS, vol. 10024, pp. 778–793. Springer, Cham (2016). doi:10.1007/978-3-319-49004-5_50
9. Yan, S., Lee, D.: Toward alternative measures for ranking venues: a case of database research community. In: JCDL. ACM (2007)

High-Pass Text Filtering for Citation Matching

Yannis Foufoulas[1]([✉]), Lefteris Stamatogiannakis[1], Harry Dimitropoulos[1],
and Yannis Ioannidis[1,2]

[1] Department of Informatics and Telecom, University of Athens, Athens, Greece
{johnfouf,estama,harryd,yannis}@di.uoa.gr
[2] "Athena" Research Center, Maroussi, Greece

Abstract. Open publications are increasing at such a rapid pace that it
is almost impossible for researchers to keep up with them. Even in terms
of computational complexity, the data are becoming bigger and bigger,
so there is a great need to provide new and faster algorithms for mining scientific articles. One such important mining task is finding citation
links between the literature, which can assist researchers looking into the
literature, finding dependencies between publications, and so on. In this
paper, we introduce a greedy citation matching algorithm, that works
with plain unstructured text and mines citations from papers regardless of the format in which the citations are presented. This research is
supported by the European Commission under projects OpenAIRE2020
(643410) and Human Brain Project (720270).

Keywords: Citation matching · High-pass filtering · Reference
extraction

1 Introduction

Scholarly communication is currently at a new phase where researcher's published results are more optimally shared, discovered, validated and re-used when
they are exposed in their full context. This means that they are best accompanied by all the relative information that provides an insight and capacity to
translate the research process and activities that have taken place. Such information may be citations. If we are able to provide links between the literature,
then this can be used for multiple purposes including literature search, finding
similar publications, analysis of research trends, etc.

Most of the time, citation extraction and parsing is not enough; there is also
a need to match the citations to metadata databases in order to enrich them
with more useful information, such as the complete author list, scientific areas,
journal information and in some cases abstracts or even fulltexts. Such info is
not always included in the citation text, while abstracts and fulltexts are never
included. Even when some of this exists, there is no algorithm that assures 100%
recall rate in the parsing and extraction phase.

Due to these reasons, the matching of the citations with a metadata database is
an important task. Currently, a user is able to download publications' metadata

© Springer International Publishing AG 2017
J. Kamps et al. (Eds.): TPDL 2017, LNCS 10450, pp. 355–366, 2017.
DOI: 10.1007/978-3-319-67008-9_28

from various sources including repositories and other systems which offer APIs (e.g. PubMed API[1], ArXiv API[2], CrossRef Search API[3], OpenAIRE[4] API)

Figure 1 presents the most common workaround to extract citations links. The first step is citation extraction and parsing. Citation extraction regards the extraction of a citation and its metadata (title, author names, journals, dates).

Fig. 1. Citation extraction, parsing and matching workflow

Titles, author names and other metadata are typed in many different ways and orders. An example of two citations that refer to the same paper, yet are cited quite differently, follows:

Friston, K. J., Holmes, A. P., Worsley, K. J., Poline, J. P., Frith, C. D., and Frackowiak, R. S. J. (1995). Statistical parametric maps in functional imaging: a general linear approach. Human Brain Mapping 2:189–210.

[Friston et al. 94] Statistical parametric maps in functional imaging: a general linear approach, Karl J Friston, Andrew P Holmes, Keith J Worsley, J-P Poline, Chris D Frith, Richard SJ Frackowiak. Human Brain Mapping Vol. 2(4), pp. 189–210.

For this reason, citation parsing is a difficult task and has already been addressed many times by the community. Mainstream citation extraction approaches use heuristics [5], machine learning techniques [6,7], knowledge-based approaches [2,4], and other methods to overcome this issue. However, due to the different ways that a document is formatted - and the different languages - this process may be time consuming. In the Related Work section, we will present more thoroughly the existing techniques.

The second step is citation matching. This phase regards the enrichment of the extracted citation. The title, the authors and the other extracted fields are matched against the repository of interest. Since the metadata of the repository are also structured, this matching seems like a simple string match. However, there are also several problems to tackle, like the different ways that author names (or other metadata) are written, title misspellings, publications with the same title and other issues.

In this work, we match the publications' plain text with the repository metadata and produce directly the enriched matched citations. We are eliminating

---

[1] https://europepmc.org/.

[2] http://arxiv.org/.

[3] http://www.crossref.org/.

[4] https://www.openaire.eu/.

the citation parsing step, replacing it with a fast text filtering step whose purpose it to keep only the sections in the text which contain references. Figure 2 presents the workaround of the proposed method.

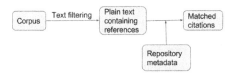

Fig. 2. Citation matching workflow

The first step uses heuristics and high pass text filtering to extract the whole reference section from the text and any other section that may contain references.

The next step is the final citation matching step. The structured repository metadata are matched against the references sections from the text using database and pattern matching techniques.

Our technique is able to extract references from anywhere in the text, including footnotes, and not only from the references section. Moreover, since the full-text is not parsed to produce a structured citations list, the algorithm does not depend on the references or the publication's format. Finally, as the experiments have shown, the presented method is able to provide citation links between a corpus consisting of publications' plain text and a specific repository up to more than an order of magnitude faster than the techniques that parse the citations before the matching.

The rest of the paper is organized as follows. In Sect. 2, we present the related work. In Sect. 3, we lay out our reference section extraction algorithm and in Sect. 4, we introduce our citation matching algorithm, and the implementation details. In Sect. 6, some experimental results are shown.

2   Related Work

While dealing with the citation matching problem, the first issue we need to address is that of data representation. The data can be either structured (e.g. XML) or unstructured (e.g. plain text). E.g., when a publication is in XML format it often has structured references, titles and authors, which can be easily matched against an existing/given database of publications. On the other hand, when provided with unstructured plain text it is important to consider that citations are presented in different formats according to each repository/publisher, making it difficult to produce an algorithm that is able to extract references from all possible repository formats.

To solve this problem: CITESEER [5] identifies the reference section and then uses heuristics and machine learning techniques to find title, author, year of publication, page numbers, and citation tags; CERMINE [12] uses some geometrical, lexical values, the format and some heuristics like the uppercase etc., to extract the references section and machine learning techniques to find reference strings; GROBID [8] and ParsCit [3] use Conditional Random Fields [9]. GROBID and CERMINE work with PDF files, whereas ParsCit works also with plain text.

All these methods target the problem of extracting and parsing citations from a paper. Having completed this step, if someone needs to match the citations against a metadata database, he has to match the extracted titles, authors and the other metadata. While this task seems simpler, it is also difficult since it includes title and metadata matching. Title matching is difficult due to possible typos. Moreover, when a title matches with another, it is a good hint that we are talking about the same paper, but this is not always the case. Matching metadata including authors, journal information etc. is very difficult due to the different ways this information is presented. The existing tools target mainly the extraction of structured citation lists, without matching them to metadata databases.

3   Reference Sections Extraction Algorithm

As already mentioned, the main challenge is the different ways that citations are presented. Thus in order to be able to locate them in a publication it is crucial to use global heuristics. Ideally, we should find a globally common characteristic. Such a characteristic is the appearance of years or URLs in the text: in all publications, the references sections are dense in dates and URLs. Using this feature, we split each publication text on its newlines, remove blank and small lines that consist of less than ten characters, and we mark all lines where at least one year (4 digit numbers, between 1900 and 20xx) or URL appears in their context, producing a list of marked or unmarked items. By considering this list as a signal in time, we can then run a high-pass filter process, keeping regions with higher than average density of dates or URLs. In this way, we are able to extract not only the references section but also text that contains citations anywhere in the body of the publication.

An illustration of how the reference extraction algorithm works can be seen in Fig. 3. In the first step, the lines containing URLs or dates are marked. The next step calculates the density of such patterns per window. In this example, a window consists of 5 lines, so the lines from 4th to 8th constitute a window. The density for the lines of this window is 0.2, since a year appears in one line (1/5). In the final step we mark the lines with density higher than average. The unmarked lines are filtered out, and the sections from the text which may contain references are extracted[5].

This step's main goal is to reduce, as fast as possible, the amount of text that will be processed without missing any valid citations. We do not care about false positives as these will be eliminated during the next steps.

---

[5] Note that this is an unreal example where the references section covers almost half of the complete text. When the main body is larger, the average density of dates and URLs is lower and possible references in footnotes are not filtered out.

text	signal	density (window size = 5)	returned lines (density > avg_density)
	0	0	0
	0	0	0
	0	0	0
	0	0.2	0
	0	0.2	0
2004	1	0.2	0
	0	0.2	0
	0	0.2	0
	0	0	0
	0	0	0
	0	0	0
	0	0.2	0
	0	0.2	0
http://google.com	1	0.2	0
	0	0.4	1
	0	0.4	1
2009	1	0.4	1
	0	0.6	1
2010	1	0.8	1
2006	1	0.8	1
2003	1	1	1
2002	1	0.8	1
1999	1	0.8	1
	0	0.8	1
2012	1	0.8	1

Fig. 3. Reference extraction algorithm with window size = 5

4 Title and Metadata Matching

Citation matching is the next step of the presented algorithm. There are several problems that we have to address. First, matching titles with plain text can be very time-consuming, because title lengths vary. Moreover, the same title does not always refer to the same paper. We also need to match with metadata. Citations may be written in different formats, the order of publication metadata varies, author names can be written in different ways (John Smith or J. Smith). Our algorithm solves the title and metadata matching problem in the following steps:

- Preprocessing phase (possibly offline)
  - Normalization of structured metadata
  - Creation of characteristic inverted index
- Matching phase
  - Title matching
  - Validation of results using metadata matching

4.1 Preprocessing

Normalization of Metadata. At first, we normalize the titles and other publication metadata by:

- Reducing spaces between words
- Replacing punctuation marks with underscores
- Converting text to lower case

For example if we have the title:
"The PageRank citation ranking: Bringing order to the web"

we transform it as follows:

"the pagerank citation ranking_ bringing order to the web"

The preprocessing phase addresses misspelling issues related to number of spaces, punctuation and case sensitivity. Exactly the same preprocessing procedure has to be applied to the publications' fulltext before the final matching.

Characteristic Inverted Index. As mentioned before, title matching is a very demanding and time-consuming task. We produce an inverted index based on all trigrams that appear in the titles. With the term trigram, we refer to any sequence of three words in the titles. We execute a JOIN operator between the text and title trigrams. When a trigram from the text matches with a trigram from a title, we examine if the whole title matches in the text. If yes, we have a matched title, if not we have a title miss. Using trigrams instead of bigrams or single words, we reduce possible title misses. On the other hand, we do not use larger N-grams, because many titles consist of just three terms. A typical trigram-based inverted index for the above example title is shown in Table 1.

Table 1. Trigram-based inverted index

Trigram	Title id
The pagerank citation	1
Pagerank citation ranking_	1
Citation ranking_ bringing	1
Ranking_ bringing order	1
Bringing order to	1
Order to the	1
To the web	1

Obviously, since this index contains all trigrams appearing in all titles, it is both memory and computationally expensive to be used in a relational JOIN operation. Moreover, common trigrams that may co-exist in many titles could lead to a huge number of matching trigrams, thus to a huge number of title misses.

A way to reduce the size of the index and the title misses is to only use *identifying trigrams* in the index. An identifying trigram, is a trigram that appears in only one title. So, an ideal inverted index would only contain one identifying trigram per title. Because the ideal inverted index is unattainable most of the time, we try to approximate it using a simple heuristic iterative method. First we build the full trigram-based inverted index that contains all trigrams appearing in all titles. From this index we "pick" the trigrams and titles that only appear once, and remove them from the full inverted index. We repeat this procedure iteratively, increasing the threshold of trigram appearance count, until we have fully covered the set of titles of the full index. This produces a characteristic inverted index containing, for each title, a trigram that appears in as few other titles as possible.

The characteristic inverted index is used in a relational EQUI-JOIN between the trigrams appearing in the text and the index trigrams. Here follows an example of the described algorithm. Let A,B,C,D,E,F,G,H be trigrams and consider the example shown in Fig. 4.

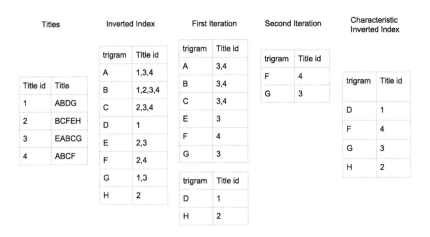

Fig. 4. Characteristic inverted index creation

The goal is to select the minimum subset of trigrams that covers the full set of titles, such that selected trigrams have minimum number of assigned titles. Figure 4 presents the steps to produce the characteristic inverted index. In the **first iteration**, we select D and H which appear uniquely in titles 1 and 2, and we remove these trigrams and titles from the index. In the **second iteration**, we are able to identify titles 3 and 4 with trigrams E,G and F. We may use E or G to identify title 3. In this situation, we select the longer trigram according to the number of characters it contains. After removing these trigrams (F,G) and titles (3,4) from the index, it turns out that no titles remain, so our characteristic inverted index is ready.

When the inverted index is complete, a query runs which scrolls a window over the publication's fulltext, extracts all the trigrams and matches them to our inverted index. If a match happens then the full title from the inverted index is matched with the context of the trigram in the publication's text. This way we ensure achieving a high recall rate since all trigrams from the text are joined.

4.2   Matching

Title Matching. A query extracts the references sections from the text, scrolls a window over the extracted text, extracts all the trigrams and matches them to the characteristic inverted index. Experimentally, the size of the window is set by default to 60 whitespace separated strings, in order to include at least the title and the metadata. If a match happens then the full title from the inverted index is matched with the corresponding window of text.

Validation of the Results. After matching a title we have to deal with the difficult problem of disambiguating and filtering out false matches. We use the following techniques:

- We create a bag of words that contains author names, publication dates, publishers, journal names for each metadata record. We pattern match this bag of words to our window of text. Each category weighs differently than the others. So, the author matches weigh more than the other metadata, the author surnames weigh more than their first names, the journal names and the publishers more than the publication dates.
- Each match also weighs differently according to its distance (in number of words with more than two characters) from the title. If a word matches, the confidence value is increased by its weight and inversely proportional to its distance from the title in words.
- The length of the title (number of words) is also considered, since a larger matched title is more possibly a true match.

The following equation shows how the confidence value is calculated:

$$Conf = \frac{\frac{10*F(AS)+3*F(AF)+3*F(PY)+5*F(JN)}{MAXVAL} + \frac{L(t)}{L(C)}}{2}, \tag{1}$$

AS, AF, PY, JN are author surnames, first names, publication years and journal names respectively. Function F sums the distance weights for all the occurrences of the input pattern. If a pattern is matched 3 words away from the title, its distance weight is 1/3. L(t) is the length of the matched title, while L(C) is the length of the window. MAXVAL is the maximum value, if all the words in the window are matched with author surnames. It is used so that the final confidence value is between 0 and 1. After experimenting with various datasets, we have defined 0.1 as an appropriate threshold. If this value is above the threshold, then the citation is marked as true positive, else as false.

The fallback of this technique is that the context used for calculating confidence value has stable length for speed purposes, so it may contain strings from previous or next citations. Nevertheless, manual curation of experimental results indicates that less than 1% of matches are false positives because of metadata that match with an adjacent reference.

4.3   Implementation Details

Our algorithm is implemented on top of madIS [11], a powerful extension of a relational DBMS with user-defined data processing functionality. MadIS is built on top of the SQLite API[6].

---

[6] https://www.sqlite.org/.

MadIS allows the creation of user-defined functions (UDFs) in Python and it uses them in the same way as its native SQL functions. Both Python and SQLite are executed in the same process, greatly reducing the communication cost between them. This is a critical architectural characteristic and has a positive impact on joint performance.

MadIS is highly scalable, easily handling 10 s of Gigabytes of data on a single machine. This benefit transparently carries over to distributed systems (e.g., Hadoop [10], Exareme [1]) which can use madIS in each node.

In madIS, queries are expressed in madQL: an SQL-based declarative language extended with additional syntax and user-defined functions (UDFs). One of the goals of madIS is to eliminate the effort of creating and using UDFs by making them first-class citizens in the query language itself.

The expressiveness and the performance of madIS along with its scalability features were compelling reasons for choosing it to implement our algorithm. Our citation matching software is open source and hosted by Github.[7]

5   Experiments

We ran three experiments to evaluate the proposed method. In the first two experiments, we test some important features of our method, whereas in the third experiment we compare our method to GROBID and ParsCit. Our experiments ran on an Intel(R) Core(TM) i7-4790 CPU @ 3.60 GHz processor with a 500 GB SSD disk and 16 GB RAM, running Ubuntu Server 14.04 LTS.

The aim of our **first experiment** was to prove the benefits of the reference extraction algorithm. We ran the citation matching algorithm to the full publications' text (without using the reference extraction algorithm) and compared the results to those that are produced if the references are extracted. For this experiment, we used 100 publications from both arXiv and PubMed repositories (50/50), matching citations with 9.6 millions publications from OpenAIRE. ArXiv and PubMed were selected, as their deposited publications do not share similar reference formats and cover many scientific areas including medicine, physics, computer science and others. The results are shown in Table 2. The precision and the recall rates are based on manual validation of the results.

In case 1, running citation matching on the fulltext, we find a total of 289 citations. The recall rate (97.40%) is high and notably the 5 citation misses are all cases that the algorithm was able to match but are filtered out due to low confidence value. In case 2, we only process 12.3% of the total text lines, so the processing is about 8 times faster. Case 2, misses 12 more correct references that were found by case 1. By increasing the window size from 5 to 7 in case 3, we only increase by 0.3% the number of processed lines but we also find all the references missed by case 2. Finally, it turns out that the use of the reference extraction algorithm is very advantageous in terms of precision, since in case 1 the precision is very low (67.10%), whereas in cases 2 and 3 there are not any

---

[7] https://github.com/madgik/recital.

Table 2. Evaluation of matched citations

Case	1. full pub text	2. ref extraction window size = 5	3. ref extraction window size = 7
High confidence matches	289	182	194
True citation misses	5	17	5
High confidence precision	67.10%	100%	100%
Recall	97.40%	91.50%	97.40%
% of total text lines	100%	12.30%	12.60%

false positives. However, while inspecting the false positives of case 1, it turned out that the titles of the publications themselves that are placed on the top of the PDF had matched. We can avoid this if we simply exclude the first lines of the plaintexts in the pre-processing phase. Thus, the impact of the reference extraction algorithm on the processing time is very important since the main body of the algorithm is only applied on a small percentage of total text lines.

Our **second experiment** concerns the citation matching algorithm. We ran our algorithm on about 450 K publications' fulltexts retrieved from the ArXiv repository. The purpose of our experiment was to find citation links to OpenAIRE publications. The first step for doing this is building the characteristic inverted index. OpenAIRE publication index, during the experiments, consisted of 9,598,093 publications in which there were 8,168,090 distinct titles. While creating the characteristic inverted index, we managed to extract identifying trigrams for 7,430,948 titles whereas for the rest we increased the threshold of trigram appearance count, until we covered the full set of titles. Then, we ran the algorithm using the text references extraction algorithm with window size set to 7. The results of our experiment are shown in Table 3.

Table 3. Citation matching in OpenAIRE

Publication count	OpenAIRE publication count	High confidence Citation matches	High confidence Precision
450.4K	9.6M	968,880	99%

In this experiment, after validating a sample of 200 citations, we found two false positives. These two cases show one main disadvantage of our method. Rarely, there are titles which are substrings of other titles. This is not a problem when the other metadata differ, because the confidence value remains low. But when the authors are also the same, then we may end up with false positives. Consider the citations below:

Caire, Giuseppe, and Daniela Tuninetti. "The throughput of hybrid-ARQ protocols for the Gaussian collision channel." IEEE Transactions on Information Theory (2001).

Caire, G., and D. Tuninetti. "ARQ protocols for the Gaussian collision channel."
IEEE INTERNATIONAL SYMPOSIUM ON INFORMATION THEORY. 2000.
These refer to two different publications with the same authors, where the
second publications' title is a substring of the first.

Our **last experiment**, compares our method to GROBID and ParsCit. Both
these tools extract and parse citations from publications. GROBID processes
PDF files while ParsCit processes plain text. These tools extract structured
citations and the matching to a given list of publications' metadata can be done
as a next step. Our algorithm (and ParsCit) works with plain text, so we had
first to convert the PDFs. We did it using the Unix util *pdftotext* without any
parameters. We compare the methods on a dataset consisting of 15K PDFs, and a
metadata database consisting of about 375 K publications from ACM[8]. GROBID
ran in multi-threaded mode (8 threads), while ParsCit and our method ran in
single-thread. Table 4 shows the execution times, and the matched citations.

Table 4. Comparison with GROBID and ParsCit

Method	PDF convertion time (minutes)	Preprocessing (minutes)	Extraction (minutes)	Matching (minutes)	Matched citations
GROBID	-	-	101	0.9	32595
ParsCit	11	-	720	0.9	31989
Proposed	11	1.9	0.5	2.2	39027

As can be seen, the proposed method is much more efficient in terms of
speed. The preprocessing time regards the normalization of the metadata and
the creation of the characteristic inverted index. The proposed method does not
extract stuctured citations, but only sections from the text that may contain
references. In this step, GROBID extracts 129252 structured citations with titles
while ParsCit extracts 125328. In the next step, the titles are matched against the
metadata database. Grobid produces 33206 title matches while ParsCit produces
32686. The presented numbers regard the citation matches where not only the
titles, but also metadata like publication year and author names have matched.

Our method matches the metadata database with the plain text and produces
39027 high confidence matched citations. All three methods extract in common
31899 matched citations. Our method matches 7089 citations more than GRO-
BID and 7105 citations more than ParsCit. The validation of a sample of 200
citations showed that 97% of them were valid and the rest were false alarms.
GROBID also extracts 657 citations that are not extracted using the proposed
method. These citations were missed mainly due to the PDF conversion. Comparing our method with ParsCit, where the same PDF conversion tool was used,
it seems that our method misses just 67 citations. This supports our claim that
our method achieves high recall rates when processing plain text.

---

[8] http://dl.acm.org.

6   Conclusions

Given a known database of publications' metadata, we propose a fast and accurate citation matching method from a corpus of publications to the metadata database. Our method does not use citation extraction but citation matching. This means that we do not extract external citations but we target extracting citations within a given dataset. That is why we can avoid using time-consuming machine learning techniques to extract full citation metadata from the publications' fulltext. Hence, the algorithm achieves the same accuracies and faster processing times regardless of the format of the publication's fulltext and in most cases even its language. Moreover, by matching every trigram in the fulltext, we ensure that our method achieves higher recall rates than methods that extract citations' metadata from fulltext before applying the match.

References

1. Chronis, Y., et al.: A relational approach to complex dataflows (MEDAL 2016) (2016)
2. Cortez, E., da Silva, A.S., et al.: Flux-CiM: flexible unsupervised extraction of citation metadata. In: Proceedings of the 7th ACM/IEEE-CS Joint Conference on Digital Libraries, pp. 215–224. ACM (2007)
3. Councill, I.G., Giles, C.L., Kan, M.Y.: ParsCit: an open-source CRF reference string parsing package. In: LREC 2008 (2008)
4. Day, M.Y., Tsai, T.H., et al.: A knowledge-based approach to citation extraction. In: International Conference on Information Reuse and Integration. IEEE (2005)
5. Giles, C.L., et al.: Citeseer: An automatic citation indexing system. In: Proceedings of the third ACM Conference on Digital Libraries, pp. 89–98. ACM (1998)
6. Han, H., et al.: Automatic document metadata extraction using support vector machines. In: Joint Conference on Digital Libraries. IEEE (2003)
7. Lawrence, S., Giles, L.C., Bollacker, K.: Digital libraries and autonomous citation indexing. Computer, 67–71 (1999)
8. Lopez, P.: GROBID: combining automatic bibliographic data recognition and term extraction for scholarship publications. In: Agosti, M., Borbinha, J., Kapidakis, S., Papatheodorou, C., Tsakonas, G. (eds.) ECDL 2009. LNCS, vol. 5714, pp. 473–474. Springer, Heidelberg (2009). doi:10.1007/978-3-642-04346-8_62
9. Peng, F., McCallum, A.: Information extraction from research papers using conditional random fields. Inf. Process. Manage. (2006)
10. Shvachko, K., et al.: The HADOOP distributed file system. In: 26th Symposium on Mass Storage Systems and Technologies (MSST), pp. 1–10. IEEE (2010)
11. Stamatogiannakis, L., et al.: madIS - extensible relational DB based on SQLite, https://github.com/madgik/madis. Accessed 5 March 2017
12. Tkaczyk, D., et al.: CERMINE - automatic extraction of metadata and references from scientific literature. In: 11th IAPR International Workshop on Document Analysis Systems, pp. 217–221. IEEE (2014)

Sentiment Analysis

Sentiment Classification over Opinionated Data Streams Through Informed Model Adaptation

Vasileios Iosifidis[1]([✉]), Annina Oelschlager[2], and Eirini Ntoutsi[1]

[1] L3S Research Center, Leibniz University Hannover, Hannover, Germany
{iosifidis,ntoutsi}@l3s.de
[2] Ludwig-Maximilians-University, Munich, Germany
annina.oelschlaeger@gmail.com

Abstract. Opinionated data streams are very popular data paradigms nowadays as more and more users share their opinions online about almost everything from products to persons, brands and ideas. One of the key challenges for opinionated stream mining is dealing with concept drifts in the underlying stream population by building learners that adapt to such concept changes. Ageing is a typical way of adapting to change in a stream environment as it potentially allows us to discard outdated information from the learning models and focus on the most recent information. Most of the existing approaches follow a fixed ageing strategy which remains the same over the whole stream; for example, a fixed window size in the sliding window model or a fixed ageing factor in the damped window model. This implies that we forget at the same rate over the whole course of the stream, which is counterintuitive given the volatile nature of the stream. What is more intuitive is to forget faster in times of change so as to adapt to new data and to forget slower, or in other words, to remember more, in times of stability. In this work, we propose an informative-adaptation-to-change approach where we first detect changes in the underlying data stream and then we tune the ageing factor of the ageing-based Multinomial Naive Bayes (MNB) classifier based on the detected change. Except for the up-to-date classifier our method also outputs the points of change in the stream, therefore offering more insights to the final users.

1 Introduction

A huge amount of opinions is available nowadays, as a result of the widespread usage of the social media and the Web. Opinions are valuable for consumers, who benefit from the experiences of others, in order to make better buying decisions [13] but also for vendors, who can get insights on what customers like or dislike [16]. Such sort of data are freely available, however due to their amount and complexity a proper analysis is required in order to gain insights.

Opinions are accumulated over time, building what we call opinionated streams, i.e., streams of documents which convey sentiment. The accumulating opinionated documents are subject to different forms of drift: the topics discussed in the stream change, the attitude of people towards certain topics

© Springer International Publishing AG 2017
J. Kamps et al. (Eds.): TPDL 2017, LNCS 10450, pp. 369–381, 2017.
DOI: 10.1007/978-3-319-67008-9_29

might change, words used to describe topics or sentiment might change and so on and so forth.

In this work, we address the issue of polarity learning over opinionated streams. That is, we want to build classifiers that can cope with the volatile nature of the stream. There are two different directions for adaptation in a stream environment [5]: *blind adaptation* methods that update the underlying models constantly over the stream and *informed adaptation* methods that adapt the model only if change has been detected. The later are computationally more expensive methods as except for the adaptation step they typically include a change detection step that looks explicitly for changes in the stream. Those methods though are more informative as except for the up-to-date classification model they provide additional information on the points of change, which comprises important knowledge for the end user and it allows the user to react to changes. We propose informed adaptation over ageing-based Multinomial Naive Bayes classifiers, which incorporate ageing through the damped window model, and in particular, an approach for the online tuning of the ageing factor *lambda* of the dumped window model based on the dynamics of the underlying stream.

The rest of the paper is organized as follows: Related work is discussed in Sect. 2. The basic concepts and motivation are presented in Sect. 3. Our informed adaptation approach is presented in Sect. 4. Experimental results are shown in Sect. 5. Conclusions and open issues are discussed in Sect. 6.

2   Related Work

Change is a key concept in data streams and refers to the fact that the distribution that generates the stream is non-stationary, rather it changes with time, causing the so-called concept drifts [18]. The ability to adapt to changes is a key property of data stream mining algorithms. There are two ways of adaptation: (i) by including new instances from the stream and updating accordingly the learning model and (ii) by discarding outdated information from the model, also known as forgetting. The forgetting mechanisms can be categorized into: abrupt forgetting and gradual forgetting. The former ones take into consideration only recent instances within a sliding window, whereas the latter ones assume that all instances can potentially contribute to the model but with a weight that is regulated by their age. The concept of drift adaptation and state-of-the-art techniques and algorithms for dealing with drift in data stream mining is nicely covered in [6], whereas forgetting has been the subject of many research works, e.g., [4, 9–12, 15, 17, 22] just to mention a few.

Multinomial Naive Bayes (MNB) [14] is a popular classifier due to its simplicity and good performance in practice, despite its naive assumption on the class-conditional independence of the features [3, 20]. Its simplicity and efficient online maintenance makes it particularly suitable for streams. Bermingham et al. [1] compared the performance of Support Vector Machines (SVM) and MNB classifiers on microblog data and reviews (not streams) and showed that MNB performs well on short-length, opinion-rich microblog messages (rather than on long texts).

In [8], popular classification algorithms were studied such as MNBs, Random Forest, Bayesian Logistic Regression and SVMs using sequential minimal optimization for the classification in Twitter streams while building classifiers at different samples. Across tested classifiers, MNBs showed the best performance for all applied data sets. In [2], MNB has been compared to Stochastic Gradient Descend (SGD) and Hoeffding Trees for polarity classification on streams. MNB approach which was used in this study is incremental, i.e., it accumulates information on class appearances and word-in-class appearances over the stream, however, it does not forget anything. Their experiments showed that MNB had the largest difficulty in dealing with drifts in the stream population, although its performance in times of stability was very good. Regarding runtime, MNB was the fastest model due to its simplicity in predictions but also due to the easy incorporation of new instances in the model. The poor performance of MNB [2] motivated the ageing-based MNB approach [21] which also considers the recency of the class and words-in-classes observations and uses this information to regulate the class priors and class-conditional word probabilities. Their approach though is a blind adaptation approach, i.e., the model is constantly tuned based on a fixed ageing factor λ without explicitly counting for change. In this work, we follow an informed adaptation approach by tuning λ upon (data) change.

3   Basic Concepts

Before we proceed we introduce some notation:

- S: the (accumulated) stream up to current timepoint.
- V: the vocabulary of S.
- S_{sl}: the current sliding window of the most recent w instances.
- V_{sl}: the vocabulary of S_{sl}.

We observe a stream S of opinionated documents arriving at distinct timepoints $t_0, \ldots, t_i, \ldots$. An opinionated document d in S is a document associated with a polarity label $c \in C$, where C is the class attribute for the polarity. In the simplest case, the polarity class has two values, positive and negative. The document d is represented through the bag-of-words model as a set of words, $d = \{w_i\}$.

Our goal is to build a polarity classifier for the prediction of the polarity of new arriving documents. Our base model is the ageing-based Multinomial Naive Bayes (*ageingMNB*) classifier [21], an MNB classifier that forgets based on the damped window model with a constant ageing factor λ. Our goal is to tune the ageing factor λ according to the dynamics of the underlying stream. That is, in times of change, the ageing should be more drastic to allow for fast adaptation to the new content received from the stream, whereas in times of stability the ageing should be kept low in order to exploit the so far learned model.

MNB is one of the most popular classifiers due to its efficiency and modest performance. The original MNB classifier works in a static setting (*staticMNB*), where the whole dataset is provided as input to the algorithm. The MNB model consists of a set of class priors and class conditional word probabilities, which are estimated from the training set. The straightforward extension of the static MNB to streams is by extending the definition of the training set to the (theoretically) never-ending stream case. In particular, the training set keeps growing by including new documents that continuously arrive from the stream. Due to its simplicity it is easy to maintain the MNB model in a stream setting; the probabilities of classes and word-class combinations are updated based on the new documents and their class labels. We refer to this model as *accumulativeMNB* [2].

The *accumulativeMNB* model includes new observations but *does not forget*. Therefore, it is difficult to adapt to changes in the stream, a fact which has been already observed in previous works [2,21]. The reason for poor adaptation is that the historical data dominate the decisions of the classifier. To overcome this issue, the *ageingMNB* model that forgets was proposed in [21].

The *ageingMNB* classifier extends the *accumulativeMNB* by including information on the recency of the observations (classes and words-in-classes observations). The recency information is derived from the original documents, which are associated with timestamps. Each class and word-class combination in the model is associated with a timestamp, the most recent timestamp where the specific class or word-class entity was observed in the stream. The recency entries are used during classification of new instances from the stream in order to downgrade the contribution of outdated observations in the model, so as more recent observations contribute more and incur model adaptation.

The (temporal) class prior for class $c \in C$ at timepoint t is [21]:

$$\hat{P}t(c) = \frac{N_c^t * e^{-\lambda \cdot (t - t_{lo}^c)}}{|\mathcal{S}^t|} \tag{1}$$

where N_c^t is the number of documents in the stream up to timepoint t belonging to class c and $|\mathcal{S}^t|$ is the total number of document in the stream up to t. The t_{lo}^c is the most recent observation of class c in the stream and $(t - t_{lo}^c)$ denotes the time lag between the last occurrence of the class label c in the stream and the current timepoint t.

The (temporal) class conditional word probability for a word $w_i \in d$ at t is given by [21]:

$$\hat{P}t(w_i|c) = \frac{N_{ic}^t * e^{-\lambda \cdot (t - t_{lo}^{(w_i,c)})}}{\sum_{j=1}^{|V^t|} N_{jc}^t * e^{-\lambda \cdot (t - t_{lo}^{(w_j,c)})}} \tag{2}$$

Again, the word-class counts N_{ic} are weighted by the recency of the observations of the specific word w_i in documents of class c. Old observations will be downgraded so their effect during classification is limited.

The *ageingMNB* approach is a *blind adaptation* method [6] as it applies a constant ageing factor λ in the MNB model over the whole course of the stream without considering whether there is an actual change or not. In the next section, we propose an adaptive ageing MNB model that tunes the ageing factor λ and therefore, the MNB model, online based on changes in the underlying stream population. There are two advantages of such an approach over the blind adaptation approach of *ageingMNB* [21]: first, it allows for ageing at different rates, which as already mentioned is more intuitive in a stream setting and second, except for the classification model, it provides additional information on the points of change, which is valuable for decision making and allows the end user to react to changes. For example, if a negative sentiment starts developing for a brand as a result of bad customer experiences, the brand can quickly address customer concerns and classify misconceptions thus transforming the negative sentiment into a winning customer experience.

4   Informed Adaptation of Multinomial Naive Bayes Classifiers over Data Streams

Our solution consists of two steps: (i) a change detection step that detects changes in the underlying stream population (Sect. 4.1), and (ii) a tuning step that adjusts the ageing factor λ, and therefore the classifier, upon detection of change (Sect. 4.2).

4.1   Detecting Change

There are several approaches for change detection, which are presented in detail in [5]. Since our focus in this work is on the adaptation of the ageing factor λ and due to lack of space, we present here the detector we used in our experiments, which showed the best performance among several methods we tried. Our detector falls into the category of monitoring the distributions in two different time-windows: such detectors compare the decision model built upon a *reference window* of past data to the decision model built over a *current window* of the most recent data points. In this work, we monitor the distance between the vocabularies of the most recent window $\mathcal{S}_{sl}$ and the reference window $\mathcal{S}$, i.e., V_{sl} vs V, for both the negative and the positive class. For the comparison, we employ precision, which equals to the fraction of the reference vocabulary words that also appear in the current vocabulary.

$$precision = \frac{V_{sl} \cap V}{|V|} \tag{3}$$

A high precision means that the current vocabulary comprises a large part of the reference vocabulary. Intuitively, this implies that the reference model, built over the reference vocabulary, could still be valid. Otherwise, the reference model is not well reflecting the current developments in the stream.

Change points are detected by comparing current precision to the moving average precision plus/minus α times the standard deviation, as follows:

$$precision < \mu - \alpha * \sigma$$
$$precision > \mu + \alpha * \sigma \tag{4}$$

where σ is the standard deviation, μ is the average precision and α is a user defined threshold that controls the trade-off between earlier detecting true alarms by allowing some false alarms. Low values of α allow faster detection, at the cost of increasing the number of false alarms.

Except for the final change points, often is also useful to detect warning points when the monitored difference between the current precision and moving average prediction exceeds some threshold $\beta \times \sigma$, with $\beta < \alpha$. Warning points are more frequent comparing to change points. Moreover, once a warning is detected a buffer of instances is maintained for model rebuild once the warning turns into an actual change point. Otherwise, the buffer is emptied.

4.2   Adapting to Change

Once a change is detected, the classifier should be updated to reflect the changing population. The most abrupt way of reacting to change is by building a new classifier over the recent data and demolishing the old one. Following a more conventional approach, one can affect the statistics of the model over the stream by tuning appropriately the ageing factor λ. We present hereafter different strategies for model adaptation to change.

Let λ_0 be an initial value of λ, set at the beginning of the stream. In the simplest case, $\lambda_0 = 0$, i.e., there is no-ageing. If $\lambda_0 > 0$, there is a constant ageing over the stream.

- **SlowIncreaseUpToALimit** - *Gradually increase λ by a constant value c up to a limit λ_{max}:*
  When a change is detected, λ is increased by a constant value c, i.e., it is set to $\lambda_i + c$, where λ_i is the value of λ before change. If there is still change, *lambda* will be further increased by c. Increasing λ after change is beneficial as the model will focus on more recent instances and the effect of old instances will be downgraded. However, the constant increase of λ might lead to high values and the total discard of historical data. To prevent this, we set an upper limit λ_{max} for the highest value of λ. If limit is reached, λ_{max} ageing is applied for the rest of the stream. Note that for efficiency issues we check for change not after each instance but after a certain number of instances, denoted by w. This implies that each λ value has an effect for at least w instances.
- **SlowIncreaseFastReset** - *Gradually increase λ by a constant value c and reset to λ_0 after λ_{max} is reached:*
  The constant increase of λ in the previous strategy implies more and more data forgetting as more changes are detected in the stream. Typically though in a stream periods of change are followed by periods of stability, therefore such a forgetting is very harsh. To count for this effect, we reset λ to its initial value λ_0 when the max value λ_{max} is reached and after a certain period at this ageing level; this period is implemented in terms of a fixed number of instances w (one could use timepoints alternatively).
- **FastSetFastReset** - *Fast set to λ_{max} upon change and fast reset to initial λ_0 after a certain period:*

When a change is detected, λ is instantly increased to an upper bound λ_{max}, i.e., $\lambda = \lambda_{max}$. The λ is reset to its initial value λ_0 after a certain period of w instances. The intuition is to forget fast (with λ_{max}) in times of change and slow (with λ_0) in times of "stability".

– **FastSetSlowDecrease** - *Fast set to* λ_{max} *upon change and slow reset to initial* λ_0 *by* $\delta\lambda\%$ *decrease at each step:*
When a change is detected, λ is instantly set λ_{max}, i.e., $\lambda = \lambda_{max}$. The λ is reset to its initial value λ_0 gradually with a $\delta\lambda\%$ step. That is, at each step, *lambda* is decreased by $\delta\lambda\%$ until it reaches λ_0. The duration of each step is w instances. This offers a more gradual adaptation of λ comparing to the previous strategy.

The above strategies aim at tuning the ageing factor λ and indirectly the MNB classifier. There are other ways to affect the classifier, which do not involve direct λ tuning though. We overview them below.

– **Rebuild** - *Constant* λ_0 *and model rebuild upon change:*
A constant λ, $\lambda = \lambda_0$, is applied over the whole stream but once a change is detected the classifier is rebuilt upon the most recent w instances. The constant ageing over the whole stream should, in times of relative stability, reduce the effect of noise and in case of drastic changes, the rebuilding implies an abrupt forgetting of old, outdated information. Rebuilding incurs the fastest adaptation to change, however it completely ignores any old knowledge.

Depending on the value of λ_0 we can, for all the above strategies, distinguish two cases: (i) $\lambda_0 > 0$ and (ii) $\lambda_0 = 0$. The former applies a constant ageing λ_0 in the stream, whereas the later does not consider ageing. Moreover, we also include the following strategies as baselines.

– **fadingMNB** - *Constant ageing, no change detection:* This is the blind adaptation approach (*fadingMNB*) [21]. There is a constant ageing, $\lambda = \lambda_0 > 0$, over the stream, but there is no change detection.
– **accumulativeMNB** - *No-ageing, no change detection:* This is the accumulative MNB approach [2], discussed in Sect. 3. It does not forget, neither invokes some change detection mechanism. The model is accumulative as it considers all instances from the beginning of the stream.

5   Experiments

5.1   Dataset

We use the TwitterSentiment dataset [19], introduced in [7]. The dataset was collected by querying the Twitter API for tweets between April 6, 2009 and June 25, 2009. The sentiment labels were derived by a Maximum Entropy classifier that was trained on emoticons [7]. The final stream consists of 1,600,000 opinionated tweets, 50% of which are positive and 50% negative. We aggregate the tweets hourly, the class distribution is shown in Fig. 1(a). The class distribution is quite stable in the beginning of the stream with the positive class slightly dominating the stream. The class distribution changes drastically towards the end of the stream as only instances of the negative class are present. The change point is instance number 1,326,000. We refer to this dataset as $DS1$.

To experiment with a more volatile stream setting, we introduced some more changes to the original stream by removing certain fractions of instances. The new dataset, denoted as $DS2$, is depicted in Fig. 1(b). The dataset is no longer balanced: it contains 1,073,065 tweets with 378,288 positive and 694,777 negative instances.

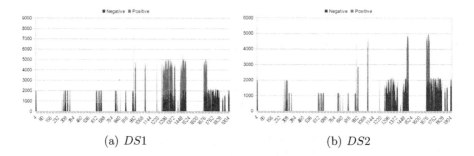

(a) $DS1$ (b) $DS2$

Fig. 1. Hourly aggregated class distribution for streams $DS1$, $DS2$.

For the evaluation, we used prequential evaluation, where each instance of the stream is first used for testing and then for training the model. As quality measures we used accuracy over an evaluation window, $evalW$. For the detection of the change points, we used $\alpha = 1.8$. We used $\beta = 0.334$ for the detection of warning points.

5.2   Classifiers Performance

We report here on the performance of the different adaptation techniques listed in Sect. 4, for both $DS1$ and $DS2$[1].

Overall performance. The overall results for $DS1$ are depicted in Fig. 2(left). The *accumulativeMNB* that does not forget achieves the worse performance, whereas *SlowIncreaseUpToALimit* with $Init - \lambda^2$ achieves the best performance, followed by *SlowIncreaseFastReset* with $Init - \lambda$. Also, for all different strategies, a constant ageing over the stream (i.e., the $Init - \lambda$ strategies where $\lambda_0 > 0$), is better than no-ageing (i.e., the $Zero - \lambda$ strategies with $\lambda_0 = 0$).

The overall results for DS_2 are depicted in Fig. 2(right). Similarly to DS1, *accumulativeMNB* achieves the worse performance, whereas *Rebuild* with Init-λ

---

[1] Parameters for DS1, DS2 are listed in Table 1.

[2] $Init - \lambda$ is the case of $\lambda_0 > 0$.

Table 1. Best parameter setting per strategy.

Strategies	λ_0		instances w (*1,000)		λ_{max}		decrease λ ratio ($\delta\lambda\%$)		increase λ value (c)	
	DS1	DS2	DS1	DS2	DS1	DS2	DS1	DS2	DS1	DS2
fadingMNB	0.2	0.3	-	-	-	-	-	-	-	-
Rebuild-Zero-λ	-	-	-	-	-	-	-	-	-	-
Rebuild-Init-λ	0.2	0.25	-	-	-	-	-	-	-	-
FastSetFastReset-Zero-λ	-	-	100	100	0.5	0.4	-	-	-	-
FastSetFastReset-Init-λ	0.1	0.15	24	22	0.5	0.5	-	-	-	-
FastSetSlowDecrease-Zero-λ	-	-	100	100	0.5	0.4	5%	5%	-	-
FastSetSlowDecrease-Init-λ	0.1	0.15	24	22	0.5	0.5	5%	5%	-	-
SlowIncreaseUpToALimit-Zero-λ	-	-	-	-	-	-	-	-	0.6	0.1
SlowIncreaseUpToALimit-Init-λ	0.2	0.2	-	-	-	-	-	-	0.1	0.1
SlowIncreaseFastReset-Zero-λ	-	-	-	-	-	-	-	-	0.4	0.2
SlowIncreaseFastReset-Init-λ	0.2	0.2	-	-	-	-	-	-	0.4	0.2

achieves the best performance, followed by *FastSetFastReset* with Init-λ, *FastSetSlowDecrease* with Init-λ and *SlowIncreaseFastReset* with Init-λ. Again, having an init λ (i.e., $\lambda_0 > 0$ is better than no-ageing (i.e., $\lambda_0 = 0$), for all cases.

We should note that $DS2$ is a very volatile stream; this might explain why rebuild ranks first for $DS2$.

Overtime performance. In Figs. 3(a), (b) we show the performance over time, for the different strategies for $DS1, DS2$, respectively.

As expected, the different strategies have an effect only after change. Before change, we can comment on the difference between approaches with an $Init - \lambda$, i.e., with ageing, and approaches with $Zero - \lambda$, i.e., no-ageing. *fadingMNB* and all the strategies with $\lambda_0 > 0$ perform better than *accumulativeMNB*, during the "stable" period. Upon change, the differences between the different methods are better manifested: The *accumulativeMNB* has the lowest performance for both datasets as it does not manage to recover after change. Methods that reset reach the poor performance of *accumulativeMNB* after a while, i.e., when they reset

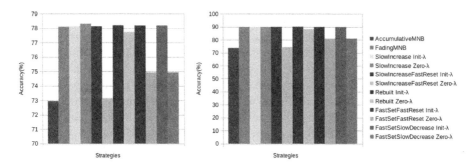

Fig. 2. Overall accuracy of different strategies $DS1$ (left), $DS2$ (right) ($Init - \lambda$ corresponds to λ_0). (Color figure online)

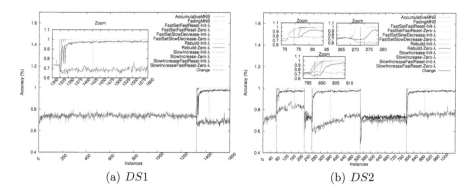

(a) $DS1$ (b) $DS2$

Fig. 3. Accuracy over time for the different strategies for $DS1$ and $DS2$

to initial lambda. Again, the $init - \lambda$ approaches perform better; this is clearly depicted from the performance of the rebuild method after change (see zoom-in figures) for both datasets (red for $\lambda_0 > 0$ vs black for $\lambda_0 = 0$).

5.3   Qualitative Evaluation

To qualitatively evaluate the change detector and the interplay with the classifier adaptation, we also experimented with a third focused dataset, collected from Twitter's public streaming API[3] for two specific entities, namely "Obama" and "Adele", during 2015. Our intention was to use very different entities, which will probably generate different words. "Obama"'s vocabulary, for example, will be related to politics, whereas "Adele"' vocabulary will be related to music with no much overlap between them. Out of the total 71,124 tweets, the majority (66,012) refers to "Obama" and the remaining (5,112) to "Adele". Figure 4 depicts the class distribution for both entities.

In the beginning of the stream, only "Obama" is present, "Adele" is introduced on instance 28,000 and remains up to instance 43,000, after that only "Obama" is present again. The vocabulary-based change detector is sensing a change at point 28,321 (recall "Adele" was introduced on instance 28,000) and raises an alarm. At point 30,000 a real change is detected and the classifier adaptation strategies take effect. The change detector starts sensing a new change at point 46,778 (recall "Adele" is removed after instance 43,000) and detects the actual change on instance 48,000. The alarms, detected changes and performance of the classifier are depicted in Fig. 5. In both cases, the change detector managed to detect the changes in the underlying stream, though with delay. The delay is due to the detector itself as even for different entities like "Adele" and "Obama" coming from different areas, the vocabulary is not completely disjoint rather common words are used in both cases.

What is interesting is that the performance of the classifier started dropping before the first actual change point. A possible explanation is that even within

---

[3] https://dev.twitter.com/streaming/overview.

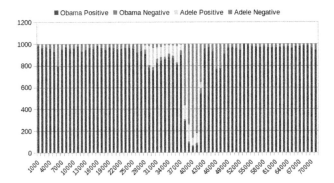

Fig. 4. Class distribution for both entities

a single topic, like "Obama" there might be changes which affect the classifier and therefore, we observe the drop. Those changes could not be detected by our vocabulary-based detector, because for example the alarm threshold α was too high or because the change itself cannot be captured by a vocabulary-based detector. The same behavior is observed after the second change point. What these incidents might indicate is that a single change detector type might not be adequate to deal with all different types of change that can occur in a stream. In practice, change might be due to different reasons like change in the class distribution, different topics discussed in the stream, internal changes within a topic etc. This calls for different types of change detectors that can be activated under different conditions. We plan to undertake this challenge of building a change detection framework of different detectors in our future work.

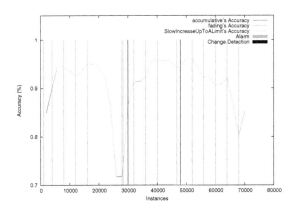

Fig. 5. Informed adaptation: change points and accuracy

6   Conclusions and Outlook

We presented an informed adaptation approach for ageing-based Multinomial Naive Bayes classifiers in order to allow adaptation at different rates over the stream based on the dynamics of the underlying stream population. Our motivation is that in times of change, ageing should be more harsh to allow for a faster adaptation of the model, however in times of stability the ageing factor should be lowered to allow for model exploitation. We proposed several adaptation techniques for the ageing factor λ. The experimental results showed that different strategies perform similarly but all of them outperform techniques that use no ageing. The same holds for harsh forgetting techniques, like model rebuild. In our experiments informed adaptation performed similarly to blind adaptation approaches. However, we should stress that the informed adaptation methods, expect for the adapted classification model, also provide the user with the points of change, which is valuable for decision making and reactions to change.

Thus far, we tune the model indirectly through the ageing factor λ. In our future work, we will also discard outdated parts of the model, i.e., outdated class priors and class conditional word probabilities to allow for faster adaptation to change and re-learning of outdated parts of the model. Moreover, as our qualitative experiment revealed, a change detector can detect a single type of change, although in practice, change might occur due to different reasons. We plan to investigate the possibility of a framework of different detectors that can be activated under different conditions and might call for different model update strategies.

Acknowledgements. The work was partially funded by the European Commission for the ERC Advanced Grant ALEXANDRIA under grant No. 339233 and by the German Research Foundation (DFG) project OSCAR (Opinion Stream Classification with Ensembles and Active leaRners).

References

1. Bermingham, A., Smeaton, A.F.: Classifying sentiment in microblogs: is brevity an advantage? In: Proceedings of the 19th ACM International Conference on Information and Knowledge Management, CIKM 2010, pp. 1833–1836. ACM, New York (2010)
2. Bifet, A., Frank, E.: Sentiment knowledge discovery in Twitter streaming data. In: Pfahringer, B., Holmes, G., Hoffmann, A. (eds.) DS 2010. LNCS, vol. 6332, pp. 1–15. Springer, Heidelberg (2010). doi:10.1007/978-3-642-16184-1_1
3. Domingos, P., Pazzani, M.: On the optimality of the simple bayesian classifier under zero-one loss. Mach. Learn. **29**(2–3), 103–130 (1997)
4. Forman, G.: Tackling concept drift by temporal inductive transfer. In: Proceedings of the 29th Annual International ACM SIGIR Conference on Research and Development in Information Retrieval, pp. 252–259. ACM (2006)
5. Gama, J.: Knowledge Discovery from Data Streams. CRC Press, Boca Raton (2010)

6. Gama, J.A., Žliobaitė, I., Bifet, A., Pechenizkiy, M., Bouchachia, A.: A survey on concept drift adaptation. ACM Comput. Surv. **46**(4), 44:1–44:37 (2014)
7. Go, A., Bhayani, R., Huang, L.: Twitter sentiment classification using distant supervision. Processing, 1–6 (2009)
8. Gokulakrishnan, B., Priyanthan, P., Ragavan, T., Prasath, N., Perera, A.S.: Opinion mining and sentiment analysis on a Twitter data stream. In: Proceedings of 2012 International Conference on Advances in ICT for Emerging Regions (ICTer), ICTer 2012, pp. 182–188. IEEE (2012)
9. Klinkenberg, R.: Learning drifting concepts: example selection vs. example weighting. Intell. Data Anal. **8**(3), 281–300 (2004)
10. Koren, Y.: Collaborative filtering with temporal dynamics. Commun. ACM **53**(4), 89–97 (2010)
11. Koychev, I.: Gradual forgetting for adaptation to concept drift. In: Proceedings of ECAI 2000 Workshop on Current Issues in Spatio-Temporal Reasoning (2000)
12. Koychev, I.: Tracking changing user interests through prior-learning of context. In: Bra, P., Brusilovsky, P., Conejo, R. (eds.) AH 2002. LNCS, vol. 2347, pp. 223–232. Springer, Heidelberg (2002). doi:10.1007/3-540-47952-X_24
13. Liu, Y., Yu, X., An, A., Huang, X.: Riding the tide of sentiment change: sentiment analysis with evolving online reviews. World Wide Web **16**(4), 477–496 (2013)
14. McCallum, A., Nigam, K.: A comparison of event models for Naive Bayes text classification. In: AAAI-98 Workshop on Learning for Text Categorization, pp. 41–48. AAAI Press (1998)
15. Pechenizkiy, M., Bakker, J., Žliobaitė, I., Ivannikov, A., Kärkkäinen, T.: Online mass flow prediction in CFB boilers with explicit detection of sudden concept drift. ACM SIGKDD Explor. Newslett. **11**(2), 109–116 (2010)
16. Plaza, L., de Albornoz, J.C.: Sentiment Analysis in Business Intelligence: A Survey, pp. 231–252. IGI-Global (2011)
17. Salganicoff, M.: Tolerating concept and sampling shift in lazy learning using prediction error context switching. Artif. Intell. Rev. **11**(1–5), 133–155 (1997)
18. Schlimmer, J.C., Granger, R.H.: Beyond incremental processing: tracking concept drift. In: AAAI, pp. 502–507 (1986)
19. Sentiment140. Sentiment140 - a Twitter sentiment analysis tool. http://help. sentiment140.com/
20. Turney, P.D.: Thumbs up or thumbs down?: semantic orientation applied to unsupervised classification of reviews. In: Proceedings of the 40th Annual Meeting on Association for Computational Linguistics, ACL 2002, Stroudsburg, PA, USA, 2002, pp. 417–424. Association for Computational Linguistics (2002)
21. Wagner, S., Zimmermann, M., Ntoutsi, E., Spiliopoulou, M.: Ageing-based multinomial Naive Bayes classifiers over opinionated data streams. In: Proceedings of the European Conference on Machine Learning and Knowledge Discovery in Databases, Porto, Portugal, pp. 401–416 (2015)
22. Widmer, G., Kubat, M.: Learning in the presence of concept drift and hidden contexts. Mach. Learn. **23**(1), 69–101 (1996)

Mining Semantic Patterns for Sentiment Analysis of Product Reviews

Sang-Sàng Tan[✉] and Jin-Cheon Na

Wee Kim Wee School of Communication and Information, Nanyang Technological University,
31 Nanyang Link, Singapore 637718, Singapore
{tans0348,tjcna}@ntu.edu.sg

Abstract. A central challenge in building sentiment classifiers using machine learning approach is the generation of discriminative features that allow sentiment to be implied. Researchers have made significant progress with various features such as n-grams, sentiment shifters, and lexicon features. However, the potential of semantics-based features in sentiment classification has not been fully explored. By integrating PropBank-based semantic parsing and class association rule (CAR) mining, this study aims to mine patterns of semantic labels from domain corpus for sentence-level sentiment analysis of product reviews. With the features generated from the semantic patterns, the F-score of the sentiment classifier was boosted to 82.31% at minimum confidence level of 0.75, which not only indicated a statistically significant improvement over the baseline classifier with unigram and negation features (F-score = 73.93%) but also surpassed the best performance obtained with other classifiers trained on generic lexicon features (F-score = 76.25%) and domain-specific lexicon features (F-score = 78.91%).

Keywords: Sentiment analysis · Semantic parsing · Pattern mining · Machine learning · Sentiment classification

1 Introduction

The preferences and decisions of internet users are increasingly influenced by peer opinions from online reviews, social networks, blogs, and other user-generated content on the web [1–3]. This growing reliance on user-generated content has triggered wide interest among stakeholders to capture these data and turn them into insightful information for various purposes including decision-making, target marketing, competitor analysis, etc. To this end, sentiment analysis has been extensively studied for gathering, extracting, and classifying users' sentiment expressed in textual content as a means to understand users' attitudes towards the targets of analysis. In the field of digital library, sentiment analysis can be employed in many ways—for example, as sentiment-based recommender system or sentiment-based searching and browsing features—for advanced retrieval of digital objects.

Sentiment analysis can be regarded as a subtask of natural language processing (NLP) that attempts to build, in machines, the abilities to imitate some cognitive abilities of human beings in interpreting human language for implying sentiment. Since a big

© Springer International Publishing AG 2017
J. Kamps et al. (Eds.): TPDL 2017, LNCS 10450, pp. 382–393, 2017.
DOI: 10.1007/978-3-319-67008-9_30

part of the complexity of sentiment analysis lies in the processing of meanings in human language, we posit that the problem should be addressed in light of semantically sound approaches. As pointed out by researchers [4, 5], progress in sentiment analysis should be made towards "(a) enriching shallow representations with linguistically motivated, rich information, and (b) focusing different branches of research and combining resources and work forces to join hands with related work in NLP" [4] (pg. 66).

In the interest of exploring the potential of semantically motivated approach for sentiment classification, we propose the use of semantic parsing and pattern mining to derive a set of semantic patterns as features for training sentiment classifiers. Given a piece of text, the primary goal of semantic parsing is to detect the events described in the text and to identify the participants and their roles in the events, as a means to answer the question "Who did What to Whom, and How, When and Where?" [6] (pg. 1). Unlike syntactic parsing that focuses on grammatical relations between the components of a sentence, semantic parsing is an important step towards the understanding of meanings. In this study, we used a PropBank-based semantic parser [7] to transform sentences from word-level representations to their semantic representations. The semantically labelled data were then fed into class association rule (CAR) mining algorithm [8] to extract the semantic patterns that were regularly associated with the positive and negative senti-ment. For instance, a semantic pattern like '*negation + buy.01 (predicate) + thing bought (argument)*' could be an evidence of negative sentiment. Since expressions that convey the similar meanings tend to share common semantic forms although they might differ in the use of words, word orders, and syntactic forms, such patterns would provide more generalizable features, resulting in a less sparse feature space for learning the classifi-cation model.

2 Related Work

Researchers have explored a wide range of features for sentiment classification. Word n-grams, which were popularized by Pang et al. [9], are among the most commonly used features that have produced acceptable performance on sentiment classification. Different types of data have their own unique characteristics that might be useful for inferring sentiment polarities. Tweets data, in particular, contain hashtags and emoticons that have shown to be closely related to the emotions expressed in tweet messages, making them well-suited as features for classifying sentiment in tweets [10, 11]. Other popular features for classifying sentiment include punctuation marks, part-of-speech tags, sentiment shifters like negators and other modifiers (e.g., very and barely), and stylistic features such as words per document and words per sentence.

As far as sentiment classification features are concerned, our stance is that semantics-based features are more likely to have a significant impact on classifiers because the inter-pretation of sentiment relies on the understanding of meanings. Generally speaking, any approach that takes into consideration the denotations and connotations of words or phrases falls into the category of semantics-based approach. One such approach that has been proven superior is the sentiment lexicon approach because an important indicator of sentiment in textual content is the use of sentiment terms that express likes and dislikes.

It is well established that combining lexical knowledge tends to show promising improvement in sentiment classification (e.g., [12–14]). The study conducted by Mohammad et al. [11] highlighted the importance of sentiment lexicons through their experiments that compared the effects of various classification features by removing one feature set at a time in the classification process. Among the feature sets compared in the experiments (including n-grams, negations, part-of-speech tags, emoticons, punctuation marks, hashtags, among others), sentiment lexicons were found to produce the most influential features for sentiment classification, to the extent that the removal of such features dropped the F-scores of the classification by more than 8.5%.

A considerable amount of literature has been published on the construction of sentiment lexicons. Earlier studies in this line of research focused on building general-purpose sentiment lexicons that only include terms of which the prior sentiment scores can be assumed with minimal uncertainty based on denotations and connotations of the terms (e.g., [15, 16]). Words like 'generosity' and 'admirer' can be considered as inherently positive whereas words like 'betrayal' and 'nauseating' can be considered as inherently negative. Generic sentiment lexicons have substantial application and research values because they are highly reusable. However, researchers have recognized that domain-specific terms are extremely crucial for interpreting opinions that require domain knowledge, especially in certain domains such as medical and chemical domains. Therefore, a wide range of studies (e.g., [10, 13]) has devoted the efforts to build domain-specific sentiment lexicons from domain corpora. One of the challenges in this line of research is that the coverage of sentiment lexicons is always a concern due to the richness of human language. Even when a corpus of enormous volume is used to learn a sentiment lexicon, it is almost certain that there will be some terms in the unseen data that the lexicon fails to cover. When no match exists for a term, the lexicon fails to provide useful information for classification. With respect to the coverage issue in lexicon-based features, the present study suggests that semantic patterns obtained from mining semantically parsed data would constitute a more generalizable feature set, in the sense that the semantic patterns are likely to match more cases in the unseen data.

Many NLP tasks that require semantic interpretation and processing could benefit from semantic parsing. One of the earliest studies that exploited semantic parsing for sentiment analysis is the work undertaken by Kim and Hovy [17]. They applied frame-based semantic parsing to identify the opinion holders and topics expressed in online news media text. The goal of their study was to find out which semantic roles could be used to identify opinion holders and topics. Another study that also used frame-like schemas for detection of opinion holders and topics was carried out by Gangemi et al. [18]. Their study used VerbNet (https://verbs.colorado.edu/~mpalmer/projects/verbnet.html) to find verb classes and thematic roles of verb arguments that indicated the presence of opinion holders and topics. As far as we know, the integration of semantic parsing and pattern mining for feature generation in sentiment analysis has not been fully explored. It would seem, therefore, that further investigations are desirable to find out whether semantic patterns mined from PropBank's verb-oriented semantic labels would have a positive impact on the performance of sentiment classifiers.

3 Method

3.1 Semantic Parsing

The development of most semantic parsing tools relies on human-annotated resources that provide annotations for verbs and their arguments. Recent years have seen increasingly rapid advances in semantic parsing due to the continuing efforts devoted to the development and maintenance of high-quality resources like PropBank [19]. PropBank has known to be an extremely influential resource in semantic parsing. It was created by the Proposition Bank project to provide predicate-argument information on top of Penn Treebank's syntactic layer.

In the present study, Punyakanok et al.'s [7] PropBank-based semantic parser was used for semantic processing. This semantic parser is able to identify the following arguments of verb predicates:

- Core arguments (*A0–A5* and *AA*) which are labelled based on the semantics of verb predicates specified in PropBank. The numbers and types of arguments vary across predicates. For instance, the arguments of predicate '*break*' in PropBank are: *A0 - breaker*, *A1 - thing broken*, *A2 - instrument*, *A3 - pieces*, and *A4 - argument 1 broken away from what*.
- Thirteen adjunct arguments which are labelled as *AM-adj* where *adj* is the type of adjunct. The adjunct types are listed in Table 1.

Table 1. Thirteen adjunct types and their descriptions

Label	Description	Label	Description
AM-ADV	Adverbial modification	AM-NEG	Negation
AM-DIR	Direction	AM-PNC	Proper noun component
AM-DIS	Discourse marker	AM-PRD	Secondary predicate
AM-EXT	Extent	AM-PRP	Purpose
AM-LOC	Location	AM-REC	Reciprocal
AM-MNR	Manner	AM-TMP	Temporal
AM-MOD	General modification		

- Continued arguments which extend other arguments (core or adjunct arguments). This type of arguments is labelled as *C-arg* where *arg* is the label of the argument for which the continuity needs to be indicated. For instance, *C-A1* and *C-AM-TMP* indicate that the current arguments are continued from core argument *A1* and adjunct argument *AM-TMP* respectively.
- Referential arguments which represent relative pronouns. Referential arguments are labelled as *R-arg* where *arg* is the label of the core argument or adjunct argument to which the relative pronoun refers.

Figure 1 shows the semantic labels generated by the semantic parser for the sentence '*Also, pieces of this liner would break off very easily while using it.*'. The semantic parser has identified two predicates—'*break*' and '*use*'—in the sentence, with each predicate and its arguments forming a sequence of semantic labels that would become the inputs to the association rule mining algorithm.

	Predicate *'break'* and its arguments	Predicate *'use'* and its arguments
Also	discourse marker [AM-DIS]	
,		
pieces		
of		
this	thing broken [A1]	
liner		
would	general modification [AM-MOD]	
break	V: break.01	
off	V: break	
very	manner [AM-MNR]	
easily		
while		
using	temporal [AM-MNR]	V: use.01
it		thing used [A1]
.		

Fig. 1. Semantic labels generated by Punyakanok et al.'s [7] semantic parser for the sentence *'Also, pieces of this liner would break off very easily while using it.'*

3.2   Class Association Rule (CAR) Mining

Association rule mining is a machine learning method introduced by Agrawal et al. [20] for discovering interesting patterns of purchases in large-scale transaction data of supermarkets. The results of the analysis are a set of association rules or statements of regularities.

Let $T = \{t_1, t_2, t_3, \ldots, t_m\}$ be a set of transactions and $I = \{i_1, i_2, i_3, \ldots, i_n\}$ be a set of items such that each transaction t_k consists of one or more items from set I. An association rule discovered from the transaction data is a statement in the form $X \Rightarrow Y$, where X and Y are some items or itemsets that appear in the transactions, with $X \subset I$, $Y \subset I$, and $X \cap Y = \emptyset$. Various measures of significance and certainty can be applied to select the interesting rules, the most widely accepted measures being the support level and the confidence level. The goal of association rule mining is then to find the rules that satisfy the user-specified minimum support threshold (*minsup*) and minimum confidence threshold (*minconf*). The support level of a rule indicates how frequent the condition (i.e. the itemset X) appears in the transactions whereas the confidence level indicates how often the appearance of the condition actually leads to the consequence (i.e. the itemset Y). A confidence level of 1.0 indicates that Y always appears in the same transaction whenever X appears. The two measures are calculated as follows:

$$support(X) = \frac{|\{t \in T; X \subseteq t\}|}{|T|} \tag{1}$$

$$confidence(X \Rightarrow Y) = \frac{support(X \cup Y)}{support(X)} \tag{2}$$

The variant of association rule mining adopted in the present study is the one proposed by Liu et al. [8]. The original algorithm introduced by Agrawal et al. [20] does not impose any restriction to the targets or the consequences of rules. Liu et al.'s algorithm, on the other hand, integrated classification and association rule mining to acquire CARs for predetermined targets. Both algorithms have wide-ranging applications but the latter is better-suited for the purpose of pattern mining in the present study, i.e. to mine regular patterns of semantic labels that are commonly associated with the positive and negative sentiment. The resulted set of rules would be in the form $X \Rightarrow$ positive and $X \Rightarrow$ negative where X is the semantic pattern of which the appearance increases the probability of positive sentiment and negative sentiment respectively.

4 Experiments

4.1 Dataset

This study used the same cosmetic dataset we collected for our earlier study [21] that assessed the significance of multi-word sentiment terms in a lexicon-based, supervised sentiment classification task. The cosmetic dataset consists of 1100 positive sentences and 1100 negative sentences collected from MakeupAlley (www.makeupalley.com), a popular beauty website that provides consumers' reviews on beauty products. After the collected reviews were tokenized into sentences using Stanford Parser, sentences were labelled by two annotators; the first annotator labelled all the sentences whereas the second annotator labelled 300 sentences. Inter-rater reliability test using Cohen's kappa showed high agreement between the two annotators ($\kappa = .829, p < .0005$).

4.2 Feature Engineering, Classifier, and Validation

Evaluation of the semantic pattern features was performed by comparing the classification results produced by support vector machine (SVM) classifiers trained on different

Table 2. Feature sets for training the four groups of SVM classifiers (C1–C4)

Feature Sets	Classifiers			
	C1	C2	C3	C4
Baseline features (BL)	√	√	√	√
Generic lexicon features (GL)		√	√	√
Domain lexicon features (DL)			√	√
Semantic pattern features (SP)				√

feature sets, as shown in Table 2. Results were obtained using 10-fold cross validation, with each partition containing the same number of positive and negative sentences. SVMs with linear kernel and parameter C = 0.1 were used in all experiments.

The baseline features consist of binary features that indicated the presence or absence of unigrams and numeric features that indicated the number of negations in a sentence. Since lexicon features have shown promising results in sentiment classification [11], we compared the semantic pattern features to lexicon-related features generated using Hu and Liu's generic sentiment lexicon [16] and a set of domain-specific lexicons. In our earlier work [21], bigram domain lexicons were found to produce the best-performing lexicon features for sentiment classification. Therefore, we adopted the same approach in the present study to generate domain-specific lexicons consisting of bigram entries selected from nine of ten partitions (except the test partition) of the data using Pointwise Mutual Information (PMI) [10]. The following lexicon-related features were generated from the generic lexicon and the domain-specific lexicons:

- **Sum of Sentiment Scores**. For each sentence, sums of sentiment scores were obtained for all terms, the positive terms, and the negative terms. With Hu and Liu's sentiment lexicon, the positive terms were given prior scores of +1 whereas the negative terms were given prior scores of −1. The sentiment scores of terms in the domain-specific lexicons are positive and negative values calculated using PMI. For instance, the term '*stay power*' has a score of +1.435.
- **Count of Terms**. Three features were generated to indicate the number of positive, negative, and neutral terms in each sentence.

For each round of the 10-fold cross validation, semantic patterns were mined from nine of ten partitions, with the test partition excluded. The semantic patterns obtained from CAR mining formed a set of binary features, of which the values were determined based on the presence or absence of the patterns in each sentence. We considered the presence of a semantic pattern as the co-occurrence of all semantic labels of the pattern in a sentence, regardless of the order in which they occurred.

Considering the size of the dataset, the *minsup* threshold of CAR mining in the experiments was set to a very low value (0.0003), allowing any pattern that occurred more than twice to be considered as frequent itemset. Several thresholds of *minconf* in the range of 0.60 and 0.85 were tested in the experiments. Besides *minsup* and *minconf*, another crucial factor that might affect the outcomes of CAR mining is the form of semantic labels used in the mining process. The semantic patterns (X) that will be generated, the number of rules, as well as the support level and confidence level of each rule might vary according to the semantic labels used in the process. For instance, the labels for the core arguments of the verb predicates can take many forms:

- The more general form like *A0–A5* and *AA*. It is general in the sense that all predicates have the same set of labels for their core arguments.
- The specific form that ties an argument to its predicate, like [*A0: break.01*]. The predicate and its sense number (*break.01*) are appended to the core argument label (*A0*) so that the label will not be confused with the core argument labels of other predicates.

- The intermediate form that provides the definition of the argument per se (based on the definition in PropBank). This label is not strictly tied to the predicate because the same definition can describe the arguments of different but related predicates. For instance, the definition for argument 2 of predicate '*purchase*' is '*seller*', yet the same definition is also used to describe argument 2 of predicate '*buy*' as well as argument 1 of predicate '*sell*'.

Table 3 shows the different forms of semantic labels for the example given in Fig. 1. The labels of the core arguments are highlighted in bold. Different forms of labels would appear at different frequencies in the data, thus the choice of labels has a huge impact on the outcomes of the pattern mining algorithm. The intermediate form is neither too general nor too specific, so it was chosen as the approach in our experiments to generate inputs for CAR mining.

Table 3. Different forms of semantic labels for the sentence '*Also, pieces of this liner would break off very easily while using it.*'

Label Form		Semantic Labels
General	t1	*[AM-DIS], [A1], [AM-MOD], [break.01], [break], [AM-MNR], [AM-TMP]*
	t2	*[use.01], [A1]*
Specific	t1	*[AM-DIS], [A1: break.01], [AM-MOD], [break.01], [break], [AM-MNR], [AM-TMP]*
	t2	*[use.01], [A1: use.01]*
Intermediate	t1	*[AM-DIS], [thing broken], [AM-MOD], [break.01], [break], [AM-MNR], [AM-TMP]*
	t2	*[use.01], [thing used]*

Semantic parsing provides the abstraction of meanings but the semantic labels might discard some information that could be quite distinctive in sentiment analysis. For example, '*really badly*' and '*really nicely*' can both be labelled as [AM-MNR] (manner), even though they are obviously related to different sentiment. To address this problem, for every word encompassed by a semantic label, the sum of polarity scores of the words was appended to the label so that each label is a two-tuple of [*semantic label:sum of polarity scores*]. For instance, the sum of polarity scores for the phrase '*really badly*' is -1, then the enhanced semantic label would be [*AM-MNR: -1*]. The polarity scores were obtained from Hu and Liu's generic sentiment lexicon [16]. Note that the sums of polarity scores are not shown in Table 3 but in the subsequent discussion, all semantic labels would be presented with the appended scores.

5   Results and Discussion

As can be seen from the results presented in Table 4, the sentiment classifiers have obviously benefited from the semantic pattern features. Results were obtained for five *minconf* thresholds. The size of the feature set (i.e. the number of semantic patterns) decreased as *minconf* increased. It has been observed that the performance of the sentiment classifiers increased marginally as *minconf* increased in the range of 0.60 and 0.75.

A reasonable explanation is that higher *minconf* values caused some spurious rules with low confidence levels to be excluded from the feature set. For example, these two ambiguous rules were excluded when *minconf* was set to 0.70:

- [*get.01:0*] + [*thing gotten:+ 1*] ⇒ negative (conf. level = 0.62)
- [*apply.02:0*] + [*applied to:0*] ⇒ negative (conf. level = 0.67)

Table 4. Classification results with p values indicated (* $p \leq .05$, ** $p \leq .01$, *** $p \leq .001$). BL: Baseline features. GL: Generic lexicon features. DL: Domain lexicon features. SP: Semantic pattern features. #SP: Number of semantic patterns.

Classifiers	Accuracy	Precision	Recall	F-Measure
C1: BL	73.95	74.04	73.94	73.93
C2: BL + GL	76.27	76.41	76.27	76.25
C3: BL + GL + DL	78.91 **	78.95 **	78.91 **	78.91 **
C4: BL + GL + DL + SP				
minconf = 0.60, #SP = 2913	81.59 ***	81.68 ***	81.60 ***	81.58 ***
minconf = 0.65, #SP = 2663	81.73 ***	81.82 ***	81.73 ***	81.71 ***
minconf = 0.70, #SP = 2136	82.14 ***	82.20 ***	82.15 ***	82.14 ***
minconf = 0.75, #SP = 2032	82.32 ***	82.38 ***	82.32 ***	82.31 ***
minconf = 0.85, #SP = 1579	81.91 ***	82.00 ***	81.92 ***	81.91 ***

Nevertheless, as demonstrated by the classifier's performance at *minconf* = 0.85, further increments of the *minconf* values would likely backfire due to the elimination of potentially useful rules. Based on this observation, a value around 0.75 seems to be a reasonable threshold for *minconf*. Despite the effects of the *minconf* values on the classifiers' performance, independent t-tests performed on the accuracy, precision, recall, and F-measure of the classification results showed that the performance of the C4 classifiers at all *minconf* thresholds was statistically significantly better than the performance of the baseline classifier (C1). Furthermore, with the inclusion of semantic pattern features, the C4 classifiers also outperformed classifier C2 and classifier C3, which were trained only on the baseline features and the lexicon-related features.

At higher confidence levels, the pattern mining algorithm was able to derive some interesting and meaningful semantic patterns, some of which have revealed implicitly expressed sentiment that is usually not detectable or distinguishable using other features. It is generally agreed that the classification of implicitly expressed sentiment is more challenging due to the absence of prominent evidence that signifies the manifestation of sentiment. As shown in the sentences that matched semantic pattern (1) and semantic pattern (2) in Table 5, without the use of sentiment-laden words, reviewers might express their sentiment implicitly by sharing experiences, describing scenarios, giving advice, and so forth. Unlike explicitly expressed sentiment that can be more easily detected from certain sentiment keywords, the classification of implicitly expressed sentiment often requires the overall meanings of the sentences to be taken into consideration. Sometimes, the presence of sentiment keywords might also mislead the sentiment classifier. For instance, the sentences that matched semantic pattern (3) both contain positive sentiment words like '*well*' and '*love*', and thus can easily be mistaken as positive sentences. However, the semantic patterns were able to recognize the two sentences as expressing negative sentiment. This finding suggests that semantic patterns might take us a step closer to perfecting sentiment analysis by improving the classification of implicit sentiment, which is expressed in a subtle manner.

Table 5. Semantic patterns with high confidence levels

(1) [*thing remaining:0*] + [*AM-NEG:0*] + [*stay.01:0*] ⇒ negative (conf. level = 1.00)
• *I have oily eyelids and the liner would not stay on after two to four hours.* • *It does not stay in the waterline at all!*
(2) [*AM-NEG:0*] + [*need.01:0*] + [*thing needed:0*] ⇒ positive (conf. level = 1.00)
• *I don't need to keep reapplying coats to get color.* • *The pots I have now will last me a very long time (as I said, you really do not need to use a lot).*
(3) [*want.01:0*] + [*thing wanted:+1*] ⇒ negative (conf. level = 0.93)
• *I really wanted this to work **well**.* • *Believe me, I wanted to **love** this product.*

6 Conclusion

In this paper, we explored the potential of semantic patterns in sentiment classification. The proposed method used PropBank-based semantic parsing and class association rule (CAR) mining to detect discriminative semantic patterns that constituted the features for building sentiment classifiers. Compared to other features, the semantic pattern features were able to improve the classifiers' performance to a greater extent. However, our experiments also revealed several issues in using PropBank-based semantic parsing to extract the sentiment-related semantic patterns. First, PropBank-based semantic parsing is verb-oriented so no semantic label was generated in the absence of verbs. Second, arguments of verb predicates might encompass large chunks of text, causing the details within the arguments to be completely omitted in the labelling process. As described earlier, this issue was partially solved in this study by appending the sum of polarity scores to each label. Such approach provided a pragmatic solution but might not be optimal. Despite the limitations, the proposed method has discovered some interesting semantic patterns that allowed subtly expressed sentiment to be recognized. This study thus suggests a potentially rewarding research direction for tackling the implicit sentiment problem that has known to be one of the highly challenging problems in sentiment analysis.

References

1. Constantinides, E., Fountain, S.J.: Web 2.0: Conceptual foundations and marketing issues. J. Direct Data Digit. Mark. Prac. **9**, 231–244 (2008). doi:10.1057/palgrave.dddmp.4350098
2. Ye, Q., Law, R., Gu, B., Chen, W.: The influence of user-generated content on traveler behavior: an empirical investigation on the effects of e-word-of-mouth to hotel online bookings. Comput. Hum. Behav. **27**, 634–639 (2011). doi:10.1016/j.chb.2010.04.014
3. Zhang, Z., Ye, Q., Law, R., Li, Y.: The impact of e-word-of-mouth on the online popularity of restaurants: a comparison of consumer reviews and editor reviews. Int. J. Hosp. Manage. **29**, 694–700 (2010). doi:10.1016/j.ijhm.2010.02.002
4. Ruppenhofer, J.: Extending FrameNet for sentiment analysis. Revista Veredas **17**, 66–81 (2013)
5. Ruppenhofer, J., Rehbein, I.: Semantic frames as an anchor representation for sentiment analysis. In: 3rd Workshop in Computational Approaches to Subjectivity and Sentiment Analysis, pp. 104–109. ACL, Stroudsburg (2012)
6. Palmer, M., Gildea, D., Xue, N.: Semantic role labeling. Synth. Lect. Hum. Lang. Technol. **3**, 1–103 (2010). doi:10.2200/S00239ED1V01Y200912HLT006
7. Punyakanok, V., Roth, D., Yih, W.T.: The importance of syntactic parsing and inference in semantic role labeling. Comput. Linguist. **34**, 257–287 (2008). doi:10.1162/coli.2008.34.2.257
8. Liu, B., Hsu, W., Ma, Y.: Integrating classification and association rule mining. In: Fourth International Conference on Knowledge Discovery and Data Mining, pp. 80–86. AAAI Press, New York (1998)
9. Pang, B., Lee, L., Vaithyanathan, S.: Thumbs up?: sentiment classification using machine learning techniques. In: ACL-02 Conference on Empirical Methods in Natural Language Processing, pp. 79–86. ACL, Stroudsburg (2002). doi:10.3115/1118693.1118704
10. Kiritchenko, S., Zhu, X., Mohammad, S.M.: Sentiment analysis of short informal texts. J. Artif. Intell. Res. **50**, 723–762 (2014). doi:10.1613/jair.4272
11. Mohammad, S.M., Kiritchenko, S., Zhu, X.: NRC-Canada: building the state-of-the-art in sentiment analysis of tweets. In: Second Joint Conference on Lexical and Computational Semantics, pp. 321–327. ACL, Stroudsburg (2013)
12. Cambria, E., Olsher, D., Rajagopal, D.: SenticNet 3: a common and common-sense knowledge base for cognition-driven sentiment analysis. In: Twenty-Eighth AAAI Conference on Artificial Intelligence, pp. 1515–1521. AAAI Press, New York (2014)
13. Ding, X., Liu, B., Yu, P.S.: A holistic lexicon-based approach to opinion mining. In: International Conference on Web Search and Data Mining, pp. 231–240. ACM, New York (2008). doi:10.1145/1341531.1341561
14. Kaji, N., Kitsuregawa, M.: Building lexicon for sentiment analysis from massive collection of HTML documents. In: Joint Conference on Empirical Methods in Natural Language Processing and Computational Natural Language Learning, pp. 1075–1083. ACL, Stroudsburg (2007)
15. Wilson, T., Wiebe, J., Hoffmann, P.: Recognizing contextual polarity in phrase-level sentiment analysis. In: Conference on Human Language Technology and Empirical Methods in Natural Language Processing, pp. 347–354. ACL, Stroudsburg (2005). doi:10.3115/1220575.1220619
16. Hu, M., Liu, B.: Mining and summarizing customer reviews. In: 10th ACM SIGKDD International Conference on Knowledge Discovery and Data Mining, pp. 168–177. ACM, New York (2004). doi:10.1145/1014052.1014073

17. Kim, S.M., Hovy, E.: Extracting opinions, opinion holders, and topics expressed in online news media text. In: Workshop on Sentiment and Subjectivity in Text, pp. 1–8. ACL, Stroudsburg (2006)

18. Gangemi, A., Presutti, V., Recupero, D.R.: Frame-based detection of opinion holders and topics: a model and a tool. IEEE Comput. Intell. Mag. **9**, 20–30 (2014). doi:10.1109/MCI.2013.2291688

19. Palmer, M., Gildea, D., Kingsbury, P.: The proposition bank: an annotated corpus of semantic roles. Comput. Linguist. **31**, 71–106 (2005). doi:10.1162/0891201053630264

20. Agrawal, R., Imieliński, T., Swami, A.: Mining association rules between sets of items in large databases. In: ACM SIGMOD International Conference on Management of Data, pp. 207–216. ACM, New York (1993). doi:10.1145/170036.170072

21. Tan, S.-S., Na, J.-C.: Expanding sentiment lexicon with multi-word terms for domain-specific sentiment analysis. In: Morishima, A., Rauber, A., Liew, C.L. (eds.) ICADL 2016. LNCS, vol. 10075, pp. 285–296. Springer, Cham (2016). doi:10.1007/978-3-319-49304-6_34

A Comparison of Pre-processing Techniques for Twitter Sentiment Analysis

Dimitrios Effrosynidis$^{(\boxtimes)}$, Symeon Symeonidis, and Avi Arampatzis

Database and Information Retrieval Research Unit,
Department of Electrical and Computer Engineering,
Democritus University of Thrace, 67100 Xanthi, Greece
{dimievfr,ssymeoni,avi}@ee.duth.gr
http://www.nonrelevant.net

Abstract. Pre-processing is considered to be the first step in text classification, and choosing the right pre-processing techniques can improve classification effectiveness. We experimentally compare 15 commonly used pre-processing techniques on two Twitter datasets. We employ three different machine learning algorithms, namely, Linear SVC, Bernoulli Naïve Bayes, and Logistic Regression, and report the classification accuracy and the resulting number of features for each pre-processing technique. Finally, based on our results, we categorize these techniques based on their performance. We find that techniques like stemming, removing numbers, and replacing elongated words improve accuracy, while others like removing punctuation do not.

Keywords: Sentiment analysis · Text pre-processing · Machine learning · Text classification

1 Introduction

In the last decade, Sentiment Analysis in microblogging has become a very popular research area. People share their daily life with messages in platforms such as Twitter, and posts of users are related with many topics. Many studies present interesting approaches for classification methods in sentiment analysis, e.g. [1,9], and refer to the important role of pre-processing before and during the feature selection process.

Pre-processing in this context is the procedure of cleansing and preparation of texts that are going to be classified. It is a fact that unstructured texts on the Internet —and in our case on Twitter— contain significant amounts of noise. By the term noise, we mean data that do not contain any useful information for the analysis at hand, i.e. sentiment analysis in our case.

According to [4], the total percentage of noise in a dataset reaches 40%, a fact that causes confusion in machine learning algorithms. Twitter users are prone to spelling and typographical errors and to the use of abbreviations and slang. They may also use punctuation signs to emphasize their emotions,

© Springer International Publishing AG 2017
J. Kamps et al. (Eds.): TPDL 2017, LNCS 10450, pp. 394–406, 2017.
DOI: 10.1007/978-3-319-67008-9_31

like many exclamation marks. Usually, it is not necessary to include all terms of the initial form of a text in the machine learning step and some of them can be ignored, replaced, or merged with others. Thus, it arises the need of cleansing and normalizing the data, as their quality is a key factor to the success of the machine learning that follows pre-processing.

The purpose of this study is to gather many common pre-processing techniques from other previous studies, plus a few novel ones such as replacing contractions and replacing negations with antonyms, and examine their significance in feature selection by measuring their accuracy in sentiment classification and their resulting number of features on two well-known datasets. In the end, based on our results, we suggest to future researchers which techniques are more suitable for Twitter sentiment analysis and which have to be avoided.

The rest of this paper is structured as follows. The following section includes a review of the related literature. Section 3 presents the pre-processing techniques that we will compare. Section 4 describes the datasets, the machine learning algorithms, and the evaluation methodology. Results and conclusions are discussed in Sects. 5 and 6, respectively.

2 Related Work

In Sentiment Analysis, especially on microblogging texts, the role of pre-processing techniques is significant as a part of text classification. Many research efforts have been made to demonstrate the difference between these techniques and their contribution to the final result of classification.

In [19], the authors examine the effects of pre-processing on twitter data for the fortification of sentiment classification. They focus on tweets which are full of symbols, abbreviations, folksonomy, and unidentified words. They remove URLs, hashtags, user mentions, punctuation, and stopwords, and they identify the importance of slang words and spelling correction. They use an SVM classifier in their experiments.

The role of pre-processing is also investigated by [18] on movie reviews. They use pre-processing techniques such as expansion of abbreviations, removal of non-alphabetic signs, stopword removal, negation handling with the addition of the prefix 'NOT_', and stemming. They also use an SVM classifier and correlate the number of features to its accuracy. They show that appropriate text pre-processing methods, including data transformation and filtering, can significantly enhance the classifier's performance.

Pre-processing techniques are also explored by [21] for two languages on e-mails and news. They use stopword removal, lowercase conversion, and stemming, and they evaluate with micro-F_1 score using an SVM classifier. They show that there is no unique combination of pre-processing techniques that improves accuracy on any domain or language and that researchers should carefully analyze all possible combinations.

There is also a workshop named 'Workshop on Noisy User-generated Text'[1], that is running since 2015 and focuses on natural language processing applied to noisy user-generated text that is found online. In 2015, they introduced a lexical normalization task, in aiming to normalise non-standard words in English Twitter messages to their canonical forms.

Thus, many studies have examined the role of pre-processing, generally and specifically in sentiment analysis, however, none of them has gathered in a comparative study the total number of techniques which will be presented in Sect. 4.

3   Common Pre-processing Techniques

Below we describe the 15 pre-processing techniques we will experiment with.

Remove Numbers. It is a common tactic to remove numbers from text, because they do not contain any sentiment. However, some researchers argue that keeping the numbers may improve classification effectiveness [6].

Replace Repetitions of Punctuation. We distinguish three punctuation signs, whose repetitions concern us. These are the exclamation, question, and stop marks. The use of these punctuation marks signals the existence of intense emotion. If we find more than one in a row, we replace it with a representative tag. For example the token '???' will be replace with 'multiQuestionMark'.

Handling Capitalized Words. Same as before, capitalized words may imply intense emotion, so we detect all the words that are longer than two characters with all of their characters capitalized. We prefix them with 'ALL_CAPS_' like [16] did, so they can be identified in machine learning.

Lowercasing. One of the most common pre-processing techniques is to lowercase all words. By doing so, many words are merged and the dimensionality of the problem is reduced.

Replace Slang and Abbreviations. Social media users usually write in an informal way and their texts contain a lot of slang and abbreviations. These words, in order to be interpreted correctly, have to be replaced to impute their meaning. We manually constructed a lookup table consisting of 290 such words and their replacements. Some examples are the words 'ty', 'qq' and 'omg', which respectively mean and replaced by 'thank you', 'crying', and 'oh my god'.

---

[1] http://noisy-text.github.io/.

Replace Elongated Words. Elongated is a word when it contains a character that is repeating more than two times, like the word 'greeeeat'. It is important to replace words like this with their source words, so they can be merged. Otherwise, the classifier will treat them as different words, and probably the elongated ones will be ignored because of their low frequency of occurrence. Detecting and replacing elongated words have been examined by researchers before, e.g. in [8].

Replace Contractions. One technique that can be used in pre-process is the replacement of contractions, i.e. words like 'won't' and 'don't', that will be replaced with 'will not' and 'do not', respectively.

Replace Negations with Antonyms. It is an approach that has not been used by many researchers and is presented in [14]. We search in each sentence for the word 'not' and then, we check if the next word has an antonym. If yes, we replace both words with the antonym. For example, the phrase 'not good' will be replaced with the word 'bad', using WordNet [7].

Handling Negations. When text analysis is performed in a word level, it is very challenging to handle negation. One method that is widely used by researchers is the detection of words that imply negation and the addition of the prefix 'NOT_' in every word after them until the first punctuation mark.

Remove Stopwords. Stopwords are function words with high frequency of presence across all sentences. It is considered needless to analyze them, because they do not contain much useful information. The set of these words is not completely predefined and it can be changed by removing or adding more to it, depending on the application. In our implementation, we used the standard stopwords provided by NLTK [2].

Stemming. It is the process of removing the endings of the words in order to detect their root form. By doing so, many words are merged and the dimensionality is reduced. It is a widely used method that generally provides good results; we used the Porter Stemmer [15].

Lemmatizing. Another method of merging many words to one is Lemmatization. In this method, we remove the endings of the words in order to detect their lemmas, i.e. their root forms in a dictionary.

Replace URLs and User Mentions. In Twitter texts, almost every sentence contains a URL and a user mention. Their presence does not contain any sentiment and one approach is to replace them in pre-processing with tags as [1] did. We used the tags 'URL' and 'AT_USER'.

Spelling Correction. It is very common in informal texts for users to make spelling errors that might make classification harder. By using tools that automatically correct these errors, it is possible to improve classification effectiveness [10]. While no corrector is perfect, they have some —usually high— accuracy of success. We used Norvig's spelling corrector.[2]

Remove Punctuation. In many works, it is common to remove punctuation signs in pre-processing [6]. However, many times the presence of punctuation marks denotes the existence of some sentiment. For example, an exclamation mark may mean an intense positive or negative sentiment. So if we remove them we might decrease the accuracy of classification.

Table 1. Correspondence of pre-processing techniques

Number	Pre-processing Technique	Number	Pre-processing Technique
0	Basic (Remove Unicode strings and noise)	8	Replace negations with antonyms
1	Remove Numbers	9	Handling Negations
2	Replace Repetitions of Punctuation	10	Remove Stopwords
3	Handling Capitalized Words	11	Stemming
4	Lowercase	12	Lemmatizing
5	Replace Slang and Abbreviations	13	Other (Replace urls and user mentions)
6	Replace Elongated Words	14	Spelling Correction
7	Replace Contractions	15	Remove Punctuation

Table 1 summarizes and assigns numbers (for later use) to all the aforementioned techniques.

4 Experimental Setup

Hitherto, several datasets for supervised Twitter sentiment analysis have been published. Each of them consists of tweets manually labeled by human annotators in one sentiment category. The most common labels are positive, negative, and neutral, but there are also some datasets which provide numeric labels that correspond to sentiment strengths.

Eight widely-used Twitter sentiment analysis datasets are presented in [17]. We chose to examine the three-point classification problem with the predefined classes of positive, negative, and neutral. For this task, we used two datasets, the first being the Sentiment Strength Twitter dataset and the second the SemEval dataset, both described next.

---

[2] http://norvig.com/spell-correct.html.

4.1   The Sentiment Strength Twitter Dataset

The Sentiment Strength Twitter or SS-Twitter dataset contains 4,242 tweets and was developed by [20] in order to evaluate SentiStrength[3], a lexicon-based method for sentiment strength detection. The tweets are labeled with positive and negative strengths: a positive strength is a number between 1 ("not positive") and 5 ("extremely positive"), and a negative strength is a number between -1 ("not negative") and -5 ("extremely negative").

By re-annotating this dataset, we created a new one with three sentiment labels (positive, negative, neutral), suitable for our task. Hence, we apply two rules, as done in [17]. Firstly, we compute the positive to negative strength ratio of each tweet. If its absolute value is equal to 1, then we label the tweet as neutral. If the positive strength ratio is 1.5 times greater than the negative one, the tweet is considered positive, and negative otherwise. After these transformations, the final dataset consists of 1,252 positive, 1,037 negative and 1,953 neutral tweets. Some statistics related to the dataset are shown in Table 2.

Table 2. Statistics of the datasets

	SS-Twitter	SemEval
Total sentences	4,242	65,854
Total words	80,246	1,454,723
Average words/sentence	18.91	22.09
Total unique tokens	22,496	176,578
Total emoticons	3,467	34,979
Total slangs	622	5,815
Total elongated words	1,543	17,355
Total multi exclamation marks	325	2,834
Total multi question marks	152	750
Total multi stop marks	1,118	14,115
Total all capitalized words	2,854	52,141

4.2   The SemEval Dataset

This dataset was constructed for the International Workshop on Semantic Evaluation (SemEval)[4]. SemEval consists of many tasks and one of them is about sentiment analysis in three-point classification. Each tweet was manually annotated by Amazon Mechanical Turk workers or CrowdFlower users, depending on the year. This task is running each year since 2013 [12], and every year more data are added. By collecting the datasets of all years (2013–2017), we gathered 65,854 tweets, i.e. 23,197 positive, 12,510 negative, and 30,147 neutral. Some statistics related to this dataset are also shown in Table 2.

---

[3] http://sentistrength.wlv.ac.uk.
[4] http://alt.qcri.org/semeval2017/.

4.3 Machine Learning Algorithms

Out of the many available supervised machine learning algorithms, we chose one algorithm for each of the three most used categories. These are, the Generalized Linear Models (GLM), the Naïve Bayes (NB), and the Support Vector Machines (SVM). From the GLM family we chose the Logistic Regression algorithm, from the NB we chose the Bernoulli Naïve Bayes, and from the SVMs we chose the Linear SVC algorithm.

Logistic Regression. It is a popular algorithm that belongs to the Generalized Linear Models methods —despite its name— and it is also known as Maximum Entropy. In this model, the probabilities describing the possible outcomes of a single trial are modeled using a logistic function [13].

Bernoulli Naïve Bayes. Naïve Bayes algorithms are the simplest probabilistic classification algorithms [5] that are widely used in sentiment analysis. They are based on the Bayes Theorem, which assumes a complete independence of variables. The Bernoulli algorithm is an alternative of Naïve Bayes, where each term is equal to 1 if it exists in the sentence and 0 if not. Its difference from Boolean Naïve Bayes is that it takes into account terms that do not appear in the sentence. It is a fast algorithm that deals well with high dimensionality.

Linear SVC. One of the most popular machine learning methods for classification of linear problems are SVMs [3]. They try to find a set of hyperplanes that separate the space into dimensions representing classes. These hyperplanes are chosen in a way to maximize the distance from the nearest data point of each class. The Linear SVC is the simplest and fastest SVM algorithm assuming a linear separation between classes.

All the models that have been selected are in fact linear. Naïve Bayes is a generative approach, whereas logistic regression and SVMs are discriminative approaches. Logistic Regression varies from SVMs in the fact that it provides a probabilistic interpretation for the results.

4.4 Feature Extraction and Evaluation

There are several ways to assess the features in a bag-of-words representation. We chose to use Term Frequency – Inverse Document Frequency (TF.IDF) which is given by

$$\text{TF.IDF} = f \log(N/df),$$

where f is the number of occurrences in the document, N is the number of documents, and df is the number of documents that contain this feature [11].

The metric that was used to evaluate the classification results is accuracy, which is the number of the correct classifications out of all classifications. Accuracy is a good metric for balanced datasets like in our case. Finally, we used uni-grams, and compare the numbers of resulting features across pre-processing methods.

5 Results

In this section, we present the results of the use of every pre-processing technique among the two datasets and between the three classifiers.

With a dataset as input, we used Python's NLTK [2] and created a new file as output for each pre-processing technique. Depending on the technique, the final file had more or less total and unique tokens than the initial as can be seen in Fig. 1.

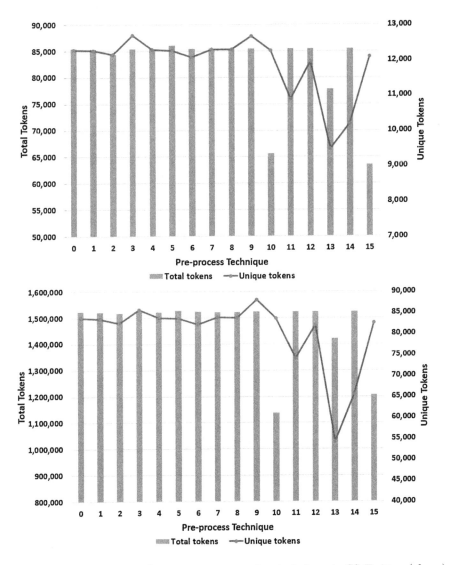

Fig. 1. Total and unique tokens per pre-processing technique in SS-Twitter (above) and SemEval (below) datasets

As we can see, handling negations by adding the 'NOT_' prefix in front of the words (technique 9), results in an augmentation of both the total and the unique tokens, with the latter being clearly visible. The other technique that increases the total tokens is the slang and abbreviation replacement (technique 5), but it decreases the unique tokens. Removing stopwords (technique 10) and punctuation (technique 15) results in a great reduction for the total tokens, but not a remarkable reduction on the unique tokens. The three techniques that reduce a lot the unique tokens are stemming, the replacement of URLs and user mentions, and spelling correction (techniques 11, 13, and 14). Both datasets present the same proportions in the total and unique tokens.

The number of unique tokens defines the number of features that will be used in a uni-gram bag-of-words representation. The quality and number of these features play a key role in the accuracy of the classifiers. Keeping a significant number of words/features will increase the temporal and spatial complexity of classifiers. This also favors the appearance of overfitting. Although, an increase in the number of features may not always result in better classification, because the quality of features also matters.

After the creation of the new pre-processed files, we apply machine learning algorithms using Sklearn [13]. For vectorization we used the tf-idf transformation and as features we utilized uni-grams, so we can see if and how the number of features has an impact on the classification results. As said before, we chose three representative algorithms (Linear SVC, Bernoulli Naïve Bayes, and Logistic Regression) and we did not make any changes to their parameters. The results for both datasets are presented in Fig. 2. For each pre-processing technique we compare the accuracy and the number of features per three classifiers.

For the SS-Twitter dataset we observe that the techniques which result in increased accuracy in all classifiers are 1, 2, 6, 7, 11 and 12. The highest results were 61.4% for the Linear SVC which was achieved by replacing the elongated words, 60.6% for the Bernoulli Naïve Bayes which was achieved by stemming, and 61% for the Logistic Regression which was achieved by using lowercase. The lowest accuracy for all classifiers occurs when we remove punctuation signs (technique 15), showing their importance in sentiment classification. Other poorly performing techniques were 3, 5, 8, 10, and 14, which only resulted in a small increase in one classifier. Finally, the techniques 4, 9, and 13, resulted in an increment in two classifiers and can be considered good techniques.

For the SemEval dataset, the techniques which provide better accuracy than the initial for all classifiers are 1, 2, 11 and 13. Especially the latter, which is the replacement of URLs and user mentions, gives the highest results with 59% for Linear SVC, 60.6% for Bernoulli Naïve Bayes, and 60.7% for Logistic Regression. The lowest accuracy is noticed when we apply the techniques 5, 10, 14, and 15 for all classifiers. The poorly performing techniques in this dataset are 3, 4, 6, 7 and 8. Finally, other highly performing techniques which result in improved accuracy in two classifiers are 8, 9, and 12.

Based on the results, we can discern 5 categories depending on the accuracy. These categories describe how the SS-Twitter and the SemEval datasets reacted

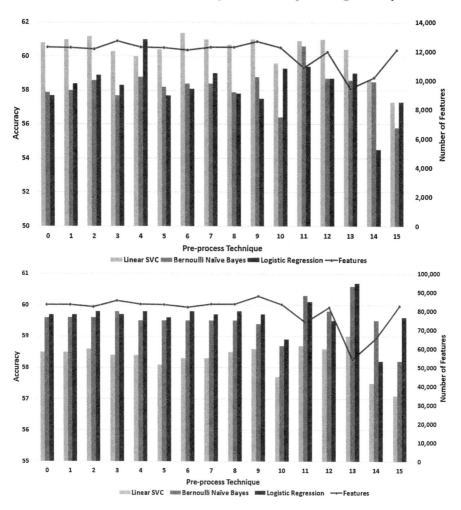

Fig. 2. Accuracy percentage and number of features for all pre-processing techniques per three machine learning algorithms in SS-Twitter (above) SemEval (below) datasets

to the 15 pre-processing techniques for three-point Twitter sentiment analysis and are presented in Table 3.

We note that there is no significant association between the number of features and the accuracy. The techniques that increase the number of features are 3, 4, 7, 8, and 9 and only one of them has high performance. The techniques which achieve a great reduction of features like 11, 13, and 14, give better results. The rest reduce the features by a few, and their accuracy varies.

Table 3. Accuracy performance categories for all pre-processing techniques on both datasets

Performance	Description	Techniques
Best	High accuracy in all classifiers and all datasets	1, 2, 11
High	High accuracy in most classifiers and all datasets	9, 12, 13
Poor	Low accuracy in most classifiers and all datasets	3, 5, 8, 10, 14
Worst	Lowest accuracy in all classifiers and all datasets	15
Varying	High or poor accuracy in most classifiers depending on the dataset	4, 6, 7

6   Conclusion and Future Work

Pre-processing is the first step in text sentiment analysis and the use of appropriate techniques can improve classification effectiveness. We examined a significant number of pre-processing techniques, which were not evaluated in a comparative study in the past, and tested them in two datasets. Each technique was evaluated in three representative machine learning algorithms on accuracy. Finally, we distinguish some performance categories based on the results and count the number of features for each technique.

Our experiments show that on Twitter sentiment analysis some techniques provide better results in classification for both of the datasets used, while others decrease the accuracy. The recommended techniques are stemming, replacement of repetitions of punctuation, and removing numbers. The non-recommended techniques include removing punctuation, handling capitalized words, replacing slang, replacing negations with antonyms, and spelling correction.

Depending on the classifier, the results vary, and if we combine these techniques we may get different results. Thus, in future work, we will extend our analysis with more machine learning algorithms and we will try to combine these techniques to achieve better results. Moreover, another future approach is to test these techniques on datasets from different domains such as news articles and product or movie reviews.

References

1. Agarwal, A., Xie, B., Vovsha, I., Rambow, O., Passonneau, R.: Sentiment analysis of twitter data. In: Proceedings of the Workshop on Languages in Social Media, LSM 2011, Association for Computational Linguistics, Stroudsburg, PA, USA, pp. 30–38 (2011). http://dl.acm.org/citation.cfm?id=2021109.2021114
2. Bird, S.: NLTK: the natural language toolkit. In: Calzolari, N., Cardie, C., Isabelle, P. (eds.) ACL 2006, 21st International Conference on Computational Linguistics and 44th Annual Meeting of the Association for Computational Linguistics, Proceedings of the Conference, Sydney, Australia, 17–21 July 2006. The Association for Computer Linguistics (2006). http://aclweb.org/anthology/p06-4018

3. Cherkassky, V.: The nature of statistical learning theory. IEEE Trans. Neural Netw. **8**(6), 1564 (1997). doi:10.1109/TNN.1997.641482

4. Fayyad, U.M., Piatetsky-Shapiro, G., Uthurusamy, R.: Summary from the KDD-03 panel: data mining: the next 10 years. SIGKDD Explor. **5**(2), 191–196 (2003). doi:10.1145/980972.981004

5. John, G.H., Langley, P.: Estimating continuous distributions in bayesian classifiers. In: UAI 1995: Proceedings of the Eleventh Annual Conference on Uncertainty in Artificial Intelligence, Montreal, Quebec, Canada, 18–20 August 1995, pp. 338–345 (1995). https://dslpitt.org/uai/displayArticleDetails.jsp?mmnu=1&smnu=2&article_id=450&proceeding_id=11

6. Lin, C., He, Y.: Joint sentiment/topic model for sentiment analysis. In: Proceedings of the 18th ACM Conference on Information and Knowledge Management, CIKM 2009, Hong Kong, China, 2–6 November 2009, pp. 375–384 (2009). http://doi.acm.org/10.1145/1645953.1646003

7. Miller, G.A.: WordNet: a lexical database for english. Commun. ACM **38**(11), 39–41 (1995). doi:10.1145/219717.219748

8. Mohammad, S., Kiritchenko, S., Zhu, X.: NRC-Canada: building the state-of-the-art in sentiment analysis of tweets. In: Proceedings of the 7th International Workshop on Semantic Evaluation, SemEval@NAACL-HLT 2013, Atlanta, Georgia, USA, 14–15 June 2013, pp. 321–327 (2013). http://aclweb.org/anthology/S/S13/S13-2053.pdf

9. Mohammad, S.M., Zhu, X., Kiritchenko, S., Martin, J.D.: Sentiment, emotion, purpose, and style in electoral tweets. Inf. Process. Manage. **51**(4), 480–499 (2015). doi:10.1016/j.ipm.2014.09.003

10. Mullen, T., Malouf, R.: A preliminary investigation into sentiment analysis of informal political discourse. In: Computational Approaches to Analyzing Weblogs, Papers from the 2006 AAAI Spring Symposium, Technical Report SS-06-03, Stanford, California, USA, 27–29 March 2006, pp. 159–162 (2006). http://www.aaai.org/Library/Symposia/Spring/2006/ss06-03-031.php

11. Na, J.C., Sui, H., Khoo, C., Chan, S., Zhou, Y.: Effectiveness of simple linguistic processing in automatic sentiment classification of product reviews. In: Conference of the International Society for Knowledge Organization (ISKO), pp. 49–54 (2004)

12. Nakov, P., Rosenthal, S., Kozareva, Z., Stoyanov, V., Ritter, A., Wilson, T.: SemEval-2013 task 2: sentiment analysis in twitter. In: Second Joint Conference on Lexical and Computational Semantics (*SEM), Proceedings of the Seventh International Workshop on Semantic Evaluation (SemEval 2013), vol. 2, pp. 312–320. Association for Computational Linguistics, Atlanta, Georgia, USA, June 2013. http://www.aclweb.org/anthology/S13-2052

13. Pedregosa, F., Varoquaux, G., Gramfort, A., Michel, V., Thirion, B., Grisel, O., Blondel, M., Prettenhofer, P., Weiss, R., Dubourg, V., VanderPlas, J., Passos, A., Cournapeau, D., Brucher, M., Perrot, M., Duchesnay, E.: Scikit-learn: machine learning in python. J. Mach. Learn. Res. **12**, 2825–2830 (2011). http://dl.acm.org/citation.cfm?id=2078195

14. Perkins, J.: Python Text Processing with NLTK 2.0 Cookbook. Packt Publishing, Birmingham (2010)

15. Porter, M.F.: An algorithm for suffix stripping. Program **14**(3), 130–137 (1980). doi:10.1108/eb046814

16. Prasad, S.: Micro-blogging sentiment analysis using baycsian classification methods. Technical report (2010)

17. Saif, H., Fernández, M., He, Y., Alani, H.: Evaluation datasets for twitter sentiment analysis: a survey and a new dataset, the STS-gold. In: Proceedings of the First International Workshop on Emotion and Sentiment in Social and Expressive Media: Approaches and Perspectives from AI (ESSEM 2013) A Workshop of the XIII International Conference of the Italian Association for Artificial Intelligence (AI*IA 2013), Turin, Italy, 3 December 2013, pp. 9–21 (2013). http://ceur-ws.org/Vol-1096/paper1.pdf

18. Shi, Y., Xi, Y., Wolcott, P., Tian, Y., Li, J., Berg, D., Chen, Z., Herrera-Viedma, E., Kou, G., Lee, H., Peng, Y., Yu, L. (eds.): Proceedings of the First International Conference on Information Technology and Quantitative Management, ITQM 2013, Dushu Lake Hotel, Sushou, China, 16–18 May 2013, Procedia Computer Science, vol. 17. Elsevier (2013). http://www.sciencedirect.com/science/journal/18770509/17

19. Singh, T., Kumari, M.: Role of text pre-processing in twitter sentiment analysis. Proc. Comput. Sci. **89**, 549–554 (2016). http://www.sciencedirect.com/science/article/pii/S1877050916311607

20. Thelwall, M., Buckley, K., Paltoglou, G.: Sentiment strength detection for the social web. JASIST **63**(1), 163–173 (2012). doi:10.1002/asi.21662

21. Uysal, A.K., Günal, S.: The impact of preprocessing on text classification. Inf. Process. Manage. **50**(1), 104–112 (2014). doi:10.1016/j.ipm.2013.08.006

Employing Twitter Hashtags and Linked Data to Suggest Trending Resources in a Digital Library

Ioannis Papadakis[1], Konstantinos Kyprianos[1(✉)], Apostolos Karalis[2], and Christos Douligeris[2]

[1] Department of Archives, Library Science and Museum Studies, Ionian University, Ioannou Theotoki 72, 49100 Corfu, Greece
papadakis@ionio.gr, k.kyprianos@gmail.com
[2] Department of Informatics, University of Piraeus, Karaoli & Dimitriou 80, 18534 Piraeus, Greece
{akaralis,cdoulig}@unipi.gr

Abstract. It is common truth that social web sites have dominated the web during the past few years. This results in the creation of vast amounts of information that is being produced by the corresponding user activities. Traditional information organization tools originating from the library domain are not applicable to the social web due to its overwhelmingly dynamic nature. Along these lines, hashtags have become an information organization tool of growing popularity among social web sites.

In this paper, it is argued that digital libraries may exploit information deriving from hashtags bringing this way two fundamentally different worlds closer to each other. Thus, a methodology is proposed, where popular hashtags are expanded through semantic web technologies and are ultimately matched against the subject index of a digital library. Successful matches are promoted to the homepage of the digital library to suggest trending resources to the end-users.

Keywords: Social web · Twitter · Hashtags · DBpedia · Semantic web · Linked data

1 Introduction

Digital libraries and libraries in general, are traditionally interested in providing up-to-date information to their users. Thus, libraries most frequently analyse usage statistics [16] in an effort to support their decisions in collection development. Such decisions concern the future of the underlying collections and are based on transactions that occurred in the past.

During the past few years, the social web has emerged as a communication channel capable of facilitating instant information sharing and collaboration. One of the most widespread features of the social web are the hashtags. Hashtags are employed by social web tools to classify messages, propagate ideas and promote specific topics and people [8]. They have been introduced by Twitter, but nowadays are commonly met in many social web tools. Hashtags have evolved into a powerful

© Springer International Publishing AG 2017
J. Kamps et al. (Eds.): TPDL 2017, LNCS 10450, pp. 407–418, 2017.
DOI: 10.1007/978-3-319-67008-9_32

classification tool for the social web. They constitute a peculiar kind of dynamic vocabulary that is controlled by the same people that employ it (i.e. the end-users). Hashtags are searchable through Twitter, Google and specialized sites [18].

In this paper, a methodology is proposed that evolves around hashtags and suggests resources within a digital library about trending topics. In contrast to traditional library practices, the proposed approach aims in observing current trends in the society and instantly suggesting relevant resources to its users. The proposed methodology is implemented and deployed as a service within the context of an academic digital library. Finally, the service is evaluated and interesting remarks are shown.

The remainder of this paper is structured as follows: In the following section, a short description is presented about the relation between social networks, Twitter, hashtags and digital libraries. Then, related research regarding the employment of hashtags across the social web is mentioned. Next, an effort is made to highlight the convergence and divergence points between controlled vocabularies and hashtags. In Sect. 5, the proposed methodology is realized as an online service and its deployment in a digital library is presented. In the following section, the service is accordingly assessed and the corresponding results are shown. Finally, Sect. 7 concludes the paper and points directions for future work.

2   Social Networks, Twitter, Hashtags, Digital Libraries

Since their initial appearance, Social Network Sites (SNS), such as Facebook, Twitter, and Myspace, have attracted millions of users globally. In fact, many users have rendered such sites an integral part of their lives [4]. The widespread of collaborative technologies has led to the formulation of instant online communities, thus facilitating the communication among people rapidly and conveniently [10].

Twitter is a microblogging service that allows users to follow other users or to be followed. Unlike most other SNS, the relation between following and being followed is not symmetric. A user may follow other users but it is not necessary to be the other way around [13]. Twitter allows users to broadcast brief text updates about things that are happening to their lives. Users refer to Twitter when they want to find information about breaking news, real-time events, people and topical information [14, 19]. Such features establish Twitter as a tool that may provide timely information quicker than any other mass media (e.g. television, radio, etc.). According to January 2017 report by Statista[1], Twitter is among the 10 most famous SNS, having more than 317 million active users. Moreover, based on Alexa website ranking[2], Twitter possesses the 16th place in the global rank regarding the most famous sites across the world.

One popular feature of Twitter is the employment of hashtags. A hashtag is a convention among Twitter users to create and follow a thread of discussion by prefixing

---

[1] Most famous social network sites worldwide as of January 2017, r. (2017). *Global social media ranking 2017 | Statistic. Statista*. Retrieved 28 March 2017, from https://www.statista.com/ statistics/272014/global-social-networks-ranked-by-number-of-users/.

[2] *Keyword Research, Competitor Analysis, & Website Ranking | Alexa*. (2017). *Alexa Internet*. Retrieved 28 March 2017, from http://www.alexa.com.

a word with a '#' character [13]. Tweets containing a hashtag are visible not only to the followers of a user that employed the specific hashtag but to anyone on the social network. Twitter provides an API through which it is possible to identify the most popular hashtags so that users can witness trending topics. A trending topic does not last forever nor does it disappear so as to never come back. Usually, it lasts for a couple of days and can have many active periods [13].

Thus, hashtags, to some extent, reflect the trending topics that users are talking about. Through the employment of hashtags, tweets can be organized, indexed, shared and discovered by anyone [12]. Consequently, hashtags can be seen as a powerful vocabulary that is created, employed and controlled by the users themselves.

In this paper, it is argued that digital libraries may exploit information deriving from hashtags in favor of their users by pinpointing trending resources that exist in their collections. Thus, digital libraries get valuable feedback from external entities in a timely fashion.

3 Using Hashtags Across the Social Web

As already mentioned, since their emergence on Twitter[3], hashtags have been used extensively in social networks and micro-blogging services. Their wide acceptance urged many researchers to study them thoroughly.

Efron [9] proposed a language modelling approach to hashtag retrieval based on the assumption that when a user is interested in a specific topic he/she might like to find hashtags that are often applied to posts about such topic. In a similar fashion, Bansal, Jain and Varma [3] proposed a method of semantic enrichment of microblogs for a particular type of entity search that ends up in retrieving a ranked list of the top-k hashtags relevant to a user's query. Such a methodology may help users to track posts of general interest. The aforementioned approaches aim in providing end users with personalized information from tweets based on hashtags.

In a different approach, Sedhai and Sun [17] introduced an entity-hashtag graph for tweets with hyperlinks. More specifically, they grouped together the hashtags of tweets containing links to various web pages and recommended them to future users that posted tweets containing the same web pages. Thus, information sharing and organization within the Twitter ecosystem can be facilitated.

Another indicative case of research work about hashtags that evolves around the Twitter ecosystem is TweetPos, which was proposed by Wijants et al. [21]. TweetPos is a versatile web-based tool that facilitates the analytical study of geographic tendencies in crowd-sourced Twitter data feeds. Hashtags play a crucial role in this tool, since they constitute the service's essential ingress parameters. When a user addresses a topic query to TweetPos, the system creates a compilation of tweets about this topic. The user may geographically and temporally filter such tweets.

---

[3] #OriginStory - Carnegie Mellon University. (2017) #OriginStory. Retrieved 29 March 2017, from http://www.cmu.edu/homepage/computing/2014/summer/originstory.shtml.

In a slightly different line of research, the following two approaches exploit hashtags to retrieve real-time and popular events that people are discussing in Twitter. More specifically, Wang et al. [20] proposed an adaptive crawling model that detects emerging popular hashtags and monitors them to retrieve high volume of relevant data for events of interest. The model analyzes the traffic patterns of the collected hashtags to update subsequent collection queries. Cui et al. [7] aim in discovering breaking events with the employment of popular hashtags in Twitter.

There are also some approaches that facilitate the search and retrieval of topic-related tweets with the employment of hashtags. More specifically, the methodology proposed by Llewellyn et al. [15] focuses on the formulation of a corpus of tweets about a specific topic based on popular hashtags, hand-selected hashtags and topic modelling. In a similar approach, Cotelo, Cruz and Troyano [6] proposed a general, dynamic and graph-based model to capture related but unknown topics in tweets based on hashtags and users. Bansal, Bansal and Varma [2] presented a machine learning methodology to segment the hashtags and link the entities in hashtags to Wikipedia, an approach that helps in finding latent semantic information about hashtags.

To conclude, it seems that a great deal of the relevant literature about hashtags emphasizes on finding ways to aid users in discovering additional hashtags and tweets regarding their initial information needs. Consequently, the rich semantic information that lies into hashtags is mostly exploited within the strict boundaries of the Twitter ecosystem.

4   Controlled Vocabularies vs. Hashtags

Controlled vocabularies can be seen as collections of terms defined by experts that are employed to index and, ultimately, to retrieve information through browsing or searching [11]. Controlled vocabularies typically include preferred, non-preferred and related terms. In many cases, these terms have hierarchical relationships among them, meaning that navigation is possible from a generic term to a more specific one and vice-versa. The purpose of controlled vocabularies is the organization of information and the provision of terminology to catalogue and retrieve information [11].

On the other hand, hashtags can be seen as vocabularies that are defined by common users and not by experts [5]. Moreover, instead of referring to formal collections, hashtags are employed to index messages or tweets on the microblog sphere. Such terms are incorporated into a tweet by the author of the specific tweet, meaning that there is no limitation or control over the term that will be created as a hashtag. After their publication, the decisive factor that transforms a hashtag to something like a controlled vocabulary term is popularity. Popular hashtags tend to be employed in many tweets thus becoming even more popular. Therefore, it seems that hashtags go under a constant control and evaluation by the users themselves. A user that decides to employ an existing hashtag, promotes this hashtag and ultimately contributes to its establishment as an authority. To sum up, it is evident that formal controlled vocabularies created by experts and popular hashtags created by users have signs of convergence.

5   Proposed Approach

In this section, we propose a digital library service, which harvests trending hashtags from Twitter to identify relevant resources within a digital library. Such resources are promoted to the homepage and then suggested to end-users through an interactive query suggestion service. The service is based on a technique that was introduced at [16].

Initially, the most popular tweets are harvested and their corresponding hashtags are stored for further process. Since hashtags rarely look like normal terms that usually exist within library indices, such hashtags undergo a spell-checking control. The controlled hashtags are tunneled towards DBpedia [1], a linked data provider containing structured content deriving from Wikipedia, in an effort to enrich the term collection with even more relevant terms. The structured information can be queried online through the employment of appropriate semantic web technologies (e.g. SPARQL). The enriched set of terms originating from popular tweets is matched against the subject index of a digital library. The successful matches are again stored and the more recent ones are ranked at the top. The top-n terms populate an HTML division element (i.e. <div>) at the homepage of a digital library. The end-users are able to interact with such terms and accordingly retrieve resources from the digital library about trending topics on Twitter.

The next section provides a detailed analysis of how the proposed service offers the aforementioned functionality.

5.1   Implementation Details

To suggest resources about trending topics in the digital library, the underlying engine goes incrementally through the phases below (see Fig. 1)[4]:

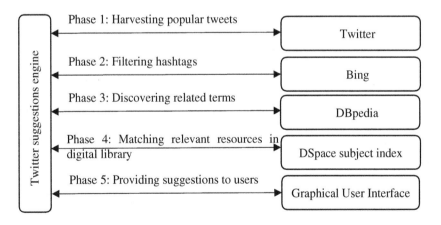

Fig. 1. Information workflow

---

[4] The source code of the service is available at bitbucket: https://bitbucket.org/akaralis/twitter-inlibraries.

Phase 1: The n most popular tweets are harvested in Twitter. For this purpose, Twitter's[5] Search API is used.

Phase 2: The hashtags of each tweet are filtered using a spelling suggestion service, namely the Bing Spell Check API[6]. For example, the phrase "Cloud Computing" is produced when the hashtag #CloudComputing is filtered using the aforementioned spelling suggestion service.

Phase 3: For the resulting set of spelling suggestions and the unfiltered hashtags, related terms and categories are requested via the SPARQL endpoint of DBpedia[7].

Phase 4: The entire set of the spelling suggestions, the unfiltered hashtags and the related terms and categories are matched against the subject index of a digital library. Obviously, the successful matches of this phase constitute the set of trending topics that are ultimately suggested to end-users.

Phase 5: Each time a client (i.e. a browser on which the homepage of the digital library is loaded) requests suggestions, the proposed service retrieves from the underlying database and returns a predefined number of the most recent ones that have the highest number of "clicks" (by the users) in a specified timeframe (e.g. the last 5 days). This way, the service promotes recent suggestions that are popular among the population of the digital library.

5.2 Deployment of the Proposed Service

The proposed service is implemented and deployed in Dione, the academic digital library of the University of Piraeus in Greece[8]. Dione mainly contains theses and dissertations from students of the four Schools of the University (i.e. School of Economics, Business and International Studies, School of Finance and Statistics, School of Maritime and Industrial Studies and School of Information and Communication Technologies).

The initial goal for the service was to suggest queries that would apply to all the scientific disciplines that are relevant to the Schools of the University. However, it was soon realized that popular hashtags contained too much noise and ambiguity that led to meaningless suggestions. Therefore, it was decided to filter the initial tweets and select just the ones that contain the word 'programming' to focus on tweets about technology.

The alpha version of the service was attached to Dione in January 2016 and the beta version was attached to the digital library on January 15th, 2017. The beta version introduced a new algorithm for populating the HTML division element of Dione's homepage. During the alpha version, the five most recent terms of the database appeared in the

---

[5] *The Search API.* (2017). *Twitter Developer Documentation.* Retrieved 20 April 2017, from https://dev.twitter.com/rest/public/search. More specifically, the parameter result_type='mixed' is employed in an effort to get a mix of recent and popular tweets.

[6] *Bing Search API | Microsoft Azure Marketplace.* (2017). *Datamarket.azure.com.* Retrieved 20 April 2017, from http://datamarket.azure.com/dataset/bing/search.

[7] The corresponding SPARQL query would be: SELECT ?related WHERE { <http://dbpedia.org/resource/Data_science><http://purl.org/dc/terms/subject>?related}. DSpace REST API, avail. at: https://wiki.duraspace.org/display/DSDOC5x/REST+API [accessed: 12/03/2017].

[8] *Dione* (2017). *Dione's Homepage.* Retrieved 29 March 2017, from http://dione.lib.unipi.gr.

division element. When a new term emerged in the database, it took the place of the oldest one in the division element. In the beta version, the division element contains seven terms. Each time a user selects a term, a counter is increased. When a new term emerges in the database, it substitutes the last term of the division element, which contains terms ranked by both the number of times they have been selected and their age. No term can stay in the division element more than five days.

As shown in Fig. 2, the proposed service is visualized as a division HTML element (i.e. <div>) containing suggested queries about trending topics at the top right of the digital library's homepage. Upon selection, a query is addressed to the underlying search engine and the matched resources are returned to the user.

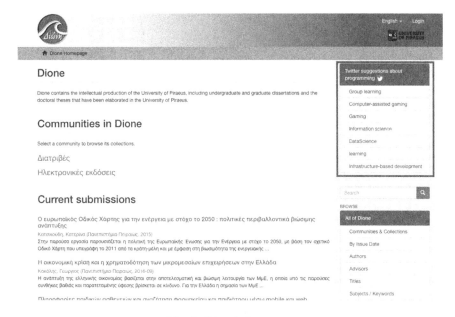

Fig. 2. Dione's homepage

6   Evaluation

The proposed service has been evaluated in order to assess the impact to the user community of the digital library as well as the performance of the various modules that constitute the service. The evaluation is based on a log file analysis of Dione's usage from January 15th, 2017 until March 15th, 2017.

6.1   Impact to the User Community

The log files of Dione provide the opportunity to compare the number of times the proposed service has been employed to the number of times the rest of the six browsing interactions of Dione have been used by the community. As shown in Fig. 3, the most popular navigational interaction is against the subject index (61.07%), followed by the author index (16.2%). The advisor index ranks third (15.44%) and Twitter suggestions appear in the fourth place (4.75%). The remaining interactions are not very popular since they cover less than 2% of the total number.

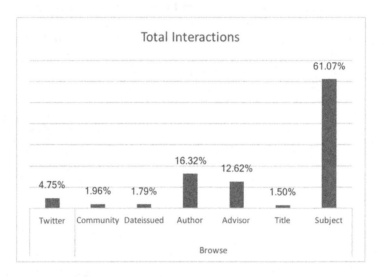

Fig. 3. Total Interactions

It is apparent that the number of Twitter suggestions is significantly lower than the top-three interactions (i.e. 'Subject', 'Author' and 'Advisor') provided by Dione. However, the proposed service ranks higher than the rest of the interactions that lie below. This is particularly important if one takes into consideration the fact that the scope of Twitter suggestions is limited to technology (through the employment of the word 'programming'), whereas the other interactions refer to the total of the disciplines covered by Dione.

Moreover, 22.9% of all the distinct Twitter suggestions that appeared in the <div> element were actually selected by the user community (63 out of 275 distinct suggestions). Thus, to a great extent, the service did not manage to filter out terms that went later unnoticed by the users of Dione. This could be attributed to the fact that too much Twitter noise managed to penetrate into the proposed service.

Since the business logic of the proposed service dictates that each suggestion may appear in the <div> element of the homepage once for each five days' timeframe, it becomes apparent that a single suggestion may appear in the <div> element many times in different timeframes. Thus, it would be interesting to see whether the number of times each hashtag appears in the <div> element influences the number of times the hashtag is selected by the users. According to Fig. 4, hashtags that make it often to the <div> element have a higher chance of being selected by users. Such a conclusion certainly comes as no surprise, since frequently appearing hashtags have more chances to get selected. Thus, it seems that the users of Dione follow the trending topics as they have been recorded in Twitter.

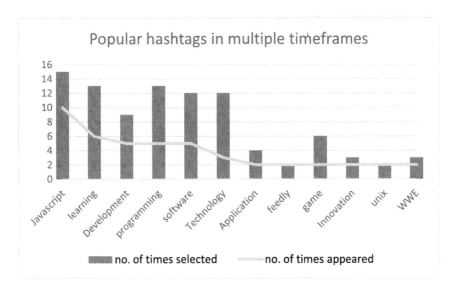

Fig. 4. Popular hashtags that have been selected by users in multiple timeframes

6.2 Under the Hood: Performance of Individual Modules

In the previous section, the impact of the proposed service to the user community of Dione was assessed. In this section, further analysis is performed to assess the various interactions between the core modules of the service and the remote online services that have been employed (namely: Twitter, Bing and DBpedia).

The vast majority (i.e. 71%) of the keywords that made it to the <div> element of the homepage come from DBpedia, whereas 17% of them originate from Twitter alone and just 2% have taken advantage of Bing's spelling suggestion service as seen in

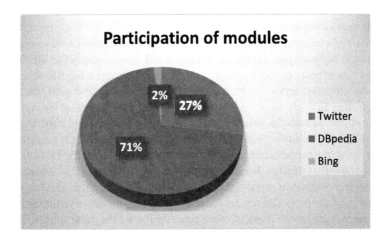

Fig. 5. The impact of the three remote modules that are employed by the proposed service

Fig. 5. It is apparent that DBpedia is very influential to the proposed service since it manages to provide useful keywords to the user community. On the other hand, Bing's spelling suggestion service has a minimal effect on the service.

From another point of view, it would be interesting to assess the impact of the three modules to the users of Dione. Thus, Fig. 6 calculates the number of times each suggestion has been selected from Dione's population, grouped by the module each suggestion originates from. This time, it is apparent that suggestions from Twitter are more popular than suggestions from DBpedia, despite the fact that most of them originate from DBpedia.

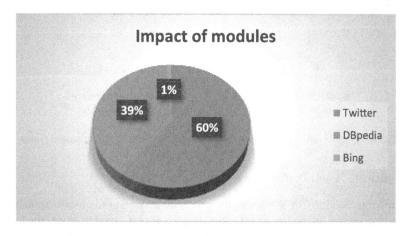

Fig. 6. The impact of the three modules to the users of Dione

7   Conclusions and Future Work

In this paper, a methodology was proposed that takes advantage of information that exists outside a digital library in favor of a query suggestion service within a digital library. The service suggests resources about trending topics that have been harvested from Twitter and expanded through DBpedia. The proposed service was integrated in the institutional repository of an academic library and was thoroughly evaluated.

The assessment process was based on quantitative methods. A log file analysis indicates that the proposed service attracts a considerable number of digital library users. Moreover, further analysis on the functionality of the service shows that the employment of DBpedia improves both the quantity and the quality of the provided term suggestions. Future work focuses on minimizing the inherent "noise" of the hashtags and on finding ways of applying the proposed approach to the entire scope of the digital library.

This work is a first step towards the integration of traditional digital library services with information originating from popular, crowd-sourcing sites that exist on the Web. The evaluation of the proposed service exhibits promising results in terms of the appreciation of the service by the users of the digital library. At the same time, it is evident that the lack of control to the information that is being accumulated in such sites contradicts the strict organization principles that traditionally govern digital libraries. Therefore, further detailed analysis is required prior to the entrance of such information to any digital library. The semantic web technologies could play a crucial role in minimizing the inherent noise of crowd-sourced information and become the catalyst for the creation of new, value-added services in the digital library domain.

References

1. Auer, S., Bizer, C., Kobilarov, G., Lehmann, J., Cyganiak, R., Ives, Z.: DBpedia: a nucleus for a web of open data. In: Aberer, K., Choi, K.-S., Noy, N., Allemang, D., Lee, K.-I., Nixon, L., Golbeck, J., Mika, P., Maynard, D., Mizoguchi, R., Schreiber, G., Cudré-Mauroux, P. (eds.) ASWC/ISWC -2007. LNCS, vol. 4825, pp. 722–735. Springer, Heidelberg (2007). doi:10.1007/978-3-540-76298-0_52
2. Bansal, P., Bansal, R., Varma, V.: Towards deep semantic analysis of hashtags. In: Hanbury, A., Kazai, G., Rauber, A., Fuhr, N. (eds.) ECIR 2015. LNCS, vol. 9022, pp. 453–464. Springer, Cham (2015). doi:10.1007/978-3-319-16354-3_50
3. Bansal, P., Jain, S., Varma, V.: Towards semantic retrieval of hashtags in microblogs. In: Proceedings of the 24th International Conference on World Wide Web, Florence, Italy, 18–22 May, pp. 7–8 (2015)
4. Boyd, D.M., Ellison, N.B.: Social network sites: definition, history, and scholarship. J. Comput.-Mediated Commun. **13**(1), 210–230 (2007)
5. Bradley, P.: Social Media for Creative Libraries, 2nd edn. Facet Publishing, London (2015)
6. Cotelo, J.M., Cruz, F.L., Troyano, J.A.: Dynamic topic-related tweet retrieval. J. Assoc. Inf. Sci. Technol. **65**(3), 513–523 (2014)
7. Cui, A., Zhang, M., Liu, Y., Ma, S., Zhang, K.: Discover breaking events with popular hashtags in Twitter. In: Proceedings of the 21st ACM International Conference on Information and Knowledge Management (CIKM), Maui, HI, 29 October–02 November, pp. 1794–1798 (2012)
8. Cunha, E., Magno, G., Comarcel, G., Almeida, V. Goncalves, M.A., Benevenuto, F.: Analyzing the dynamic evolution of hashtags on Twitter: a language-based approach. In: Proceedings of the Workshop on Language in Social Media (LSM 2011), pp. 58–65 (2011)
9. Efron, M.: Hashtag retrieval in a microblogging environment. In: Proceedings of the 33rd International ACM SIGIR Conference on Research and Development in Information Retrieval, Geneva, Switzerland, 19–23 July, pp. 787–788 (2010)
10. Fu, F., Liu, L., Wang, L.: Empirical analysis of online social networks in the age of web 2.0. Phys. A: Stat. Mech. Appl. **387**(2–3), 675–684 (2008)
11. Harpring, P.: Introduction to Controlled Vocabularies: Terminology for Art, Architecture, and Other Cultural Works. Getty Research Institute Publications, Los Angeles (2010)
12. Ivankova, M.: Understanding Microblogging Hashtags for Learning Enhancement. Form@re **11**(74), 17–23 (2011)
13. Kwak, H., Lee, C., Park, H., Moon, S.: What is Twitter, a social network or a news media? In: Proceedings of the 19th International Conference on World Wide Web, Raleigh, North Carolina, USA, 26–30 April, pp. 591–600 (2010)
14. Lau, C.H., Li, Y., Tjondronegoro, D.: Microblog retrieval using topical features and query expansion. In: Text REtrieval Conference (2011)
15. Llewellyn, C., Grover, C., Alex, B., Oberlander, J., Tobin, R.: Extracting a topic specific dataset from a twitter archive. In: Kapidakis, S., Mazurek, C., Werla, M. (eds.) TPDL 2015. LNCS, vol. 9316, pp. 364–367. Springer, Cham (2015). doi:10.1007/978-3-319-24592-8_36

16. Papadakis, I., Kyprianos, K., Karalis, A.: Highlighting timely information in libraries through social and semantic web technologies. In: Garoufallou, E., Subirats Coll, I., Stellato, A., Greenberg, J. (eds.) MTSR 2016. CCIS, vol. 672, pp. 297–308. Springer, Cham (2016). doi: 10.1007/978-3-319-49157-8_26

17. Sedhai, S., Sun, A.: Hashtag Recommendation for hyperlinked tweets. In: Proceedings of the 37th International ACM SIGIR Conference on Research & Development in Information Retrieval, Gold Coast, Queensland, Australia, 06–11 July, pp. 831–834 (2014)

18. Small, T.A.: What the hashtag?: a content analysis of canadian politics on Twitter. Inf. Commun. Soc. **14**(6), 872–895 (2011)

19. Teevan, J., Ramage, D., Morris, M.R.: #TwitterSearch: a comparison of microblog search and web search. In: Proceedings of the Fourth ACM International Conference on Web Search and Data Mining (WSDM 2011), Hong Kong, 9–12 February, pp. 35–44 (2011)

20. Wang, X., Tokarchuk, L., Cuadrado, F., Poslad, S.: Exploiting hashtags for adaptive microblog crawling. In: IEEE/ACM International Conference on Advances in Social Networks Analysis and Mining, Niagara Falls, Canada, 25–28 August, pp. 311–315 (2013)

21. Wijnants, M., Blazejczak, A., Quax, P., Lamotte, W.: Geo-spatial trend detection through Twitter data feed mining. In: Monfort, V., Krempels, K.-H. (eds.) WEBIST 2014. LNBIP, vol. 226, pp. 212–227. Springer, Cham (2015). doi:10.1007/978-3-319-27030-2_14

Information Behavior

Social Tagging: Implications from Studying User Behavior and Institutional Practice

Õnne Mets[1]([✉])[iD] and Jaagup Kippar[2]

[1] University of Milano-Bicocca, Milan, Italy
o.mets@campus.unimib.it
[2] Tallinn University, Tallinn, Estonia
jaagup.kippar@tlu.ee

Abstract. This paper aims to describe users' tagging behavior in catalogues and in Flickr. Six platforms of two institutions and one consortium are analyzed: the main catalogue Discovery and Flickr page of the National Archives of the United Kingdom, the main catalogue Explore, the catalogue Archives and Manuscripts and Flickr page of the British Library, and the consortial search engine of the pan-European eBooks on Demand Library Network. The results of the document and user data analysis point to differences between archival and library collections, between catalogues and Flickr, illustrate the impact of different authorization and procedural rules, and confirm previous studies as regards to the small size of the active user group. Based on the data analysis, we offer eight recommendations for social tagging in libraries and archives concerning the issues of interface functionality and management, data collection, reflection of tags and maintaining the community.

Keywords: Crowdsourcing · Social tagging · Folksonomies · Catalogues · Flickr

1 Introduction

Metadata is considered a key feature to discoverability of the collections of cultural institutions [1, 2]. Traditional indexing techniques are costly and labor-intensive and may not provide the only or best way to meet user needs in online resource discovery [3–5]. Yet the information that needs to be described is diverse and voluminous. Numerous digitization projects in libraries [6] reveal undescribed aspects of the content, and archives may lack of item-specific information, while organizing their records by collections [7]. Then crowdsourcing is applied in the form of social tagging.

Crowdsourcing projects have become popular in the cultural heritage sector. Projects can be characterized by making collections available in smaller sets to achieve systematically sub-goals, progress of task completion is monitored and communicated, volunteers are motivated and sometimes specialized skills, knowledge or equipment is required [8]. The conclusions of Transcribe Bentham project illustrate the overall picture: majority of work is done by minority of users; volunteers have interest in the subject, crowdsourcing or the technology and sense of altruism; lack of time and issues with technology might limit participation whereas media attention increases it;

© Springer International Publishing AG 2017
J. Kamps et al. (Eds.): TPDL 2017, LNCS 10450, pp. 421–433, 2017.
DOI: 10.1007/978-3-319-67008-9_33

the project resulted in increasing the digital literacy skills of participants, contribution to scholarship and widen access to the material, adjusting workflows, and exploiting investments for digitization, software development and staff salaries [9].

Examples of ongoing tagging activity without the project-type framework provide interesting findings, but in a single institutional context: e.g. most tagged objects in a catalogue of Powerhouse Museum in Sydney were not on public display [10]; 67000 tags by 2518 people were attributed in the first 10 months in the Flickr page of the Library of Congress - a pioneer in Flickr, whose collaboration led to development of the Flickr Commons [11]. But the (collaborative) projects or ongoing activities in a single institution report different aspects of the results, which are difficult to compare.

This paper takes a comparative insight into the action taken by volunteers on six platforms of two institutions and one consortium: the main catalogue Discovery of the National Archives of the United Kingdom (TNA), the main catalogue Explore and catalogue Archives and Manuscripts of the British Library (BL), the Flickr pages of both organizations, and EOD Search, the consortial catalogue of the pan-European eBooks on Demand (EOD) Library Network. All cases represent social tagging as ongoing linear activity unlike the crowdsourcing projects.

The research questions are: What characterizes the tagging behavior, if it is not part of a crowdsourcing project? What are its affecting factors, outcomes, and implications? It contributes to research on online user behavior and folksonomies, and has a consultancy value for cultural organizations. However, the use of social tags by wider user community requires access to different data and is not studied in this paper.

The next section introduces the applied methods. Based on document analysis the platforms are described and the results of user data analysis are presented in Sect. 3. The findings are discussed and recommendations are proposed to archives and libraries for further consideration in Sect. 4. The paper is concluded in Sect. 5 with an outlook to the future work, where the current findings will be embedded.

2   Methods

2.1   Document Analysis

First, the six platforms were described referring to the interfaces, help articles alongside on the websites or linked pages. Some information was received directly from the institutions with delivery of data or by special enquiry.

The document analysis looked at 14 parameters: type of the platform (catalogue, social network site or other), the collection available for tagging (records, textual or non-textual items), online access to the items (full, restricted, partial or no access), collection size (number of items), pre-existing metadata, existence of application programming interface (API), releasing collections by small sets for tagging, time of launching social tagging, authorization of taggers (procedures of registration and sign in), publishing of social tags (immediate, verified), representation of tags to view or browse, procedure for deletion of tags, instructions to tag, syntax (separators of tags).

2.2   User Data Analysis

The acquired parameters for the user data were: tags, user IDs (anonymous for catalogues), item IDs, time of tag attribution (if recorded). The datasets with social tagging information in catalogues of BL, TNA, and EOD were composed by the respective institutions on request and delivered as separate CSV files. The dataset for BL Flickr account was composed earlier by the institution, and delivered as TSV file. The data for TNA Flickr account was extracted by using the Flickr API.

The data were imported to R [12] for analysis. All in all 25 parameters were calculated, including total and unique tags and tagged items per person, total and unique tags and contributing users per item, users and items per tag, returns and tagging activity per person across catalogues (for BL), correlations between parameters. Calendar converter[1] was used for calculating periods of returns.

3   Results

3.1   Document Analysis

Overview of the catalogues. BL main catalogue Explore[2] searches around 70 million items (records for books, journals, newspapers, maps, articles, Sound Archive items, Web Archive links etc.), being the biggest dataset in the comparison of the six platforms. TNA main catalogue Discovery[3] holds over 32 million descriptions of records held by TNA (available for tagging) and more than 2500 archives across UK. BL catalogue Archives and Manuscripts[4] includes unpublished documents, prints, drawings etc., the number of records in the catalogue is unknown. The EOD Search[5] is a multi-lingual consortial catalogue, which runs on open source platform VuFind and searches over 7 million records of public domain literature from 35 libraries in Europe. The records link to institutional repositories for free full-text or display a button to request digitization for a fee [13]. Other catalogues in this comparison provide mostly limited or restricted access to view items. All are traditional catalogues with pre-existing metadata. No APIs are available for users.

Social tagging settings in the catalogues. Social tags were enabled first in Explore in November 2008, followed by EOD Search in the beginning of 2011, Archives and Manuscripts in January 2012, and Discovery in October 2012. Yet TNA may also be called a pioneer in this comparison due to launching a wiki site Your Archives[6] in April 2007. A button was placed on the Document Details page of the catalogue taking to Your Archives to see if there was any additional information; otherwise, it created a

---

[1] https://www.timeanddate.com/date/duration.html .

[2] http://explore.bl.uk .

[3] http://discovery.nationalarchives.gov.uk .

[4] http://searcharchives.bl.uk .

[5] https://search.books2ebooks.eu .

[6] http://yourarchives.nationalarchives.gov.uk .

special page inviting the user to add content [14]. By 2012 the functionality was developed for Discovery, social tags were imported and the wiki was closed.

In the BL catalogues tagging requires sign in, which is only available to registered readers (registration can be completed in person at the Library[7]) and registered document supply customers (frequent users of the service, purchasing over 100 documents a year[8]). In Discovery anyone can register online, providing Reader's ticket number is optional. The EOD Search also enables anyone to register online.

Instructions about tagging are given briefly from each record's page in the BL catalogues. More detailed information is available from the opening page behind two clicks under 'Help articles'. The same information can be found behind the tab 'Tags', which is visible throughout navigation. Comma is required to separate multiple tags. In Discovery the record's page offers a link to sign in to add a tag, but detailed information about tagging is only available from the opening page. Another link after that page gives short tips about useful and appropriate tags, including instruction: "Simply enter a tag and click 'submit'. You can add as many tags as you like"[9]. In the EOD Search there is a note on a field in the record: "No Tags, Be the first to tag this record!". The only instruction in EOD Search for tagging appears after clicking 'Add Tag' button: "Spaces will separate tags. Use quotes for multi-word tags".

In all cases social tags are published immediately without verification, mostly next to the record, in EOD Search in a field in record. Tags can be deleted by the users, who attributed them and by the institutions. In the BL catalogues all tags can be browsed by most recent and most given. Logged in users can select to view only their own tags. In addition to tags BL enables to add notes, which are not indexed and not searchable, but moderated [15, 16]. TNA enables users to flag inaccurate tags, which are then checked by the staff. A spam and profanity filter to manage spam words is also in use. All tags can be browsed alphabetically, by most given and most recent.

Flickr. Both BL and TNA use Flickr to expose their selected collections of images. The BL Flickr account[10] was established in August 2007 for corporate promotion. In December 2013 the BL Labs project added over 1 million undescribed images cropped from 65 000 volumes of digitized works from 17th to 19th century [17]. The experiment was meant for anyone to use, remix and repurpose and to spread new ways to navigate and display the content; and to stimulate the research concerning the materials [18]. First offered to Wikimedia, but rejected because of the lack of metadata, Flickr was chosen next because of tagging option, API existence, and attributing a unique URL for every image. BL imported to Flickr the metadata of the books, where the images came from, but there was no metadata about the images. Additionally, geotags are imported for maps from the BL crowdsourcing platform Georeferencer [19]. TNA joined Flickr in October 2008 and started to expose their thematic image collections since the beginning of 2011 "to give a flavour of their massive holdings"[11].

---

[7] http://www.bl.uk/help/how-to-get-a-reader-pass .

[8] http://www.bl.uk/reshelp/atyourdesk/docsupply/help/register/regularcustomers/index.html .

[9] http://discovery.nationalarchives.gov.uk/tags/index/howtotag .

[10] http://www.flickr.com/people/britishlibrary .

[11] http://www.flickr.com/photos/nationalarchives .

Anyone can sign up as a Flickr user and tag the images. Tags are displayed alongside the images as is the link 'Add tags'. The tags added by Flickr robots are visible together with community tags, but distinguished by their white background. Next to 'Tags' under '?' is a short description about tags and a link to some more information, including the instruction "Separate single word tags with spaces and add phrases in quotes". Users can remove both tags they create and ones Flickr has added for them[12]. The Flickr API[13] enables anyone to write a program to present public Flickr data (photos, video, tags, profiles or groups) in different ways, and make their applications available to other users.

3.2   User Data Analysis

General overview. According to availability of the data, the period of observation varies as follows: 28 months (Dec 2013–Mar 2016) for BL Flickr page, 52 months (Oct 2012–Jan 2017) for Discovery, 60,5 months (Jan 2012–Feb 2017) for Archives and Manuscripts, 74 months (Jan 2011–Feb 2017) for EOD Search, 75 months (Jan 2011–March 2017) for TNA Flickr page, and 99 months (Nov 2008–Feb 2017) for Explore. Total numbers are presented for social tags and taggers on Fig. 1, because there is no significant distinction in the proportions compared to the results in average per month. The figure excludes 15% of total tags in Discovery (i.e. mostly tags attributed by users to Your Archives, then imported to Discovery by a single institutional user account), 43% of total tags in BL Flickr page (i.e. mostly tags attributed by users in Georeferencer, then imported to Flickr by a single account); 96% of total tags in TNA Flickr page (i.e. collection names etc. attributed as tags by TNA).

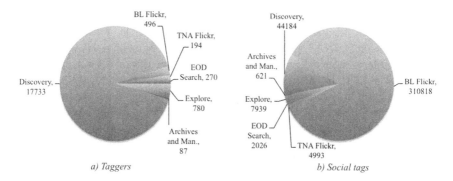

a) Taggers   *b) Social tags*

Fig. 1. Distribution of (a) engaged taggers, and (b) social tags attributed by them.

---

[12] http://help.yahoo.com/kb/flickr/tag-keywords-flickr-sln7455.html .

[13] https://www.flickr.com/services/apps/about .

In all cases majority of users attribute up to 10 tags, and about 8 people form a group of top taggers (by attribution of total and unique tags or tagged items) (Fig. 2). If we exclude the tags by institutional accounts, the median value is 6 tags per person for BL Flickr page, 3 for TNA Flickr, 2 tags per person for Explore and EOD Search and 1 for others. BL Flickr page is also first by the sum of unique tags, but both Flickr pages have least unique tags out of total tags (TNA 5% and BL Flickr page 8%, Explore 26%, Archives and Manuscripts 28%, EOD Search 46%, Discovery 60%).

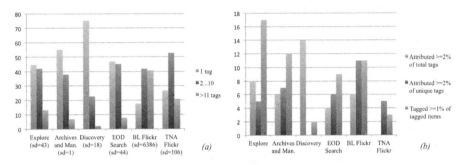

Fig. 2. (a) Number of tags per person (percentage of total taggers); (b) division of top taggers by total and unique tags and tagged items (absolute numbers).

Items. In average per month the most items were tagged in BL Flickr page (6290 items per month, 176 133 in total), less in Discovery (794 per month, 38 106 in total), in TNA Flickr page (209 per month, 15 679 in total), in Explore (56 per month, 5548 in total), in EOD Search (18 per month, 1360 in total) and in Archives and Manuscripts (9 per month, 528 in total). The tagged items gained mostly one tag per item in the catalogues (mean, median, mode < 1,5; sd < 2,2), more in Flickr (BL Flickr: mean = 3, sd = 68,5; TNA Flickr: mean = 7,2; median = 8; mode = 11, sd = 4,38). In all cases the items were tagged mostly by one person (mean < 1,42; median = 1; mode = 1; sd < 1,69). In Archives and Manuscripts no records were tagged by at least 2 different people. In Explore the maximum number of taggers per record was 4 (one occasion), in BL Flickr 5 people per image in maximum (18 occasions), in TNA Flickr page 8 people, but in Discovery there were 104 people per record in maximum (one occasion, followed by 100 people per record once, 80 people per record once etc.).

Correlation between tagged items and attribution of total tags is very strong in all cases (r >= 0,82). Correlation between attribution of total and unique tags is strong in case of archival content (TNA Flickr r = 0,98, Discovery r = 0,88, Archives and Manuscripts r = 0,72), but remains moderate in other cases (r <= 0,46). Similarly correlation between tagged items and attribution of unique tags is strong or medium for archival content (0,8 >= r >= 0,57), low for others (r <= 0,27).

Time interval. Dates of tag attribution were analyzed for Explore, Archives and Manuscripts and EOD Search. In other cases the time parameter of tag attribution was not available. In Explore there were 956,5 total tags per year in average (sd = 529), in EOD Search 305 tags (sd = 315), and in Archives and Manuscripts 123 tags in average per year (sd = 152). The number of engaged users and time interval for their returns in average per person are illustrated on Fig. 3. In all three cases users tagged mostly on one day, in average per person on 1,15 ≤ dates ≥ 1,58 (maximum in EOD Search: 112 dates per user, sd = 6,79). The users, who returned to tag on a different date, are 165 people (21% of total taggers) in Explore; 13 people (4,81% of total taggers) in EOD Search, and only 11 people (1,4% of taggers) in Archives and Manuscripts.

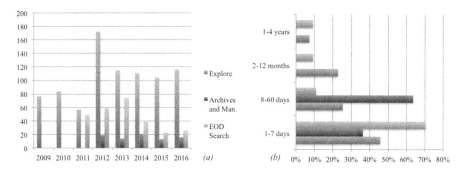

Fig. 3. Distribution of (a) engaged taggers by years, and (b) time interval for returns.

The period from first to last day of tag attribution varies as follows: 260 days in average per user (sd = 373, max. 3,7 years) in Explore; 252 days (sd = 489, max. 4,5 years) in EOD Search; 24 days (sd = 23, max. 2,3 months). The correlation between number of days from first to la st day and days with tagging activity is stronger in Archives and Man. (r = 0,68) and Explore (r = 0,54), and weak in EOD Search (r = 0,23).

Correlation between the number of returns and attributed tags is modest (Explore r = 0,46, EOD Search r = 0,36, Archives and Manuscripts r = 0,25). For instance, the 18 top taggers (by total tags) in Explore divide into 3 people, who returned on >10 dates, 12 people on 2 to 9 dates, and 3 top taggers gave all their tags on one day. In Archives and Manuscripts one person added tags on 4 dates within 2 months. All other 10 of 11 users tagged on 2 dates and returned within the same or subsequent month, incl. the top contributor (attributed 55% of tags) on two subsequent dates.

Data for Explore and Archives and Manuscripts allows looking at tagging across catalogues. 16 people have attributed tags both to Explore and Archives and Manuscripts, i.e. 2% of taggers in Explore and 18% in Archives and Manuscripts. 2 people appear as top taggers in both catalogues.

Tags. The most attributed tags refer often to import of many tags at a time (e.g. 'a level - korean war' imported to Discovery from the previously run wiki site or 'geo:continent = europe' in BL Flickr page through its API), whereas tags attributed by most people differ in content, are sometimes private by their nature (e.g. 'to read') or point to the misuse of the syntax rules (e.g. not using quotes for phrases results in prepositions as tags, like in EOD Search) (Fig. 4).

Tags attributed most in total	Tags attributed by most people	Tags attributed most in total	Tags attributed by most people
Explore, BL		Flickr, BL	
silent cinema	dissertation	map	map
business directory	fantasy fiction	georefphase2	portrait
war of 1812	history	togeoref	church
telecommunications industry guide	read	rotate90	bird
this	to read	hasgeoref	london
anarchist newspapers & periodicals (english language)	biography	wp:bookspage=geography	castle
propaganda 2013 exhibition	check	geo:continent=europe	river
construction	music	coatofarms	dog
renewable energy industry guide	1	portrait	boat
Discovery, TNA		colorful	horse
a level - korean war	grandad	colourful	ship
a level - chartism	grandfather	people	woman
movcon	dad	lettert	cathedral
cycle	father	wp:bookspage=synopticindexusa	bridge
haiti list	ww1	rotate270	letter
EOD Search		geo:country=uk	man
incunable	1	geo:country=unitedkingdom	egypt
1	2	music	split
CS	of	Flickr, TNA	
officium	12	thenationalarchivesuk	woman
http://search.books2ebook	Geschichte	tna:departmentreference=co	horse
2	a	tna:seriesreference=co1069	railway
Band	des	tna:divisionreference=cod32	road
Ainsworth	in	africathroughalens	school
Hubertusburg	the	asiathroughalens	street
alphabets	von	asia	

Fig. 4. Tags attributed most in total and tags attributed by most people.

In Archives and Manuscripts only one tag 'john' was attributed by at least 2 people. If we exclude geotags by BL in Flickr, the list of top tags remains similar. And if we manipulate that dataset further by losing the computational parts of tag strings, more geographical locations appear as most attributed tags in total.

4 Discussion

The comparison includes two main catalogues of the memory institutions, their Flickr pages, a smaller archival catalogue of the Library and a consortial catalogue. The difference in the size of collections available for tagging does not seem to impact directly the tagging activity. The difference in the type of collections may have increased the tags for BL Flickr collection of cropped images as it had no pre-existing metadata and users had freedom to add anything, even if the image discovery may have been more serendipitous compared to described collections.

The two institutions have different user authorization procedures in the catalogues. The low number of taggers in BL catalogues may be caused by the registration procedure, which requires personal presence or being a frequent customer of document delivery service, compared to Discovery where anyone can register online. Also Flickr offers free online registration, but engaged less users than the catalogues. The devoted subject communities may not be as used to use Flickr than catalogues.

Even though the datasets are different by the number of participants and tags, they result in similar number of top taggers by attribution of total tags (up to 14 people), unique tags (up to 11 people) and tagged items (2 to 17 people).

It is common in catalogues, that most users add only one tag. The Flickr users tend to add more tags per person, and there are more tags per item in Flickr than in catalogues - both trends possibly affected by the use of Flickr API.

It is evident in all cases that people, who add more tags tag also more items. Also given that only a few taggers per item we may conclude that users tend to describe items briefly and choose different items from each other. The phenomenal 104 users per a record in Discovery is an exception that may be explained by the biggest number of taggers in total - the higher the number of users the higher the probability of tagging the same object; or it may refer to users' mistake to attribute tags on a collection rather than a specific item level. It could have been explained by the collection size - the smaller the size of collection the higher the probability of people tagging the same item - unless Discovery exceeds the number of images in Flickr about 30 times.

The common feature for all three platforms with archival content is the attribution of unique tags compared to libraries' content, which can be described more by universal tags. It may refer to the perception of users to make a distinction of unique archival content compared to the published materials.

Comparison between most attributed tags and tags attributed by most people reveals interesting dichotomy. In four cases the most attributed tags refer more to applying computational techniques for tag attribution, mass import of tags from other sources or to the form of the tagged object, whereas tags attributed by most people tend to be more telling in content. E.g. Discovery suggests that the core interest of most taggers is genealogy. And even if tags like 'grandad' is a noise for others, it tells us to update the instructions with the most common examples. Additionally tags like 'to read', or 'alan's summer project' refer to marking up for individual need, suggesting to develop the functionality for private tags. Interestingly these examples occur in catalogues and not in Flickr, which has more liberal or noisy image than the controlled and verified one of the catalogues. It may also rise from the nature of the collections exposed in Flickr - selected images instead of records of materials.

The misuse of syntax for multiple tags or for multi-word tags may turn useful tags into noise. E.g. commas must be used in the BL catalogues, but if users follow the record and add 'Last Name, First Name' without quotes, it results in having a tag 'john' instead of full name. Similarly not using quotes for phrases in EOD Search resulted in having prepositions as tags. Flickr has the same rules, but these mistakes are not common there. The crowdsourcing projects usually avoid this kind of mistakes by having separate text boxes for different descriptive data.

The data of three catalogues, where time factor was available, point to rather surprising finding that not only most users, but also most top taggers by total tags make their contributions within a short time-frame, less than 10 days and in some cases only once. The timescale does not imply that tagging has become more popular over time, rather it is unstable in terms of total tags and a bit steadier by participating users.

Additional content analysis of the social tags and analytics of the usage of the social tags could shed light to understand better the tagging under these conditions.

4.1   Considerations for Institutional Practice

Enable and instruct

(1) Enabling tagging for everyone upon an online registration seems to have a positive impact on the number of contributors compared to ID verified users only. Institutional practices refer that volunteers come from different countries [8], so are the users of the catalogues, who are potential taggers. About 90% of the visits to the EOD search engine [13] and 30% of visits to Your Archives were from Google searches [14]. The description of the record may become more varied and more reliable, if tagged by more people.

(2) Once the taggers come, they might not come through the first door, i.e. opening page of the catalogue, what was in one occasion the only place where to find instructions on tagging. EOD Search has experienced only 3% of the visits landing on the opening page, and majority of visits came directly to single records [13]. Help pages should be cross-linked.

(3) Available and clear instructions proved to be vital especially for the use of separators. In some case comma was intuitively used instead of space, quotes were not used for phrases or users did not understand the instruction to insert one tag at a time. If user behavior suggests that the requirements for separators must be adjusted, it will likely not affect many users, who were used to different separators and will likely not return to tag in the future as suggested by the data.

Advance and reflect

(4) Some tags even within top tags do not contribute to increasing discoverability, but are initiated from another motive - individual need. It is not clear, if users intentionally break the rule of 'making a useful contribution' or they have overseen the notion that all tags are made publicly available. Still users seem to need an alternative option to add some tags visible only to themselves, because these tags are not meaningful to other people ('to read' etc.).

(5) As the digital skills of users improve (use of APIs, running software libraries, image recognition tools etc.), providing an API justifies the effort and significantly increases the amount of tags. It requires full availability of items, which in turn might reduce the risk of tagging based on assumptions when the item was not

fully seen. It may also lead us to an intriguing option to enable social (computational) tagging not in catalogues, but repositories.

(6) Displaying tags, which are given by most people might be more telling than showing the most attributed tags by no matter how many people or technique. It is not the case when the amount of contributors is too low.

Monitor and maintain

(7) Recording the time of tag attribution is important for monitoring the returns of the contributors. That data were available for 3 platforms of 6, all suggesting that not only majority, but even top taggers occur in short time.

(8) If the goal is to keep the top taggers, their contribution should be detected quickly, and if seemed valuable, the dialogue should be started and maintained quickly before they leave.

5   Conclusion and Future Work

The variety of tags and especially the variety of tagged items, which becomes evident from this study, illustrates the importance of social tagging for the whole collection. The overall user activity confirmed the previous studies on the small proportion of active users. The current study also showed that driving tagging activity onto a social network site does not guarantee that the activity goes viral. The power of Flickr in this case lies in its API, which was used for mass tagging, but not in increasing engaged audiences, which could have been expected due to the social nature of the platform.

The take-aways for organizations suggest reviewing sign up procedures, making instructions clear and available, considering individual need for tagging, developing tools for computational tagging by users, defining "top tags" not only by their sum, but also by the number of people attributing them, monitoring the activity in time and cherishing the valuable contributors.

The results will be used for further research on determining the relations between social interaction and discoverability of resources. Next the research proceeds with social network analysis; thematic analysis of the interviews with BL and TNA for better knowledge of participatory practices [20, 21] and institutional activity systems according to Activity Theory [22]; and thematic analysis of the questionnaires to the users for better persona creation and understanding individual activity systems.

Acknowledgements. The author is grateful to the British Library and the National Archives of UK for the data, Baseer Baheer for his help with Flickr API, and Prof. David Lamas and Dr. Marco Gui for their feedback to this paper.

References

1. Higgins, S.: Digital curation: the emergence of a new discipline. Int. J. Dig. Curation **6**(2), 78–88 (2011)
2. Westbrook, R. N., Johnson, D., Carter, K., Lockwood, A.: Metadata clean sweep: a digital library audit project. D-Lib Mag. **18**(5–6) (2012). doi:10.1045/may2012-westbrook
3. Matusiak, K.K.: Towards user-centered indexing in digital image collections. OCLC Syst. Serv. Int. Dig. Libr. Perspect. **22**(4), 283–298 (2006). doi:10.1108/10650750610706998
4. Macgregor, G., McCulloch, E.: Collaborative tagging as a knowledge organisation and resource discovery tool. Libr. Rev. **55**(5), 291–300 (2006)
5. Daly, E.K., Ballantyne, N.: Ensuring the discoverability of digital images for social work education: an online tagging survey to test controlled vocabularies. Webology **6**(2), Article 69 (2009). http://www.webology.org/2009/v6n2/a69.html
6. Galloway, E., DellaCorte, C.: Increasing the discoverability of digital collections using Wikipedia: the pitt experience. Pennsylvania Libr. Res. Pract. **2**(1), 84–96 (2014). doi:10. 5195/palrap.2014.60
7. Szajewski, M.: Using Wikipedia to Enhance the Visibility of Digitized Archival Assets. D-Lib Mag. **19**(3/4) (2013). doi:10.1045/march2013-szajewski
8. Ridge, M. (ed.): Crowdsourcing Our Cultural Heritage. Series: Digital Research in the Arts and Humanities. Routledge, London (2017)
9. Causer, T., Terras, M.: 'Many hands make light work. many hands together make merry work': transcribe bentham and crowdsourcing manuscript collections. In: Ridge, M. (ed.) Crowdsourcing Our Cultural Heritage. Series: Digital Research in the Arts and Humanities. Routledge, London (2017)
10. Chan, S.: Tagging and searching – serendipity and museum collection databases. In: Trant, J., Bearman, D. (eds.) Museums and the Web 2007: Proceedings, Toronto, Archives and Museum Informatics, 1 March 2007. http://www.archimuse.com/mw2007/papers/chan/chan. html. Accessed 24 Mar 2017
11. Oomen, J., Aroyo, L.: Crowdsourcing in the cultural heritage domain: opportunities and challenges. In: Proceedings of the 5th International Conference on Communities and Technologies (C&T 2011), pp. 138–149. ACM, New York (2011). doi:http://dx.doi.org/10. 1145/2103354.2103373
12. R Core Team. R: A language and environment for statistical computing. R Foundation for Statistical Computing, Vienna, Austria (2016). https://www.R-project.org
13. Mets, Õ., Gstrein, S., Gründhammer, V.: Increasing the visibility of library records via consortial search engine. In: Proceedings of the 14th ACM/IEEE-CS Joint Conference on Digital Libraries, pp. 169–172. IEEE Press, Piscataway (2014). doi:10.1109/JCDL.2014. 6970164
14. Grannum, G.: Harnessing user knowledge: the national archives' your archives Wiki. In: Theimered, K. (ed.) A Different Kind of Web: New Connections Between Archives and Our Users. Society of American Archivists, Chicago (2011)
15. Explore the British Library. British Library (2014). http://www.bl.uk/catalogues/search/pdf/ tags.pdf
16. Explore the British Library. British Library (2014). http://www.bl.uk/catalogues/search/pdf/ notes.pdf
17. O'steen, B.: BL Flickr image dataset: User Submitted Tags (till March 2016). Figshare. https://dx.doi.org/10.6084/m9.figshare.3126481.v1. Accessed 15 Oct 2016. https://figshare. com/articles/BL_Flickr_image_dataset_User_Submitted_Tags_til_March_2016_/3126481

18. O'steen, B.: A million first steps. Digital scholarship blog. http://blogs.bl.uk/digital-scholarship/2013/12/a-million-first-steps.html. Accessed 12 Dec 2013

19. Mahey, M.: Interview, Skype. 15 October 2016, 27 October 2016

20. Bonney, R., Ballard, H., Jordan, et al.: Public participation in scientific research: defining the field and assessing its potential for informal science education. A CAISE Inquiry Group Report, Washington, D.C.: Center for Advancement of Informal Science Education (CAISE) (2009). http://www.birds.cornell.edu/citscitoolkit/publications/CAISE-PPSR-report-2009.pdf. Accessed 24 Jan 2017

21. Simon, N.: The Participatory Museum. Museum 2.0. (2010)

22. Kaptelinin, V., Nardi, B.: Acting with Technology: Activity Theory and Interaction Design. The MIT Press, Cambridge (2006/2009)

The Ghost in the Museum Website: Investigating the General Public's Interactions with Museum Websites

David Walsh[1,2]([envelope]) [ORCID], Mark Hall[2] [ORCID], Paul Clough[1] [ORCID], and Jonathan Foster[1] [ORCID]

[1] University of Sheffield, Sheffield, UK
[2] Edge Hill University, Ormskirk, Lancashire, UK
david.walsh@edgehill.ac.uk

Abstract. Museums are increasing access to their collections via web-based interfaces, but are seeing high numbers of users looking at only one or two pages within 10 s and then leaving. To decrease this rate, a better understanding of the type of user who visits a museum web-site is required. Existing models for museum web-site users tend to focus on a small number of groups or provide little detail in their definitions of the groups. This paper presents the results of a large scale museum user survey in which data on a wide range of user characteristics was collected to provide well founded definitions for the user group's motivations, tasks, engagement, and domain knowledge. The results highlight that the general public and non-professional users make up the majority of users and allow us to clearly define these two groups.

Keywords: Digital cultural heritage · Museum web-site · Users · Survey

1 Introduction

Museums have expanded their web-based offerings, providing access not only to general information about the museums, but also direct access to their holdings. This has opened up museums to a wider public and led to a significant rise in the number of visitors to museum websites [1]. However, museums have been struggling with large numbers (more than 50%) of users visiting their sites, looking at one or two pages, and then leaving within a very short period of time (generally less than 10 s) [2,3].

This raises two questions: who are these users and what could be done to keep them on the museum's site for longer? Where digital cultural heritage (DCH) users have been studied in the past, the focus has primarily been on user groups that are easier to access, such as experts, researchers, and museum staff. The general user and the non-professional user generally receive less attention, but we hypothesise that it is from these groups that the majority of users that bounce off museum websites come. Understanding these user groups and how their needs and behaviours differ from the user groups that have been studied

© Springer International Publishing AG 2017
J. Kamps et al. (Eds.): TPDL 2017, LNCS 10450, pp. 434–445, 2017.
DOI: 10.1007/978-3-319-67008-9_34

more frequently will enable museum web-sites to adapt their content and style of presentation to better support them.

To this end we present the first large-scale study of users from the National Museums Liverpool's web-site. National Museums Liverpool (NML) is a collection of seven museums that cover a wide range of areas from art galleries to natural history and slavery. Similar to the studies previously cited data from their transaction logs indicates that approximately 60% of their users leave within 10 s. They thus form an appropriate case study, particularly as their wide spread of subject areas leads to wide range of museum visitors. Within this context the study addresses the following research questions:

RQ1 Which user groups use NML's web-site?
RQ2 How can we define the general public/non-professional user groups?
RQ3 Is there a difference between the general public and non-professional groups?

The remainder of this paper is structured as follows: in Sect. 2 we discuss existing work to understand and classify digital cultural heritage users. Section 3 describes the study we undertook; Sects. 4 and 5 present and discuss the results; and Sect. 6 presents our conclusions and directions for future work.

2 Background

Visitors to physical museums have been studied for a long time. Indeed one of the first studies was conducted in 1884 in Liverpool Museum, identifying four groups of users: students, observers, loungers, and German and Scandinavian immigrants [4]. Following on from this, studies have investigated museum visitors in various contexts including their motivations [5], who they visited with [6,7], the role taken [5], and their engagement with the museum [8]. While it is tempting to apply physical visitor models to the digital world, there is no certainty that the two entirely overlap [9] and on-line visitors should be studied in their own context [10,11].

2.1 User Expertise

The user's expertise is one of the most common facets for distinguishing different user groups. The simplest distinction here is between generic groups, such as *novice* and *expert* [1]. Vilar et. al [12, p.150] define professional users as those who act within the formal part of a profession, having good knowledge of the task, being trained and usually having experience with it and deep understanding of its context. More generally [13] defines *experts* as "specialists in the field of cultural heritage.", while [14] introduces the *Museum Information Professional* as someone working with information resources and a desire for meeting user needs whether users are inside or outside the museum.

In contrast the *lay user*, *non-expert*, or *novice* are typified as having no formal or only limited training [12,15] in relation to DCH or as being completely new to the entire environment [13]. [16] list "knowledge of the task, information needs and system expectations" as the main distinction from the *expert*.

Between these two extremes lies the *hobbyist* or *non-professional* user [7, 17–20] who shares with the *expert* the knowledge of DCH, but has the *lay user*'s focus on personal reasons. Related to both the *novice* and *hobbyist* are the *casual leisure users* who are often "first- and short-time visitors" [21, p.74], who have "just stumbled across [the digital] collection in the same way that they would wander into the CH institution's physical space" [22, p.1].

2.2   Information Needs, Motivation, and Role

An analysis of the London Science Museum's physical and virtual visitors defined three groups based on their information needs [23]: *general visitors* who require general information, such as opening hours or prices; *educational visitors* who require additional, detailed information to plan their visit; and *specialist visitors* who require more detailed information on collections and offer more expertise.

Similarly, [24] describe Library of Congress' National Digital Library users: groups were defined by combining their motivations, domain knowledge, system knowledge, task focus, and time allocation. This lead to nine different groups: staff, hobbyists, scholars, professional researchers, rummagers (browsers), object seekers, surfers, Teachers K-16, Students K-16. Similarly, the CULTURA project identified the following groups: professional researchers, apprentice investigators, informed users, and the general public [25].

2.3   Definitions of User Groups and the General Public

As this brief review shows, there are a large number of potential classification systems, which in some cases overlap; in some use different terminology for equivalent or very similar groups; and in some cases use the same terminology very differently. Additionally, many of the user groups identified above are defined via a single sentence or phrase, such as "specialists in the field of cultural heritage." When it comes to deciding how to support these user groups, this low level of detail in the definition limits the usefulness of the groupings.

Additionally, the closer the user group is to the *general public*, the less clearly defined the user groups become. Frequently, the *general public* or *general visitor* are treated as catch-alls for those users who do not fit into any of the more well-defined groups. However, it is our hypothesis that these groups are actually the most common type of visitor to museum web-sites and thus require closer attention than they have received so far.

The study reported in this paper addresses these shortcomings by acquiring on-line museum visitor responses for a wide range of criteria derived from the literature and in particular provides a detailed view onto the general public/non-professional user.

3   Methodology

To study visitors of the NML web-site an on-line survey was created based on user group definitions identified in the existing literature. The on-line survey consisted of 22 questions and was delivered via the PollDaddy system[1]. Six questions covered standard demographics (age, gender, education, employment status). The remaining questions were derived from user group definitions, or surveys, found in the literature. These were grouped into seven categories around aspects previously used to define groups: motivation [7], task [19,26–28], content types, sharing [28], engagement [26,29,30] , domain knowledge [19], usage [31], and technical expertise [30]. Table 1 shows those questions where significant differences between the "general public", "non-professional" and other user groups were found.

Table 1. Survey questions that show significant differences between the "general public", "non-professional" and other user groups. Question #7 allowed the participant to select multiple responses.

#	Category	Question
1	Motivation	Today I am visiting the NML website: [personal, study, pass time, work]
2	Task	What is the primary purpose of your visit to the NML website today?
3	Engagement	How frequently do you visit the NML website?
4	Domain knowledge	In the context of cultural heritage and your current visit to the NML website please select the appropriate statement: [novice, some experience, highly experienced, don't know]
5	Domain knowledge	Rate your general Cultural Heritage knowledge
6	Demographics	Where in the world are you at the moment?
7	User group	Which of the following groups would you place yourself in for this visit to the NML website?

3.1   Recruitment of Participants

Participants were recruited from NML's web-sites via a small banner pop-up, which appeared after a 10 s delay. Visitors were only invited once, regardless of whether they chose to participate or ignore the pop-up.

In the survey, visitors first had to confirm that they agreed to participate. They then answered the 14 questions focusing on the aspects defining user

---

[1] https://polldaddy.com/.

groups, before providing demographics data. Finally, on the last page they self-classified into a set of user groups identified from the literature (question #7). Participants were then thanked and provided with a link back to the NML site.

The survey was available for a four week period (1/2/2017 to 14/2/2017) on the Museum-focused areas (World Museum[2], International Slavery Museum[3], Sudley House[4], and the Maritime Museum[5]) and (15/2/2017 to 1/3/2017) on the Gallery areas (Walker Art Gallery[6] and Lady Lever Art Gallery[7]).

3.2   Participants

1118 participants were recruited, of which 573 completed the survey (51% completion rate). Of these, 9 were aged below 18 and subsequently filtered out (to avoid safeguarding issues), resulting in a final data-set of 564 participants.

348 participants were female (61%) and 211 male (37%) (14 unspecified). The majority of participants (204, 36%) were in the 35–54 age group, 147 (26%) were between 55 and 64, 110 (20%) between 18 and 34, 84 (15%) between 65 and 74, and 19 (3%) over 75. 61 (11%) were educated to secondary school level, 134 (24%) to further-education level, 193 (34%) had a degree, 116 (21%) had a masters-level qualification and 33 (6%) held a doctoral qualification, 21 (4%) participants chose no standard qualifications. Additionally 170 held a professional certification.

The majority (55%) of participants were employed, either full-time (208, 37%) or part-time (107, 19%). 122 (22%) were retired, 56 (10%) students, and the remainder not in employment. Participants were recruited from across the globe: 196 (35%) from the Liverpool/Merseyside area, 129 (23%) from the north-west of England, 102 (18%) from the rest of England, 35 (6%) from the rest of the UK, and 102 (18%) from the rest of the world.

The wide range and distribution of participants indicates that while participants self-selected for participation, the data-set is highly likely to be representative of the range of users of the NML web-sites.

4   Results

To address the first research question we look at how participants self-classified themselves for question #7. Participants could select any number of responses and were provided with an free-text "other" option as well. Table 2 shows the ten most frequently selected responses, which cover 90% of the participant responses. The remaining 10% are covered by multiple-selection responses, where no individual set of responses covers more than 1%. The majority of

---

[2] www.liverpoolmuseums.org.uk/wml.

[3] www.liverpoolmuseums.org.uk/ism.

[4] www.liverpoolmuseums.org.uk/sudley.

[5] www.liverpoolmuseums.org.uk/maritime.

[6] http://www.liverpoolmuseums.org.uk/walker.

[7] http://www.liverpoolmuseums.org.uk/ladylever/.

responses are for a single group only, strongly supporting the idea that participants had clearly defined views on how the groups were delineated and where they saw themselves. The exceptions to this are participants who classified themselves as "non-professional/general public", "academic/teacher", and "non-professional/teacher/general public".

To simplify further analysis we first investigated whether these multi-selection groups could be merged into the single selection groups. Our hypothesis was that "non-professional/general public" should be merged with "non-professional", "academic/teacher" with "academic", and "non-professional/teacher/general public" with "teacher". The multi-selection groups were compared to each of the single-selection groups using χ^2 tests. For the "non-professional/general public" group there were significant differences to the "general public" group $(p < 0.05)$[8] and no significant differences to the "non-professional" group. Likewise the "academic/teacher" showed no differences from "academic", but differed from the "teacher" group $(p < 0.05)$. For the "non-professional/teacher/general public" group there were no differences to the "teacher" group, but significant differences $(p < 0.05)$ to "non-professional" and "general public" groups. The multi-select groups have thus been merged following the hypothesis, resulting in the set of seven primary user groups (Table 2), which will be used for the further analysis (the "other" group has not yet been analysed in more detail and is not taken into account for the further analysis).

Table 2. Most frequently selected user groups, before applying the rules merging the multi-selection responses (**pre-merging**) and after (**merged**). In both cases the "Other" group has not been subjected to further analysis.

Group	Pre-merging	Merged
General Public	253	253
Non-professional	89	137
Non-professional/General public	48	-
Student	33	33
Other	26	26
Teacher	18	25
Academic	16	25
Museum staff	10	10
Academic/Teacher	9	-
Non-professional/Teacher/General public	7	-

To investigate research questions 2 and 3, a series of χ^2 tests were used to compare the merged groups' responses to all questions. Based on these the

---

[8] reporting individual p-values and detailed χ^2 statistics for grouped results exceeds the available space, but we intend to report them in detail in a future publication.

Table 3. Responses to the question "Today I am visiting the NML website:"

	Personal	Pass time	Study	Work
General public	**200**	**43**	6	4
Non-professional	**112**	15	5	5
Student	7	2	**23**	1
Academic	8	1	**9**	**7**
Teacher	11	4	2	8
Museum staff	2	2	0	**6**

Table 4. Responses to the question "What is the primary purpose of your visit to the NML web-site today?". MO - Museum Overview (gain an overview over the museums' content), CO - Collection Overview (gain an overview over a collection), KC - Known Collection (look at the content of a known collection), KI - Known Item (look for a known item).

	Pre-Visit	MO	CO	KC	KI	Shop	News	Unknown	Other
General public	**154**	23	13	4	8	12	1	1	37
Non-professional	49	17	9	1	12	0	2	3	35
Student	11	6	6	3	1	0	0	1	5
Academic	5	0	2	3	**5**	1	1	0	**8**
Teacher	**15**	3	1	0	1	0	2	0	3
Museum staff	4	1	1	2	0	0	0	0	2

questions that provide significant differences between the "general public" and "non-professional" were identified (see Table 1).

For question #1 (motivation) Table 3 clearly shows that the main distinction is the focus on personal reasons for the visit (differences to all groups are significant at $p < 0.001$). Interestingly, there is a significant number of "general public" users who have visited the web-site purely to pass some time; a group that is commonly identified in the physical museum.

For question #2 (Table 4), results show a slightly different picture. Preparation for a visit is a major characteristic for both the "general public" and the "non-professional" groups. However, here the "general public" group is significantly different to both the "academic" ($p < 0.001, \chi^2 = 41.3, df = 8$) and "museum staff" ($p = 0.04, \chi^2 = 16.5, df = 7$) groups but the "non-professional" group is only significantly different ($p = 0.03, \chi^2 = 16.7, df = 8$) to the "academic" group. In fact the "teacher" group is almost identical in its purpose to the "general public" group. At the same time there is a significant difference ($p < 0.001, \chi^2 = 32.1, df = 7$) between the "general public" and "non-professional" groups.

Table 5. Responses to the question "How frequently do you visit the NML website?"

	First visit	Yearly	Monthly	Weekly	Daily
General public	**133**	**82**	32	6	2
Non-professional	**78**	**40**	13	6	0
Student	22	7	2	2	0
Academic	10	10	2	1	**2**
Teacher	11	8	3	3	0
Museum staff	0	2	5	1	**2**
Professional	5	0	0	0	0

Table 6. Responses to the question "In the context of cultural heritage and your current visit to the NML website, please select the appropriate statement"

	Novice	Intermediate	Expert	Unknown
General public	**78**	**153**	16	6
Non-professional	29	**98**	10	0
Student	**14**	15	3	1
Academic	0	10	**15**	0
Teacher	5	10	8	2
Museum staff	0	5	5	0

A similar picture emerges for the frequency of visit (Table 5), with significant differences to the "academic" and "museum staff" groups ($p < 0.001$), but no significant differences to the "teacher" and "student" groups.

While in the previous questions the "general public" and "non-professional" groups have been similar, when it comes to domain knowledge, there are some differences between the two. For the domain knowledge about NML (Table 6) the "general public" is significantly different to all other groups at $p < 0.001$, except for the "student" group where there is no significant difference. On the other hand, the "non-professional" group is significantly different at $p < 0.01$ to all groups including the "student" group and the "general public" group.

For general CH knowledge (Table 7), the pattern is the same for the "general public", but here the " non-professional" group is only significantly different from the "academic" and "museum staff" groups ($p < 0.05$). The difference to the "general public" is borderline, but not significant ($p = 0.66$).

Finally, the results for location (Table 8) show some differences. The "general public" is significantly different from the "academic" and "student" groups ($p < 0.03$), while the "non-professional" group also differs significantly from the "teacher" group ($p = 0.05$, $\chi^2 = 8.83$, $df = 4$). The difference clearly being that both the "general public" and "non-professional" groups are much more local than the other groups.

Table 7. Responses to the question "Rate your general Cultural Heritage knowledge" (Likert-like scale, 1 - low, 5 - high)

	Low	2	3	4	High
General public	8	47	112	70	16
Non-professional	3	14	56	49	15
Student	1	7	15	7	3
Academic	1	0	2	10	15
Teacher	0	1	11	6	7
Museum staff	0	0	1	5	4

Table 8. Location: Distance from the physical museum.

Group	Merseyside	Northwest	England	UK	World
General public	95	73	44	18	23
Non-professionals	47	30	25	10	25
Students	8	8	7	1	9
Academics	4	3	4	1	13
Teachers	13	3	4	1	4
Museum staff	8	1	1	0	0

5   Discussion

The results clearly show that the "general public" and "non-professional" groups are the primary audience of NML's web-sites. These two groups have significantly lower experience with DCH and an early analysis of their 'other' responses indicates that they are less likely to visit repeatedly and less likely to remain engaged with the web-site if they do not immediately find what they are looking for. Based on this, it is likely that a significant fraction of those 60% of users who bounce from the web-sites within 10 s also belong to those two groups (particularly the "general public" group). A better understanding of these two groups, that a more detailed analysis of the survey responses will allow, should enable museums to provide more appropriate services and reduce the bounce rate.

Six questions have been identified that show significant differences between the "general public" and "non-professional" user groups and the other groups. From these four areas have been isolated that define the groups in relation to each other (Table 9). Both groups are generally more motivated by personal reasons and will have a lower amount of domain knowledge. The "general public" also has a strong interest in information for preparing a visit. However, there is also interest in the digital collections, where the personal focus and the lower domain knowledge might mean that current offerings, which are generally structured around the search box (requiring domain knowledge for the search terms), are not providing these user groups with the appropriate type of access and guidance.

Table 9. Defining characteristics for the main user groups. Characteristics marked "-" indicate no clear preference for that characteristic/group. P - Personal, PT - Pass Time, S - Study, W - Work, PV - Pre-Visit, C - Digital collections, N - Novice, I - Intermediate, E - Expert, L - Local, D - Distant. Defining differences between the "general public" and "non-professional" groups are in bold.

Group	Motivation	Task	Domain knowledge	Location
General public	P/PT	**PV**/C	**N/I**	L
Non-professionals	P	C	**I**	L/**D**
Students	S	-	N/I	-
Academics	S/W	C	I/E	D
Teachers	-	PV	I	L
Museum staff	W	-	I/E	L

The responses also enable separating the "general public" from the "non-professional" users based on their task, domain knowledge, and location. While both groups primarily come for personal reasons and are mostly from the local area, "non-professional" users are also drawn from further afield, while the "general public" has a stronger pre-visit information need. However, the main distinction between the two is the amount of domain knowledge the two groups possess. While the "general public" contains a mix of novice and intermediate users, the "non-professional" users generally see themselves as intermediate users. Services to support the two groups will thus have to take into account and support these varying levels.

The results presented here are derived from NML's visitors, but the wide distribution of study participants provides strong support that they will generalise to other DCH web-sites that have both a physical and virtual presence. To what degree they also apply to purely virtual DCH sites, such as Europeana, requires further study.

6 Conclusions and Future Work

The majority of research into the users of DCH web-sites has focused on those user groups that are easier to access ("academics", "museum staff", "students", and "professionals"). However, as the results of the survey reported here show, they form only a small fraction of the total number of web-site visitor. The main user groups are the "general public" and "non-professional" visitors, who make up nearly 70% of all visitors. In addition to identifying these groups as the main user groups, the survey data also allowed us to define those criteria (motivation, task, engagement, domain knowledge, and location) that distinguish these two groups from the other groups and also the criteria (domain knowledge and location) that distinguish the two groups from each other.

Due to the lower degree to which these two groups have been studied, it is also highly likely that current DCH web-site offerings are not as suitable for these groups as ideally desired. This would also explain why DCH web-sites suffer such high bounce rates, as based on the survey results, those users who leave immediately are more likely to belong to the "general public" and "non-professional" user groups, as the initial analysis of their responses indicates that they are more likely to give up quickly.

The analysis presented here provides an initial view onto the responses; however, significant work remains to investigate exactly how the different user groups interact with the site, whether patterns emerge and how users' interactions can be better supported across a range of tasks and goals.

Acknowledgements. We would like to thank National Museums Liverpool for giving us access to their users by allowing us to run the survey on their web-sites.

References

1. Johnson, A.: Users, use and context: supporting interaction between users and digital archives. What Are Archives?: Cultural and Theoretical Perspectives: A Reader, 145–164 (2008)
2. Hall, M.M., Clough, P.D., de Lacalle, O.L., Soroa, A., Agirre, E.: Enabling the discovery of digital cultural heritage objects through wikipedia. In: Proceedings of the 6th Workshop on Language Technology for Cultural Heritage, Social Sciences, and Humanities, Association for Computational Linguistics, pp. 94–100 (2012)
3. Ciber: Europeana 2012–2013: usage and performance update. Technical report, CIBER Research, July 2013
4. Hein, G.E.: Learning in the Museum, vol. 47 (1998)
5. Falk, J.H.: Identity and the museum visitor experience. Left Coast Press (2009)
6. Dierking, L.D., Falk, J.H.: Family behavior and learning in informal science settings: a review of the research. Sci. Educ. **78**(1), 57–72 (1994)
7. Spellerberg, M., Granata, E., Wambold, S.: Visitor-first, mobile-first: Designing a visitor-centric mobile experience. In: Museums and the Web (2016)
8. Templeton, C.A.: Museum visitor engagement through resonant, rich and interactive experiences (2011)
9. Cunliffe, D., Kritou, E., Tudhope, D.: Usability evaluation for museum web sites. Museum Manage. Curatorship **19**(3), 229–252 (2001)
10. Peacock, D., Brownbill, J.: Audiences, visitors, users: Reconceptualising users of museum on-line content and services. In: Museums and the Web. Citeseer (2007)
11. Marty, P.F.: Museum websites and museum visitors: digital museum resources and their use. Museum Manage. Curatorship **23**(1), 81–99 (2008)
12. Vilar, P., Šauperl, A.: Archival literacy: different users, different information needs, behaviour and skills. In: Kurbanoğlu, S., Špiranec, S., Grassian, E., Mizrachi, D., Catts, R. (eds.) ECIL 2014. CCIS, vol. 492, pp. 149–159. Springer, Cham (2014). doi:10.1007/978-3-319-14136-7_16
13. Pantano, E.: Virtual cultural heritage consumption: a 3d learning experience. Int. J. Technol. Enhanced Learn. **3**(5), 482–495 (2011)
14. Marty, P.F.: Meeting user needs in the modern museum: profiles of the new museum information professional. Libr. Inf. Sci. Res. **28**(1), 128–144 (2006)
15. Hogg, C., Williamson, C.: Whose interests do lay people represent? towards an understanding of the role of lay people as members of committees. Health Expect. **4**(1), 2–9 (2001)

16. Cifter, A.S., Dong, H.: User Characteristics: Professional vs. Lay Users (2009)
17. Kelly, L.: The interrelationships between adult museum visitors' learning identities and their museum experiences. chap. 3. Methodology, pp. 3–46 (2007)
18. Skov, M., Ingwersen, P.: Exploring information seeking behaviour in a digital museum context. In: Proceedings of the Second International Symposium on Information Interaction in Context, IIiX 2008, pp. 110–115. ACM, New York (2008)
19. Skov, M.: The reinvented museum: Exploring information seeking behaviour in a digital museum context. PhD thesis, Københavns Universitet'Københavns Universitet', Faculty of Humanities, School of Library and Information Science, Royal School of Library and Information Science unpublished thesis (2009)
20. Elsweiler, D., Wilson, M.L., Lunn, B.K.: Chapter 9 understanding casual-leisure information behaviour. In: New Directions in Information Behaviour (Library and Information Science), vol. 1, pp. 211–241. Emerald Group Publishing Limited (2011)
21. Ardissono, L., Kuflik, T., Petrelli, D.: Personalization in cultural heritage: the road travelled and the one ahead. User Model. User-Adap. Inter. $22(1–2)$, 73–99 (2012)
22. Gäde, M., Hall, M., Huurdeman, H., Kamps, J., Koolen, M., Skov, M., Toms, E., Walsh, D.: Supporting complex search tasks. In: Hanbury, A., Kazai, G., Rauber, A., Fuhr, N. (eds.) ECIR 2015. LNCS, vol. 9022, pp. 841–844. Springer, Cham (2015). doi:10.1007/978-3-319-16354-3_99
23. Booth, B.: Understanding the information needs of visitors to museums. Museum Manage. Curatorship $17(2)$, 139–157 (1998)
24. Marchionini, G., Plaisant, C., Komlodi, A.: The people in digital libraries: Multifaceted approaches to assessing needs and impact. In: Digital Library Use: Social Practice in Design and Evaluation, pp. 119–160 (2003)
25. Sweetnam, M., Siochru, M., Agosti, M., Manfioletti, M., Orio, N., Ponchia, C.: Stereotype or spectrum: designing for a user continuum. In: The Proceedings of the First Workshop on the Exploration, Navigation and Retrieval of Information in Cultural Heritage, ENRICH (2013)
26. Europeana: Results of the europeana user survey 2014 - europeana professional, June 2014. http://pro.europeana.eu/blogpost/results-of-the-europeana-user-survey-2014. Accessed 14 Mar 2016
27. Stiller, J.: From curation to collaboration. PhD thesis, Humboldt-Universität zu Berlin, Philosophische Fakultät I (2014)
28. Goldman, K.H., Schaller, D.T., Adventures, E.W., et al.: Exploring motivational factors and visitor satisfaction in on-line museum visits. In. In: Bearman, D., Trant, J. (eds.) Museums and the Web 2004. Citeseer (2004)
29. Russell-Rose, T., Tate, T.: Designing the Search Experience: The Information Architecture of Discovery. Newnes (2012)
30. Taylor, E.: The national archive, UK government web archive discovery project: User survey. An unpublished technical report (2015)
31. Lapatovska, I., O'Brien, H.L., Rieh, S.Y., Wildemuth, B.: Capturing the complexity of information interactions: measurement and evaluation issues. In: Proceedings of the ASIST Annual Meeting, vol. 48 (2011)

Evaluating the Usefulness of Visual Features for Supporting Document Triage

Dagmar Kern[1(✉)], Maria Lusky[2], and Dirk Wacker[3]

[1] GESIS Leibniz-Institute for the Social Sciences, Cologne, Germany
dagmar.kern@gesis.org
[2] RheinMain University of Applied Sciences, Wiesbaden, Germany
maria.lusky@hs-rm.de
[3] Fernuniversitaet in Hagen, Hagen, Germany
dirk.wacker@studium.fernuni-hagen.de

Abstract. One of many challenges the users of digital libraries face is the quick and easy identification of relevant documents as well as their effective use – a process which is called *document triage*. While there is a variety of means for supporting the actual search process in digital libraries, only few address document triage. In this paper, we investigate the usefulness of user initiated support features that might assist users in this process. Therefore, we implemented SortBoard, a visual workspace for assisting document triage in a digital library for the social sciences. The results of a user study with 16 participants show that the features are highly useful for comparing and organizing documents as well as finding and examining similarities between documents.

1 Introduction

Finding relevant literature is essential for researchers of all disciplines. A challenge of literature search in digital libraries is the quick and easy identification of relevant documents, including actions like investigating documents, structuring and organizing a set of documents and using their information effectively. This process is also referred to as "document triage". According to Toms et al., two-thirds of the total time for finding relevant documents are used for "reviewing documents that had already been found" [21], while only one-third is spent on searching. There is a variety of means that support searching in digital libraries, such as term recommenders or re-ranking techniques [10]. However, the support for document triage is often limited to some kind of favorite lists or bookmarks. Although there are promising research activities regarding visual support for document triage, e.g. [1,15,18], this is still almost non-existent in digital libraries. In our work, we investigate the usefulness of user initiated support features for assisting the document triage process visually as well as in a transparent way for the user. We implemented SortBoard, a prototypical application and integrated it into Sowiport[1], a digital library for the social sciences. In SortBoard

---

[1] http://sowiport.gesis.org/.

© Springer International Publishing AG 2017
J. Kamps et al. (Eds.): TPDL 2017, LNCS 10450, pp. 446–458, 2017.
DOI: 10.1007/978-3-319-67008-9_35

the user can move, arrange and structure documents as well as investigate their properties and similarities.

In a user study with 16 participants, we compare SortBoard regarding its usefulness to a common method for document triage: opening relevant documents in new tabs and making notes. The results show that SortBoard's user initiated support features are more useful than the commonly used method regarding comparing and organizing documents as well as finding and examining similarities. The most useful features are the display of a metadata box, reference and citation relations and colored highlighting of documents with similarities.

2   Background and Related Work

In their model of the information searching process, Ellis et al. [7,8] describe the stages academic researchers go through while searching for information. In focus of our work is the *differentiating* stage where the user views and investigates documents and decides about their relevance. This process is also often called "document triage". In the literature, there are different definitions of this term [2,5,13]. We follow the definition of Marshall et al. [15], who describe "document triage" as "sorting through relevant materials and organizing them to meet the needs of the task at hand". Loizides [13] distinguishes between three levels of document triage: *surrogate triage stage, within document triage stage* and *further reading triage stage*. In our work, we focus on the *surrogate triage stage*. In this stage, the user investigates documents, represented by surrogates, based on their metadata. It is the first encounter of a user with a document, typically after searching. The user judges the relevance of documents without looking at the full texts.

One opportunity to support users in sorting through documents is showing these documents represented as movable objects on a 2- or 3-dimensional workspace. On such a workspace the objects can be arranged (1) manually [15,16], (2) manually with automatic support by suggesting the belonging of new objects to already formed groups on the space [3,4,17], (3) automatically according to predefined categories [11,19,22] or (4) regarding their similarity to used search terms [9]. The expert tool TRIST (Rapid Information Scanning Tool) [11] allows, additionally to automatic clustering of search results, to highlight documents' commonalities (such as the same year of publication) across the clusters' borders. A detailed up-to-date summary of different approaches for visual support for the document triage process in digital libraries is provided by Loizides et al. [14].

In our work, we take up basic principles of an early support tool for document triage, called VIKI [15]. VIKI is a spatial hypertext system and allows users to structure and organize documents on a visual workspace like analysts used to do so with physical papers. VIKI has been further developed into the visual knowledge builder (VKB) [18]. In a more recent version of VKB [1] additionally to user-generated visualizations (like used in VIKI) system-generated visualizations are used that organize documents implicitly according to users' search interests.

This information is also used for highlighting new information on the workspace that might also be relevant for the user. In this paper, we focus solely on user-generated visualization. The user gets system support while exploring similarities and relations between preselected documents on a 2-dimensional workspace, but has full control over it. A filter list next to the workspace shows metadata like keywords and topics and their frequency of occurrence. This list can be used to get an overview of the documents on the workspace, to get inspiration for further search terms and as a filter to find and highlight documents with commonalities. We are especially interested in the usefulness of user initiated support features and compare them to the common practice of document triage during publication search in digital libraries.

3   SortBoard

In this section, we introduce SortBoard, a web-based application we implemented to evaluate user initiated support features for visual document triage. SortBoard is implemented as a prototype in Sowiport[2], a digital library with more than 9.5 million literature references. Taking up basic principles from VIKI [15] and VKB [18] the main intention of SortBoard is: (1) to offer a 2-dimensional workspace where a user can drop relevant documents in a digital library by just one click for further triage, (2) to sort and structure documents graphically, (3) to support users in finding similarities in a set of documents and additionally (4) to derive further search terms. All system support in SortBoard is user initiated, thus transparent to the user.

The user interaction in SortBoard is as follows: While searching for literature in a digital library, the user can drop documents directly from the result list to SortBoard by clicking on a link "Add to SortBoard". The actual task of scanning the result list for relevant literature is not affected until the user explicitly switches to SortBoard by clicking on the corresponding link. After switching to SortBoard the user sees all the dropped documents represented as movable object cards (A in Fig. 1). The user can drag the cards to arrange them according to her preferences.

To support finding similarities and to provide an overview of the documents on the workspace we adapted the concept of faceted search, which is already used in Sowiport to filter search results in the result list. We show selected metadata of the documents as filters at the left side of SortBoard (B in Fig. 1). They are divided into five categories: author, classification, topic, document type, and notes. Next to each filter the frequency of property occurrence in the documents on SortBoard is shown (C in Fig. 1). By selecting one of the filters, all documents that match this criterion are highlighted with the same colored border, which is called colored highlighting (D in Fig. 1). This visualization of similarities should support users by structuring documents on the workspace. Selected filters are listed at the top of the list so that the user can easily remove one or all of them (E in Fig. 1). By dragging and dropping a document on the recycle bin, it is

---

[2] http://sowiport.gesis.org/.

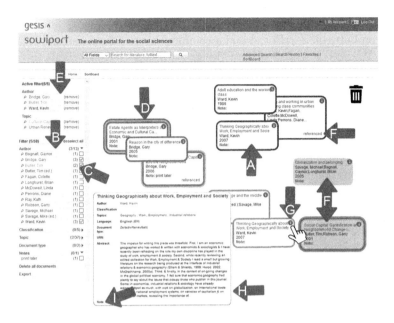

Fig. 1. SortBoard: movable object card (A), filter area (B), frequency of property occurrence (C), colored highlighting (D), selected filters (E), reference and citation relations (F), information icon (G), which triggers on mouse hover the appearance of the metadata box (H). The user can add notes (I) to an object.

removed irrevocably from SortBoard. To support users in exploring co-citation relations [20] between the documents on SortBoard, the card gets the additional note "referenced" or "cited by". A mouse-over triggers the highlighting of the documents that are referenced in the selected object or cited by the selected object respectively (F in Fig. 1). Each movable object card provides the following information about a document at a glance: title, authors, year of publication and notes, as well as further information on demand by mouse-hovering over the information icon (G in Fig. 1). This triggers the appearance of a metadata box (H in Fig. 1) containing the following metadata: author, title, classification, topics, language, document type, URN, abstract. This metadata box opens in the direction that provides the freest space and will close automatically again after the cursor is positioned outside the box. By clicking on "notes", the user can add notes to a movable object card (I in Fig. 1).

4 Evaluation

We performed a user study to evaluate the usefulness of SortBoard's user initiated support features for assisting document triage. Our study focuses on the *surrogate triage stage* as defined by Loizides [13], particularly on the stage after the user has made a first relevance decision and the relevant document is stored for further investigation. Following Kelly [12], we designed our lab study as a comparison between SortBoard and a common method for document triage as a baseline. We used a within-subject design with two triage methods (SortBoard and the baseline) as well as two different tasks as independent variables. We conducted our lab study in single sessions with a duration of about one hour each. Having the two triage methods and the two tasks as independent variables, we rotated them in a Graeco-Latin square design and distributed the four experimental conditions randomly but counterbalanced among 16 participants.

4.1   Setup

Triage Methods. The first triage method is represented by SortBoard described above. To find a baseline as the second triage method, we performed an online survey and asked 102 researchers about their preferred strategies to store relevant documents during the search process in digital libraries. The results show that storing a document on a PC is the most preferred method (used by 92%), followed by opening an interesting document in a new tab (used by 91%) and making notes (used by 81%). As our digital library Sowiport does not provide full texts to all literature references for storing them on a PC, we chose a combination of the second and third most common methods "open in new tab" and "make notes". For supporting the baseline triage method, functionalities of a common internet browser can be used. Sowiport provides a detailed result page for each document showing the same information as the metadata box of the movable object cards in SortBoard. While examining the result list a user can open a new tab by middle-click or right-click and select "open in new tab". A word processing program or a sheet of paper can be used to take notes.

Tasks. We defined two identical tasks (A and B) that differed only in the topic they dealt with. While task A addressed "urban research", task B dealt with "scientific organizations". For each of these tasks, we defined a set of 18 documents and created a sample solution on how this documents could be grouped. The selection of the documents for the sets as well as the sub-topic groups in the sample solutions was based on reviewed document collections on each of the topics composed by social scientists. The participant's task was to get an overview of the documents, to group the documents by sub-topics, to delete documents that are not relevant and to extract search terms that seem to be useful for further searches. Participants were requested to write these search terms in an excel sheet. The time for solving the task was limited to 15 min but could be ended earlier, if the participant stated to have finished the task.

Apparatus. All participants used the same laptop with a 15.6" screen and a resolution of 1366×768. For ease of use, an external mouse was provided as a controlling device. We recorded the screen activities as well as the think-aloud comments during the sessions using the screen capture software Camtasia. SortBoard was equipped with 18 documents of set A (respectively B) that had to be examined. For taking notes, SortBoard's note function should be used. In the baseline triage method, 18 documents of set B (respectively A) were opened in tabs of a web browser and participants were allowed to take notes either in a word document or on a sheet of paper.

4.2 Procedure

The procedure of each session followed a detailed protocol to ensure the same conditions for each participant. Solely the combination and sequence of the two document triage methods and the tasks differed as described above. The procedure mainly consisted of three parts: (1) In an introduction, the purpose of the study was disclosed to the participants, and a short explanation of the stages of document triage was given. Then they were informed about the procedure of the session as well as of the recording that was taken. They signed a consent form, filled out the pre-questionnaire including questions about demographics and general search experience and then the recording was started. (2) In the second part, the participants had to solve the tasks one after another. They were introduced to the first triage method they used by a video tutorial. After that, they were handed out the first task and started solving it. After finishing this first task, they were shown a video tutorial of the second triage method. The second task was given to the participants and carried out by them. The participants noted down their sub-topics in an excel sheet and saved their groups of documents in SortBoard. 10 min after starting each task a notification was given that five minutes were left. The tasks ended after 15 min. If the participants finished a task before the 15 min expired, they signaled it. (3) In the third and final part of the session, the participants filled out a post-questionnaire including questions about task performance and usefulness.

4.3 Participants

Participants were recruited via email and personal request at three universities and one research institute. The group was composed of 16 German speaking researchers from different fields of the social sciences, of which six were female, and ten were male. Their age ranged from 25 to 47 (M = 31.31, SD = 5.91). Three participants held a bachelor's degree, six a master's degree, four a diploma and three were postdoctoral researchers. All of our participants stated that they used the internet daily and that they searched for information online daily or at least once a week. Regarding the participants' experience with digital libraries, 56% stated to have much or very much experience in searching digital libraries. Their experience regarding the two topics did not differ significantly.

4.4 Measured Variables

Task Performance. For participants' task performance we examined: (1) how well participants identified topic clusters and non-relevant documents (correctness), (2) if they needed the whole 15 min to solve the task (time) and

(3) how many new terms for future search could be found. Since the participants were allowed to finish earlier than 15 min, the time needed to accomplish it was measured, with a maximum of 15 minutes. Furthermore, in the post-questionnaire, the participants had to state if they felt they have finished the task or not.

Usefulness. For investigating the usefulness we followed two approaches: (1) For evaluating the usefulness of the two triage methods in order to compare them, we used the evaluation model based on the main goal and sub-goal levels introduced by Cole et al. [6]. We formulated the main goal, sub-goals, and information seeking strategies (ISSs) for our task (see Table 2). After becoming familiar with both triage methods, we asked participants on five-point Likert scales about the usefulness and support of the respective triage method, based on our main goal, sub-goals, and ISS levels. At the level of the main goal (investigating documents) and the sub-goals (obtaining an overview, identifying sub-topics, separating relevant and non-relevant documents, identifying starting points for further searching) they were asked how useful (1= "not at all useful" to 5= "very useful") the triage method was for accomplishing these goals. At the level of the ISSs (see Table 2) we asked how well the triage method supported (1= "not at all" to 5 = "very") them in this strategies. (2) For evaluating the usefulness of the user initiated support features, we asked participants on a five-point Likert scale how useful (1= "not at all useful" to 5= "very useful") they considered the filter list, the number of occurrence of these filters, the colored highlighting of documents that share commonalities, the opportunity to add notes as well as the display of reference and citation relations between documents. In addition, we asked how useful (1= "not at all useful" to 5= "very useful") they considered SortBoard in general for supporting their document triage process. Finally, to gather qualitative feedback, participants were asked to note down improvements for the features of SortBoard as well as suggestions for new features that they found the tool lacking in.

5   Evaluation Results

We analyzed the task performance, and the results of the post-questionnaire, comprising the comparison of the two triage methods at the different levels of main goal, sub-goals, and ISSs as well as the usefulness of the user initiated support features. To determine the statistical significance, Wilcoxon Signed Rank tests were used with $\alpha = 0.05$.

5.1   Task Performance

Table 1 summarizes all findings that focus on the task performance. The number of deleted documents was significantly higher while using SortBoard for document triage than with the baseline method. The same is true for the accordance of documents that participants considered not relevant compared to the ones that

Table 1. Task performance with the two triage methods, mean (SD). *p<0.05; **p<0.01.

	Baseline	SortBoard
Number of identified groups of sub-topics	4.81 (1.87)	4.50 (0.97)
Number of deleted documents	1.00 (0.97) *	**1.93 (1.06)** *
Number of deleted documents matching with the sample solution	0.69 (0.70) *	**1.38 (0.72)** *
Number of identified search terms	**17.75 (6.84)** **	11.5 (9.26) **
Time needed	14:35 (00:42)	14:04 (01:21)
Number of participants that finished earlier	8	7
Number of participants that stated to have finished the task	13	10

are not relevant in the sample solution. Furthermore, there was a significant difference regarding identifying new search terms. Participants elicited significantly ($p < 0.01$) more search terms for further searching with the baseline method. There was no significant difference in the number of sub-topics that were found and in the time needed to accomplish the task.

5.2 Usefulness

Table 2 shows participants' evaluation of the usefulness of the two triage methods at the different main goal, sub-goal and ISS levels. The mean usefulness for accomplishing the main goal was rated significantly higher for SortBoard than for the baseline. Additionally, the ratings for identifying sub-topics and all its ISSs

Table 2. Mean (SD) usefulness ratings (1= "not at all useful" to 5= "very useful") at the main goal, sub-goal, and ISS levels for each triage method. *p<0.05; **p<0.01.

	Baseline	SortBoard
Main goal: Investigating documents	3.19 (0.83) *	**3.94 (0.68)** *
1. Sub-goal: Obtaining overview	3.13 (1.09)	3.69 (1.09)
ISS 1.1: Extracting main information	3.63 (0.89)	3.44 (0.73)
ISS 1.2: Making notes	2.81 (1.22)	2.94 (1.29)
2. Sub-goal: Identifying sub-topics	3.31 (0.70) *	**4.06 (0.85)** *
ISS 2.1: Comparing documents	2.56 (0.89) **	**4.13 (0.81)** **
ISS 2.2: Finding similarities	2.50 (0.52) **	**3.81 (0.75)** **
ISS 2.3: Forming groups of sub-topics	3.13 (0.62) *	**4.19 (1.05)** *
ISS 2.4: Assigning documents to the groups	3.00 (0.82) *	**4.00 (0.89)** *
3. Sub-goal: Separating documents	3.13 (0.81)	3.81 (0.83)
ISS 3.1: Evaluating documents	3.13 (0.72)	3.50 (0.73)
ISS 3.2: Deleting documents	3.94 (0.93)	4.06 (1.00)
ISS 3.3: Keeping or saving documents	3.06 (1.06) *	**3.81 (0.83)** *
4. Sub-goal: Identifying starting points for further searching	3.19 (0.98)	3.75 (0.93)
ISS 4.1: Examining similarities	2.75 (0.68) **	**4.00 (0.52)** **
ISS 4.2: Finding new search terms	3.31 (0.95)	3.50 (0.63)

as well as for keeping or saving documents and examining similarities differed significantly in favor of SortBoard.

Table 3 shows participants' usefulness ratings for the different features of SortBoard. The tool in general was rated as "useful". The metadata box gained the highest ratings, followed by reference and citation relations and the colored highlighting, whereas adding notes and filtering — yet still rated as "rather useful"— received the lowest values. While all of the participants used the movable object cards of SortBoard and the metadata box, only 38% added notes. 69% browsed through the filter list and 56% of all participants used them to highlight documents with similarities.

Table 3. Mean (SD) usefulness of SortBoard's features (1= "not at all useful" to 5= "very useful").

Feature	Mean usefulness (SD)
Metadata box	4.50 (0.52)
Reference and citation relations	4.44 (0.96)
Colored highlighting	4.25 (0.68)
Frequence of property occurrence	4.00 (0.89)
Adding notes	3.88 (0.89)
Filters	3.88 (0.89)
SortBoard in general	4.06 (0.93)

5.3 Suggestions for SortBoard

We gained extensive qualitative user feedback from the participants. Most of the suggestions referred to the metadata box that should either be opened via clicking instead of hovering over a symbol or be aligned in the center of the screen. Another difficulty that occurred was the high number of filters displayed on the left side of SortBoard. The most popular feature suggested by the participants was the creation of labeled groups on SortBoard, e.g. by drawing a rectangle around a group or creating sub-folders. Further ideas dealt with highlighting the cards of documents that are highly relevant, such as documents with a very high number of citations, as well as possibilities for import and export from SortBoard into literature management software or the direct export of full texts.

6 Discussion

The results showed that participants experienced SortBoard in general as a useful tool that supports the document triage process through user-initiated support features. Participants found SortBoard significantly more useful for the following tasks (compared to using browser tabs and making notes): investigating documents, identifying sub-topics, comparing documents, finding similarities, forming groups of sub-topics, assigning documents to groups, keeping or saving documents and examining similarities. These findings support participant's need for visual document triage tools found in previous user studies [1, 4, 11, 22]. As we concentrate solely on user-initiated support features, we showed that even without any system-generated visualizations user assistance during document triage process could be improved. Especially the results regarding the deleted documents tell us that SortBoard's features do support the user's relevance decision and the "quality check" that is part of the *differentiating* stage [7, 8]. A reason may be that on a 2-dimensional workspace the user sees all documents at a glance and gets a better overview, while with browser tabs, individual documents can be overseen easily. Additionally, the user has to switch between browser tabs to inspect and compare titles and authors, while SortBoard's movable object cards provide this important information directly. However, for seeing the abstract and further details of a document the user has to open the metadata box, which was consequently rated as the most useful feature. Participants found significantly more search terms in the baseline condition; this might be due to the fact that on the result view of a document opened in a tab the user sees all information at once while on SortBoard the metadata box has to be open explicitly. We had assumed that the filter list is very useful for the user in finding further search terms, by providing an overview of commonly occurred keywords in the documents on SortBoard. However, the filter list was not used extensively. One explanation for that was given by our participants in their qualitative feedback. The number of filters was far too high, and the terms were sorted alphabetically instead of based on their frequency of occurrence. Most of the terms provided as filters, only appear in one document and therefore were less helpful in finding new search terms. By decreasing the number of filters, e.g. by excluding those that only relate to one document, or changing their order, we assume that the filter feature, as well as the colored highlighting feature, would be used more extensively.

One of the most popular suggestions for new features was the creation of labeled groups on SortBoard. That further differentiation and labeling of grouped documents can support document triage was also found in the studies on VIKI and VKB [1, 15, 18]. The wish for colored highlighting of most relevant documents in the document set on SortBoard is in line with findings in [1]. We plan to provide additional filters (like most cited papers) to address this need. In doing so, the process is still transparent, and user initiated. Import and export of documents seems to be an important matter for many participants. It suggests that user initiated support features would be useful beyond Sowiport as well and in a broader context of the whole information searching process.

At the end of this section, we would like to summarize some limitations of our study. Since the triage methods we compared are quite different from each other, one could assume that the comparison was inappropriate. Nevertheless, working with browser tabs is the most common method used during document triage, according to our online survey which was also confirmed by the participants during the study. Also, the sub-topics the participants found with both triage methods do not differ significantly. In turn, several participants remarked that SortBoard felt new to them and that they had needed some time to get accustomed to it. The comparatively high usefulness rankings of features that have not been used by all participants support the conclusion that they had not enough time, but see the potential of these features. Apparently, the video tutorial was not sufficient, and it might have been better to additionally give them a few minutes to make themselves familiar with SortBoard or to generally remove the time constraints of the scenarios.

7   Conclusions and Future Work

In this paper, we investigate the usefulness of user initiated support features for the document triage process in a digital library. Therefore, we presented Sort-Board, a visual workspace for storing relevant documents, for sorting and structuring documents graphically, for finding and examining similarities between documents, and for deriving search terms for further search processes. A user study with 16 participants was conducted and showed that the user initiated support features were rated as highly useful for supporting the document triage process. We showed that SortBoard's features provide the user with a better usefulness than the baseline triage method regarding half of all goals, sub-goals, and ISSs. The user initiated support features, as well as SortBoard in general, were rated as useful on average, though there is still room for improvement.

In future work, we plan to conduct a long-term user study with an improved version of SortBaord, including the user feedback given in our study. We are interested in how a visual workspace can support the user beyond the document triage process, addressing questions like, might SortBoard be helpful for resuming a previous task, or might it be helpful to use SortBoard for different information types, e.g. to find connections between literature, research data, research projects, and websites.

As Ellis et al. [7, 8] showed in their research, the information searching behavior of researchers from different disciplines differs only minimally. We are sure that the features evaluated in this work are also appropriate for researchers from other fields of study and that digital libraries could benefit from implementing similar concepts.

Acknowledgments. We thank all volunteers that participated in our study and the focus group on interactive information retrieval at GESIS.

References

1. Bae, S., Kim, D., Meintanis, K., Moore, J.M., Zacchi, A., Shipman, F., Hsieh, H., Marshall, C.C.: Supporting document triage via annotation-based multi-application visualizations. In: Proceedings of the 10th Annual Joint Conference on Digital Libraries, pp. 177–186. ACM (2010)
2. Bae, S., Marshall, C.C., Meintanis, K., Zacchi, A., Hsieh, H., Moore, J.M., Shipman, F.M.: Patterns of reading and organizing information in document triage. Proc. Am. Soc. Inf. Sci. Technol. **43**(1), 1–27 (2006)

3. Buchanan, G., Blandford, A., Thimbleby, H., Jones, M.: Integrating information seeking and structuring: exploring the role of spatial hypertext in a digital library. In: Proceedings of the Fifteenth ACM Conference on Hypertext and Hypermedia, pp. 225–234. ACM (2004)
4. Buchanan, G., Blandford, A., Thimbleby, H., Jones, M.: Supporting information structuring in a digital library. In: Heery, R., Lyon, L. (eds.) ECDL 2004. LNCS, vol. 3232, pp. 464–475. Springer, Heidelberg (2004). doi:10.1007/978-3-540-30230-8_42
5. Buchanan, G., Loizides, F.: Investigating document triage on paper and electronic media. In: Kovács, L., Fuhr, N., Meghini, C. (eds.) ECDL 2007. LNCS, vol. 4675, pp. 416–427. Springer, Heidelberg (2007). doi:10.1007/978-3-540-74851-9_35
6. Cole, M., Liu, J., Belkin, N., Bierig, R., Gwizdka, J., Liu, C., Zhang, J., Zhang, X.: Usefulness as the criterion for evaluation of interactive information retrieval. In: Proceedings of the Third Workshop on Human-Computer Interaction and Information Retrieval, HCIR 2009, pp. 1–4 (2009)
7. Ellis, D.: Modeling the information-seeking patterns of academic researchers: a grounded theory approach. Libr. Q. **63**, 469–486 (1993)
8. Ellis, D., Cox, D., Hall, K.: A comparison of the information seeking patterns of researchers in the physical and social sciences. J. Doc. **49**(4), 356–369 (1993)
9. Fowler, M., Bellis, C., Perry, C., Kim, B.: Exploring web-based visual interfaces for searching research articles on digital library systems (2016)
10. Hienert, D., Sawitzki, F., Mayr, P.: Digital library research in action: supporting information retrieval in sowiport. D-Lib Mag. **21**(3), 8 (2015)
11. Jonker, D., Wright, W., Schroh, D., Proulx, P., Cort, B., et al.: Information triage with TRIST. In: International Conference on Intelligence Analysis, pp. 2–4 (2005)
12. Kelly, D.: Methods for evaluating interactive information retrieval systems with users. Found. Trends Inform. Retrieval **3**(1–2), 1–224 (2009)
13. Loizides, F.: Understanding and conceptualising the document triage process through information seekers' visual and navigational attention. Ph.D. thesis, City University (2012)
14. Loizides, F., Buchanan, G., Mavri, K.: Theory and practice in visual interfaces for semi-structured document discovery and selection. Inf. Serv. Use **35**(4), 259–271 (2016)
15. Marshall, C.C., Shipman III., F.M.: Spatial hypertext and the practice of information triage. In: Proceedings of the Eighth ACM Conference on Hypertext, pp. 124–133. ACM (1997)
16. Robertson, G., Czerwinski, M., Larson, K., Robbins, D.C., Thiel, D., Van Dantzich, M.: Data mountain: using spatial memory for document management. In: Proceedings of the 11th Annual ACM Symposium on User Interface Software and Technology, pp. 153–162. ACM (1998)
17. Schwarzkopf, E.: Enhancing the interaction with information portals. In: Proceedings of the 9th International Conference on Intelligent User Interfaces, pp. 322–324. ACM (2004)
18. Shipman III., F.M., Hsieh, H., Maloor, P., Moore, J.M.: The visual knowledge builder: a second generation spatial hypertext. In: Proceedings of the 12th ACM Conference on Hypertext and Hypermedia, NY, USA, pp. 113–122 (2001). http://doi.acm.org/10.1145/504216.504245
19. Shneiderman, B., Feldman, D., Rose, A., Grau, X.F.: Visualizing digital library search results with categorical and hierarchical axes. In: Proceedings of the Fifth ACM Conference on Digital Libraries, pp. 57–66. ACM (2000)

20. Small, H.: Co-citation in the scientific literature: a new measure of the relationship between two documents. J. Am. Soc. Inf. Sci. **24**(4), 265–269 (1973). doi:10.1002/asi.4630240406

21. Toms, E.G., Villa, R., McCay-Peet, L.: How is a search system used in work task completion? J. Inform. Sci. **39**, 15–25 (2013)

22. Wong, B.L.W., Choudhury, S.T., Rooney, C., Chen, R., Xu, K.: INVISQUE: technology and methodologies for interactive information visualization and analytics in large library collections. In: Gradmann, S., Borri, F., Meghini, C., Schuldt, H. (eds.) TPDL 2011. LNCS, vol. 6966, pp. 227–235. Springer, Heidelberg (2011). doi:10.1007/978-3-642-24469-8_24

Building User Groups Based on a Structural Representation of User Search Sessions

Wilko van Hoek and Zeljko Carevic[✉]

GESIS – Leibniz Institute for the Social Sciences,
Unter Sachsenhausen 6–8, 50667 Cologne, Germany
{wilko.hoek,zeljko.carevic}@gesis.org

Abstract. Identifying user groups is an important task in order to personalise search results. In Digital Libraries, visited resources and the sequential search patterns are often used to measure user similarity. Whereas visited resources help to understand what users want, they do not reveal how users prefer to search. In contrast, sequential patterns allow to decode the way in which users search, but they are very strict and do not allow changes in the order of the search. A third alternative and compromise could be the analysis of the structure of a search session. In this paper, we aim to obtain some insights into the potential of analysing search sessions on a structural basis. Therefore, we will investigate a structural representation of search sessions based on tree graphs. We will present a novel method to merge multiple session trees into a combined tree. Based on combined tree taken from similar sessions, we will build archetypical trees for different user groups.

Keywords: Retrieval sessions · User behaviour · Session trees · Exploratory search

1 Introduction

To improve the user experience in information retrieval systems, understanding user needs has become more and more important. Methods in information retrieval have ventured from strictly text based models like TF-IDF and language models [13] to learning models based on user behaviour [1]. Along this process, personalisation has become an ever growing field [7]. The more user information is collected and evaluated, the better search results can be tailored to a specific user or user group. Therefore, it is important to understand the users' search behaviour. The methodologies to analyse search behaviour range from descriptive counts and user feedback (e.g. [19]), qualitative feedback and interviews (e.g. [2]), gaze-data (e.g. [12]) to mixed-methods (e.g. [18]). The goal is to identify specific signals in the usage behaviour that indicate the users' needs. In live systems, most systems rely on measurable signals, like click through rates, search terms, the set of visited resources, or sequential patterns.

Sequential pattern analysis has proven to be a very useful approach in personalisation [7,16]. It can be used to measure the similarity between different users [15].

© Springer International Publishing AG 2017
J. Kamps et al. (Eds.): TPDL 2017, LNCS 10450, pp. 459–470, 2017.
DOI: 10.1007/978-3-319-67008-9_36

An important advantage of sequential patterns is the ability to keep the temporal order in which user activity has been executed. This helps to understand if there is a relevance to whether an action or information should be presented before or after another. Sequential pattern analysis stands opposite to bag-of-words-like approaches, where user activities are treated irrespectively of the order in which they were conducted. Using such approaches, user similarity can be measured, based on the objects the users have visited [7].

However, there is another aspect of user activity which is the structure of the user behaviour. Instead of analysing which objects have been visited in which order, the question would be, how have the objects been accessed? A structural representation shows the connections between various objects regardless of the order in which they have been accessed. It is less detailed than a sequential and more detailed than a bag-of-words representation. In this paper, we will investigate a structural representation of search sessions based on tree graphs. We will present a novel method to merge multiple session trees into a combined tree. Based on the combined tree of session groups, we will build archetypical trees for different user groups. We exemplify our approach based on the results of a user study with 32 participants performing an exploratory search task in a digital library. Our goal is to look if we can discover groups that show (a) economic, (b) exhaustive-active and (c) exhaustive-passive behaviour similar to the groups described in [4].

2 Related Work

In [14], a study was conducted in which the search intention given by the participants could be identified automatically. The authors found evidence that there was a connection between search pattern and task type. [6] were able to distinguish between low-level tasks, based on the activity patterns and introduce a novel technique that allowed to detect aspects of tasks. Going beyond the connection between patterns and tasks, [4] found that the user's task influences the result page examination behaviour. They analysed queries, clicks, mouse cursor movement, scrolling, and text highlighting that was collected from the usage of the Bing search engine during a time period of 13 days. By using a set of features derived from the logged data, they were able to derive six clusters. By clustering only data from non-navigational tasks, they were able to distinguish three types of search engine result pages (SERPs) examiners: the economic, the exhaustive-active and the exhaustive-passive user. While economic users do not spend much time on SERPs, show more mouse movements, and abandon SERPs more often, users from the exhaustive groups investigate their SERPs more intensely.

Similar groups have been found in [3]. Here a lab study with 28 participants was conducted. Based on the eye tracking data, specific examination patterns were identified and manually clustered into the two groups economic and exhaustive evaluation styles. For both groups significant differences in the search behaviour could be found. White and Drucker [17] also focus on patterns in the search behaviour. They collected five months of live data from 3290 users and extracted the users' search trails. Based on those trails, they identified differences in the interaction patterns, which led to two identifiable user groups, navigators and explorers. Navigators showed more consistent interaction patterns. They showed few deviations in their behaviour, tackled problems sequentially and revisited former pages more often. In contrast, explorers used a variety of different patterns, they branched frequently, submitted more queries and visited new websites more often. Only one of the studies mentioned above investigated the search session on a structural level. In [17], the search sessions were transferred into a web graph representation, which allowed to identify structural aspects of the user sessions. However, as the authors were interested in other aspects of the user behaviour, they derived sequential patterns from the graph representation and then analysed the sequences instead of the structural information.

In other fields, we can find research on the benefit of analysing structural information. In biology for example, structural information is used to detect common cell developments. The cells are represented as trees. By measuring pairwise similarities of those trees, similar groups of cell developments can be found. In [9], an overview is given on different approaches utilising structural analysis of trees in the field of biology.

3   Merging User Sessions

In this section, we will introduce our data set which we took from a user study on exploratory search. Furthermore, we will explain how to create a structural representation of a user's search session as a tree. At last, we will show how this representation can be used to merge multiple user sessions into a combined tree.

3.1   User Study

The user study involved 32 participants from the social sciences – 16 postdoctoral researchers and 16 students – who were asked to search for related work to a given topic. All participants started with the same document titled *Ethnical education inequality at start of school* and had a limit of ten minutes to solve the task. Having to use Sowiport [11] for their literature search, the participants had access to about 9 million social scientific documents. The seed document (see Fig. 1) was published by two authors, had five keywords and one classification and was published in a German journal for sociology and social psychology. All metadata fields could be utilised for further exploration within the system via a hyperlink. Additionally, participants could browse through citations, references or read the abstract or the full text. Besides this information, the participants were provided with ten recommended documents. Five documents of these were provided using the SOLR *more like this function* and the remaining five were documents published in the same journal. The system comprised 18 different databases and thus duplicates could have been recommended.

A more detailed description of the user study and the procedure can be found in [5]. In this paper, we only focus on the participants' activities and will therefore not go into any more detail regarding the study procedure.

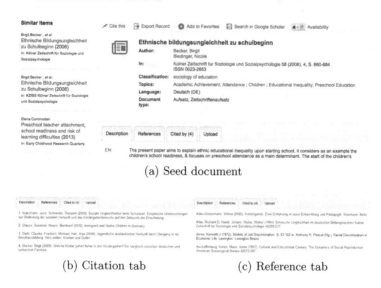

(a) Seed document

(b) Citation tab   (c) Reference tab

Fig. 1. Seed document for the user study.

3.2   Tree Representation of Search Sessions

We use the screen casts and the notes taken during the experiments to create a tree representation of the user behaviour and transformed them into a JSON format. To illustrate how we create such session trees, Fig. 2 shows three fictitious examples of session trees and the corresponding search patterns. In the first session, starting from the seed document, the document's citation list was clicked and one of the cited documents visited. This is represented by the left two nodes in Fig. 2a. The user then returned to the citation list and to the seed document. This user activity only involved already visited pages. Thus, it did not result in additional nodes. Finally, the search conducted at the session's end accounts for the right node in the tree. The other trees are created accordingly.

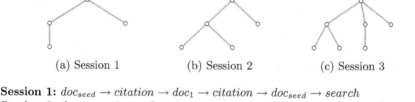

(a) Session 1   (b) Session 2   (c) Session 3

Session 1: $doc_{seed} \rightarrow citation \rightarrow doc_1 \rightarrow citation \rightarrow doc_{seed} \rightarrow search$
Session 2: $doc_{seed} \rightarrow journal \rightarrow doc_{seed} \rightarrow author \rightarrow doc_1 \rightarrow author \rightarrow doc_2$
Session 3: $doc_{seed} \rightarrow search \rightarrow doc_1 \rightarrow search \rightarrow doc_2 \rightarrow search \rightarrow doc_{seed} \rightarrow journal \rightarrow doc_{seed} \rightarrow citation \rightarrow doc_3$

Fig. 2. Session trees (a–c) for three example search sessions (1–3)

For this representation we ignore the type of user activity. Whether a node represents a document, a search, or a citation list is discarded. Furthermore, we discard the order of the activities. Instead, we sort the tree by subtree size from left to right. On one level, nodes with more subnodes are sorted further to the left than those with less subnodes. This can be observed in Fig. 2b. The author search happens after the journal search, but the corresponding nodes are on the left side, because there are more subsequential actions involved.

To extend the analysis from individual sessions to session groups, we need a way to combine multiple session trees into one conjoint tree[1]. Instead of combining all trees at once, we merge pairs of trees iteratively. We start with an empty tree and merge it with the first session tree. The resulting tree is then merged with the next tree and so forth. Figure 3a shows the merged tree of the example session tree 1 (cf. Fig. 2a) and 2 (cf. Fig. 2b). When two nodes are merged, the weights of their edges are summed up. After merging session tree 1 and 2, we merge the result with the tree of session 3 (cf. Fig. 2c), shown in Fig. 3b.

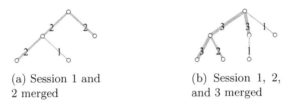

(a) Session 1 and 2 merged

(b) Session 1, 2, and 3 merged

Fig. 3. Combined session trees of the example sessions from Fig. 2.

When merging two trees, one has to decide which nodes are merged. As each node and its child nodes can be interpreted as an individual tree, this decision can be made recursively for each node. We create all possible combinations of subtrees for each pair of nodes and select the best combination along two conditions: First, the number of nodes in the resulting subtree has to be minimal. Second, the weight distribution of the subtree has to be optimised.

For the second condition we needed to define what a weight of a subtree is and what an optimum for the subtree weight is. We define the subtree weight $W(p)$ as follows.

Definition 1. *Let $p \in N_T$ be a node in the tree T, $w(p)$ be the weight of the edge leading to the node p, and C_p the set of child nodes of p. The weight $W(p)$ of the subtree with the root node p is then defined as:*

$$W(p) = \begin{cases} \lg(2 \cdot w(p)) & \text{if } C_p = \emptyset \\ \lg(\sum_{q \in C_p} w(q) \cdot W(q)) & \text{else} \end{cases}$$

---

[1] The Java and R based tool is available under: https://github.com/wilkovanhoek/amur-session-graph/tree/tpdl2017.

We sum the product of the weights of the edges leading to child nodes and the subtree weight of those nodes and calculate the logarithm of this sum. We use the logarithm only to keep the subtree weight from increasing exponentially, as we only need it to compare the subtrees, not to assess an actual summed subtree weight. After defining the subtree weight, we need to define what an optimal subtree weighting is. Because we want the weight of the resulting subtrees to increase from left to right and we do not want the weight to be distributed equally, we favour building maximal heavy subtrees. Therefore, the merged nodes with the heaviest subtree is considered to be optimal.

Following the described procedure, we receive exactly one resulting tree, when combining two session trees. However, when merging multiple trees iteratively, the resulting tree depends on the order in which the trees are merged. To find the optimal merging order, we would need to calculate all possible permutations in which the session trees could be merged. This would result in $32! \approx 2,63 \times 10^{35}$ merging orders for the complete set of trees in our study. As this exceeded our computational capacities for this paper, we decided to use another approach. Before merging the trees, we have sorted the session trees in ascending order with respect to their root node's subtree weight. Now, when merging them, the resulting tree of each merge slowly increases in weight (in general trees with a smaller subtree weight tend to be more compact trees). In this way, common structures that exist in many trees are merged very early, whereas outliers are merged later. Note that the sorting does not guarantee an optimal merging order. However, it ensures that there is only one resulting tree for a set of trees and that all combined trees are created with the same procedure.

3.3   Building Subtrees

Figure 4a shows the combined session tree for all participants. In most sessions, starting from the root node, at least three different actions were conducted. In addition, a larger group of users followed a longer trail of multiple consecutive actions (cf. trail of nodes on the left). Overall, the combined session tree is not very compact. There is a certain density within the first two levels, but behind that, only a few longer trails exist in the tree. Instead, there are many edges that are introduced by a few intense sessions. We consider this to be noise, because it inflates the combined session tree. Therefore, we will introduce a way to reduce this noise, without removing the session trees that are responsible for the noise.

Edges with a low weight as displayed in Fig. 4a represent an activity that has happened in a minority of sessions. Figure 4b illustrates how many nodes would remain in the tree, if we remove all edges (and their nodes) below an increasing weight threshold. After the strong decline in the beginning we can see a first 'plateau' where no bigger drop in the number of remaining nodes for the threshold of 6 and onwards is observed. At a threshold of 11 we can observe a similar development.

Figure 5 shows three subtrees with different thresholds, extracted from the combined session tree in Fig. 4a. Figures 5a and b show the subtrees for the thresholds of 6 and 11. In addition, we include the tree for a threshold of 17

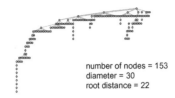

number of nodes = 153
diameter = 30
root distance = 22

(a) Combined session tree for all participants.

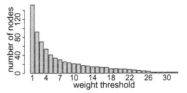

(b) Distribution of the number of nodes in the combined tree, after removing all nodes with a weight below a given threshold value.

Fig. 4. Combined session tree for all participants (a) and thresholds of graph nodes (b).

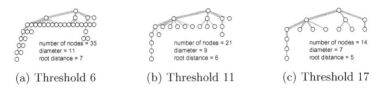

number of nodes = 35
diameter = 11
root distance = 7

(a) Threshold 6

number of nodes = 21
diameter = 9
root distance = 6

(b) Threshold 11

number of nodes = 14
diameter = 7
root distance = 5

(c) Threshold 17

Fig. 5. Subtrees extracted from the combined session tree (cf. Fig. 4). Each tree is created by removing all edges (and nodes) with a weight below the specified threshold.

which represents all nodes that appear at least in half of all session trees. In Fig. 5c we can now see our former observation more clearly. Most users start at least three independent activities from the root document and follow at least one longer trail. Comparing all three subtrees in Fig. 5, we can observe that with an increasing threshold mainly the number of nodes per level decreases, whereas the overall structure does not change decisively.

4   Grouping the User Behaviour

Based on the session trees, we tried to divide the user sessions into groups. We grouped similar session trees. We did this based on our visual impression and aspects like the number of nodes on the same level, the number of parallel subtrees, the overall depth of the tree, the number of subtrees with a similar depth, and the branchiness (how often single nodes are followed by multiple nodes).

Figures 6 and 7 show sessions with very intense activity in which many actions have been conducted. The sessions in Fig. 6, however, nearly exclusively show activity close to the seed document. This activity represents SERPs that are closely examined. We characterise this as an highly exhaustive behaviour with a focus on breadth. In contrast, the sessions displayed in Fig. 7, show more activity venturing away from the seed document. It seems as if a trail is being followed. The participants investigated deeper into a specific direction. We characterise this as an exhaustive behaviour with focus on depth.

Fig. 6. Exhaustive breadth group

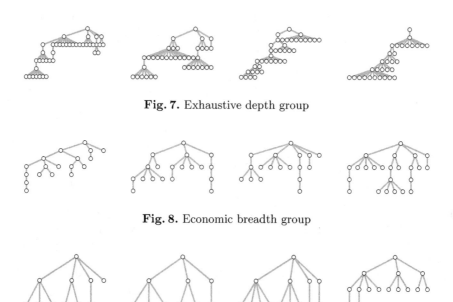

Fig. 7. Exhaustive depth group

Fig. 8. Economic breadth group

Fig. 9. System support group

Figures 8 and 9 display sessions with a lower number of actions. The trees are less dense and appear more balanced. In Fig. 8, different trails are followed which are inspected to some extend, but not very deeply. It appears to be a more swift examination of the area around the seed document. We characterise this as an economic behaviour with focus on breadth. The sessions in Fig. 9 display a very particular behaviour. When building this group, we realised, that they have strong familiarities with the subtree of the complete data set for a threshold of 17 (Fig. 5c). Structurally speaking, these sessions are similar to a common behaviour in the majority of sessions. To understand this, we watched the sessions' screen casts again. We could see that the participants almost solely relied on information that was provided by the system (recommendations, references and citations) and rarely conducted own searches. We characterise this group as behaviour with a focus on system support. However, we don't know in how far the study task and situation has triggered this behaviour. Possibly, these participants could not identify with the task.

5   Building Archetypical Session Trees per Group

Now that we have four different groups of behaviour, the next step is to build archetypical trees for each group. We will use the method described in Sect. 3.3. At first, we will merge all session trees within each individual group into one combined tree. After that, we will create subtrees based on a suitable edge weight threshold. We will define the resulting subtree as the archetypical tree for the different groups. Figure 10a shows the results for the group of sessions with exhaustive behaviour focused on breadth. We can see that with an increased threshold, the number of remaining nodes decreases constantly. However, for a threshold of 3, still half of the nodes are left in the subtree. Therefore, this subtree represents behaviour that is shared by at least three participants. Looking at the resulting subtree, it seems that this tree does represent the group. We feel that this is a reasonable threshold for creating an archetypical session tree.

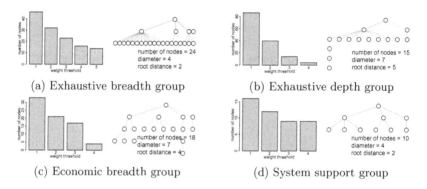

(a) Exhaustive breadth group   (b) Exhaustive depth group

(c) Economic breadth group   (d) System support group

Fig. 10. Distribution of number of nodes and combined session trees for the four different groups

Using the same method to create an archetypical session tree for the exhaustive depth group leads to bad results. The decline of number of nodes left in the subtree, when increasing the weight threshold is a lot faster than with the previous group (cf. Fig. 10b). In addition, the resulting subtree for a threshold of 3 does not look similar to any of the group's individual sessions. Only around one fifth of the combined tree's nodes remain in the subtree. It seems that our method cannot create a suitable archetypical session tree for this group. Although this tree does not represent the underlying group sufficiently we will keep it for further analysis as a negative example.

The results seem to be more sound with respect to the economic breadth group. In Fig. 10c, we can see that a subtree for a threshold of 3 contains more than half of the nodes. Only, when we raise the threshold up to 4 we see a strong decrease in the number of nodes. When comparing the resulting subtree and the group's individual session trees, strong similarities can be observed. For the last group, (system support) our method works best. Figure 10d shows that the number of nodes remaining in all subtrees stay on a higher level than for the other groups. Also, the subtree for threshold 3 resembles the group's session trees reasonably. Analogous to our previous observation, this group's archetype has a strong resemblance to the threshold 17 subtree created from the combination of all 32 session trees (cf. Fig. 5).

Based on the results of our method on creating subtrees for groups of session trees, we propose a following definition of an archetype of a group:

Definition 2. *For a group of session trees $G = \{t_1, t_2, \ldots, t_n\}$, their combined tree t_{comb} and the set of its subtrees $G_{t_{comb}} = \{sub_{t_{comb},1}, sub_{t_{comb},2}, \ldots, sub_{t_{comb},n}\}$, an archetypical session tree t_{arch} is the session tree, that satisfies the following conditions:*

- *it is included in the set of subtrees of the combined tree: $t_{arch} \in G_{t_{comb}}$*
- *the number of nodes is higher than half of the number of nodes in the combined tree: $|t_{arch}| > |t_{comb}|/2$*
- *the weight of each node is higher than half the number of sessions in the group: $\forall n \in t_{arch} : w(n) > n/2$*

For all groups except the exhaustive depth group, the subtrees for threshold 3 are archetypical session trees. Our approach was not able to merge the trees in the exhaustive depth group very well. As a result, we cannot derive an archetypical tree from this group.

6   Discussion

Utilising our approach based on a tree representation of user activity, we were able to visually identify similar structural patterns. By merging the similar sessions, we created a combined tree per group. By removing edges, with a low edge weight, we created a subtree of the combined tree that represents nodes that exist in most of the individual trees. This has led to the idea that there could be an archetypical tree for each group of users. We proposed a Definition 2 for such an archetypical tree. We applied this method to individual sessions of different users. However, this could also be done for multiple sessions of a user to create a user specific graph. Comparing and merging multiple user specific graphs could then be used to derive user groups in a live system. This could be achieved by merging a user specific tree with different archetypes and measure the number of unmatched nodes. A more rigid version of Definition 2 is imaginable but should be based on a larger data set. Especially the weighting constraint should be investigated more closely in future work. Also, we have not investigated the 15 user session that were not grouped. As future work we plan to assess the most suitable archetypes of those sessions.

So far, we have ignored the types of search activities and objects. It remains unclear how the present approach performs when only edges or nodes of the same type are merged. Untyped trees do not allow us to identify different search tactics. A group of users that frequently utilises a specific tactic, like a journal run, cannot be identified in this way. However, this would require a significantly larger data set. One promising study that involves a larger data set can be found in [10] and could be suitable for future work.

Another aspect we disregarded is the time spent on a specific object. Thus, there is no information about the intensity with which activities are followed. A user who only inspects the list of citations briefly is treated equally to a user that inspects the list in more detail. This could be addressed by including the action duration per node. However, we lack a suitable way to include time into the merging method. In addition, further effort should be invested in evaluating our approach with respect to more profound models in information seeking behaviour like the model by Ellis [8]. One example could be to investigate whether the present approach is capable of identifying the different stages (e.g. *Starting*, *Chaining*) of information seeking.

7 Conclusion

In this paper, we have discussed the structural analysis of patterns in user search behaviour based on a tree representation of the user sessions. We divided the different user sessions manually into groups similar to those in [4]. When grouping, we considered the graphs visual resemblance and graph attributes like the number of nodes. We proposed a novel method to merge multiple user session into one combined tree. We merged the sessions for each of our groups individually and could see that for most groups a relatively high number of nodes can be merged in a combined tree. Furthermore, we defined criteria to derive an archetypical tree from a combined tree. In this way, our method could potentially be used to assign user groups for new users of a system, based on the way in which they perform their searches. By comparing different user sessions, it could also help to identify different search strategies.

Acknowledgments. This work was partly funded by DFG, grant no. MA 3964/5-1; the AMUR project at GESIS. We thank all participants of our user study and Maria Lusky.

References

1. Agichtein, E., Brill, E., Dumais, S., Ragno, R.: Learning user interaction models for predicting web search result preferences. In: Proceedings of the 29th Annual International ACM SIGIR Conference on Research and Development in Information Retrieval, pp. 3–10. ACM (2006)
2. Athukorala, K., Hoggan, E., Lehtiö, A., Ruotsalo, T., Jacucci, G.: Information-seeking behaviors of computer scientists: challenges for electronic literature search tools. Proc. Am. Soc. Inf. Sci. Technol. **50**(1), 1–11 (2013)
3. Aula, A., Majaranta, P., Räihä, K.-J.: Eye-tracking reveals the personal styles for search result evaluation. In: Costabile, M.F., Paternò, F. (eds.) INTERACT 2005. LNCS, vol. 3585, pp. 1058–1061. Springer, Heidelberg (2005). doi:10.1007/11555261_104

4. Buscher, G., White, R.W., Dumais, S., Huang, J.: Large-scale analysis of individual and task differences in search result page examination strategies. In: Proceedings of the Fifth ACM International Conference on Web Search and Data Mining, WSDM 2012, pp. 373–382. ACM, New York (2012)
5. Carevic, Z., Lusky, M., Van Hoek, W., Mayr, P.: Investigating exploratory search activities based on the stratagem level in digital libraries. Int. J. Dig. Libr. **18**(1), 1–3 (2017). https://arxiv.org/abs/1706.06410
6. Cole, M.J., Hendahewa, C., Belkin, N.J., Shah, C.: User activity patterns during information search. ACM Trans. Inf. Syst. (TOIS) **33**(1), 1 (2015)
7. Eirinaki, M., Vazirgiannis, M.: Web mining for web personalization. ACM Trans. Int. Technol. (TOIT) **3**(1), 1–27 (2003)
8. Ellis, D.: A behavioural approach to information retrieval system design. J. Documentation **45**(3), 171–212 (1989)
9. Fangerau, J.: Interactive Similarity Analysis for 3D+t Cell Trajectory Data. Ph.D. thesis (2015)
10. Hagen, M., Potthast, M., Völske, M., Gomoll, J., Stein, B.: How writers search: analyzing the search and writing logs of non-fictional essays. In: Proceedings of the 2016 ACM on Conference on Human Information Interaction and Retrieval, pp. 193–202. ACM (2016)
11. Hienert, D., Sawitzki, F., Mayr, P.: Digital library research in action: supporting information retrieval in sowiport. D-Lib Mag. **21**(3), 8 (2015)
12. Kules, B., Capra, R., Banta, M., Sierra, T.: What do exploratory searchers look at in a faceted search interface?. In: Proceedings of the 9th ACM/IEEE-CS Joint Conference on Digital Libraries, pp. 313–322. ACM (2009)
13. Manning, C.D., Raghavan, P., Schütze, H., et al.: Introduction to information retrieval, vol. 1. Cambridge University Press, Cambridge (2008)
14. Mitsui, M., Shah, C., Belkin, N.J.: Extracting information seeking intentions for web search sessions. In: Proceedings of the 39th International ACM SIGIR Conference on Research and Development in Information Retrieval, SIGIR 2016, pp. 841–844. ACM, New York (2016)
15. Mobasher, B., Cooley, R., Srivastava, J.: Automatic personalization based on web usage mining. Commun. ACM **43**(8), 142–151 (2000)
16. Srivastava, J., Cooley, R., Deshpande, M., Tan, P.N.: Web usage mining: discovery and applications of usage patterns from web data. ACM SIGKDD Explor. Newsletter **1**(2), 12–23 (2000)
17. White, R.W., Drucker, S.M.: Investigating behavioral variability in web search. In: Proceedings of the 16th International Conference on World Wide Web, WWW 2007, pp. 21–30. ACM, New York (2007)
18. Wildemuth, B.M.: The effects of domain knowledge on search tactic formulation. J. Am. Soc. Inf. Sci. Technol. **55**(3), 246–258 (2004)
19. Wilson, M.L., Schraefel, M.: A longitudinal study of exploratory and keyword search. In: Proceedings of the 8th ACM/IEEE-CS Joint Conference on Digital Libraries, JCDL 2008, pp. 52–56. ACM, New York (2008)

Information Retrieval

Multiple Random Walks for Personalized Ranking with Trust and Distrust

Dimitrios Rafailidis[1(✉)] and Fabio Crestani[2]

[1] Department of Computer Science, University of Mons, Mons, Belgium
dimitrios.rafailidis@umons.ac.be
[2] Faculty of Informatics, Università Della Svizzera Italiana, Lugano, Switzerland
fabio.crestani@usi.ch

Abstract. Social networks with trust and distrust relationships has been an emerging topic, aiming at identifying users' friends and foes when sharing information in social networks or purchasing products online. In this study we investigate how to generate accurate personalized rankings while considering both trust and distrust user relationships. This paper includes the following contributions, first we propose a social inference step of missing (indirect) trust relationships via multiple random walks, while considering users' direct trust and distrust relationships during the inference. In doing so, we can better capture the missing trust relationships between users in an enhanced signed network. Then, we introduce a regularization framework to account for (i) the *structural properties* of the enhanced graph with the inferred trust relationships, and (ii) the user's trust and distrust *personalized preferences* in the graph to produce his/her personalized ranking list. We evaluate the performance of the proposed approach on a benchmark dataset from Slashdot. Our experiments demonstrate the superiority of the proposed approach over state-of-the-art methods that also consider trust and distrust relationships in the personalized ranking task.

Keywords: Personalized ranking · Signed graphs · Social inference

1 Introduction

With the advent of social networks such as Epinions[1] and Slashdot[2], users share various information through interactions with other users while expressing their positive and negative opinions. Based on their feedback, users establish trust and distrust relationships on each other, forming a signed graph with positive and negative links, respectively [13]. Node ranking with trust and distrust relationships has several applications including community detection [18], collaborative filtering [2], trust prediction [3], sign prediction [13] and troll detection [17], just to name a few. Popular ranking models like PageRank and HITS do not consider both the positive and negative links at the same time and thus, they do

---

[1] http://www.epinions.com/.
[2] https://slashdot.org/.

© Springer International Publishing AG 2017
J. Kamps et al. (Eds.): TPDL 2017, LNCS 10450, pp. 473–484, 2017.
DOI: 10.1007/978-3-319-67008-9_37

not perform well on the personalized ranking task in signed graphs. Recently, in an attempt to perform node ranking in graphs with trust and distrust relationships Shahriari and Jalili [13] review different ranking algorithms in signed graphs. In this study, they propose a modified PageRank algorithm to account for both the trust and distrust relationships in signed networks. They compute the PageRank values in a positive and a negative subgraph separately and then subtract negative PageRank values from positive ones to compute an aggregated PageRank value for each node. In the study reported in [17], a variant of PageRank is introduced to model the probability of trustworthiness of individual data sources as an interpretation for the underlying ranking values. Jung et al. [4] extend the Random Walk with Restart algorithm, namely Signed Random Walk with Restart, to generate personalized rankings in signed graphs. They introduce a random walker based on the balance theory that considers both the positive and negative links by changing the walker's sign when performing random walks on the signed graph and producing a user's ranking list.

The shortcomings of the above state-of-the art methods are that if a user/node does not have enough interaction information, the aforementioned ranking methods face difficulties in calculating the probability of trustworthiness between two users. In addition, these ranking methods in signed networks mainly rely on the trust and distrust relationships between users based on direct signed relationships e.g., direct friend or foe, and depend on the number of common neighbours to capture the indirect relationships. So, provided that the social relationships are sparse, the challenge that we face is how can we perform personalized ranking in graphs with trust and distrust relationships to infer the missing (indirect) relationships between the users? Inferring social relationships of trust and distrust users is a challenging task [14]. Given explicit (direct) social relationships, the goal is to infer the indirect relationships of trust and distrust users. Trust relationships show strong transitivity, which means that inferring trust relationships can be computed in a network of trust users, mainly because if two users a and b are friends and a third user c is friend with a, then user c might be a friend of b as well. However, recent studies showed that distrust is certainly not transitive [1, 15, 16]. Therefore, distrust cannot be considered as the negative of trust when inferring users' distrust relationships. Accounting for the transitivity of trust relationships, a few prediction models have been proposed to infer the implicit trust relationships, while exploiting explicit distrust relationships in their predictions [3, 14]. Nonetheless, these models are designed to predict missing trust relationships and not to generate personalized rankings. Therefore, a pressing challenge resides on how to infer trust relationships of users with their distrust relationships to improve the accuracy of users' personalized rankings.

1.1   Contribution and Outline

As generating accurate personalized rankings becomes more and more important in social networks, in this study our contributions are the following:

- We introduce a social inference step via multiple random walks, aiming to solve the sparsity of social relationships and consequently better predict the

indirect relationships between the users. In our social inference step, we predict missing (indirect) trust relationships, while considering users' direct trust and distrust relationships during the inference.
- We propose a regularization framework for personalized ranking to consider both (i) the *structural properties* of the enhanced signed graph with the inferred relationships and (ii) the input query vector of a user with his/her *personalized preferences* of trust and distrust relationships in the signed network.

Our experiments on a benchmark dataset with trust and distrust relationships show that the proposed approach significantly outperforms other competitors in the personalized ranking prediction task in signed networks. The remainder of the paper is organized as follows, in Sect. 2 we formally define our problem, Sect. 3 details the proposed model, Sect. 4 presents the experimental results and finally, Sect. 5 concludes the study.

2 Problem Formulation

In this study, we consider a directed graph $\mathcal{G}$ with $n = |\mathcal{V}|$ nodes and $i, j \in \mathcal{V}$. Two nodes are connected with edges in the form $(i, j) \in \mathcal{E}$. The edges are considered directed and weighted, and in our setting we consider positive and negative weights to express trust and distrust relationships, respectively. Both positive and negative weights are stored in a $(n \times n)$ weighting matrix $\mathbf{W}$. In our approach we generate two different graphs, a graph $\mathcal{G}_+$ which contains only the positive edges and a second graph $\mathcal{G}_-$ with the negative ones. Given $\mathcal{E} \equiv \mathcal{E}_+ \cup \mathcal{E}_-$, we compute two different $(n \times n)$ weighting matrices $\mathbf{W}_+$ and $\mathbf{W}_-$, corresponding to the weights of the positive $(i, j)_+ \in \mathcal{E}_+$ and negative edges/relationships $(i, j)_- \in \mathcal{E}_-$. Notice that $\forall (i, j)_- \in \mathcal{E}_-$ we set $(\mathbf{W}_-)_{ij} = |\mathbf{W}_{ij}|$, storing the absolute values if the weights of the edges are negative. In our setting, we consider a *query vector* $\mathbf{y} \in \mathbb{R}^n$ for a node/user $m \in \mathcal{V}$, expressing his/her personalized preferences of trust and distrust in the signed graph. The query vector $\mathbf{y}$ is formally defined as follows:

Definition 1 (Query vector). *A query vector $\mathbf{y} \in \mathbb{R}^n$ of a user m expresses his/her personalized preferences of trust and distrust relationships in the signed graph, computed as follows, $\mathbf{y}_i = 1$, if $i = m$ and $\mathbf{y}_i = \mathbf{W}_{mi}$ otherwise.*

With these settings, the problem of personalized ranking with trust and distrust relationships is formally defined as follows:

Definition 2 (Problem). *"Given (i) the weighting matrices $\mathbf{W}_+$ and $\mathbf{W}_-$, and (ii) a query vector $\mathbf{y} \in \mathbb{R}^n$ of node/user m with his/her personal preferences of trust and distrust in the signed graph, the goal of the proposed approach is to generate an optimized ranking vector $\mathbf{r} \in \mathbb{R}^n$ for ranking the n nodes, accounting for both the structural properties of the signed graph and the personalized preferences on trust and distrust relationships of user m."*

In the first step of our approach we run multiple random walks on the graph $\mathcal{G}_+$, while considering the distrust relationships in graph $\mathcal{G}_-$. In doing so, we enhance and better capture the relationships between the trusted and distrusted nodes, inferring the missing (indirect) trust links between the nodes. Then, in the second step we consider the enhanced graph with the inferred trust and the direct (explicit) trust and distrust relationships and calculate an optimized ranking vector $\mathbf{r}$ per node/user to produce the personalized ranking list.

3   Proposed Approach

3.1   Social Inference via Multiple Random Walks

To infer the missing (indirect) trust relationships, we perform random walks on the n nodes in graph $\mathcal{G}_+$, by taking into account the distrust relationships in graph $\mathcal{G}_-$ during the inference. In particular, the proposed approach runs multiple random walks on the graph $\mathcal{G}_+$ with the trust relationships and then filters out the inferred trust relationships by considering the distrust relationships in graph $\mathcal{G}_-$. The main reason that we avoid to perform random walks on graph $\mathcal{G}_-$ is that *distrust is not transitive*, as opposed to trust [3,14–16]. Next, we present the case of performing a single random walk on graph $\mathcal{G}_+$ and we show how to perform multiple random walks to better infer the implicit trust relationships.

Single Random Walk. Given a source node *sou* and a target node *tar*, with $(sou, tar) \notin \mathcal{E}_+$, the goal is to start a random walk from *sou* to reach *tar* to infer their trust relationships, denoted as $(\overline{\mathbf{W}}_+)_{sou,tar}$. We assume that the walk moves from one node to a neighbourhood node at each step, and at time t the walk has moved to node i. The walk chooses whether to move to another node with probability ξ_t or terminate the walk with probability $1 - \xi_t$. In the case of terminating the walk, the value $(\overline{\mathbf{W}}_+)_{sou,tar}$ is returned only if edge $(i, tar) \in \mathcal{E}_+$, and 0 otherwise. The *transition probability* of moving from a current node i to another node j is calculated as follows:

$$p_+(j|i) = (\mathbf{W}_+)_{ij}/d_i$$

where $d_i = \sum_j (W_+)_{ij}$ is the degree of i. The $(n \times n)$ *transition matrix* of a random walk is given by

$$\mathbf{T}_+ = \mathbf{D}_+^{-1}\mathbf{W}_+$$

where $(\mathbf{D}_+)_{ii} = d_i$ is the $(n \times n)$ degree diagonal matrix. A vector $\mathbf{p}_+^{(t)} \in \mathbb{R}^n$ represents the visiting distribution over all n nodes at a certain time t. With these settings, if the walk continues at the next time $t + 1$, the distribution vector will be updated as follows:

$$\mathbf{p}_+^{(t+1)} = \mathbf{p}_+^{(t)} \times \mathbf{T}_+$$

Multiple Random Walks. Instead of performing a single walk, we run multiple random walks from a source node in graphs $\mathcal{G}_+$ and $\mathcal{G}_-$ to better infer the missing trust relationships. The main reason that we can achieve better inference is that *multiple random walks start from the source user sou to seek more alternatives for the implicit (indirect) relationship to the target user tar*. Considering the graph $\mathcal{G}_+$, we define s as the total length of a single walk for which we recursively update the distribution vector $\mathbf{p}^{(s)}$. For a target node tar we consider all its in-linked edges, denoted by $(\mathbf{W}_+)_{*tar}$, that is the tar-th column vector of $\mathbf{W}_+$. With these settings, the returned value for a random walk terminated at time s is:

$$(\overline{\mathbf{W}}_+)_{sou,tar}|s = \mathbf{p}^{(s)}(\mathbf{W}_+)_{*tar} \tag{1}$$

Theoretically, we can perform random walks with infinite lengths from the source node. Aggregating the multiple random walks from the source node we have:

$$(\overline{\mathbf{W}}_+)_{sou,tar} = \sum_{t=1}^{\infty} \omega_+(t)\mathbf{p}_+^{(0)}\mathbf{T}_+^t(\mathbf{W}_+)_{*tar} \tag{2}$$

where $\mathbf{p}_+^{(0)}$ is the starting distribution of a walk on $\mathcal{G}_+$ and $\omega_+(t)$ expresses the probability that a random walk will terminate at a certain time t:

$$\omega_+(t) = p_+(s = t|\xi) = \xi_t \prod_{i=1}^{t-1}(1 - \xi_i) \tag{3}$$

Therefore, the weighting matrix $\overline{\mathbf{W}}_+$ with the inferred trust relationships is calculated as follows:

$$\overline{\mathbf{W}}_+ = \sum_{t=1}^{\infty} \omega_+(t)\mathbf{T}_+^t\mathbf{W}_+ \tag{4}$$

In our implementation, we avoid long (infinite) walks on the graph, following the idea of the "six degrees of separation", that is most nodes can be reached with a six step walk length [2]. This means that if a walk has reached more than six steps, then the walk is terminated. In practice, we observed that random walks do not reach more than four steps in our experiments with $\xi_t = 0.85$, equal to the dampening factor of PageRank [6].

When performing multiple random walks on graph $\mathcal{G}_+$, the distrust relationships in graph $\mathcal{G}_-$ are ignored. Consequently, an inferred trust relationship between a source user sou and a target user tar in $\overline{\mathbf{W}}_+$, might have a conflict of a distrust relationship between sou and tar in graph $\mathcal{G}_-$. To avoid this conflict, we recompute matrix $\overline{\mathbf{W}}_+$ by setting $\overline{\mathbf{W}}_+ \leftarrow 0$, if $(\overline{\mathbf{W}}_+)_{ij} > 0 \wedge (\mathbf{W}_-)_{ij} > 0$, $\forall i, j = 1 \ldots n$. Finally, the filtered trust relationships and their positive weights are stored into the initial adjacency matrix with the trust relationships, by setting $\mathbf{W}_+ \leftarrow \overline{\mathbf{W}}_+$.

3.2   Node Ranking with Trust and Distrust Relationships

A Regularization Framework for Node Ranking. In a plain graph with a single type of edges that considers only positive weights, we have a weighting matrix $\mathbf{W} > 0$ and a query vector $\mathbf{y} \in \mathbb{R}^n$ (Definition 1), where the i-th element $\mathbf{y}_i$ is the initial query score of node i. The goal is to find a new optimized ranking vector $\mathbf{r} \in \mathbb{R}^n$ that is smooth and close to $\mathbf{y}$, formulating the following cost function C as a minimization problem:

$$\min_{\mathbf{r}} C(\mathbf{r}) = S(\mathbf{r}) + \theta \widehat{E}(\mathbf{r}; \mathbf{y}) \tag{5}$$

The first term $S(\mathbf{r})$ is a smoothness function to consider the *structural cost* of the graph. The second term $\widehat{E}$ is the *ranking error* between the vectors $\mathbf{r}$ and $\mathbf{y}$ to express how well the optimized ranking vector fits the input query vector. Parameter θ controls the influence of the second term when minimizing $C(\mathbf{r})$. According to the regularization framework of [19] the smoothness function can be calculated as:

$$S(\mathbf{r}) = \mathbf{r}^\top (\mathbf{I} - \mathbf{A})\mathbf{r}$$

where $\mathbf{A} = \mathbf{D}^{-1/2}\mathbf{W}\mathbf{D}^{-1/2}$. The ranking error is computed as follows:

$$\widehat{E}(\mathbf{r}; \mathbf{y}) = ||\mathbf{r} - \mathbf{y}||^2 = (\mathbf{r} - \mathbf{y})^\top (\mathbf{r} - \mathbf{y})$$

To compute the optimized ranking vector $\mathbf{r}$, we take the gradient of $C(\mathbf{r})$ and set it to zero:

$$\frac{\partial(C)}{\partial \mathbf{r}} = (\mathbf{I} - \mathbf{A})\mathbf{r} + \theta(\mathbf{r} - \mathbf{y}) = 0 \Rightarrow$$
$$\mathbf{r} = (1 - \lambda)(\mathbf{I} - \lambda A)^{-1}\mathbf{y} \propto (\mathbf{I} - \lambda A)^{-1}\mathbf{y} \tag{6}$$

with $\lambda = 1/(\theta + 1) \in (0, 1)$.

Ranking with Trust and Distrust Relationships. In our setting the graph contains the initial trust and distrust relationships and the inferred trust relationships in the weighting matrices $\mathbf{W}_+$ and $\mathbf{W}_-$. As in the case of the plain graphs we have to calculate an optimized ranking vector $\mathbf{r}$, given a query vector $\mathbf{y}$ with the user's personalized preferences of trust and distrust in the signed graph. In this respect, we have to reformulate the smoothness function $S(\mathbf{r})$ in Eq. (5) when minimizing the objective function $C(\mathbf{r})$, considering the weighting matrices $\mathbf{W}_+$ and $\mathbf{W}_-$.

We combine $\mathbf{W}_+$ and $\mathbf{W}_-$ into a global weighting matrix $\mathbf{W} = \mathbf{W}_+ - \mathbf{W}_-$. As the trust and distrust relationships are directed, for a node i we calculate the out-degree $d_i^{out} = \sum_{j\neq i} \mathbf{W}_{ij}$ and the in-degree $d_i^{in} = \sum_{j\neq i} \mathbf{W}_{ji}$, as well as the $(n \times n)$ diagonal degree matrices $(\mathbf{D}^{out})_{ii} = d_i^{out}$ and $(\mathbf{D}^{in})_{ii} = d_i^{in}$.

The overall smoothness of the ranking vector is the summation of all *local variations* [19]:

$$S(\mathbf{r}) = \sum_{i \in \mathcal{V}} ||\nabla_i \mathbf{r}||^2 \tag{7}$$

where the *local variation* at each node n in our graph with trust and distrust relationships is calculated as follows:

$$||\nabla_i \mathbf{r}|| \sqrt{\frac{1}{2}\left[\sum_{j\neq i}\partial\mathbf{r}/\partial(i,j) + \sum_{j\neq i}\partial\mathbf{r}/\partial(j,i)\right]} \tag{8}$$

with

$$\partial\mathbf{r}/\partial(i,j) = \sqrt{\mathbf{W}_{ij}/d_i^{out}}\,\mathbf{r}_i - \sqrt{\mathbf{W}_{ij}/d_j^{in}}\,\mathbf{r}_j$$
$$\partial\mathbf{r}/\partial(j,i) = \sqrt{\mathbf{W}_{ji}/d_i^{in}}\,\mathbf{r}_i - \sqrt{\mathbf{W}_{ji}/d_j^{out}}\,\mathbf{r}_j \tag{9}$$

Provided that $\partial\mathbf{r}/\partial(i,j) = -\partial\mathbf{r}/\partial(j,i)$, according to Eqs. (8) and (9) it is easy to verify that the overall smoothness in Eq. (7) can be reformulated as follows:

$$S(\mathbf{r}) = \frac{1}{2}\sum_{i=1}^{n}\mathbf{r}_i^2 + \sum_{j=1}^{n}\mathbf{r}_j^2\sum_{i=1}^{n}\sum_{j=1}^{n}\frac{\mathbf{W}_{ij}\mathbf{r}_i^2\mathbf{r}_j^2}{\sqrt{d_i^{out}d_j^{in}}}$$
$$= \mathbf{r}^\top\mathbf{r} - \mathbf{r}^\top\mathbf{D}^{out^{-1/2}}\mathbf{W}\mathbf{D}^{in^{-1/2}}\mathbf{r} = \mathbf{r}^\top(\mathbf{I}-\mathbf{B})\mathbf{f} \tag{10}$$

with $\mathbf{B} = \mathbf{D}^{out^{-1/2}}\mathbf{W}\mathbf{D}^{in^{-1/2}}$. Based on the formulation of the ranking problem in plain graphs in Eq. (5) and the smoothness function in Eq. (10), the ranking problem in our graph with trust and distrust relationships becomes:

$$\min_{\mathbf{r}} C(\mathbf{r}) = \mathbf{r}^\top(\mathbf{I}-\mathbf{B})\mathbf{f} + \theta(\mathbf{r}-\mathbf{y})^\top(\mathbf{r}-\mathbf{y}) \tag{11}$$

Similar to the case of a single graph, we derive the following closed-form solution of the optimized ranking vector:

$$\mathbf{r} = (\mathbf{I}-\lambda\mathbf{B})^{-1}\mathbf{y} \tag{12}$$

To generate a *personalized ranking* for a node m, we set the query vector $\mathbf{y}_i = 1$, if $i = m$ and $\mathbf{y}_i = \mathbf{W}_{mi}$ otherwise, expressing user's m personalized preferences of trust and distrust in the signed graph (Definition 1). Having computed the query vector $\mathbf{y}$ for the node m then we perform the personalized ranking based on Eq. (12).

4   Experimental Evaluation

4.1   Experimental Setup

As there is no groundtruth available to evaluate directly the performance of the ranking models in graphs with trust and distrust relationships, we examine the ranking performance on the troll detection task [4]. Trolls are users that can intentionally post misleading information, either having malicious intent, profit motives, or simply behaving in a disruptive way [17]. The goal of the troll detection task is to identify trolls in a user's personalized ranking list. In our experiments we used the "Slashdot Zoo" dataset[3] [5], which consists of 77,985 users, 388,190 friend (trust) links and 121,967 foe (distrust) links. Following [5], we use the foes of a user, called No-More Trolls in the "Slashdot Zoo" dataset. As we investigate the case of personalized ranking in signed graphs, we generate a personalized distrust ranking list, aiming to detect trolls high at a user's ranking list.

In our experiments we used the ranking-based metrics precision, recall and Normalized Discounted Cumulative Gain (NDCG). Precision is defined as the ratio of the relevant items in the top-N list, and recall is defined as the ratio of the relevant items in the top-N ranked list over all the relevant items. The NDCG metric considers the ranking of the relevant items in the top-N list. For each user the Discounted Cumulative Gain (DCG) is defined as:

$$DCG@N = \sum_{j=1}^{N} \frac{2^{rel_j} - 1}{\log_2 j + 1}$$

where rel_j represents the relevance score of item j, that is binary relevance in our case. As we focus on the troll detection task, we consider an item as relevant if a user is a troll, and irrelevant otherwise. NDCG is the ratio of DCG over the ideal iDCG value for each user, that is the DCG value given all trolls in the users' personalized list. We report precision, recall and NDCG at the top-$N = 100$ results of the user's ranking list. The reason that we consider the top-100 ranked results is that in total there are 96 trolls. We repeated our results five times and in each run we averaged the evaluation metrics over all users.

4.2   Compared Methods

We evaluate the performance of the following methods:

- **MPR** [13]: a Modified PageRank algorithm for ranking nodes in signed graphs. MPR computes the PageRank values in the positive $\mathcal{G}_+$ and negative graph $\mathcal{G}_-$ separately, and then subtracts negative PageRank values from positive ones to calculate an aggregated PageRank value per node.
- **Troll-Trust** [17]: a model that performs personalized ranking with trust and distrust relationships. Troll-Trust first uses a Bernoulli distribution to characterize each user as either being trustworthy or being a troll, and then constructs a probabilistic model based on the users' trust and distrust relationships with an iterative algorithm.
- **SRWR**[4] [4]: a Signed Random Walks with Restart method for personalized ranking in signed graphs. SRWR starts a signed random surfer so that she considers negative edges by changing her sign for walking. In particular, SRWR first considers the sign of the surfer either positive or negative, that is favorable or adversarial to a node respectively, and then when a random surfer

---

[3] http://dai-labor.de/IRML/datasets.
[4] http://datalab.snu.ac.kr/srwr.

encounters a negative edge, she changes her sign from positive to negative, or vice versa. Otherwise, she keeps her sign.

– **MRW-TD***: a variant of the proposed method to evaluate the effect on the ranking performance of our model when we do not perform the inference of trust relationships with the inference step of Sect. 3.1. To achieve this, in the MRW-TD* variant we feed the second step of our approach in Sect. 3.2 with the initial (direct) trust and distrust relationships of the original graph in the weighting matrices $\mathbf{W}_+$ and $\mathbf{W}_-$, respectively.

– **MRW-TD**: the proposed method of Multiple Random Walks with Trust and Distrust relationships.

4.3 Balancing Personalized Preferences of Trust and Distrust with Graphs' Structural Properties

Figure 1 shows the effect on NDCG, when varying the θ parameter of Eq. (11) in the proposed MRW-TD approach and its MRW-TD* variant. While considering the structural properties of the signed graph, higher θ values indicate that the personalized preferences of trust and distrust will influence more the objective function in Eq. (11). The results in Fig. 1 demonstrate that a selection of $\theta = 1e-2$ achieves the best NDCG value for both methods. Setting $\theta = 1e-1$ results in the model's overfitting, degrading the NDCG metric for both methods. On the other hand, lower values $\theta \leq 1e-3$ consider less user's preferences of trust and distrust in the signed network, thus reducing the NDCG metric. Compared to the variant MRW-TD*, MRW-TD achieves a relative improvement of 35.54%. This indicates the importance of the proposed social inference step of missing trust relationships in Sect. 3.1, which solves the sparsity in user's social relationships. As a consequence, MRW-TD can better capture the missing relationships than its MRW-TD* variant, hence MRW-TD produces more accurate personalized rankings, expressed by the higher NDCG values.

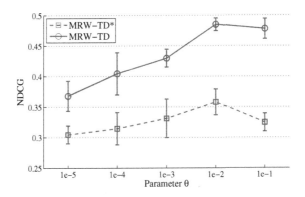

Fig. 1. Effect on NDCG when varying parameter θ.

Table 1. Methods comparison in terms of NDCG, precision and recall. Bold values denote the best scores ($p < 0.05$).

	NDCG	Precision	Recall
MPR [13]	.2963 ± .0243	.0812 ± .0131	.0775 ± .0107
Troll-Trust [17]	.3646 ± .0058	.1146 ± .0129	.0904 ± .0096
SRWR [4]	.4165 ± .0155	.1492 ± .0087	.1363 ± .0055
MRW-TD*	.3579 ± .0211	.1046 ± .0115	.0805 ± .0088
MRW-TD	**.4851 ± .0104**	**.1608 ± .0062**	**.1588 ± .0047**

4.4 Comparison with State-of-the-Art

In Table 1, we report average NDCG, precision and recall, to compare the proposed MRW-TD method with the state-of-the-art methods for personalized ranking in graphs with trust and distrust relationships. Compared to the second best method of SRWR, in this set of experiments we achieve a relative improvement of 16.47, 7.76 and 16.51% in terms of NDCG, precision and recall, respectively. Using the paired t-test we found that MRW-TD outperforms its competitors in all runs, with the results being statistically significant at $p < 0.05$. This occurs because MRW-TD performs the social inference step, and as a consequence can better capture the missing (indirect) relationships than other methods. At the same time MRW-TD balances the structural properties of the signed graph with the user's preferences of trust and distrust in the ranking regularization framework of Eq. (11). Notice that the competitors face difficulties in the presence of sparsity in the social relationships. For instance, the second best method SRWR changes the sign of the walker based on users' direct trust and distrust relationships and then generates the personalized rankings accordingly. However, SRWR does not predict the missing (indirect) trust relationships when producing the personalized ranking lists. Instead, our proposed MRW-TD method infers the missing trust relationships, while considering users' direct trust and distrust relationships. Clearly, the competitors do not capture well the missing relationships, which negatively affects their ranking performance. In our approach, it is the combination of the two steps of (i) social inference of missing relationships and (ii) node ranking in the regularization framework of Eq. (11) that makes MRW-TD significantly outperform the baseline signed ranking techniques, by balancing well the structural properties of the enhanced graph with the personalized preferences of users' trust and distrust relationships in the graph.

5   Conclusions

We presented an accurate personalized ranking method in signed graphs with trust and distrust relationships. As users' social relationships are sparse, in the first step of our approach we infer missing trust relationships, while considering users' explicit trust and distrust relationships during the inference. In addition, we introduce a ranking regularization framework to balance users' personalized preferences of trust and distrust with the structural properties of the signed graph. Our experiments show that the proposed approach wins all the competitors by correctly inferring the missing relationships, and taking into account the graph's structural properties and user's social preferences.

Recently, collaborative ranking has gained much attention for generating personalized recommendations with trust relationships [7,8]. However, the distrust relationships are not considered in these studies. As future work we plan to extend our approach for designing a collaborative ranking model with both trust and distrust relationships, while considering evolving users' social relationships. This is a challenging task for online social networks, as users' preferences evolve over time as well [9–12].

Acknowledgments. Dimitrios Rafailidis was supported by the COMPLEXYS and INFORTECH Research Institutes of University of Mons.

References

1. Guha, R.V., Kumar, R., Raghavan, P., Tomkins, A.: Propagation of trust and distrust. In: Proceedings of the 13th ACM International Conference on World Wide Web, New York, NY, USA, pp. 403–412 (2004)
2. Jamali, M., Ester, M.: Trustwalker: a random walk model for combining trust-based and item-based recommendation. In: Proceedings of the 15th ACM SIGKDD International Conference on Knowledge Discovery and Data Mining, Paris, France, pp. 397–406 (2009)
3. Jang, M., Faloutsos, C., Kim, S., Kang, U., Ha, J.: PIN-TRUST: fast trust propagation exploiting positive, implicit, and negative information. In: Proceedings of the 25th ACM International on Conference on Information and Knowledge Management, Indianapolis, IN, USA, pp. 629–638 (2016)
4. Jung, J., Jin, W., Sael, L., Kang, U.: Personalized ranking in signed networks using signed random walk with restart. In: Proceedings of 16th IEEE International Conference on Data Mining, Barcelona, Spain, pp. 973–978 (2016)
5. Kunegis, J., Lommatzsch, A., Bauckhage, C.: The slashdot zoo: mining a social network with negative edges. In: Proceedings of the 18th International Conference on World Wide Web, Madrid, Spain, pp. 741–750 (2009)
6. Page, L., Brin, S., Motwani, R., Winograd, T.: The PageRank citation ranking: Bringing order to the web. Technical report, Stanford Digital Libraries SIDL-WP-1999-0120 (1999)
7. Rafailidis, D., Crestani, F.: Collaborative ranking with social relationships for top-n recommendations. In: Proceedings of the 39th International ACM SIGIR Conference on Research and Development in Information Retrieval, Pisa, Italy, pp. 785–788 (2016)
8. Rafailidis, D., Crestani, F.: Joint collaborative ranking with social relationships in top-n recommendation. In: Proceedings of the 25th ACM International on Conference on Information and Knowledge Management, Indianapolis, IN, USA, pp. 1393–1402 (2016)
9. Rafailidis, D., Kefalas, P., Manolopoulos, Y.: Preference dynamics with multimodal user-item interactions in social media recommendation. Expert Syst. Appl. **74**, 11–18 (2017)

10. Rafailidis, D., Nanopoulos, A.: Modeling the dynamics of user preferences in coupled tensor factorization. In: Proceedings of the 8th ACM Conference on Recommender Systems, Foster City, Silicon Valley, CA, USA, pp. 321–324 (2014)
11. Rafailidis, D., Nanopoulos, A.: Repeat consumption recommendation based on users preference dynamics and side information. In: Proceedings of the 24th International Conference on World Wide Web Companion, Florence, Italy, pp. 99–100 (2015)
12. Rafailidis, D., Nanopoulos, A.: Modeling users preference dynamics and side information in recommender systems. IEEE Trans. Syst. Man Cybern.: Syst. **46**(6), 782–792 (2016)
13. Shahriari, M., Jalili, M.: Ranking nodes in signed social networks. Social Netw. Analys. Min. **4**(1), 172 (2014)
14. Tang, J., Chang, Y., Aggarwal, C., Liu, H.: A survey of signed network mining in social media. ACM Comput. Surv. **49**(3), 42:1–42:37 (2016)
15. Tang, J., Hu, X., Chang, Y., Liu, H.: Predictability of distrust with interaction data. In: Proceedings of the 23rd ACM International Conference on Conference on Information and Knowledge Management, Shanghai, China, pp. 181–190 (2014)
16. Victor, P., Cornelis, C., Cock, M.D., Teredesai, A.: Trust- and distrust-based recommendations for controversial reviews. IEEE Intell. Syst. **26**(1), 48–55 (2011)
17. Wu, Z., Aggarwal, C.C., Sun, J.: The troll-trust model for ranking in signed networks. In: Proceedings of the 9th ACM International Conference on Web Search and Data Mining, San Francisco, CA, USA, pp. 447–456 (2016)
18. Zhang, J., Zhan, Q., He, L., Aggarwal, C.C., Yu, P.S.: Trust hole identification in signed networks. In: Proceedings of the European Conference on Machine Learning and Knowledge Discovery in Databases, Riva del Garda, Italy, pp. 697–713 (2016)
19. Zhou, D., Schlkopf, B.: A regularization framework for learning from graph data. In: ICML Workshop on Statistical Relational Learning and Its Connections to Other Fields, pp. 132–137 (2004)

Plagiarism Detection Based on Citing Sentences

Sidik Soleman[✉] and Atsushi Fujii

Tokyo Institute of Technology, Tokyo, Japan
soleman.s.aa@m.titech.ac.jp, fujii@cs.titech.ac.jp

Abstract. Plagiarism, which is one of the forms of academic misconducts, is problematic. It results in discouraging innovation, and losing trust in the academic community. We modeled the plagiarism for academic publications, by means of the similarity between textual contents, and citation relations. Furthermore, we adopted the model in our proposed method for plagiarism detection. We evaluate our method using two types of dataset, namely auto-simulated and manually judged dataset. Our experiment shows that our method outperforms the baseline, which only uses the similarity between textual contents, on the auto-simulated dataset and the manually judged one for the ACL sub-dataset.

Keywords: Plagiarism detection · Information retrieval · Citation analysis

1 Introduction

Digital archives for academic publications have enabled us to efficiently access a large volume of scientific information. However, its misuse and misconduct have of late become a crucial problem. Plagiarism is "the act of using another person's words or ideas without giving credit to that person"[1], which results in discouraging innovation and losing trust in the scientific research community. To alleviate this problem, a number of methods for detecting plagiarisms specifically for academic publications have been proposed.

In a broad sense, plagiarism detection (PD) is a task to identify whether a document in question is produced by means of plagiarism, and is often requested to present one or more source documents as evidences for the plagiarism. However, in this paper we consider only cases where an input document is a plagiarized one and focus only on identifying one or more source documents for the input document.

As with an adversarial information processing like filtering spam e-mails, a person who conducts plagiarism, or a plagiarist for short, usually intends to hide the plagiarism, for example, by means of editing and summarizing source documents. As a result, PD is a cat-and-mouse game between plagiarists and people who develop PD systems.

---

[1] https://www.merriam-webster.com/dictionary/plagiarism.

© Springer International Publishing AG 2017
J. Kamps et al. (Eds.): TPDL 2017, LNCS 10450, pp. 485–497, 2017.
DOI: 10.1007/978-3-319-67008-9_38

Whereas the above scenario is associated with intentional plagiarism, detecting unintentional plagiarism is also important to avoid innocent mistakes. Fang et al. [1] investigated approximately 2 000 papers that were once indexed by PubMed[2] but retracted later and found that 9.8% of them were retracted due to being judged as a plagiarized paper. Irrespective whether those papers are associated with intentional or unintentional plagiarism, effective methods for plagiarism detection will have a significant impact on our society.

One of the crucial steps in PD is to measure the similarity between two documents. In the field of citation analysis, it is well-known that the number of same citations between two documents can be a good indicator whether they are related/similar or not, i.e. bibliographic coupling [2]. The more same citations two documents have, the more related they are. In this paper, we proposed a model for plagiarism that combines the similarity between textual contents and citation relations. More precisely, our model combines the similarity between textual contents in citing and non-citing sentences. We further applied this model to our PD system, which identifies source documents.

In this paper, our contribution is twofold. First, we modeled plagiarism by means of the similarity between textual contents and citation relations, and applied this model to PD system. Second, we evaluated the effectiveness of our PD system.

2   Related Work

Generally, the existing PD systems that focus on identifying source documents can be classified into two categories as shown in Fig. 1. These categories are search engine-based and direct comparison-based PD system.

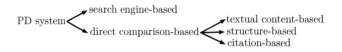

Fig. 1. The categories of PD systems

The search engine-based PD system, which was introduced in PAN workshop[3], utilizes a search engine to identify source documents, because plagiarists are likely to use a search engine to find source documents when plagiarizing document in the Web [3]. The PD systems generate a number of queries from input document, and submit to a search engine for retrieving source documents. Therefore, the system should produce queries that represent the source documents in order to be able to retrieve them. However, the performances of the

---

[2] https://www.ncbi.nlm.nih.gov/pubmed.
[3] a competition for plagiarism detection.

systems are often limited due to the capabilities of the search engine, e.g. query length, and document-query weighting scheme.

Unlike the previous category, the direct comparison-based PD systems compare input and target documents[4] directly, one by one. In this category, the PD systems can be divided into three types based on the aspects that the systems use for comparing documents, namely textual content, structure, and citation-based.

In the textual content-based type, the PD systems compare textual contents of input and target documents whether they have significant similarity. The systems use various textual comparison strategies, e.g. word n-gram [4].

In the structure-based type, the PD systems take the structure of document into consideration when comparing input and target documents since there are some parts of the document that may be less important than the others. For instance, Alzahrani et al. [5] used section-based component to represent the structure of document, such as *introduction, method,* and *conclusion section* as the components. They put different weight for each component, thus the important components have heavier weight than the less important ones, e.g. *method section* has heavier weight than *introduction section* has. They used these weights to re-weight terms in input and target documents when comparing them.

In the citation-based type, the PD systems consider citation relations when comparing input and target documents. There are two kind of citation considerations in the existing PD systems. First, the existence of citations is the sign of innocent case, such as in the system developed by Alzahrani et al. [5]. Thus, input document is not a plagiarized one and target documents that are similar to the input one are not source documents, as long as the input one cites them.

The second consideration is that the existence of citation relations are used to measure the similarity between input and target documents, which is motivated by the phenomena in citation, i.e. bibliographic coupling [2]. HaCohen-Kerner et al. [6] compared reference lists between input and target documents whether they have a significant degree of reference overlap. However, their system resulted in producing many false positives. It means that the innocent documents are labeled as plagiarized ones. One possible explanation is that these innocent ones cite the same documents with others, but their contents may be different.

Different from HaCohen-Kerner et al. [6], Gipp et al. [7] used the pattern of citation anchors[5] in input and target documents. They generated a number of chunks of citation anchors from the input and the target documents to compare whether the documents have a significant degree of chunk overlap or not. Since their system is likely to fail when there is no citation anchor, recently this work was extended by Pertile et al. [8], where they combined the similarity of textual content in document level, the similarity of reference list, and the pattern of citation anchors.

---

[4] Target documents are a set of documents in a collection where source documents exist.

[5] Citation anchors refer to characters in citing sentences that point to documents in reference list.

In summary, the recent works in PD consider citation relations to measure the similarity between input and target documents. However, these works may fail when there is no citation relations, or produce false positives. Thus, comparing citation relations alone is not sufficient.

To alleviate this problem, we proposed a model for plagiarism that combines the similarity between textual contents and citation relations, and adopted this model to our PD system. More precisely, we combined the similarity between textual contents in citing and non-citing sentences. Hence, a document is likely to be a plagiarized document, when it has a significant amount of citing and/or non-citing sentences that are similar to the other documents.

3   Proposed Approach

3.1   Model for Plagiarism

As mentioned previously, we model plagiarism by means of the similarities between textual contents in citing and non-citing sentences. Thus, given input (X) and target document (Y), their similarity score is calculated as follows:

$$Score(X,Y) = \alpha \ Sim(Cite(X), Cite(Y) + \\ (1 - \alpha) \ Sim(NCite(X), NCite(Y)) \qquad (1)$$

with

- $Cite$: a function that returns citing sentences from a document.
- $NCite$: a function that returns non-citing sentences from a document.
- α: a weighting parameter with value [0,1]. Thus, by tuning this value, we are able to prioritize between the similarity of citing and non-citing sentences.
- Sim: a function that measures the similarity of textual content.

Next, given d_1 and d_2 as vectorized text fragments generated by using bag-of-word method (i.e. word as the dimension of the vector), we define Sim, which calculates the similarity of textual content, by the following equation:

$$Sim(d_1, d_2) = \frac{d_1 \cdot d_2}{\|d_1\| \, \|d_2\|} \qquad (2)$$

In order to transform a text fragment to its vector representation, we calculate a weight for each word in the text fragment based on the frequency of that word in the text fragment, and inverted document frequency of that word in a document collection, by the following equation:

$$w_t = f_t \ log \frac{N}{n_t} \qquad (3)$$

with

- f_t: total number of word t that appears in the text fragment.
- N: total number of documents in document collection.
- n_t: total number of documents in document collection that contain word t.

Unlike the PD systems that only consider citation anchors/reference lists, our model is still able to perform PD when citation relations are not available since the model compares non-citing sentences. In addition, when citation relations are available, our model considers them by means of the similarity between the textual contents in citing sentences. Therefore, our model is different from the textual content-based PD system, which does not consider the citation relations.

Regarding our task of PD that identifies source documents, the similarity score in our model is used to rank the target documents. Thus, the source documents ideally should be located at the top of the target document list.

3.2 PD System

Here, we describe our PD system, given an input document and a set of target documents in a collection. The system outputs a ranked document list, which in ideal situation, the source documents should be located at the top of the document list. Our PD system consists of three components as described in Fig. 2, namely sentence classification, preprocessing, and document comparison.

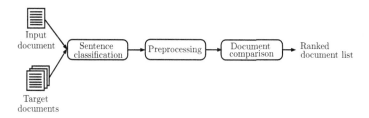

Fig. 2. The components in PD system

Sentence Classification. Since our model combines the similarity between the textual content in citing and non-citing sentences, all sentences in a document should be classified into two classes, i.e. citing and non-citing sentence. This component performs the classification based on the condition whether a sentence contains citation anchor or not. Thus, a sentence containing citation anchor is classified as citing sentence, otherwise it is non-citing sentence.

We employed regular expression to recognize citation anchors for the following formats:

- Combination of author name and publication year, e.g.: *(name, 2010), (name, 2010; name, 2010a), name (2010), [name, 2010b],* and *[name, 2010; name, 2010b].*

– Combination of author name, publication year, and page/paragraph number, e.g.: *(name, 2010, p.1)*, *(name, 2010, para.1)*, and *(name, 2010, p.i)*.
– Citation anchor is a sequence of characters that refers to a document in reference list, e.g.: *[1]*, *[LIZ2]*, and *(1)*.
– Combination of author name, publication year, and a document identification in reference list, e.g.: *[name, 2010 (1)]*

Preprocessing. This component performs some modifications to a text fragment, which is its input. First, the text fragment is lowercased, and any numerical character is removed. The next step is to remove any word that is considered as stopwords[6], and lastly words are stemmed using stemmer[7] for English language.

Document Comparison. This component measures similarity between input and target documents by applying our model for plagiarism, which is described in Sect. 3.1. An input document is compared with target documents one by one, and the target documents are sorted in descending order according to their similarity scores. The list of ordered target document is the output of this component, which is also the output of our system (i.e. the ranked document list).

4   Experiment

4.1   Dataset

To evaluate our system, we need dataset that is suitable for our PD task. Since identifying source documents is our goal, the dataset should consist of input documents and document collection containing their source documents. As we have mentioned earlier, we only use plagiarized documents as the input of our system. Additionally, because our model of plagiarism combines the textual similarity between content in citing and non-citing sentences, the dataset should contain citation relations. In this experiment, we used two types of dataset, namely auto-simulated and manually judged dataset.

The auto-simulated dataset was produced by Alzahrani et al. [5] by constructing plagiarized documents automatically since it is difficult to obtain verified plagiarized documents. In this dataset, they controlled the length and the obfuscation level of the plagiarized text fragment. They performed obfuscation by using several text modification techniques, such as verbatim copy-paste, word shuffling, synonym replacement, back-translation, and auto-summarization.

To construct the plagiarized documents, Alzahrani et al. [5] used document collection from Directory of Open Access[8]. First, they divided documents in the collection into two groups, namely plagiarized and target group. Second, they inserted text fragments from any document in the target group to any

---

[6] http://snowball.tartarus.org/algorithms/english/stop.txt.

[7] https://opennlp.apache.org/.

[8] http://doaj.org.

document in the plagiarized group after the text fragment is obfuscated. Thus, documents in the plagiarized group that are inserted with the text fragments are the plagiarized documents, and the ones where these text fragments come from are the source documents.

The manually judged dataset was created by Pertile et al. [8] by identifying documents that are suspected as the result of plagiarism in two document collections, namely ACL anthology[9] and PubMed[10]. First, they compared all documents in each document collection by using some similarity methods. Second, they pooled pairs of document from the top 30 ranked pairs for each similarity method. Lastly, they asked 10 annotators to judge these pairs of document by using the definition of plagiarism from ACM[11] and IEEE[12]. Thus, the identified pair of document is the pair of input and source document. The complete information about these datasets is described in Table 1.

Table 1. The statistics of the datasets

Type	Manually judged		Auto-simulated
	ACL	PubMed	
Topic	Computation linguistics	Biomedical and life science	Science and technology
Target document	4 685	1 440	8 657
Input document	40	60	3 950
Avg. word (target)	2 557.7	2 868.8	4 417
Avg. word (input)	2 797	3 732	5 263
Source/input document	1.025	1.05	2.5
Kappa	.675	.524	—
Agreement rate	84%	80%	—

4.2   Evaluation Method

Since our system outputted a ranked list of document, we used Mean Average Precision (MAP), which measured the ranking quality of a document list for evaluation method. In addition, we evaluated our system by measuring recall (R), precision (P), and F1. We calculated those methods by the following equations:

$$MAP(n) = \frac{1}{|D|} \sum_{d=1}^{|D|} \frac{1}{|src_d|} \sum_{i=1}^{n} P(L_{d,i}) \qquad (4)$$

$$P(L_{d,i}) = \frac{|\{s \in src_d \cap L_{d,i}\}|}{i} \qquad (5)$$

$$R(L_{d,i}) = \frac{|\{s \in src_d \cap L_{d,i}\}|}{|src_d|} \qquad (6)$$

$$F1(L_{d,i}) = \frac{2 \times P(L_{d,i}) \times R(L_{d,i})}{P(L_{d,i}) + R(L_{d,i})} \qquad (7)$$

with

- n: cut-off value for ranked document list.
- $L_{d,i}$: top i documents of ranked document list for input document d.
- src_d: set of source documents for input document d.
- D: set of input documents.

MAP, R, P, and F1 produce score between [0,1]. When MAP score is 1, it means that all source documents for a given input document are located at the top of the ranked document list, consecutively. Thus, the higher the MAP score, the better the system performs.

When R score is equal to 1, it means that all source documents are contained in the ranked document list. While P score is 1, it suggests that all documents in the ranked document list are source document. Both R and P are combined as F1, thus the higher the F1 score, the better the system performs.

4.3   Experiment Result

Baseline. In our experiment, we compared our system with baseline that measured the similarity between the textual content of input and target documents without distinguishing citing and non-citing sentences. Thus, the baseline belongs to the category of textual content-based PD system. The processes in this baseline are similar to our system, except it does not perform sentence classification, and it uses Eq. 2 to compute the similarity scores.

Citation-Based PD. We also compared our system with the citation-based PD methods in Pertile et al. [8]. Generally, the methods compared list of reference and citation anchors between two documents, i.e. the number of similar references (BC), the overlap of references divided by its union (JR), the co-occurrence of citation anchors (CC), and the summation of a weighted similar reference, thus a reference cited by fewer documents has heavier weight (CF).

Result and Discussion. First, we present and discuss the results when our system only use the similarity of textual content in citing sentences (CS) or in non-citing ones (NS). Thus, we can identify which factor is the best and should be prioritized in this experiment. In addition, we compare our system with the methods in Pertile et al. [8]. Second, we show and discuss the results of our system with the best weight (α) and its improvement, which is achieved. Lastly, we discuss the errors that happen in this experiment.

Table 2 presents the results when we only use the similarity of textual content in citing sentences (CS) or non-citing ones (NS) on the auto-simulated dataset. While the results of similar experiment on the manually judged dataset are shown in Tables 3 and 4. In Table 2, CS outperforms the baseline and NS at every cut-off based on their MAP scores. On average, the MAP scores of CS and NS are about .067 and .005 higher than the baseline, respectively. Moreover, we conducted 2-tailed paired t-test among them using their MAP scores (cut-off = 100), we found that their differences are significant at level 1%. Thus, these results suggest that CS should be prioritized on this dataset.

Table 2. The performance of baseline, NS, and CS on the auto-simulated dataset

Cut-off	Baseline				NS				CS			
	MAP	F1	R	P	MAP	F1	R	P	MAP	F1	R	P
10	.308	.100	.361	.058	.313	.107	.375	.063	**.379**	**.136**	**.435**	**.080**
30	.314	.052	.442	.028	.320	.054	.452	.029	**.384**	**.061**	**.488**	**.032**
100	.318	.024	.574	.012	.324	**.025**	**.582**	**.013**	**.386**	.023	.553	.012

Based on F1, R, and P scores in Table 2, CS outperforms the baseline and NS at cut-off 10 and 30, while NS outperforms CS and the baseline at cut-off 100. These results indicate that CS may be unable to identify some source documents that have a significant amount of similar non-citing sentences but not citing ones with input documents, and they are identified by NS. Thus, combining CS and NS is likely better, since both of them may complement each other.

In Table 3, which shows the MAP scores on the manually judged dataset, the performance of the baseline is pretty good. Since the ratio of input and source documents is approx. 1 (Table 1), and the R scores in Table 4 are close or equal to 1, the MAP score about .9 means the majority of the source documents are located at the first position in the ranked document lists for each input document. We found 7 out of 100 input documents (1 in the PubMed and 6 in ACL sub-dataset), which their source documents are not located at the first position in their ranked document lists. Consequently, to outperform the baseline on this dataset may be difficult. This happened probably due to the limitation of this dataset since Pertile et al. [8] only focused on pairs of document that have large amount of similar textual contents to be annotated. Thus, the source documents are mostly located at the top of the ranked document lists in our baseline.

The above reason may also explain why the performance of our system is different in both datasets. Since Alzahrani et al. [5] controlled the length of plagiarized text fragments in the auto-simulated dataset from the short to the

Table 3. The MAP scores of baseline, NS, CS, and methods in Pertile et al. [8] on the manually judged dataset

Sub-dataset	Cut-off	Baseline	NS	CS	CF	BC	JR	CC
PubMed	10 or more	**.993**	.978	.978	.900	.900	.900	.430
ACL	10	.908	.908	**.963**	—	—	—	—
	30	.908	.910	**.964**	—	—	—	—
	100	.910	.911	**.964**	.800	.810	.810	.620

long one, thus it is more difficult to identify source documents in this dataset. Moreover, they controlled the obfuscation level of the plagiarized text fragments from the light one to the heavy one, unlike the manually judged dataset.

On the PubMed sub-dataset, the MAP scores of CS and NS are about .015 lower than the baseline in Table 3, although their F1, R, and P scores in Table 4 are the same. While on the ACL sub-dataset, on average, the MAP scores of CS are .054 higher than the baseline and NS as shown in Table 3. In addition, CS outperforms the baseline and NS on this sub-dataset at cut-off 10 and 30 according to their F1, R, and P scores in Table 4. These results suggest that CS may be better if it is given heavier weight in the manually judged dataset.

In Table 3, CS also outperforms all the methods in Pertile et al. [8] based on their MAP scores on the manually judged dataset. These results indicate that comparing list of reference and citation anchors alone is not sufficient for PD.

Tables 5 and 6 present the experiment results of our system with the best weight (α) on the auto-simulated and manually judged dataset, respectively. As shown in Table 5, our system with $\alpha = .9$ achieves the best MAP scores at every cut-off on the auto-simulated dataset, which is about .07 higher than the baseline on average. We also conducted 2-tailed paired t-test between our system ($\alpha = .9$) and the baseline by using their MAP scores (cut-off = 100), we found that their difference is significant at level 1%.

Table 4. The F1, P, and R of the baseline, NS, and CS on the manually judged dataset

Sub-dataset	Cut-off	Baseline			NS			CS		
		F1	R	P	F1	R	P	F1	R	P
PubMed	10 or more	.220	1.000	.123	.220	1.000	.123	.220	1.000	.123
ACL	10	.181	.950	.100	.181	.950	.100	**.190**	**.975**	**.105**
	30	.064	.950	.033	.068	.988	.035	**.069**	**1.000**	**.036**
	100	.021	1.000	.011	.021	1.000	.011	.021	1.000	.011

Table 5. The best performance of our system on the auto-simulated dataset

Cut-off	Baseline				$\alpha = .5$				$\alpha = .9$			
	MAP	F1	R	P	MAP	F1	R	P	MAP	F1	R	P
10	.308	.100	.361	.058	.360	.135	.435	.080	**.383**	**.137**	**.440**	**.081**
30	.314	.052	.442	.028	.368	**.065**	**.518**	**.035**	**.388**	.062	.495	.033
100	.318	.024	.574	.012	.372	**.028**	**.633**	**.014**	**.390**	.024	.567	.012

In addition, our system ($\alpha = .9$) achieves the best F1, R, and P on the auto-simulated dataset in Table 5 at cut-off 10. While at cut-off 30 and 100, our system with $\alpha = .5$ achieves the best F, R, and P on this dataset, which is .007 higher than our system with $\alpha = .9$ based on their F1 scores.

In Table 6, our system does not outperform the baseline on the PubMed sub-dataset, although its best MAP score ($\alpha = .9$) is about .007 lower than the baseline and their F1, R, and P scores are the same. While on the ACL sub-dataset, we find that our system ($\alpha = .9$) outperforms the baseline about .067 higher on average based on their MAP scores, and also achieves the best F1, R, and P scores at cut-off 10 and 30.

Based on the MAP scores, our system achieves its best performance when $\alpha = .9$ on both datasets. Thus, it confirms our previous finding that it is better to give heavier weight on the similarity between textual content in citing sentences in this experiment. The results also suggest that the similarity between textual content in citing and non-citing sentences complement each other, since the MAP scores become worse when only using one of them. Moreover, we conducted 2-tailed paired t-test among the MAP scores (cut-off $= 100$) on the auto-simulated dataset when our system uses $\alpha = 0$ (NS), $\alpha = .9$, and $\alpha = 1$ (CS). We found that their differences are significant at level 1%.

Table 6. The best performance of our system on the manually judged dataset

Sub-dataset	Cut-off	Baseline				$\alpha = .9$			
		MAP	F1	R	P	MAP	F1	R	P
PubMed	10 or more	**.993**	.220	1.000	.123	.986	.220	1.000	.123
ACL	10	.908	.181	.950	.100	**.975**	**.190**	**.975**	**.105**
	30	.908	.064	.950	.033	**.976**	**.069**	**1.000**	**.036**
	100	.910	.021	1.000	.011	**.976**	.021	1.000	.011

Error Analysis. In this experiment, we conducted error analysis on the manually judged dataset using the best results of our system ($\alpha = .9$). We found 4 input documents (3 in the PubMed and 1 in the ACL sub-dataset), which their MAP scores are not as high as the others. We suspected that there are three reasons why these errors happened.

First, some of the target documents that are not annotated as source document have similarity scores higher than the source documents in our system. Moreover, we observed a significant similarity of textual contents between them and the input document. This happened probably because Pertile et al. [8] used cut-off (i.e. the top 30) instead of similarity threshold on their method to pool document pairs from the document collections to be annotated. Since their method might rank these target documents lower than the cut-off, these target documents are ignored and not annotated. We found two input documents from the PubMed sub-dataset associated with this error. Thus, determining similarity threshold to decide whether an input and a target document are a plagiarized and source document, respectively is another crucial issue in PD.

Second, the target documents mentioned above have similar topic with the input document. For instance, they discuss the same research problem, use the same learning algorithm, and evaluate their methods using the same dataset. However, their proposed methods are different. We also observed that they and the input document cite some documents together, and also use similar terminologies and descriptions. We identified one input document from each sub-dataset associated with this error.

Lastly, we suspect that our similarity method (see Eq. 2) may be sensitive to the length of text fragments when one of them is longer. We identified one input document from the ACL sub-dataset may associate with this error.

5   Conclusion

Plagiarism, which is one of the forms of academic misconducts, is problematic. It results in discouraging innovation and losing trust in the scientific research community. We proposed a method for plagiarism detection (PD) based on our model of plagiarism, which combines the similarity between textual contents in citing and non-citing sentences.

Given a plagiarized document as the input, our system identifies its source documents. We evaluated our system using two types of dataset, namely auto-simulated and manually judged dataset. In the evaluation, we compares our system with the baseline, which measures the similarity of textual contents between two documents without distinguishing citing and non-citing sentences.

According to the experiment results, our system does not outperform the baseline for the PubMed sub-dataset on the manually judged dataset, although the difference of their MAP (Mean Average Precision) scores is about .007. However, our system outperforms the baseline on the auto-simulated and the manually judge dataset for the ACL sub-dataset about .07 and .067 higher (MAP) than the baseline, respectively.

As for future work, we may extract more features from citation relations, and integrates them with the current system. Additionally, to reduce the vector sparsity of text fragments when measuring their similarities, we may use an algorithm to learn their vector representations.

References

1. Fang, F.C., Steen, R.G., Casadevall, A.: Misconduct accounts for the majority of retracted scientific publications. Proc. Nat. Acad. Sci. **109**(42), 17028–17033 (2012). doi:10.1073/pnas.1212247109. NAS
2. Kessler, M.M.: Bibliographic coupling between scientific papers. Am. Documentation **14**(1), 10–25 (1963). doi:10.1002/asi.5090140103. Wiley
3. Potthast, M., Gollub, T., Hagen, M., Graßegger, J., Kiesel, J., Michel, M., Oberländer, A., Tippmann, M., Barrón-Cedeño, A., Gupta, P., Rosso, P., Stein, B.: Overview of the 4th international competition on plagiarism detection. In: Forner, P., Karlgren, J., Womser-Hacker, C. (eds.) Working Notes Papers of the CLEF 2012 Evaluation Labs (2012)
4. Gupta, P., Rosso, P.: Text reuse with ACL: (upward) trends. In: Proceedings of the ACL-2012 Special Workshop on Rediscovering 50 Years of Discoveries, pp. 76–82. ACL (2012)
5. Alzahrani, S., Palade, V., Salim, N., Abraham, A.: Using structural information and citation evidence to detect significant plagiarism cases in scientific publications. J. Am. Soc. Inf. Sci. **63**(2), 286–312 (2012). doi:10.1002/asi.21651. Wiley
6. HaCohen-Kerner, Y., Tayeb, A., Ben-Dror, N.: Detection of simple plagiarism in computer science papers. In: Proceedings of the 23rd International Conference on Computational Linguistics, pp. 421–429. ACL (2010)
7. Gipp, B., Meuschke, N.: Citation pattern matching algorithms for citation-based plagiarism detection: greedy citation tiling, citation chunking and longest common citation sequence. In: Proceedings of the 11th ACM Symposium on Document Engineering, pp. 249–258. ACM (2011). doi:10.1145/2034691.2034741
8. Pertile, S.D.L., Moreira, V.P., Rosso, P.: Comparing and combining content-and citation-based approaches for plagiarism detection. J. Assn. Inf. Sci. Tec. **67**(10), 2511–2526 (2016). doi:10.1002/asi.23593. Wiley

Lexicon Induction for Interpretable Text Classification

Jérémie Clos$^{(\boxtimes)}$ and Nirmalie Wiratunga

Robert Gordon University, Garthdee Road, Aberdeen, UK
{j.clos,n.wiratunga}@rgu.ac.uk

Abstract. The automated classification of text documents is an active research challenge in document-oriented information systems, helping users browse massive amounts of data, detecting likely authors of unsigned work, or analyzing large corpora along predefined dimensions of interest such as sentiment or emotion. Existing approaches to text classification tend toward building black-box algorithms, offering accurate classification at the price of not understanding the rationale behind each algorithmic prediction. Lexicon-based classifiers offer an alternative to black-box classifiers by modeling the classification problem with a trivially interpretable classifier. However, current techniques for lexicon-based document classification limit themselves to using either hand-crafted lexicons, which suffer from human bias and are difficult to extend, or automatically generated lexicons, which are induced using point-estimates of some predefined probabilistic measure in the corpus of interest. This paper proposes LEXICNET, an alternative way of generating high accuracy classification lexicons offering an optimal generalization power without sacrificing model interpretability. We evaluate our approach on two tasks: stance detection and sentiment classification. We find that our lexicon outperforms baseline lexicon induction approaches as well as several standard text classifiers.

Keywords: Text classification · Lexicon induction · Sentiment analysis · Stance classification

1 Introduction

Text classification is a core task in natural language processing, with applications ranging from web search to author detection. For example, support vector machines [11], a common and extremely powerful classification algorithm [10] have helped improve document navigation tasks by categorizing web search results [4], analyzed corpora to identify anonymous authors [7], and are used to identify spam e-mails [8] at large scale. However, supervised classification algorithms suffer from not providing predictions that can be explained. Understanding the reason behind a classification allows us to establish trust in further predictions, which can have far-reaching consequences in algorithms deployed in production systems such as search engines and document categorization pipelines.

© Springer International Publishing AG 2017
J. Kamps et al. (Eds.): TPDL 2017, LNCS 10450, pp. 498–510, 2017.
DOI: 10.1007/978-3-319-67008-9_39

Lexicons attend to this need by offering a white-box approach to text mining. They do so by using an additive model, where the probability of an instance belonging to a class is a weighted sum of the probabilities of each term belonging to that class.

A lexicon prediction can thus be interpreted by observing the terms that are contained in the instance and the terms which have contributed the most to the prediction, and it is possible for a human agent to modify the model in order to correct a mistake without restarting the learning process entirely. Figure 1 illustrates the explanation step with an example.

However, current techniques used to build those lexicons are lacking in many respects compared to standard supervised text classifiers. This paper attempts to conciliate lexicon-based classification and traditional classification models by defining a simple and effective training procedure that can generate lexicons with a classification accuracy that is competitive with modern classification algorithms. Firstly, they use point estimates of text statistics (such as raw co-occurrence or mutual information) in order to build a lexicon that is susceptible to overfitting. Secondly, they perform significantly worse than black-box models.

Example 1. In a binary sentiment classification setting, for a given sentence "*I love horror books*", a lexicon $\mathcal{L}$ referred on the figure, the lexicon could find an aggregated score of $f(love) \times 1.0 + f(horror) \times 0.3 + f(books) \times 0.5 = 1.8$ for the *Positive* class, and $f(love) \times 0 + f(horror) \times 0.7 + f(books) \times 0.5 = 1.2$ for the *Negative* class, where f is a function measuring some notion of local term frequency. The decision function $\mathcal{D}$ would then return the class with the maximum value, i. e., *Positive*. A human reader can read the sentence and identify that the term "*love*" is responsible for tipping the classification towards the *Positive* class.

Example Lexicon		
Term	*Positive*	*Negative*
love	1.0	0.0
horror	0.3	0.7
books	0.5	0.5

Fig. 1. Classification and explanation with a sentiment lexicon

We first formalize the concept of lexicons and explore the state of the art in the domain of lexicon-based classification. We then detail our contribution, formalizing lexicon-based classification as a form of computational graph and inducing optimal weights using a regularized objective function. We then detail our evaluation protocol on two classification tasks: stance detection and sentiment classification. We perform an evaluation against standard lexicons and baselines found in the literature and report that our approach significantly outperforms standard text classification techniques. Finally, we analyze and discuss our results, before exploring the next steps of our work.

2 Related Works

Despite its widespread use in real-world applications, text classification heavily relies on black-box models offering little if any explanation on their predictions [21]. Lexicon-based classifiers overcome this limitation by constraining the classification to a simple model: each term/class pair is linked to a score, and a new instance gets assigned a score for each class corresponding to a sum of those scores weighted by the frequency of the corresponding term.

Those scores get weighted according to the frequency of that term in the instance and then added together, and finally the class with the highest total score for a given instance is chosen as the prediction. Such a classification model offers the flexibility of transparency: each prediction can be explained trivially by analyzing the terms that were present in the text, and any domain expert could revise the model manually with a simple text editing software. This transparency however comes at the cost of some classification accuracy, due to the simplistic nature of its inference scheme.

2.1   Lexicon-Based Classification

Lexicons are linguistic tools for the automated analysis of text. Their most notorious uses are classification and feature extraction [2,5]. They can take many forms, the most common of which is a simple list of terms associated to a certain class of interest. Classification is done by counting the number of terms belonging to each list in a given unlabeled instance, and returning the class associated to the list with the most occurrences. Optionally, the terms can be weighted according to their strength of association with a given class. Some lexicons also contain additional contextual information in order to help their users build more complex models [17], but they all share the same architecture:

Definition 1 (formal classification lexicon). *A classification lexicon Lex is a tuple* $Lex = \langle \mathcal{L}, \mathcal{A}, \mathcal{D} \rangle$ *where:*

$$\mathcal{L} : T \times C \mapsto \mathbb{R}$$
$$\mathcal{A} : \mathbb{R}^n \mapsto \mathbb{R}$$
$$\mathcal{D} : \mathbb{R}^n \mapsto \mathbb{R}$$

For a given dictionary of terms T and set of classes of interest C, $\mathcal{L}$ is a mapping function that assigns an unbounded value to each pair (t, c) where term $t \in T$ and class $c \in C$. The function $\mathcal{A}$ is an aggregation function that accumulates scores and returns one value, and $\mathcal{D}$ is a decision function that selects and returns a single one of these aggregated values. Concretely, the mapping determines an evidence score for each term using a look-up list (the lexicon), propagates it to the aggregation function which aggregates the evidence into one cumulative score per class. Finally, the decision function evaluates each score to select the one that is the most likely. Figure 1 provides an example of the classification process.

We therefore define a core challenge in lexicon-based classification: the lexicon induction problem. The next section reviews techniques traditionally used to solve the lexicon induction problem.

Definition 2 (lexicon induction problem). *The **lexicon induction problem** is the estimation, given aggregation function $\mathcal{A}$ and decision function $\mathcal{D}$, of the optimal function $\mathcal{L}$ so that the resulting lexicon $Lex = \langle \mathcal{L}, \mathcal{A}, \mathcal{D} \rangle$ minimizes its classification errors on unseen data.*

2.2 Lexicon Induction Techniques

Research in lexicon induction outlines multiple families of techniques that can be used to produce a computational lexicon. Those techniques are either built on an extensive lexical resource such as an ontology, or on an estimation of strength of association between each term and a class in a reference corpus. Research has shown that merging multiple lexicons produces a reliable feature extractor to augment an existing classifier [27], but using those lexicons for direct classification was not explored.

Traditional hand-crafted lexicons (THCL). Due to the computational cost of building a lexicon from text, early lexicons were hand-crafted by domain experts [24] and while higher performance in automated classification tasks has been shown using modern techniques, there still exist handcrafted lexicons in use to this day such as the Linguistic Inquiry and Word Count lexicon [20]. The strengths of these approaches are that they generalize well and are highly interpretable due to their human (and not algorithmic) origin. Conversely their weakness are that they tend to be small due to the human labor involved in generating them, and less effective than other methods due to their focus on human interpretability. However they can provide a commonsense knowledge back-up in hybrid lexicons [16] with some degree of success.

Ontology-based lexicons (OBL). OBL learning techniques use a few human-provided seed words for which the class is known, and leverage some external relationship (typically synonymy, antonymy and hypernymy) in a semantic graph such as WordNet [15] to propagate class values along that graph [9]. Because this family of techniques is extremely foreign to the one we are proposing, we do not evaluate against it and only refer to it for the sake of exhaustiveness.

Corpus statistic-based lexicons (CSBL). CSBL learning techniques use a labeled corpus of interest in order to learn a domain-specific lexicon. The two main statistics used for this purpose are the conditional probability (Eq. 1) of observing a term given a class, and the pointwise mutual information (PMI, Eq. 2) between the observation of a term and the observation of a class. These approaches are flawed in that they can overemphasize spurious correlations between terms and classes. For example, if a non-class specific term such as "Monday" accidentally co-occurs too often within one class, it will be misconstrued as being indicative of that class, and the lexicon will overfit. Bandhakavi et al. [1] describe a method for building conditional probability-based lexicons and Turney [25] an approach using PMI and an external search engine to compute lexicon scores.

Other works [6] have shown some improvement using the normalized PMI measure (NPMI, Eq. 3) on a stance classification task.

$$P(t;c) = \frac{p(t|c)}{\sum_{i=0}^{|C|} p(t|c_i)} \tag{1}$$

$$PMI(t;c) = \frac{\log(p(t;c))}{p(t)p(c)} \tag{2} \qquad NPMI(t;c) = \frac{\frac{\log(p(t;c))}{p(t)p(c)}}{-\log\left[p(t;c)\right]} \tag{3}$$

3   Lexicon Induction by Backpropagation

CSBL learning techniques traditionally use point estimates of some statistical values on a corpus. Assuming $F^{n \times 1}$ is a $n \times 1$ matrix containing the frequencies of each of the n terms in instance x and $W^{c \times n}$ is a $c \times n$ matrix indexed by a class c and containing an association score for each term-class pair, we define the classification step of a lexicon in the following way:

$$Prediction(x) = ArgMax_c\,(W_c \cdot F) \tag{4}$$

We can observe that a standard lexicon is a computational graph, i.e., a composition of functions, as shown in Fig. 2 illustrating the network topology of a binary classification lexicon. This allows us to use gradient-based learning techniques such as backpropagation in order to solve the lexicon induction problem. The details of the network topology and the training protocol are explained in the following sections.

3.1   The Lexicon Network Topology

The lexicon network topology corresponds to a shallow network with linear units (the lexicon layer), where one regressor is trained per class and the output of each regressor (the aggregation layer in Fig. 2) is fed into a SoftMax normalization layer (the decision layer in Fig. 2) so as to produce a probability distribution as a final output, which is necessary to backpropagate the error gradient to find the optimal lexicon weights. In this section we review each layer of the neural lexicon and their function.

The vocabulary input layer. The input layer feeds term frequencies into the network. The output of this layer is a $n \times 1$ matrix F (see Eq. 4) where n is the number of terms in the lexicon. The inputs can be logarithmically scaled to smooth out the differences in input length using the ScaledFrequency function where RawFrequency corresponds to the number of times a term has appeared in the current input. More complex scaling functions are typically applied in text classification.

$$ScaledFrequency(t) = Log(1 + RawFrequency(t)) \tag{5}$$

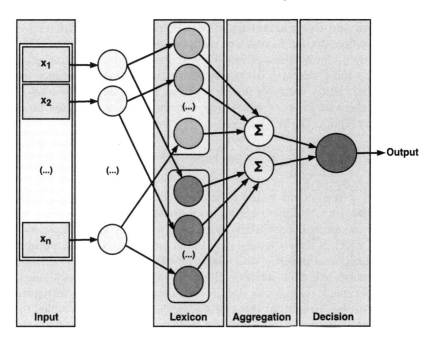

Fig. 2. The LEXICNET network topology

The lexicon layer. The lexicon layer maps a term to its respective class-dependent scores. This layer is represented by a $c \times n$ matrix W (see Eq. 4) where n is the number of terms in the lexicon and c the number of classes.

The aggregation layer. The aggregation layer adds up evidence towards a class from a list of units in the previous layer by performing an inner product between $n \times 1$ matrix F and $c \times n$ matrix W. The output of this layer is a $1 \times c$ row vector O containing the aggregated scores for each of the classes.

The decision layer. The decision layer transforms the row vector O into a probability distribution using the SoftMax function [3] and returns it as the output of the network. At testing time, the decision layer returns the $ArgMax$ of the probability distribution in the output in order to compute the accuracy of the current model. At training time, it returns only the probability distribution since the $ArgMax$ function is not differentiable, which is a required feature for the backpropagation algorithm.

3.2   Lexicon Network Training

We train the LEXICNET network using an Elastic Net-regularized average cross-entropy error function. In this section we detail our training procedure and cost function.

Cost Function and Regularization. The backpropagation algorithm relies on reverse-mode differentiation in order to train the network in a computationally efficient way, by updating the weights of the units based on the error gradient with respect to those weights. We transform the label of each instance into a probability distribution vector of length n where $Y_i = 1$ for the relevant class and $Y_i = 0$ otherwise in order to use the average cross-entropy cost function (function E detailed in Eq. 6).

$$E(Y, \hat{Y}) = -\frac{1}{m} \cdot \sum_{i=1}^{m} \sum_{j=1}^{n} \left(Y_{i,j} \times log(\hat{Y}_{i,j}) + (1 - Y_{i,j}) \times log(1 - \hat{Y}_{i,j}) \right) \quad (6)$$

Given a set of predictions $\hat{Y}$ and their corresponding ground truth labels Y, the average cross-entropy function iterates over each pair $(y \in Y; \hat{y} \in \hat{Y})$ where both y and $\hat{y}$ are probability distributions over n classes and computes their cross-entropy, which is then averaged over the m instances. However, optimizing over a direct function of the error with a large amount of free parameters (number of classes $\times$ number of lexicon entries) will lead to overfitting on the training data and poor performance on the test data, which indicates the need to regularize our training process. To counter that effect a regularization term is added to the optimization process using Elastic Net regularization [30] which has been shown to work on neural networks architectures [13].

The cost function J resulting is shown in Eq. 7.

$$J(\theta, X, Y) = E(Y, h(X)) + \lambda * \left(\alpha * \sum_{j=0}^{m} |w_j| + (1 - \alpha) * \sqrt{\sum_{j=0}^{m} w_j^2} \right) \quad (7)$$

Here we can observe the presence of two different regularization parameters: λ corresponds to a regularization weight, which modulates the importance that we are putting on obtaining a generalizable lexicon against having a low error in the training set and is selected empirically, α corresponds to the elastic weight which weights the importance put on minimizing respectively the L_1-norm or the L_2-norm of the lexicon weights, w_i corresponds to the weight of unit i in the lexicon layer.

Optimization. We train our network using Backpropagation and the full batch Gradient Descent algorithm with Nesterov momentum [18] as described in Eq. 8, where μ is the velocity scaling parameter and γ is the learning rate. Nesterov momentum works by updating each weight in two steps: firstly using a scaled version of their previous update (conditioned by a fixed velocity parameter), followed by a course correction step using the error gradient after the first update, mimicking the effect of momentum in physical objects.

$$w_i \leftarrow w_i - \mu \times u_i$$
$$u_i \leftarrow \frac{\partial J}{\partial w_i} \quad (8)$$
$$w_i \leftarrow w_i - \gamma \times u_i$$

4   Experiments

We evaluated our approach using a 10-fold cross-validation [23] on multiple tasks. We used standard TF-IDF term weighting [22] for the supervised classifiers because of it is shown to be competitive with more complex weighting methods for text classification where there is little to no term filtering [29]. We measured the performance of our algorithm against standard supervised classifier baselines and lexicon induction techniques using the accuracy performance metric (percentage of test instances correctly classified). The rest of this section details the datasets and baseline algorithms used in our experiments.

4.1   Datasets

We performed our evaluation on two tasks, containing a total of 4 datasets for stance classification and sentiment analysis. Class statistics for each dataset can be found in Table 1. We describe the tasks and datasets associated in the rest of this section:

– **Stance detection** is the study of local stance of a document with respect to a topic or another stance. For example, if the topic of discussion is "death penalty" and a document d_1 is for the death penalty, then a document d_2 that is against the death penalty is said to be in disagreement with document d_1, while a document d_3 that is also for the death penalty is said to be in agreement with document d_1. Stance classification is the classification of unseen documents with respect to a topic or an existing stance. In this work, we consider a reduced version of the stance classification task with an unobserved topic, which means that the classifiers do not have any contextual information.
  - **The IAC dataset** is a subset of the Internet Argument Corpus [26] containing forum comments crawled from 4FORUMS[1] on different topics: e. g., politics, ... and labeled on a scale from −5 to 5. A subset of comments that ensured disjoint class membership (with an average score far from 0) and containing more than 3 words was binned into 2 classes (agreement and disagreement) and used for our experiments.
  - **The CD dataset** is a dataset collected from the CREATEDEBATE forum[2] dedicated to social argumentation on political and religious topics and labeled using 2 classes (agreement and disagreement).
– **Sentiment classification** is the study of the sentiment (positive or negative) contained within a piece of text. While many datasets propose finer-grained sentiment classes (positive, negative, and neutral or numerical gradation of sentiment) we chose to use a binary sentiment classification task as the object of our study.

---

[1] http://www.4forums.com.
[2] http://www.createdebate.com.

Table 1. Class statistics of datasets IAC, CD, AYI and AMZ.

	Dataset			
	Stance		Sentiment	
	IAC	CD	AYI	AMZ
Number of instances	3,910	4,902	3,000	8,000
Frequency of the *agreement/positive* class	1,955	2,912	1,500	4,000
Frequency of the *disagreement/negative* class	1,955	1,989	1,500	4,000
Minimum instance length (in words)	10	10	43	10
Maximum instance length (in words)	3245	6,685	12,220	79
Average instance length (in words)	68.41	74.10	588.84	13.55

- **The AYI dataset** was collected from Amazon[3], Yelp[4] and IMDB[5] and was built from individual sentences from product, location and movie reviews (respectively), labeled with a binary positive/negative judgment.
- **The AMZ dataset** was collected from Amazon user reviews and labeled with a binary positive/negative judgment. The dataset is provided with preprocessed unigrams and bigrams. Only the unigrams are used for our experiments, so as to be more similar to the other tasks and datasets.

4.2   Baselines

We used two families of baselines as comparison points with our approach:

Lexicons: Two lexicons used as a baseline are the CPBLEX and the PMILEX, which are standard methods for building lexicons for other purposes, such as sentiment lexicons [12]. Section 2.2 on corpus-based lexicons details their implementation. These algorithms were provided tf-idf normalized raw frequencies, which has been shown to be a competitive term weighting scheme for text classification [14];

Standard classifiers: SVM (with a RBF kernel), which has been shown to perform well in stance detection tasks by Yin et al. [28] and is a regular top performer in general classification tasks [10], and NAIVEBAYES and DECISIONTREE which are two popular baselines for text classification. Parameters for the classifiers were taken from the default recommendations of the SCIKIT-LEARN[6] [19] library. Tf-idf was also used here due to the disparity in document length shown in Table 1.

---

[3] http://www.amazon.com.
[4] http://www.yelp.com.
[5] http://www.imdb.com.
[6] http://scikit-learn.org.

5   Results and Discussion

Table 2 shows that LEXICNET outperforms all the baselines by a statistically significant margin (on a one-tailed paired T-test, with $p < 0.05$) except for the AYI dataset where decision trees are the top performing algorithm and the AMZ dataset where the performance improvement comes close to statistical significance but does not reach it. The parameter values for LEXICNET were determined empirically on a hold-out dataset, leading to the choice of a lexicon size of 400 words (the 400 most frequent words in the corpus) and a regularization coefficient (λ) of 0.5, an elastic net weight (α) of 0.4 and a momentum velocity coefficient of 0.7.

Table 2. Experimental results

	Method	Accuracy			
		Stance		Sentiment	
		IAC	CD	AYI	AMZ
Baseline lexicons	CPBLEX	0.524	0.441	0.528	0.519
	PMILEX	0.557	0.529	0.571	0.554
Baseline classifiers	NAIVEBAYES	0.536	0.474	0.665	0.637
	SVM	0.589	0.594	0.689	0.671
	DECISIONTREE	0.582	0.573	**0.742**	0.707
Approach	**LexicNet**	**0.647**	**0.642**	0.729	**0.718**

We note from manual examination of the results on stance classification tasks that a large proportion of the classification errors made by LEXICNET was made on documents where stance was heavily implied and there is no term used that is heavily indicative of the class, e.g., *"Thus, an important point is raised"* (agreement). While algorithms like SVM manage to draw on their complexity to target higher-order (and sometimes even coincidental) relationships between sets of words and classes, LEXICNET lacks the expressiveness to do so. On sentiment analysis tasks, we note that a likely explanation of the poor performance on the AYI dataset is due to the multiple sources of data: Amazon, Yelp and IMDB. Such diversity would tend to cause a large standard deviation of input length, something that a lexicon-based classifier is more sensitive to than a decision tree. We conduct another experiment in order to confirm this hypothesis, and display the results in Table 3.

We can see from Table 3 that when separating the AYI dataset in the three domains a performance improvement can be observed in LEXICNET. However the improvement does not completely bridge the gap with DECISIONTREE, meaning that other factors at play influence the algorithm and will be studied in further work.

Table 3. Experimental results after separating domains in the AYI dataset

	Method	Accuracy		
		Amazon	Yelp	IMDB
Baseline lexicons	CPBLex	0.545	0.511	0.562
	PMILex	0.557	0.593	0.612
Baseline classifiers	NaiveBayes	0.691	0.704	0.728
	SVM	0.677	0.692	0.686
	DecisionTree	**0.755**	**0.704**	0.740
Approach	**LexicNet**	0.738	0.691	**0.742**

6 Conclusion and Future Work

In this work we showed the viability of using elastic net-regularized backpropagation to learn effective lexicon weights, producing a lexicon that is competitive with standard classifiers and significantly outperforms baseline lexicon learning techniques in several datasets.

However we identified two major flaws in the LexicNet which will be subject to future work: (1) its lack of expressiveness to capture higher-order relationships between sets of terms and classes, as well as (2) its deficiencies when dealing with inputs of varying length, due to the way it models the classification process as a weighted sum of term-class association strength, which means that a term that only appears in very long instances will be assigned a lower weight than one appearing only in short texts.

Our future work will focus on improving the expressiveness of LexicNet to counter those issues while keeping it in the form of a human-readable lexicon. Our objective is to do so by incorporating lexical context such as modifier words (e.g., adverbs) and topical context (words with multiple meanings with different class values) in the network topology, thus improving accuracy without hurting interpretability.

References

1. Bandhakavi, A., Wiratunga, N., Deepak, P., Massie, S.: Generating a word-emotion lexicon from# emotional tweets. In: Proceedings of the 3rd Joint Conference on Lexical and Computational Semantics (*SEM 2014) (2014)
2. Bandhakavi, A., Wiratunga, N., Deepak, P., Massie, S.: Lexicon based feature extraction for emotion text classification. Pattern Recogn. Lett. **93**, 133–143 (2016)
3. Bishop, C.M.: Pattern recognition. Mach. Learn. **128**, 1–58 (2006)
4. Chen, H., Dumais, S.: Bringing order to the web: automatically categorizing search results. In: Proceedings of the SIGCHI Conference on Human Factors in Computing Systems, pp. 145–152. ACM (2000)
5. Clos, J., Bandhakavi, A., Wiratunga, N., Cabanac, G.: Predicting emotional reaction in social networks. In: Jose, J.M., Hauff, C., Altıngovde, I.S., Song, D., Albakour, D., Watt, S., Tait, J. (eds.) ECIR 2017. LNCS, vol. 10193, pp. 527–533. Springer, Cham (2017). doi:10.1007/978-3-319-56608-5_44

6. Clos, J., Wiratunga, N., Massie, S., Cabanac, G.: Shallow techniques for argument mining. In: ECA': Proceedings of the ECA, vol. 63, p. 2 (2016)
7. Diederich, J., Kindermann, J., Leopold, E., Paass, G.: Authorship attribution with support vector machines. Appl. Intell. **19**(1), 109–123 (2003)
8. Drucker, H., Wu, D., Vapnik, V.N.: Support vector machines for spam categorization. IEEE Trans. Neural Netw. **10**(5), 1048–1054 (1999)
9. Esuli, A., Sebastiani, F.: Sentiwordnet: a publicly available lexical resource for opinion mining. In: Proceedings of LREC, vol. 6, pp. 417–422. Citeseer (2006)
10. Fernández-Delgado, M., Cernadas, E., Barro, S., Amorim, D.: Do we need hundreds of classifiers to solve real world classification problems. J. Mach. Learn. Res. **15**(1), 3133–3181 (2014)
11. Hearst, M.A., Dumais, S.T., Osuna, E., Platt, J., Scholkopf, B.: Support vector machines. IEEE Intell. Syst. Appl. **13**(4), 18–28 (1998)
12. Jurafsky, D., Martin, J.H.: Lexicons for sentiment extraction, chap. 18. Pearson (2016)
13. Kang, G., Li, J., Tao, D.: Shakeout: a new regularized deep neural network training scheme. In: Proceedings of the Thirtieth AAAI Conference on Artificial Intelligence, pp. 1751–1757. AAAI Press (2016)
14. Lan, M., Tan, C.L., Low, H.B., Sung, S.Y.: A comprehensive comparative study on term weighting schemes for text categorization with support vector machines. In: Special Interest Tracks and Posters of the 14th International Conference on World Wide Web, pp. 1032–1033. ACM (2005)
15. Miller, G.A.: Wordnet: a lexical database for english. Commun. ACM **38**(11), 39–41 (1995)
16. Muhammad, A., Wiratunga, N., Lothian, R.: A hybrid sentiment lexicon for social media mining. In: IEEE 26th International Conference on Tools with AI (ICTAI), pp. 461–468 (2014)
17. Muhammad, A., Wiratunga, N., Lothian, R.: Contextual sentiment analysis for social media genres. Knowl.-Based Syst. **108**, 92–101 (2016)
18. Nesterov, Y.: Gradient methods for minimizing composite objective function (2007)
19. Pedregosa, F., Varoquaux, G., Gramfort, A., Michel, V., Thirion, B., Grisel, O., Blondel, M., Prettenhofer, P., Weiss, R., Dubourg, V., et al.: Scikit-learn: machine learning in python. J. Mach. Learn. Res. **12**, 2825–2830 (2011)
20. Pennebaker, J.W., Francis, M.E., Booth, R.J.: Linguistic inquiry and word count: LIWC 2001. Mahway: Lawrence Erlbaum Associates 71, 2001 (2001)
21. Ribeiro, M.T., Singh, S., Guestrin, C.: Why should i trust you?: explaining the predictions of any classifier. In: Proceedings of the 22nd ACM SIGKDD International Conference on Knowledge Discovery and Data Mining, pp. 1135–1144. ACM (2016)
22. Salton, G., Buckley, C.: Term-weighting approaches in automatic text retrieval. Inf. Process. Manage. **24**(5), 513–523 (1988)
23. Sammut, C., Webb, G.I. (eds.): Cross-Validation, pp. 249–249. Springer, Boston (2010)
24. Stone, P.J., Dunphy, D.C., Smith, M.S.: The General Inquirer: A Computer Approach to Content Analysis. MIT press, Cambridge (1966)
25. Turney, P.D.: Thumbs up or thumbs down?: semantic orientation applied to unsupervised classification of reviews. In: Proceedings of the 40th Annual Meeting on ACL. ACL (2002)
26. Walker, M.A., Tree, J.E.F., Anand, P., Abbott, R., King, J.: A corpus for research on deliberation and debate. In: LREC, pp. 812–817 (2012)

27. Wang, L., Cardie, C.: Improving agreement and disagreement identification in online discussions with a socially-tuned sentiment lexicon. In: ACL 2014, vol. 97 (2014)
28. Yin, J., Thomas, P., Narang, N., Paris, C.: Unifying local and global agreement and disagreement classification in online debates. In: Proceedings of the 3rd Workshop in Computational Approaches to Subjectivity and Sentiment Analysis, pp. 61–69. ACL (2012)
29. Zhang, W., Yoshida, T., Tang, X.: A comparative study of TF* IDF, lSI and multi-words for text classification. Expert Syst. Appl. **38**(3), 2758–2765 (2011)
30. Zou, H., Hastie, T.: Regularization and variable selection via the elastic net. J. Roy. Stat. Soc.: Ser. B (Stat. Methodol.) **67**(2), 301–320 (2005)

The Clustering-Based Initialization for Non-negative Matrix Factorization in the Feature Transformation of the High-Dimensional Text Categorization System: A Viewpoint of Term Vectors

Le Nguyen Hoai Nam[(✉)] and Ho Bao Quoc

VNUHCM - The University of Science, Ho Chi Minh City, Vietnam
{lnhnam, hbquoc}@fit.hcmus.edu.vn

Abstract. Due to the non-negativity of the matrix factors, Non-negative Matrix Factorization (NMF) is favorable for transforming a high-dimensional original Terms-Documents matrix into a lower-dimensional semantic Concepts-Documents matrix in the text categorization. With the iterative nature of all NMF algorithms, the NMF matrix factors need initializing. In this paper, we propose a clustering-based method for initializing the NMF according to the term vectors instead of the document vectors as the previous researches.

Keywords: Feature transformation · Matrix factorization · Text categorization

1 Introduction

Document categorization is a process that automatically classifies a given collection of documents into the predefined categories [21]. In this task, each document is converted to a Bag-of-Words (BoW) vector, so a corpus can be presented by a Terms-Documents matrix [19]. However, in a large corpus, the number of terms highly increases. This leads to the risk of irrelevant features, sparse vectors, and over-fitting producing the negative effects on the categorization [21]. In addition, the phenomenon of synonymy and polysemy has become more common when the corpus is larger.

Feature Transformation (FT) [14] is one of the main techniques to solve the problems in the high-dimensional text categorization. In an FT, all of the original terms join together to build new features, also known as semantic concepts. It is mainly concerned with a Low-Rank Approximation (LRA) to replace a high-dimensional original Terms-Documents matrix with a lower-dimensional semantic Concepts-Documents matrix. The document presentation under a new axis system being the semantic concepts helps deal with the issue of synonymy and polysemy [6]. A commonly used LRA is Singular Value Decomposition (SVD) [8]. However, the SVD cannot guarantee the non-negativity of the output matrices, although the Terms-Documents input matrix is inherently non-negative. After performing the SVD, the negative elements make the corpus presentation hard to be interpreted [9, 23]. Non-negative Matrix Factorization (NMF) [12] addresses this shortcoming by a non-negativity constraint on the matrix

© Springer International Publishing AG 2017
J. Kamps et al. (Eds.): TPDL 2017, LNCS 10450, pp. 511–522, 2017.
DOI: 10.1007/978-3-319-67008-9_40

approximation. Due to the iterative nature of all NMF algorithms, the NMF matrix factors need initializing. The previous researches [4, 24, 26] interpret the NMF matrix factors under the viewpoint of the column (document) vectors. Then, they are initialized according to this viewpoint by clustering the document vectors, i.e. the column vectors of the Terms-Documents matrix. However, an arising question is how the clustering on the document vectors initializes itself. In this case, it is very difficult to efficiently and quickly select which documents in a large corpus to be the first centroids for clustering.

To avoid this difficulty, we change viewing the NMF matrix factors from the column (document) vectors to the row (term) vectors. Under the row (term) viewpoint, we analyze the interpretation of the NMF matrix factors which is very different from that under the column (document) viewpoint. Therefore, we customize the idea of a clustering-based method to initialize the NMF matrix factors according to the row (term) vectors. The effectiveness of our method motivates further studies on the text FT based on the NMF not only at the NMF initialization stage but also the other NMF stages by interpreting the NMF matrix factors looking at their row (term) vectors.

2   Related Work

2.1   Non-negative Matrix Factorization (NMF)

NMF [12] is used to approximate a non-negative Terms-Documents matrix, called $X:R^+_{m \times n}$, to the product of two non-negative matrices, called $W : R^+_{m \times r}$ and $H : R^+_{r \times n}$, with rank $r < <min\{m, n\}$. For NMF, every original document, i.e. a column vector of the Terms-Documents matrix $X(X_{col_i} i = 1 \ldots n)$, can be reconstructed as follows:

$$X \approx W.H \Leftrightarrow X_{col_i} \approx W.H_{col_i} \Leftrightarrow X_{col_i} \approx \sum_{j=1}^{r} (H_{ji}.W_{col_j}) \tag{1}$$

Specifically, the ith original document vector (X_{col_i}) can be reconstructed by a linear combination of all the column vectors of W ($W_{col_j} j = 1 \ldots r$) with the coefficients contained in the ith column vector of H (H_{col_i}). Due to the non-negativity of W and H, the column vectors of W can be considered as the document basis vectors, and every column vector of H shows the real (only additive) coordinates of the corresponding original document vector with respect to the new axes being these document basis vectors. Therefore, under the viewpoint of the column vectors, W and H are addressed as a *document basis matrix* and a *document coordinate matrix*. This characteristic of the NMF is called the "parts-based presentation" because it shows an additive combination of the non-negative parts for constructing data [12, 23]. It makes sense to the analysis on real-world data in general and textual data in particular [23].

2.2   Text Feature Transformation Based on the NMF

A Feature Transformation (FT) [14] aims at not only reducing the feature space dimension but also creating more meaningful features by combining all the original features. It is to project the original document vectors onto a low-dimensional semantic subspace.

With W $(m \times r)$ and H $(r \times n)$ obtained from the NMF on the Terms-Documents matrix X $m \times n)$, if the semantic subspace is spanned by the basis vectors being the column vectors of W, the ith original document $(X_{col_i} : m \times 1)$ is now presented by the ith column vector of H $(H_{col_i} : r \times 1)$, i.e. the projection of X_{col_i} onto the semantic subspace. In other words, an FT based on the NMF transforms the high-dimensional original Terms-Documents matrix X $(m \times n)$ into the lower-dimensional semantic Concepts-Documents matrix H $(r \times n)$ under the semantic basis W $(m \times r)$.

Another typical FT is Latent Semantic Indexing (LSI) [7]. The LSI is based on a Singular Value Decomposition (SVD) [8] on the corpus matrix and then considers eigenvectors as the basis vectors of the semantic space for the transformation. However, the SVD cannot guarantee the non-negativity of its eigenvectors. Therefore, compared with the NMF, the SVD factors are less meaningful in text domain [23].

2.3   NMF Initialization

In NMF algorithms, W and H are iteratively updated to decrease an approximation error of X to $(W.H)$ [5, 12]. The most natural way for constructing this error function is to use Euclidean distance. Table 1 shows the general structure of a NMF algorithm.

Table 1. The general structure of a NMF algorithm.

Input: $X: R^+_{m \times n}$; NMF rank $r \ll min\{m, n\}$. **Output**: $W: R^+_{m \times r}$ and $H: R^+_{r \times n}$. **Initialize** W and H $(W^{(0)}$ and $H^{(0)})$. **While** (Not satisfying **the convergence criterion**):   **Update** W and H.

Due to the iterative characteristic, the NMF approximation error has a tendency to converge on a local minimum instead of a global minimum as expected [12]. It directly depends on the initial values of W and H, called $W^{(0)}$ and $H^{(0)}$. A good initialization leads the NMF to a faster convergence and better error at convergence [2, 4]. As presented above, all the column vectors of W $(W_{col_j}$ $j = 1...r)$ play the role of the document basis vectors. They can easily be associated with the cluster centroids obtained from clustering the original document vectors. Therefore, a *clustering-based NMF initialization* [4, 24, 26] clusters the document original vectors $(X_{col_i}$ $i = 1...n)$, and then utilizes the cluster centroids as the column vectors of $W^{(0)}$ $(W^{(0)}_{col_j}$ $j = 1...r)$. Based on Eq. (1), the initial document coordinate matrix $H^{(0)}$ showing the relation between each original document $(X_{col_i} i = 1...n)$ and each cluster centroid (initial document basis) $W^{(0)}$ $(W^{(0)}_{col_j} j = 1...r)$. Specifically, $H^{(0)}_{ji}$ is 1 or 0 indicating whether X_{col_i} belongs to the cluster $W^{(0)}_{col_j}$ or not. Furthermore, an NMF can use the output factors of other matrix factorizations for its initialization. NNDSVD [2] implements a *factorization-based NMF initialization* by two SVDs.

3   Motivation

For the clustering-based NMF initialization, the commonly used clustering algorithms are the K-means (KM) and Fuzzy C-Means (FCM). However, one of the biggest challenges these clustering algorithms face is to determine a starting centroid for each cluster. In the KM-Clustering-based NMF initialization [24] and FCM-Clustering-based NMF initialization [26], the used clustering algorithm is started with cluster centroids selected at random among the original documents. However, the randomization makes the NMF non-deterministic. Recently, [4] implements a clustering-based NMF initialization with the Subtractive Clustering (SC), i.e. a clustering algorithm of no cluster centroid initialization. For SC, every obtained cluster centroid is just one of the original documents. If an original document is used as an initial document basis, the distance between the initial document basis and the true document basis is too far.

To improve a clustering-based NMF initialization, the clustering algorithm should begin to run with the initial cluster centroids being important documents instead of random documents. In this case, it is necessary to select important documents from the corpus simply and fast. However, this task is neither highly efficient nor inexpensive, especially with a large corpus. We realize that if X, W, and H are only interpreted under the viewpoint of the column (document) vectors, and W and H are then initialized according to this viewpoint, it is too difficult to define better starting centroids of a clustering algorithm when it used to initialize the NMF. Therefore, we introduce a new interpretation of the NMF by looking at the row (term) vectors as follows:

$$X \approx W.H \Leftrightarrow X_{row_i} \approx W_{row_i}.H \Leftrightarrow X_{row_i} \approx \sum_{j=1}^{r} \left(W_{ij}.H_{row_j} \right) \qquad (2)$$

Concretely, the ith original term vector, i.e. the ith row vector of X (X_{row_i}), is constructed by a linear combination of the row vectors of H ($H_{row_j} j = 1...r$) with the weights contained in the ith row vector of W (W_{row_i}). Therefore, under the viewpoint of the row vectors, H and W are called a *term basis matrix* and a *term coordinate matrix*. Thanks to this interpretation, we propose a clustering-based method for initializing the NMF according to the row (term) vectors. The term viewpoint enables our method to utilize the researches on the term description in the text. By these ways, we overcome the challenge of a clustering-based NMF initialization when defining the starting cluster centroids of the used clustering algorithm. The NMF becomes deterministic and more effective. Sect. 4 presents our clustering-based NMF initialization.

4   A Term Clustering-Based NMF Initialization

4.1   Term Basis Matrix Initialization

Under the term (row) interpretation, a clustering-based NMF initialization becomes clustering the original term vectors ($X_{row_i} i = 1...m$), and the cluster centroids are then used as the row vectors of the initial term basis matrix $H^{(0)}$. To determine the good starting cluster centroids, it is necessary to select important term vectors from the original term vectors. In the text classification domain, the notion of important terms implicitly indicates the terms which make a big contribution to the classification. Based on the labeled training set, a wide variety of methods called supervised feature selection (FS) methods [25] is proposed for picking up the most important subset of features (terms) for the purpose of the classification.

Obviously, it is easier to address how a clustering algorithm initializes itself when it is used for initializing the NMF on a Terms-Documents matrix if we change viewing the factors from the column (document) vectors (*Document Clustering-based NMF Initialization* [4, 24, 26]) to the row (term) vector *(Term Clustering-based NMF Initialization)*. The Term Clustering-based NMF Initialization is as follows:

- A supervised FS is used to select important terms. They become the first centroids for clustering the term vectors, i.e. the row vectors of Terms-Documents matrix X.
- Thanks to a clustering algorithm, the selected important terms turn into the true cluster centroids. The cluster centroids are pushed into the row vectors of the initial term basis matrix $H^{(0)}$ ($H^{(0)}_{row_i}$, $i = 1 \ldots r$).

The supervised FS is known as an offline and relatively low-cost process [10]. It is a major tasks right after the document presentation to completely eliminate noise features without altering the information of the important features. After the FS, an FT combines the remaining important features with each other to form the new and more important features. An FS prior to an FT is to avoid the negative impacts of the noise features on the new features created by the FT as well as to decrease the computational cost of the FT. Therefore, reusing the pre-existing FS results for initializing the NMF does not impose any extra burdens on the NMF computation. For the FS, we aim at our effective supervised term selection named *DtFCFS-BRatTL*. For DtFCFS [18], a term gets a higher score if it makes both the documents in every category become closer and the categories become more separated. Based on the term scores, the BRatTL [17] selects a final term set covering all categories as well as possible.

4.2   Term Coordinate Matrix Initialization

Based on Eq. (2), $W^{(0)}_{ij}$ of the initial term coordinate matrix shows the association between the ith original term (X_{row_i}) and the jth initial term basis (cluster centroid: $H^{(0)}_{row_j}$). For a clustering-based NMF initialization, it is 1 or 0 indicating whether the term X_{row_i} belongs to the cluster $H^{(0)}_{row_j}$ or not. However, this does not show the nature of an FT in resolving the polysemy issue which allows a term to be related to many clusters. To create a better model, we compute $W^{(0)}_{ij}$ ($i = 1 \ldots m; j = 1 \ldots r$) based on the *Pointwise Mutual Information* (PMI) [22] between the original term X_{row_i} and initial term basis $H^{(0)}_{row_j}$. The PMI is one of the effective methods mainly used for measuring the semantic association between two terms as follows:

$$p_{ij} = \frac{f_{ij}}{\sum_{k=1}^{m}\sum_{p=1}^{r}f_{kp}}; p_{i*} = \frac{\sum_{p=1}^{r}f_{ip}}{\sum_{k=1}^{m}\sum_{p=1}^{r}f_{kp}}; p_{*j} = \frac{\sum_{k=1}^{m}f_{kj}}{\sum_{k=1}^{m}\sum_{p=1}^{r}f_{kp}}; W^{(0)(pmi)}_{ij} = \log_2\frac{p_{ij}}{p_{i*}\cdot p_{*j}} \quad (3)$$

where f_{kp} is co-occurrence value of X_{row_k} and $H^{(0)}_{row_p}$. However, PMI highly values rare terms [22]. [13] indicates that PMI works more effectively when raising p_{*j} to the power of α which is set to 0.75 for the most significant performance as follows:

$$p\_\alpha_{*j} = \frac{\left(\sum_{k=1}^{m}f_{kj}\right)^{\alpha}}{\sum_{p=1}^{r}\left(\sum_{k=1}^{m}f_{kp}\right)^{\alpha}}; W^{(0)(pmi\_\alpha)}_{ij} = \log_2\frac{p_{ij}}{p_{i*}\cdot p\_\alpha_{*j}} \quad (4)$$

To be compatible with the non-negativity of the NMF, we change from the PMI to Positive PMI (PPMI) [3] $\left(W^{(0)(pmi\_\alpha)}_{ij} = \max\left(0, W^{(0)(pmi\_\alpha)}_{ij}\right)\right)$. [3] points out that the PPMI is better than the PMI and many other methods.

5   Experiment

5.1   Experimental Setup

The experiments are carried out on the 10 top-sized categories of the Newsgroup of the "bydate" split [1], the Reuters of the ModApte split, and the Ohsumed of the Joachims split [11]. Our aim is to investigate the NMF initializations in the study:

- The Document Clustering-based NMF Initialization is implemented with K-Means (KM) (Doc-KM-Cluster-NMFInit [24]); with Fuzzy C-Means (FCM) (Doc-FCM-Cluster-NMFInit [26]); with the Subtractive (SC) (Doc-SC-Cluster-NMFInit [4]).
- The Term Clustering-based NMF Initialization is considered with the KM and the FCM Clustering (Term-KM-Cluster-NMFInit; Term-FCM-Cluster-NMFInit).
- The NNDSVD [2], a well-known factorization-based NMF initialization.
  Figure 1 shows the details in the experimental setup as follows:
- The training set is pre-processed by removing the stop words and word stemming. It is then presented by a Terms-Documents ($m \times n$) with TF-IDF weighting [21]. A supervised FS by the DtFCFS-BRatTL is applied on the training Terms-Documents matrix to remove noise terms as well as decrease the computation cost of the FT.
- After the FS, the L. best terms are selected. The reduced training Terms-Documents matrix ($L \times n$), called X, is taken into the NMF. The NMF is computed by the Multiplicative Update (MU) [12] or Alternate Least Square (ALS) [5] with a NMF rank r. Remember that for a term clustering-based NMF initialization, the FS results are again used. The column vectors of the output H ($r \times n$) are used for building a model using an SVM by SMO [20]. Every new document is converted to a TF-IDF vector t. ($L \times 1$) only based on the terms selected in the FS. Under the semantic

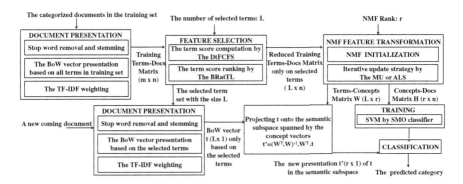

Fig. 1. The experimental setup of the NMF feature transformation for text classification.

space spanned by column vectors of the output W $(L \times r)$, its new presentation is computed by the projection t' $(r \times 1)$ where $t' = (W^T . W)^{-1} . W^T . t$. [24].

Determining the parameter L in an FS and the parameter r in an FT is a big challenge. A common trend in the FS is that the quality of the selected feature (term) set gradually moves towards a saturation point when its size increases. At this point, the rest of features have lower quality, and selecting more features (terms) does not bring high effect. In many our FS researches [15–18] on these experimental datasets, with about the 2000 best selected terms, the classification performance nearly reaches up to the peak. That is the reason why L is set to 2000. Regarding the NMF rank r, it is also the number of clusters in a clustering-based NMF initialization. With the 2000 selected terms for the FT, the maximal number of clusters (rank r) is set up to 600.

5.2   Experimental Result and Discussion

During the iterative process, an NMF aims at minimizing the approximation error. In Fig. 2A, we present the examples about the approximation errors of two NMF algorithms (MU and ALS) under the different initializations when incrementally altering the number of iterations. Firstly, we emphasize the approximation errors at the small iterations, which are heavily affected by the NMF initializations. Noticeably, at small iterations, when the K-Means Clustering (KM) is used, the approximation errors of the NMFs using the term clustering-based initialization (Term-KM-Cluster-NMFInit) are better than those using the document clustering-based initialization (Doc-KM-Cluster-NMFInit). For Fuzzy C-Means Clustering (FCM), this phenomenon is still the same. In comparison with Doc-SC-Cluster-NMFInit and NNDSVD, the NMFs using Term-KM-Cluster-NMFInit obtain more impressive errors in both MU and ALS.

At the increasing iterations, each NMF moves toward its own stability of approximation error, called the convergence point. Similar to the previous researches, at convergence, the approximation errors of NMFs by ALS using the different initializations are nearly equal. That is because an NMF by ALS is less dependent on the initialization. It only needs to initialize W, and H. is computed at the first iteration. In this case, the effectiveness of a NMF initialization is shown through the number of iterations.

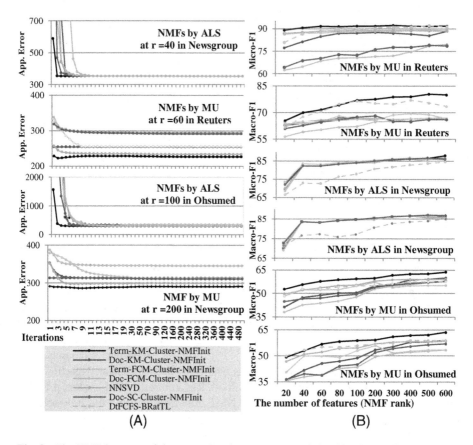

Fig. 2. The NMF in terms of the approximation error (A) and classification performance (B).

As presented in Fig. 2A, NMFs by ALS initialized by the Term-KM-Cluster-NMFInit run quickly toward the convergence. Contrary to the NMFs by ALS, the NMFs by MU have the different errors with the different number of iterations at their convergence points. Concretely, the NMFs by MU using the Doc-KM-Cluster-NMFInit require the fewest number of iterations for convergence. However, at convergence, their approximation errors are larger than those using the Term-KM-Cluster-NMFInit.

In Table 2, the approximation error and the number of iterations of the NMFs are considered under the different initializations at the point satisfying a convergence criterion for more NMF ranks. A popular convergence criterion is that the change of the error in two successive iterations is below 10^{-6} or the number of iterations reaches to 500. Interestingly, what happen here is the same as our analysis on the examples in Fig. 2A. The NMFs by ALS move the equal approximation errors regardless of their different initializations, while the best number of iterations goes to the one initialized by the Term-KM-Cluster-NMFInit. As to the NMFs by MU, the fastest in the race to the convergence is the NMFs by MU initialized by the Doc-KM-Cluster-NMFInit but

Table 2. The approximation error and the number of iterations of the NMF at the convergence point: (1) Term-KM-Cluster-NMFInit; (2) Doc-KM-Cluster-NMFInit; (3) Term-FCM-Cluster-NMFInit; (4) Doc-FCM-Cluster-NMFInit; (5) Doc-SC-Cluster-NMFInit; (6) NNDSVD.

r	20		40		60		80		100		200		300		400		500		600	
	Iter	Err	Iter	Err	Iter	Err	Iter	Err	Iter	Err	Iter	Err	Iter	Err	Iter	Err	Iter	Err	Iter	Err
NMF by MU in the Reuters dataset																				
(1)	358	**256**	412	**239**	286	**226**	314	**224**	208	219	251	**187**	383	218	500	**206**	500	176	**474**	**164**
(2)	171	293	**102**	288	**92**	254	**179**	252	**116**	243	**158**	221	**233**	231	**221**	294	500	185	500	185
(3)	500	266	500	248	500	253	389	230	411	224	424	192	500	**167**	500	215	500	132	500	176
(4)	500	317	500	303	500	291	500	285	500	270	500	245	500	241	500	266	500	181	500	193
(5)	500	320	500	305	500	291	500	284	329	276	500	242	500	218	500	296	500	176	500	180
(6)	430	316	271	244	500	235	180	232	396	**217**	500	230	500	186	500	260	500	**161**	500	172
NMF by ALS in the Newsgroup dataset																				
(1)	**162**	346	**27**	354	130	343	**185**	325	500	317	**360**	281	**421**	268	500	233	**335**	222	500	**212**
(2)	222	**345**	31	354	**136**	343	211	325	500	317	476	281	492	280	500	233	500	222	500	213
(3)	259	346	32	354	142	343	174	325	500	317	425	281	456	**255**	500	234	500	223	500	213
(4)	270	346	35	354	146	343	211	325	500	317	500	282	500	282	500	233	500	223	500	213
(5)	275	346	32	354	148	343	210	325	500	317	500	282	500	285	500	233	401	222	500	213
(6)	210	345	32	354	147	343	217	325	500	317	500	281	500	267	500	233	500	223	500	213
NMF by MU in the Ohsumed dataset																				
(1)	89	**333**	253	**322**	170	**314**	145	**306**	500	**300**	500	**279**	500	**248**	**402**	239	500	**230**	500	230
(2)	**50**	354	**82**	345	**92**	341	**101**	337	**232**	333	**307**	309	**271**	283	500	257	**449**	244	500	244
(3)	259	342	500	334	500	326	500	322	500	317	500	292	500	255	500	249	500	239	500	239
(4)	445	371	500	352	500	348	500	337	500	337	500	311	500	279	500	268	500	253	500	253
(5)	394	355	500	349	83	344	351	340	410	335	500	317	500	292	500	283	468	274	257	274
(6)	438	345	500	335	500	326	207	320	482	314	500	293	500	268	500	261	500	253	500	253

due to the randomization at starting cluster centroids, they can easily fall down a local minimum worse than that of the NMFs by MU initialized by the Term-KM-Cluster-NMFInit. Compared with the other initializations, the Term-KM-Cluster-NMFInit leads the NMFs by MU to a faster convergence and better overall error at convergence. For another used clustering algorithm, i.e. the Fuzzy C-Means Clustering (FCM), the term clustering-based NMF Initialization (Term-FCM-Cluster-NMFInit) is better than the document clustering-based NMF Initialization (Doc-FCM-Cluster-NMFInit). However, it is not more effective than the Term-KM-Cluster-NMFInit.

Next, the NMF initializations are evaluated by the classification performance on the transformed feature set. In order to consider the overall performance of a multi-category classification, two well-known measures, namely Macro-F1 [21] and Micro-F1 [21], are used. Figure 2B shows the Micro-F1 and Macro-F1 results for the NMF FTs. Concerning NMFs by ALS initialized by the different methods, similar to the approximation errors, the Micro-F1 and Maco-F1 results are almost identical. However, as the remarks above on Fig. 2A and Table 2, in order to attain these Micro-F1 and Maco-F1 results, the NMFs by ALS initialized Term-KM-Cluster-NMFInit require the fewest number of iterations. It can be seen from Fig. 2B that the NMFs by MU initialized by the Term-KM-Cluster-NMFInit outperform those initialized by the other methods at most sizes of the feature set in both the Micro-F1 and Macro-F1, while NMFs by MU initialized by the Term-FCM-Cluster-NMFInit produce better Micro-F1 and Macro-F1 than those initialized by the Doc-FCM-Cluster-NMFInit and show competitive and even superior performance to the good methods.

Another emphasis in Fig. 2B is the superiority of the classification on the transformed feature set of the NMFs initialized by the Term-KM-Cluster-NMFInit over that on the non-transformed feature set only simply selected by the DtFCFS-BRatTL. This further confirms the effectiveness of the NMFs initialized by the Term-KM-Cluster-NMFInit. At some sizes of output feature set, the classification performance on the transformed feature set of the NMFs by MU initialized by the document clustering-based methods (Doc-KM-Cluster-NMFInit and Doc-FCM-Cluster-NMFInit) falls even lower than that on the non-transformed feature set selected by the DtFCFS-BRatTL. Therefore, the randomization in starting a document clustering-based NMF initialization has a strong negative influence on the classification performance.

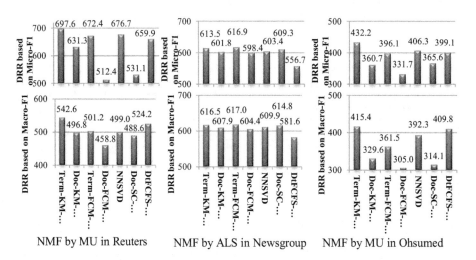

Fig. 3. The NMF performance in terms of the Dimension Reduction Rate (DRR).

Finally, for explicit comparison, we consider Dimension Reduction Rate (DRR) [15, 16, 18] of the NMFs in Fig. 3. The DRR is computed as follows:

$$DRR = \frac{1}{k}\sum_{i=1}^{k} \frac{Dim_N}{Dim_i} R_i \tag{5}$$

where k is the number of experiments; Dim_i is the number of output features (rank) in the ith experiment; R_i is the Micro-F1 or Macro-F1 in the ith experiment; and Dim_N is the maximal number of output features in all experiments. With the Micro-F1 and Macro-F1 results in Fig. 2B, for every of the clustering algorithms, i.e. the K-means and Fuzzy C-Means Clustering, the NMF using the term clustering-based initialization produces more impressive DRRs than that using the document clustering-based initialization. Especially, when the K-means Clustering is used, the term clustering-based NMF initialization (Term-KM-Cluster-NMFInit) shows superior DRRs to the others including NNDSVD, a well-known factorization-based NMF initialization.

Another dominant strength of the term clustering-based NMF initialization is to make the NMF become deterministic. Furthermore, in the experiments on the document clustering-based NMF initialization, we usually face the issue of empty clusters. In this case, to achieve the best NMF initialization, we must rerun the document clustering process with other random starting cluster centroids. For example, in the Reuters dataset, the document clustering process in the Doc-KM-Cluster-NMFInit is rerun 3 times with 200 clusters; 15 times with 300 clusters; 46 times with 400 clusters; 119 times with 500 clusters; 523 times with 600 clusters. For the Term-KM-Cluster-NMFInit, the issue of empty clusters does not happen even with 600 clusters. This demonstrates the goodness of selecting the starting term cluster centroids by using a supervised FS (DtFCFS-BRatTL) in the term clustering-based NMF initialization.

6   Conclusion

This study may pave the way for further studies on the text FT based on the NMF not only at the NMF initialization stage but also the other NMF stages by interpreting the NMF matrix factors according to their row (term) vectors. Under the viewpoint of term vectors, it is possible to exploiting the researches on term description in the text to further improve the NMF FT. For instance, in this paper, we utilize the DtFCFS-BRatTL, which is a recent supervised term selection of ours, and the PPMI, which is an effective semantic term association, to propose a new clustering-based method for initializing the NMF matrix factors according to the row (term) vectors instead of the column (document) vectors. And it is called a term clustering-based NMF initialization. This facilitates settling how a clustering algorithm defines the starting cluster centroids when it is used for the NMF initialization. Therefore, one of the dominant strengths of the term clustering-based NMF initialization is to make the NMF become deterministic. We investigate the performance of the document clustering-based NMF initializations and the proposed term clustering-based NMF initializations. The results show that the NMFs by ALS obtain the nearly equal approximation errors, classification performance, and dimension reduction rate regardless of their different initializations. However, when the K-means clustering is used, the term clustering-based NMF initializations leads the NMFs by ALS to a faster convergence. For the NMFs by MU, the term clustering-based NMF initialization is better than that using the document clustering-based initialization in terms of the approximation error, the classification performance, and the dimension reduction rate. Especially, with K-Means Clustering, the term clustering-based NMF initialization is superior to the others.

References

1. Asuncion, A., Newman, D.: UCI machine learning repository (2007)
2. Boutsidis, C., Gallopoulos, E.: SVD based initialization: a head start for nonnegative matrix factorization. Patt. Recogn. **41**(4), 1350–1362 (2008)
3. Bullinaria, J.A., Levy, J.P.: Extracting semantic representations from word co-occurrence statistics: a computational study. Behav. Res. Methods **39**(3), 510–526 (2007)
4. Casalino, G., Del Buono, N., Mencar, C.: Subtractive clustering for seeding non-negative matrix factorizations. Inf. Sci. **257**, 369–387 (2014)

5. Cichocki, A., Zdunek, R., Phan, A.H., Amari, S.I.: Nonnegative Matrix and Tensor Factorizations: Applications to Exploratory Multi-way Data Analysis and Blind Source Separation. Wiley, Chichester (2009)
6. Correa, R.F., Ludermir, T.B.: Improving self-organization of document collections by semantic mapping. Neurocomputing **70**(1), 62–69 (2006)
7. Deerwester, S., Dumais, S.T., Furnas, G.W., Landauer, T.K.: Indexing by latent semantic analysis. J. Am. Soc. Inf. Sci. **41**(6), 391 (1990)
8. Golub, G.H., Van Loan, C.F.: Matrix Computations, vol. 3. JHU Press, Baltimore (2012)
9. Hosseini-Asl, E., Zurada, Jacek M.: Nonnegative matrix factorization for document clustering: a survey. In: Rutkowski, L., Korytkowski, M., Scherer, R., Tadeusiewicz, R., Zadeh, Lotfi A., Zurada, Jacek M. (eds.) ICAISC 2014. LNCS, vol. 8468, pp. 726–737. Springer, Cham (2014). doi:10.1007/978-3-319-07176-3_63
10. Janecek, A., Gansterer, W.N., Demel, M., Ecker, G.: On the relationship between feature selection and classification accuracy. In: FSDM, pp. 90–105 (2008)
11. Joachims, T.: Text Categorization with Support Vector Machines: Learning with Many Relevant Features. Springer, Heidelberg (1998). pp. 137–142
12. Lee, D.D., Seung, H.S.: Algorithms for non-negative matrix factorization. In: Advances in Neural Information Processing Systems, pp. 556–562 (2001)
13. Levy, O., Gold, Y.: Improving distributional similarity with lessons learned from word embeddings. Trans. Comput. Linguist. Assoc. **3**, 211–225 (2015)
14. Liu, H., Motoda, H. (Eds.): Feature Extraction, Construction and Selection: A Data Mining Perspective. Springer, New York (1998)
15. Nam, L.N.H., Quoc, H.B.: A comprehensive filter feature selection for improving document classification. In: Proceedings of 29th Pacific Asia Conference on Language, Information and Computation 2015, pp. 169–177 (2015)
16. Nam, L.N.H., Quoc, H.B.: A combined approach for filter feature selection in document classification. In: 2015 IEEE 27th International Conference on Tools with Artificial Intelligence (ICTAI), pp. 317–324. IEEE (2015)
17. Nam, L.N.H., Quoc, H.B.: The ranking methods in the filter feature selection process for text categorization system. In: Proceedings of the 20th Pacific Asia Conference on Information Systems (PACIS 2016) (Paper 159) (2016)
18. Nam, L.N.H., Quoc, H.B.: The hybrid filter feature selection methods for improving high-dimensional text categorization. Int. J. Uncertainty Fuzziness Knowl.-Based Syst. **25** (02), 235–265 (2017)
19. Pinheiro, R.H., Cavalcanti, G.D.: Data-driven global-ranking local feature selection methods for text categorization. Expert Syst. Appl. **42**(4), 1941–1949 (2015)
20. Platt, J.C.: 12 fast training of support vector machines using sequential minimal optimization. In: Advances in Kernel Methods, pp. 185–208 (1999)
21. Sebastiani, F.: Machine learning in automated text categorization. ACM Comput. Surv. (CSUR) **34**(1), 1–47 (2002)
22. Turney, P.D., Pantel, P.: From frequency to meaning: vector space models of semantics. J. Artif. Intell. Res. **37**(1), 141–188 (2010)
23. Wang, Y.X., Zhang, Y.J.: Nonnegative matrix factorization: a comprehensive review. IEEE Trans. Knowl. Data Eng. **25**(6), 1336–1353 (2013)
24. Xue, Y., Tong, C.S., Chen, Y.: Clustering-based initialization for non-negative matrix factorization. Appl. Math. Comput. **205**(2), 525–536 (2008)
25. Yang, Y., Pedersen, J.O.: A comparative study on feature selection in text categorization. In: ICML, vol. 97, pp. 412–420, July 1997
26. Zheng, Z., Yang, J., Zhu, Y.: Initialization enhancer for non-negative matrix factorization. Eng. Appl. Artif. Intell. **20**(1), 101–110 (2007)

Short Paper

Analysis of Interactive Multimedia Features in Scientific Publication Platforms

Camila Wohlmuth da Silva$^{(\boxtimes)}$ and Nuno Correia

FCT, Universidade Nova de Lisboa,
Campus Universitário, 2829-516, Caparica, Portugal
camila.wohlmuth@gmail.com

Abstract. Some platforms provide another dimension in published articles, using interactive enrichments to explore related content and help readers test their understanding. In the case of scientific publication platforms, incorporating these features and providing the user with additional content remain a challenge. Therefore, this study based on the taxonomy of Paul, analyzes how interactive multimedia resources are used on different scientific platforms with hypermedia systems. The paper discusses the usage of these features and how current practices could be improved for other scientific publication platforms.

Keywords: Hypermedia · Interactivity · Information behaviours

1 Introduction

Scientific publications are essential elements in the dissemination and evolution of science and new technology. They are used as a support for retrieving information as well as staying updated, providing cooperation and integration among researchers. Through scientific publications, research findings on various subjects are presented, contributing to the recognition of scientific discoveries, and confirming the competence of the researcher in the scientific community.

Digital libraries must improve their efforts to provide users with effective and flexible access to materials which will, in turn, empower them to make new observations and discoveries. As an example, recent studies have shown that the technologies currently available incorporate hypermedia systems as a powerful tool for the development of new scientific communication models, offering multimedia content (video, audio, animation, etc.), which create new interactive expectations in the reader [1, 2].

Hypermedia presents the possible integration of interactive systems with multimedia [3, 4], creating the prospect of application in articles hosted within the scientific publication platforms. Nevertheless, unlike cinema, it has a vocabulary all its own to define its specific features. The digital environments, as in scientific publication platforms, are still adapting their framework from other media. However, it is necessary for new practices and techniques along with a new vocabulary [5]. In order to create a specific vocabulary for scientific publications, this study first understands how interactive multimedia resources are used on different scientific platforms with hypermedia systems, the analysis adopts the benchmarks that were developed by Paul [5, 6].

© Springer International Publishing AG 2017
J. Kamps et al. (Eds.): TPDL 2017, LNCS 10450, pp. 525–530, 2017.
DOI: 10.1007/978-3-319-67008-9_41

The present study seeks to identify the specifics of the digital environments of Paul [5, 6]. Scientific platforms have been selected which represent a sample of the three types currently available, open source platforms, closed source platforms and collaborative platforms. Thus, there is a general understanding of how different platforms use interactive multimedia resources in their scientific publications.

The paper is organized into four main sections, section one is the introduction; section two is the analysis of scientific publication platforms; and section three is presented as the discussion and conclusions of this study.

2   Analysis

Paul [5, 6] states that interactivity is essential for digital environments. In fact, interactivity is important for the scientific publication platforms, but it is a concept that still lacks a precise definition. To address the issue of an absence of specificity and overcome the imprecision of the term "interactivity", they have developed a taxonomy for digital narratives, which can be applied to the narratives of scientific texts. In the taxonomy developed by Paul [5, 6], five main elements are discussed: (1) media, (2) action, (3) relationship, (4) context and (5) communication.

Media – The material or information can have different properties. Paul [5, 6] discusses properties such as media type (e.g. text, video, and image), the configuration - single media, multiple media or multimedia, and the rhythm (or flow) - whether the content is asynchronous or synchronous.

Action – This element concerns the actions for two distinct aspects: the movement of the content itself and the action required by the user to access the content. This element refers to the digital environment as being dynamic or static, and in regards to the need for the user to participate in order to move content, active or passive.

Relationship – This discusses the relationship between the user and the content. It manifests itself in a combination of five types: linear or non-linear, customizable or standard, calculable or non-calculable, manipulative or fixed, and expandable or limited.

Context – Refers to how much the content of the narrative can be more than offered in its space, such as links or external content that relate to the narrative offered. In the case of the presence of these links, the context is considered hypermedia, having built-in or parallel links, internal or external, supplementary or duplicative, and contextual or related.

Communication – This element refers to the ability to connect with others through digital media. In this element, the configuration can be one-to-one, one-to-many, and many-to-one or many-to-many, depending on the amount of agents involved.

For instance, different types of communication (e.g. video, e-mail, chat and forum) are discussed, as well as the purpose of the communication. The study analyzes three selected scientific publication platforms: Elsevier (closed source), Plos One (open source) and Pub Pub (collaborative platform). A publication of each platform was analyzed. The scientific area and the theme of the publications used in the analysis of the platforms will not be considered. It will only consider what features the platforms provide and how is the integration of these features with the scientific publication.

2.1   Elsevier

The Elsevier platform provides web-based solutions for science, health, and technology professionals. The Elsevier project called "The Article of the Future" explores better ways of presenting online journal articles and enriching their content in regards to three key elements: presentation, content and context. Its main goal is to offer new content and tools that would help researchers in their specific scientific needs.

Conforming to the taxonomy of Paul [5, 6], the analyzed scientific publication[1] inside the Elsevier platform presents in its content, the different types of media separately, that is, static (text and images) in the center of the publication body, and dynamic (video, as additional information) in the margin. Even so, it is featured with a multimedia publication. Some Elsevier journals offer multimedia content such as videos and podcasts that are not included in the full text.

The layout features different levels of attention. Thus, in relation to the action, the content is static and active when the user navigates by clicking the left menu or links. The publication has three-pane-based content exploration. Consequently, the content movement is non-linear and non-calculated. Since each pane can be scrolled independently, it is possible to see both the text (on the middle pane) and an image (on the right pane sidebar) at the same time.

The left navigation pane displays a clickable section of content and thumbnails of images and tables. There are two options for navigation: with or without figure and table thumbnails displayed (customizable content). The right pane provides access to supplementary information (cited, metrics, related content and featured multimedia).

The content is fixed as more information to the content cannot be added, but it is expandable with links to related content. In addition, regarding presented links, the context is considered hypermedia, having parallel links, internal (dropdown menu) and external (references), supplementary and duplicative (featured multimedia - video) and recommended (recommended articles). Based on Paul [5, 6], communication is the element that refers to the ability to connect with others through digital media. In this way, the journal's configuration of communication is one-to-one, allowing the user to send emails to one of the authors to exchange information or ask questions.

2.2   Plos One

Plos One is an international journal, it presents research publications from all disciplines in the field of science and medicine. By not excluding articles based on the subject area, it is possible to discover connections between papers or across disciplines. Authors can send support files and multimedia files along with their manuscripts. The Plos One publishing platform supports several file types for the information system. Therefore, it imposes specific requirements and formats for multimedia files.

Concerning the taxonomy of Paul [5, 6], the analyzed publication[2] presents in its content, static and dynamic media, characterizing it as a multimedia publication

---

[1] Viewable in: https://www.sciencedirect.com/science/article/pii/S0022314X13002424.

[2] Viewable in: http://journals.plos.org/plosone/article?id=10.1371/journal.pone.0111730.

(text, video and animations). In relation to time and space, the content has a menu bar that affords the ability to navigate between parts of the paper.

As follows, the content is static and active, the reader is encouraged to click (on the links, slide images, buttons, icons and menus), participating in the movement of the content - from one session to another. The relationship can be characterized as non-linear, fixed and expandable, but it cannot be calculated. Relative to context, the content of the publication is more than what is offered in its space, with links and external contents. Moreover, because of the links presented, the context is considered hypermedia, having links that are built-in and parallel, internal and external, as well as complementary and related.

Communication of the publication takes place through the reader comments and media coverage sessions, sending an email to the author, in addition to other content and sharing pages (twitter). In this element, a one-to one configuration (reader comments) and one-to-many (sharing on social networks) are provided as tools of communication.

According to Plos One, the media exposure extends article reach globally. The latest research summaries, author interviews, multimedia pieces, and other content are found in the section of Plos Research News.

2.3   Pub Pub

Pub Pub[3] is a free and open platform with instant publishing and continuously reviewed journals. The platform has been designed and developed by Travis Rich and Thariq Shihipar at the MIT Media Lab. The platform enables immediate author-driven publishing. Moreover, it provides versioned histories and collaborative open-source editing, that allows for evolving content and formats. Pub Pub affords support for dynamic assets and live visualizations. It is a tool built for and by the community. Furthermore, researchers are invited to code, design and contribute to improvements in the platform.

In relation to the the taxonomy of Paul [5, 6], the Pub Pub collaborative platform has the ability to provide tools for building a multimedia publication. The platform allows the users to insert media (static and dynamic) In this way, users can insert tables, static and dynamic images (gif), video, audio and other interactive resources. Along these lines, media content is visually asynchronous, the file is linked to the text and published only when the user chooses.

The action element in this platform concerns the actions for two distinct aspects: the movement of the publication contents itself and the action required by the user to access the content. The content is dynamic because it has the possibility to add animation in the publication. It is also static because the text does not move by itself. In regards to movement, it is active, the user manipulates the scroll bar so that the text moves down.

The relationship between the user and the content is non-linear, customizable, non-calculable, manipulative and expandable. For instance, content is customizable,

---

[3] Viewable in: https://www.pubpub.org/pub/about.

users have the possibility to edit their publications. The creator of the document and invited users may include external links, images (static and dynamic), videos, and references. In addition, there is the option to give feedback, make comments, highlight, etc.

Concerning the context of the platform, the publication is considered hypermedia. Inserting built-in links to external content, supplementary (references), contextual and related. Regarding the communication element and the ability to connect with others, Pub Pub has a one-to-one configuration such as feedback or emails, one-to-many (citing and following), and many-to-many (discussion chat) (Table 1).

Table 1. Result of the analysis.

	Elsevier	Plos One	Pub Pub
Media	Static, Dynamic (separately); Multimedia; Asynchronous.	Static, Dynamic Multimedia; Asynchronous.	Static, Dynamic; Multimedia; Asynchronous.
Action	Static; Active.	Static; Active.	Dynamic, Static; Active.
Relationship	Non-linear; Fixed; Customizable; Non-calculated; Expandable.	Non-linear; Fixed; Standard; Non-calculated; Expandable.	Non-linear; Fixed; Customizable; Non-calculated; Expandable; Manipulative.
Context	Hypermedia	Hypermedia	Hypermedia
Communication	One-to-one	One-to-one; One-to-many.	One-to-one; One-to-many; Many-to-many.

3   Discussion and Conclusions

With the transition from print to online publishing, the layout and presentation of scientific publications have remained relatively unchanged in most publications, still following a static format with indications for interactive multimedia content. Based on Lancaster [7], one can say that electronic journals are in a third transition for the fourth stage of technological development, that is, they begin to be developed and distributed only in electronic media with incorporations of multimedia resources.

Meanwhile, a number of available enrichments on the scientific platform grows, these can provide variations in the intensity of their use, as it was measured in the analyzed platforms. Thus, it is based on the linear model of printed materials for its integration with multimedia using a more visual layout and directed to nonlinear reading.

The way these contents are being embedded into the publication is what to watch out for. The integration must consider the publication more visual, as the understanding of all of the scientific information within the publication is necessary to optimize the reading time, maintain publishing efficiently, and especially for increasing scientific knowledge. Nonetheless, as analyzed on Elsevier, the multimedia content is processed separately as single objects that ends up individualizing the content, which could not be extended for other scientific publication platforms.

The platforms analyzed have multimedia enhancements, such as video, interactive graphics, animation, among others. Although these enhancements are expected to influence the acquisition of the knowledge embodied in the publication, they do not have the necessary interactivity to become more dynamic. In this way, interactivity is operationally defined as a purposeful action by the user, which causes a directional change or a meaningful response by the system with regard to the content. In essence, the dynamic action and interactivity would require the user to actively intervene in the process, capturing the user's attention and expanding the understanding of the content.

Scientific platforms must improve the content access for scholars and users, merely Elsevier and Pub Pub are customizable, and only Pub Pub is manipulative - for being a collaborative platform. The layout being customizable and manipulative offers a possible relationship between the user and the content. As the layout can adapt to the page according to the resolution being viewed, increase font size and images if desired, hide unnecessary elements in smaller reading devices, adapt button sizes and links, etc.

Through responsive design for user-centered approaches as the best way to create more effective learning tools, with effective and flexible access to materials which will, in turn empower them to make new observations and discoveries. Hypermedia systems are presented as a tool for the development of scientific communication - providing new alternatives of formats for scientific publications, adding interactive multimedia resources and non-linear content in scientific publication platforms.

References

1. Breure, L., Voorbij, H., Hoogerwerf, M.: Rich Internet Publications: "Show what you tell". J. Digit. Inf. **12**(1), 1 (2011)
2. Siegel, E.R., Lindberg, D.A.B., Campbell, G.P., Harless, W.G., Godwin, C.R.: Defining the next generation journal: the NLM- Elsevier interactive publications experiment. Inf. Serv. Use **30**(1–2), 17–30 (2010)
3. Rada, R.: Hypertext, multimedia and hypermedia. New Rev. Hypermedia Multimedia **1**, 1–21 (1995)
4. Fluckiger, F.: Understanding Networked Multimedia, Chap. 6, pp. 109–121. Prentice Hall, London (1995)
5. Paul, Nora.: Elementos das narrativas digitais. In: Ferrari, P. (ed.), Hipertexto, hipermídia: as novas ferramentas de comunicação digital. Contexto, São Paulo (2007)
6. Paul, N.: The Elements of Digital Storytelling (2005) http://www.academia.edu/459532/The_Elements_of_Digital_Storytelling
7. Lancaster, F.W.: The evolution of electronic publishing. Library Trends **43**(4), 518–524 (1995)

Extending R2RML with Support for RDF Collections and Containers to Generate MADS-RDF Datasets

Christophe Debruyne[✉], Lucy McKenna, and Declan O'Sullivan

ADAPT Centre, Trinity College Dublin, College Green, Dublin 2, Ireland
{christophe.debruyne,lucy.mckenna,
declan.osullivan}@adaptcentre.ie

Abstract. It is a best practice to avoid the use of RDF collections and containers when publishing Linked Data, but sometimes vocabularies such as MADS-RDF prescribe these constructs. The Library of Trinity College Dublin is building a new asset management system backed by a relational database and wants to publish their metadata according to these vocabularies. We chose to use the W3C Recommendation R2RML to relate the database to RDF datasets, but R2RML unfortunately does not provide support for collections and containers. In this paper, we propose an extension to R2RML to address this problem. We support gathering collections and containers from different fields in a row of a (logical) table as well as across rows. We furthermore prescribe how the extended R2RML engine deals with named graphs in the RDF dataset as well as empty sets. Examples and our demonstration on a part of the Library's database prove the feasibility of our approach.

Keywords: R2RML · Linked Data Publishing · MODS · MADS

1 Introduction

The Digital Resources and Imaging Services (DRIS) department of the Library of Trinity College Dublin (TCD) hosts the Digital Collections Repository of the university, providing open access to the university's growing collection of digitized cultural heritage materials. DRIS hopes to move towards publishing the bibliographic data of their digital collections as Linked Data (LD) as to increase their materials' visibility. To this end, a bespoke tool backed by a relational database has been developed that accepted URIs to other Linked Data datasets. The Library decided[1] that records should follow the Metadata Object Description Schema (MODS) as this standard was: suitable for cataloguing DRIS resources to the required level of detail, compatible with existing MAchine-Readable Cataloging (MARC) records in other catalogues and also less complex than MARC, and available as an RDF vocabulary.

Given that the information was stored in a relational database, adopting the RDB to RDF Mapping Language (R2RML) [1], a vocabulary for declaring customized mappings from relational databases to RDF datasets, is a sensible approach. During the

---

[1] Which explains why no other models such as CIDOC-CRM were considered.

© Springer International Publishing AG 2017
J. Kamps et al. (Eds.): TPDL 2017, LNCS 10450, pp. 531–536, 2017.
DOI: 10.1007/978-3-319-67008-9_42

creation of R2RML mappings, a challenge arose: complete RDF records could not be generated as, to produce such a record, the use of RDF collections was required by the ontology, but were not support by R2RML.

We thus propose a minimal extension to the R2RML language and algorithm for the generation of RDF collections *and* containers. After elaborating on our approach, we demonstrate it on a part of the tool's database. We discuss our approach with respect to related work prior to concluding this paper and formulating future directions.

2   Background

MODS and MADS. Both Metadata Object Description Schema (MODS) and the Metadata Authority Description Schema (MADS) are XML schemas to describe bibliographic metadata and share quite a few elements. An OWL ontology was developed for both schemas; MODS-RDF and MADS-RDF. The MODS-RDF ontology, however, excluded all elements it had in common with MADS. If one relied on the MODS XML schema and wants to generate "semantically" equivalent RDF, MADS-RDF has to be adopted as well. But, unlike MODS-RDF, where properties are represented individually, in MADS-RDF properties are grouped in collections.

Resource Description Framework (RDF) provides two constructs to gather RDF terms for use in statements; **RDF Containers and Collections**. The difference between RDF containers (`rdf:Bag`, `rdf:Seq`, and `rdf:Alt`) and collections (`rdf:List`) is that the latter has an explicit terminator (`rdf:nil`, or the empty list) and is therefore immutable. One can add additional elements to the former. We note that it is generally considered a bad practice to use these constructs in Linked Data publishing, but some ontologies rely on it.

We assume the reader is familiar with **R2RML**, and otherwise refer to [1]. We have chosen to adopt R2RML as it is a W3C Recommendation (i.e., a standard) and hence supported by various tools, and also because it provides us with a scalable declarative approach. However, R2RML provides no *elegant* support for creating such mappings. In some cases, one can resort to an additional triples map for creating these provided the underlying relational database has support for pivot tables (allowing one to "pivot" a table and treat a particular row as the column). We deem this approach also too complex. Another approach is to go through several pre-, or post-processing stages, but that renders RDF generation not self-contained.

3   Approach

In this section, we describe our approach to provide support for generating RDF containers and collections in R2RML. We will exemplify our approach with a simple database (see Fig. 1), and then cover both cases of collecting RDF terms (per row, and per column)[2] But we first formulate the following requirements: (1) **Collecting RDF terms per row from various cells** with the additional requirement that one should be

---

[2] Our prototype is available at: https://opengogs.adaptcentre.ie/debruync/r2rml/src/r2rml-col.

able to specify what type of terms can be collected and that they can differ in a collection/
container. (2) **Collect RDF terms across rows:** grouping the RDF terms that are generated from an object map for each subject. (3) **Nesting:** the ability to nest containers and
collections, and both approaches. (4) **Provide support for managing empty collections
and containers**. (5) **Managing named graphs**.

BOOK	
ID	TITLE
1	Frankenstein
2	The Long Earth

AUTHOR				
ID	BOOKID	TITLE	FNAME	LNAME
1	1	*NULL*	Mary	Shelley
2	2	Sir	Terry	Pratchett
3	2	*NULL*	Stephen	Baxter

Fig. 1. Two relational tables representing books and their authors.

To collect terms for each row in a logical table, we extended R2RML's vocabulary
in the following ways: the introduction of a predicate `rrf:gather` to indicate which
RDF terms need to be gathered into a collection or a container, and allowing the predicate
`rr:termType` to refer to `rdf:Bag`, `rdf:Seq`, `rdf:Alt`, or `rdf:List`. The last
is the default when a valid term type is absent. We note that these are all part of the RDF
namespace, which we reused. The subject of `rrf:gather` must be a list of object
maps that generate RDF terms. We thus have an object map that is comprised of object
maps – which we will call a *gather map*. When none of the object maps generate a term
as prescribed by the R2RML W3C Recommendation, an empty list or container is
generated. One can also use `rrf:gatherAsNonEmpty` to avoid the generation of
empty collections/containers.

The application of this gather-object map *g* on a row will result in the application of
each object map part of *g* to create the container or collection. Using the running example
described above, the R2RML snippet in Listing 1 (top) generates the RDF shown in
Listing 1 (bottom) One can see how rows in the person table generates a bag only
containing a first- and last name when a title is non-existent.

```
rr:predicateObjectMap [
  rr:predicate ex:name;
  rr:objectMap [
    rrf:gather ([ rr:column "TITLE" ] [ rr:column "FNAME" ] [ rr:column "LNAME" ]);
    rr:termType rdf:Bag;
  ];
]
person:2 ex:name [ a rdf:Bag; rdf:_1 "Sir"; rdf:_2 "Terry"; rdf:_3 "Pratchett" ].
person:1 ex:name [ a rdf:Bag; rdf:_1 "Mary"; rdf:_2 "Shelley" ].
person:3 ex:name [ a rdf:Bag; rdf:_1 "Stephen"; rdf:_2 "Baxter" ].
```

The fifth requirement will be covered here as it necessitates prescribing how different
named graphs across rows that are gathered should be treated as the target graphs in the
subject and predicate-object maps may differ for each row. While gathering collections or
containers per row is fairly straightforward as it introduced a new object map that needs to
be applied to each row of a logical table, collecting RDF terms per column is a bit more
challenging in terms of coming up with an appropriate extension of the vocabulary and the
algorithm, especially the latter as one needs to keep track of the rows that need to be

grouped in order to generate the collection or container. We extended the algorithm as follows: the implementation keeps track of all object maps with a `rr:collectAs` (See Listing 2) statement. The algorithm generates a subject for each row in the logical table (or join in case of reference-object maps). A special data structure keeps track of the RDF terms generated by the graph- and predicate-maps for each row – and thus also subject – whilst collecting the objects for the creation of the collection or container. Since each row may generate different predicates or graphs, but objects are collected across them, we have decided to store the collection or container in all possible combinations of graphs and predicates related to a particular subject. Though we think that this would be an unlikely use case, we deemed it important to think this aspect of the extension through.

```
rr:predicateObjectMap [
  rr:predicate ex:writtenby;
  rr:objectMap [
    rr:parentTriplesMap <#AuthorsTriplesMap>;
    rr:joinCondition [ rr:child "ID"; rr:parent "BOOKID"; ];
    rrf:collectAs rdf:List;
  ];
];
book:1 ex:writtenby ( person:1 ).
book:2 ex:writtenby ( person:2 person:3 ).
```

We allow nesting in the following ways: (i) gather maps may be nested with gather maps; and (ii) one may collect (nested) gather maps with `rrf:collectAs`. Since we have created an object map of object maps to tackle the case of gathering RDF terms for each row to cover the first case, it is fairly straightforward to nest them. For obvious reasons, however, no "cycles" are permitted in nested object maps. Due to space limitations, we will not be able to provide examples and refer to the documentation instead. What we do not allow is the use of `rrf:collectAs` in nested object maps; it does not make sense to start aggregating, for each row, terms across rows.

4   Demonstration

Here, we demonstrate our approach to generate a MADS-RDF dataset from the relational database of The Library's cataloguing system. Concepts in MODS, such as `mods:Title`, are related to a collection of `mads:Element` instances. Elements, which act as an abstract concept for something that has a label, are attributed such a label with the predicate `mads:elementValue` whose range is an `xsd:string`. The concept `mads:Element` is then specialized into a number of subclasses such as `mads:TitleElement`, which itself is an abstract concept for all elements one can find in a title. One needs to use instances of "concrete" concepts such as `mads:Main-TitleElement`, and `mads:PartNameElement` in that list.

In the database, a record must have at least one TitleInfo – terminology adopted from MODS XML, which acts as a "container for all subelements related to title information. The table `TitleInfo` thus has a foreign key to a record in the table `Record`. In `TitleInfo`, all subelements are captured in the fields `nonSort`, `partName`, `partNumber`, `subtitle`, and `title`. Due to space limitations and since the structure of these mappings are the same for all subelements, we will only describe one. We also leave out the mapping for `Record` (and also how records are then related to `TitleInfo`), and focus on the creation of title elements instead. We note a mapping was created for the whole database (including other elements). The evaluation of our approach's performance was not within the scope of this study.

Our mapping is shown in Listing 3 (top), we use HTTP URIs for TitleInfo, but URNs for the individual elements. We chose URNs as we do not (yet) foresee a reason why users want to engage with these resources via resolvable HTTP URIs, but we also wanted to avoid the use of blank nodes. Since the actual value of the title elements are not suitable for creating URNs as they can contain illegal characters, we provided specific IDs for each element when they exist. These conditionals appear in the SQL query. The title info and title element are linked by reusing the same URN template (highlighted in yellow). Listing 3 (bottom) contains some RDF statements that were generated of one of TCD Library's assets.

```
<#TitleInfo>
  rr:logicalTable [
    rr:sqlQuery """SELECT *, IF(nonSort IS NULL, NULL, id) AS nId, IF(subtitle IS
NULL, NULL, id) AS sId, IF(partNumber IS NULL, NULL, id) AS nuId, IF(partName IS
NULL, NULL, id) AS naId FROM TitleInfo"""; ];
  rr:subjectMap [
    rr:template "http://data.library.tcd.ie/resource/titleinfo/{id}";
    rr:class madsrdf:Title;
  ];
  # Mapping to generate rdfs:label based on "title" omitted
  rr:predicateObjectMap [
    rr:predicate madsrdf:elementList;
    rr:objectMap [
      rrf:gather (
        [ rr:template "urn:tcd:title-nonsort-{nId}" ]
        [ rr:template "urn:tcd:title-main-{id}" ]
        [ rr:template "urn:tcd:title-subtitle-{sId}" ]
        [ rr:template "urn:tcd:title-partname-{naId}" ]
        [ rr:template "urn:tcd:title-partnumber-{nuId}" ]
      );
    ];
    rr:termType rdf:List;
  ];
.
<#TitleInfo-Title>
  rr:logicalTable [ rr:sqlQuery "SELECT id, title FROM TitleInfo"; ];
  rr:subjectMap [
    rr:template "urn:tcd:title-main-{id}"; rr:class madsrdf:MainTitleElement;
  ];
  rr:predicateObjectMap [
    rr:predicate madsrdf:elementValue; rr:objectMap [ rr:column "title"; ];
  ];
.

<http://data.library.tcd.ie/resource/titleinfo/2> a mads:Title; mads:elementList (
<urn:tcd:title-nonsort-2> <urn:tcd:title-main-2> <urn:tcd:title-partnumber-2> ).
<urn:tcd:title-main-2> a mads:MainTitleElement; mads:elementValue
  "Transactions of the Institution of Civil Engineers of Ireland".
<urn:tcd:title-nonsort-2> a mads:NonSortElement; mads:elementValue "The".
<urn:tcd:title-partnumber-2> a mads:PartNumberElement; mads:elementValue "Vol.26".
```

5   Related Work

We focus on related word of generating RDF datasets from *relational databases* only. To the best of our knowledge, xR2RML [3] is the only initiative that aimed to extend R2RML with support for containers and collections. It extends both R2RML for relational databases and RML [2], itself a superset of R2RML, to handle other source data formats such as JSON, XML, and CSV. At the time of writing, the implementation of xR2RML provides no support for named graphs, nested collections and containers, and different term types in collections and containers[3]. We consider the first a non-implemented feature rather than a real limitation. Interesting about their approach is how they handled "representation agnostic" mappings allowing one to mix representation formats. One can, for instance, treat the contents of a column as JSON. This feature allows one to generate collections or containers for tables with such columns.

6   Conclusions and Future Work

This paper provides evidence that a *minimal* extension of R2RML to support the generation of RDF collections and containers from relational databases is feasible. The Library of Trinity College Dublin, who wished to generate RDF from their relational database using MADS-RDF, provided the motivation of this study, as those vocabularies prescribe the use of RDF collections for which there is no support in R2RML.

Our approach furthermore supports a wider range of cases than the one needed for our motivating use case; nesting collections/containers, collections/containers across rows, and dealing with empty collections and containers. Though the Library did not need to gather collections across rows, this could potentially be useful to generate a collection of disjoint OWL classes when generating a taxonomy from a table, for instance. With respect to existing state of the art, our approach covers a wider range of cases, and does not intermix data representation formats. We believe that this would ease the maintenance mappings, though evidence for this needs to be gathered.

Acknowledgements. The ADAPT Centre for Digital Content Technology is funded under the SFI Research Centres Programme (Grant 13/RC/2106) and is co-funded under the European Regional Development Fund. We also express our gratitude to the Library of Trinity College Dublin (TCD) for providing us their data and Garg Abhivan who explored the development of an earlier prototype.

References

1. Das, S., Sundara, S, Cyganiak, R.: R2RML: RDB to RDF mapping language. In: W3C Recommendation (2012)
2. Dimou, A., Vander Sande, M., Colpaert, P., Verborgh, R., Mannens, E., Van de Walle, R.: RML: a generic language for integrated RDF mappings of heterogeneous data. In: Bizer, C., Heath, T., Auer, S., Berners-Lee, T. (eds.) Proceedings of the Workshop on Linked Data on the Web (LDOW 2014), CEUR Workshop Proceedings, vol. 1184 (2014). CEUR-WS.org
3. Michel, F., Djimenou, L., Faron-Zucker, C., Montagnat, J.: Translation of relational and non-relational databases into RDF with xR2RML. In: Monfort, V., Krempels, K., Majchrzak, T.A., Turk, Z. (eds.) Proceedings of the 11th International Conference on Web Information Systems and Technologies (WEBIST 2015), pp. 443–454. SciTePress (2015)

---

[3] See https://github.com/frmichel/morph-xr2rml, last accessed March 23, 2017.

Building the Brazilian Academic Genealogy Tree

Wellington Dores, Elias Soares, Fabrício Benevenuto,
and Alberto H.F. Laender$^{(\boxtimes)}$

Computer Science Departament,
Universidade Federal de Minas Gerais, Belo Horizonte, Brazil
{wellingtond,eliassoares,fabricio,laender}@dcc.ufmg.br

Abstract. Along the history, many researchers provided remarkable contributions to science, not only advancing knowledge but also in terms of mentoring new scientists. Currently, identifying and studying the formation of researchers over the years is a challenging task as current repositories of theses and dissertations are cataloged in a decentralized way through many local digital libraries. In this paper we focus our attention on building such trees for the Brazilian research community. For this, we use data from the Lattes Platform, an internationally renowned initiative from CNPq for managing information about individual researchers and research groups in Brazil.

Keywords: Academic genealogy trees · Academic mentorship · Lattes

1 Introduction

Science has evolved over the centuries as a system that not only promotes progress through the scientific method, but that is also centered on the processes of mentoring and teaching. The academic mentoring activity is a form of relationship that promotes the scientific development, as well as the formation and evolution of new researchers. Despite the complex system behind science, most of the existing efforts in the literature that aim at measuring individuals' research productivity within a scientific community usually account only for the publications produced, citations received and collaborations established [1,9], neglecting the formation of new researchers.

There has been only a limited number of initiatives, by specific academic communities, in the sense of documenting, analyzing and classifying advisor-advisee relationships. Sometimes this kind of study considers a representation usually called academic genealogy tree [2,3,7], in which nodes represent researchers and relations indicate that a researcher was the advisor of another one. However, these efforts have focused on specific scientific fields, such as Mathematics [7] and Neuroscience [3], or have been restricted to a specific community as it is the case of a career retrospect of prominent American physicists [2]. Although limited to specific contexts, overall these efforts show that the analysis of such relationships in the form of a genealogy structure contributes to a greater understanding

© Springer International Publishing AG 2017
J. Kamps et al. (Eds.): TPDL 2017, LNCS 10450, pp. 537–543, 2017.
DOI: 10.1007/978-3-319-67008-9_43

of a scientific community and of its individual values, allowing us to identify the impact generated by individuals in the formation of a community [10].

Complementary to all of them, we have started an ambitious project towards building a large network that records the academic genealogy of researchers across fields and countries [4]. Our preliminary work used data from NDLTD, the Networked Digital Library of Theses and Dissertations[1] [6], and aimed to reconstruct advisor-advisee relationships from ETD records from many institutions around the world and from distinct disciplines.

In this paper, we move one step forward by constructing academic genealogy trees from a completely different data source, the Lattes Platform[2]. Maintained by CNPq, the Brazilian National Council for Scientific and Technological Development, this platform is an internationally renowned initiative [8] that provides a repository of researchers' curricula vitae and research groups, all integrated into a single system. In order to be able to submit any research grant proposal, all researchers in Brazil, from all levels (from junior to senior), are required to keep their curricula updated in this platform, which provides a great amount of information about the researchers' activities and their scientific production that can be used for many purposes. We then crawled the entire Lattes platform and collected the curricula of all researchers holding a PhD degree. Next, we developed a basic framework to extract specific data from the collected curricula, identify and disambiguate the respective researchers, and establish their advisor-advisee relationships, from which we carried out a series of analyses that describe the main properties of the genealogy trees we were able to construct. Finally, we developed the first version of a system that allows users to browse and explore the academic genealogy trees. We believe that this is the first large-scale effort to generate a general academic genealogy tree involving as much distinct research fields as possible.

2   Building the Academic Genealogy Trees

In this section, we discuss how we built the researchers' individual academic genealogy trees (AGT's, for short) using data from the Lattes Platform. To build such AGT's, we first crawled the Lattes Platform and collected the curricula vitae (in XML format) of 222,674 researchers holding a PhD degree. Then, following the procedure described by Algorithm 1, we parsed each collected curriculum extracting the data required to build the researchers' AGT's. Such data appears basically in two specific sections of each curriculum: the Identification section, which includes the researcher's name, institution and degrees held, and the Mentorships section, which includes the researcher's list of all Master's and PhD students she has advised in her career. Thus, for each one of these two sections, we wrote specific XPath queries to extract each required piece of data (e.g., the researcher's name and the names of her advisees). Note that the output of this procedure is actually a directed acyclic graph, since in her academic life

---

[1] http://www.ndltd.org.

[2] http://lattes.cnpq.br.

a researcher might have had more than one advisor (e.g., PhD and Master's) or acted as a co-advisor for one or more students.

According to Algorithm 1, in order to build the individual AGT's, we first sort the set of all collected curricula according to the researcher's PhD degree year (line 1). This aims to establish a chronological order to build such trees, thus avoiding unnecessary name matchings when processing the advisees' curricula. Then, we set the graph G empty (line 2). Next, for each curriculum in the set C (lines 3 to 17), we execute the following three main steps: (i) search G for the respective researcher's node, creating a new node if it does not yet exist or updating it otherwise (lines 4 to 6); (ii) search G for the nodes of the researcher's PhD and Master's advisors, creating them if they do not yet exist or updating them otherwise, and then connect them to the researcher's node (lines 7 to 10); (iii) for each researcher's advisee, search G for her respective node, creating it if it does not yet exist or updating it otherwise, and then connect it to the researcher's node (lines 11 to 16).

A critical component of our algorithm is the search function in lines 4, 7 and 12. Although the Lattes Platform provides an internal identifier for each researcher with a registered curriculum, it is not always possible to use this mechanism to instantaneously identify another researcher whose name appears, for instance, in the list of mentorships of a specific researcher's curriculum, since this requires some action from the researcher when updating her curriculum, which is not always done. Thus, to overcome this problem, we have implemented a simple, but quite effective strategy to handle this typical name disambiguation problem [5], which considers the following parameters as input for a similarity function: the researchers' names, the names of their institutions, the titles of their theses or dissertations, and the respective years of defense. However, a detailed discussion of this similarity function is out of the scope of this paper.

3 Characterizing the AGT's

In this section, we briefly characterize some aspects of the AGT's we have been able to build. Our main motivation is to identify aspects that highlight the legacy of a researcher, measured in terms of formation of other researchers, and not in terms of the traditional counts of publications, impact factor, and scientific discoveries. Table 1 shows some figures about the AGT's. Besides basic figures such as number of nodes, edges and trees, the later defined by the number of "roots" found in the graph (i.e., nodes without a known advisor), the table also shows the number of components (i.e., connected trees) and the values of two important metrics: the average tree size and the average tree width. The values of these two last metrics are calculated by dividing, respectively, the number of descendants by the number of subtrees (average size) and the number of out-links of all nodes by the number of nodes (width).

We have found in total 70,610 AGT's with 40.19 nodes on average. The average width of such trees is 3.81, i.e., each advisor in our dataset has advised on average 3.81 PhD or Master's students. Despite the average size of the trees

---

Algorithm 1. AGT Bulding Procedure

Input: A set C of Lattes Curricula;
Output: A graph G with all AGT's built;
1 Sort C by the researchers' PhD degree year;
2 Set G empty;
3 **foreach** *Curriculum c in C* **do**
4   Search G for the researcher's node n;
5   **if** *there is no such a node in G* **then** Create node n;
6   **else** Update the academic attributes of n;
7   Search G for the nodes p and m of the researcher's PhD and Master advisors;
8   **if** *either p or m are not found* **then** Create them;
9   **else** Update the academic attributes of p and m;
10   Connect p and m to n;
11   **foreach** *advisee in c* **do**
12   Search G for the advisee's node a;
13   **if** *there is no such a node in G* **then** Create node a;
14   **else** Update the academic attributes of a;
15   Connect a to n
16   **end**
17 **end**

---

being 40.19, the 10 largest trees have more than 5,000 nodes, although 80% of them have less than 20 nodes, as shown in Fig. 1(left graph). On the other hand, almost half of the trees have depth 1, as also shown in the same figure (right graph). We also noted that Brazilian trees are about 6.77 times wider than deeper. This number is much higher in comparison with the same ratio for trees built from NDLTD [4], which is 2.48. We conjecture that this difference might be related to the quality of the trees we have obtained from both sources. NDLTD contains theses and dissertations from many institutions and countries, but it is unclear which scientific community it represents. On the other hand, Lattes represents an entire and complete scientific community, as basically all Brazilian researchers are forced to regularly update their academic records on the platform.

Table 1. Characterization of the AGT's

# of Nodes	903,183	# of Components	22,061
# of Edges	1,144,051	Avg. Tree Size	40.19
# of Trees	70,610	Avg. Tree Width	3.81

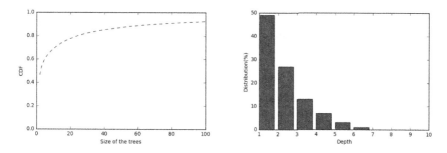

Fig. 1. Cumulative distribution function of the tree sizes and tree depth distribution

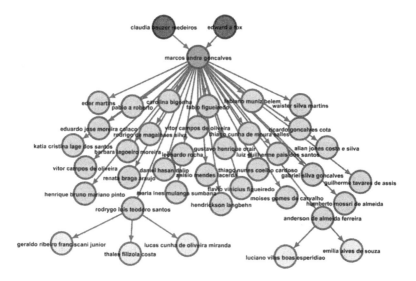

Fig. 2. Example of an academic genealogy tree built from Lattes (Color figure online)

4   Conclusions and Future Work

In this work, we used data crawled from the Lattes Platform to build a preliminary version of the Brazilian academic genealogy tree. Although still preliminary, our effort identified a number of interesting findings related to the structure of academic formation in Brazil, which highlight the importance of cataloging academic genealogy trees. Our effort, together with our previous work using data from NDLTD [4], allowed us to identify many challenges that we need to tackle towards developing a large repository that records the academic genealogy of researchers across fields and countries. More importantly, we have developed the first version of a system[3] that deploys the dataset studied here and allows users to browse the trees.

---

[3] Available at http://www.sciencetree.net.

To briefly illustrate the potential of this system, Fig. 2 shows an excerpt of the genealogy tree of Dr. Marcos André Gonçalves, a Brazilian associate professor at the Universidade Federal de Minas Gerais (UFMG), who is a well-known researcher in the digital library community. As we can see, the node colors represent the levels in his tres. The red nodes correspond to Dr. Gonçalves' advisors during his Master's and PhD studies, respectively Prof. Claudia Bauzer Medeiros, from UNICAMP, Brazil, and Prof. Edward A. Fox, from Virginia Tech, USA. The main subtree (the orange one) includes the graduate (Master's and PhD) students that have been advised by Dr. Gonçalves, which, in turn, span an additional level of subtrees (the yellow ones).

Thus, by analyzing this kind of tree we hope to better understand a research lineage. Moreover, we believe this system represents a preliminary step towards the understanding of more important questions related to science, which we will be able to answer once we have a world-wide academic genealogy tree. For example, this system would allow us to identify the important researchers within areas and the role they have played on the creation and evolution of scientific communities. It would also provide a better understanding about where research areas came from, the birth and death of research communities, the identification of one's academic lineage, and the role of interdisciplinary formation on the evolution of specific research fields. Ultimately, it would allow us to better comprehend the evolution of science and, consequently, of our society. We note, however, that the current version of our system is still beta and its development is part of our future work.

Acknowledgments. This research is funded by projects InWeb (grant MCT/CNPq 573871/2008-6) and MASWeb (grant FAPEMIG/PRONEX APQ-01400-14), and by the authors' individual grants from CAPES, CNPq and FAPEMIG. Fabrício Benevenuto and Alberto H.F. Laender are also supported by the Humboldt Foundation and IEAT, respectively.

References

1. Benevenuto, F., Laender, A.H.F., Alves, B.L.: The H-index paradox: your coauthors have a higher H-index than you do. Scientometrics **106**(1), 469–474 (2016)
2. Chang, S.: Academic genealogy of american physicists. AAPPS Bull. **13**(6), 6–41 (2003)
3. David, S.V., Hayden, B.Y.: Neurotree: a collaborative, graphical database of the academic genealogy of neuroscience. PLoS ONE **7**(10), e46608 (2012)
4. Dores, W., Benevenuto, F., Laender, A.H.F.: Extracting academic genealogy trees from the networked digital library of theses and dissertations. In: Proceedings of JCDL, pp. 163–166 (2016)
5. Ferreira, A.A., Gonçalves, M.A., Laender, A.H.F.: A brief survey of automatic methods for author name disambiguation. SIGMOD Rec. **41**(2), 15–26 (2012)
6. Fox, E.A., Gonçalves, M.A., McMillan, G., Eaton, J.L., Atkins, A., Kipp, N.A.: The networked digital library of theses and dissertations: changes in the university community. J. Comp. H. Educ. **13**(2), 102–124 (2002)

7. Jackson, A.: A labor of love: the mathematics genealogy project. Not. AMS **54**(8), 1002–1003 (2007)

8. Lane, J.: Let's make science metrics more scientific. Nature **464**(7288), 488–489 (2010)

9. Liu, X., Bollen, J., Nelson, M.L., Van de Sompel, H.: Coauthorship networks in the digital library research community. IPM **41**(6), 1462–1480 (2005)

10. Malmgren, R.D., Ottino, J.M., Amaral, L.A.N.: The role of mentorship in protégé performance. Nature **465**(7298), 622–626 (2010)

When a Metadata Provider Task Is Successful

Sarantos Kapidakis[✉]

Laboratory on Digital Libraries and Electronic Publishing,
Department of Archive, Library and Museum Sciences,
Ionian University, 72, Ioannou Theotoki Street, 49100 Corfu, Greece
sarantos@ionio.gr

Abstract. Computer services are normally assumed to work well all the time. In this work we examined the operation and the errors of metadata harvesting services and tried to find clues that will help predicting the consistency of the behavior and the quality of the harvesting. The large number of such services, the huge amount of harvested information and the possibility of meeting transient conditions makes this work hard. We studied 395530 harvesting tasks from 2138 harvesting services in 185 harvesting rounds during a period of 9 months, of which 214163 ended with error messages and the remaining tasks occasionally returning fewer records. A significant part of the OAI services never worked or have ceased working while many other serves occasionally fail to respond. It is not trivial to decide when a tasks is successful, as tasks that return without an error message do sometimes return records and also tasks that declare that complete normally sometimes return less or no records. This issue is fundamental for further analysis of the harvesting outcome and any assessment that may follow. Therefore, on this work we studied the error messages and the task outcome patterns in which they appear and also the tasks that returned no records, to decide on which is the most essential condition to decide when a task is successful. Our conclusion is that a task should be considered successful when it returns some records.

Keywords: OAI · Metadata harvesting · Reliability · Services · Temporary error · Permanent error

1 Introduction

Computer services like metadata harvesting and document retrieval do not always have the expected behavior. The users may notice delays or unavailability – but they have no idea how often these happen, and assume that each problem is rare and temporary. The big diversity of computer services, their different requirements, designs and interfaces and also network problems and user-side malfunctions, make it hard to know when the behavior of a service is normal or not and what to measure in each case. To overcome some of these restrictions, we examined the behavior of a large number of similar services of a specific type: data providers using the Open Archives Initiative Protocol for Metadata Harvesting (OAI-PMH).

Metadata harvesting is used very often, to incorporate the resources of small or big providers to large collections. The metadata harvesters, like Science Digital Library and

© Springer International Publishing AG 2017
J. Kamps et al. (Eds.): TPDL 2017, LNCS 10450, pp. 544–552, 2017.
DOI: 10.1007/978-3-319-67008-9_44

Europeana, accumulate metadata from many collections (or sources) through the appropriate services, belonging to metadata providers mostly memory institutions, by automatically contacting their services and storing the retrieved metadata locally. Their goal is to enable searching on the huge quantity of heterogeneous content, using only their locally store content.

The reliability of the services is important for ensuring current information. If the metadata harvesting service is not responding, the corresponding metadata records will not be updated at that time. Additionally, the unreliability - downtime of the metadata harvesting services usually indicate a proportional unreliability or downtime of the resource providing service, which always resides on the local sites, where both the local and the harvested metadata link to. When the resources are not available, the corresponding user requests are not satisfied, affecting the quality of the service.

In [5] Lagoze et al. discuss the NSDL development and explains why OAI-PMH based systems are not relatively easy to automate and administer with low people cost, as one would expect from the simplicity of the technology. It is interesting to investigate the deficiencies of the procedure.

The behavior and the reliability of a service, as well as the quality of its content, is important to the outcome and the satisfaction from the service. The evaluation and quality of metadata is examined as one dimension of the digital library evaluation frameworks and systems in the related literature, like [1, 6, 8]. Fuhr et al. in [1] propose a quality framework for digital libraries that deal with quality parameters. The service reliability falls under their System quality component.

In [7] Ward describes how the Dublin Core is used by 100 Data Providers registered with the Open Archives Initiative and shows that is not used to its fullest extent. In [2] Kapidakis studies the responsiveness of OAI services, and the evolution of the metadata quality over 3 harvesting rounds between 2011 and 2013. In [3] Kapidakis examines how solid the metadata harvesting procedure is, by making 17 harvesting rounds, over three years, from 2014 to 2016, and exploiting the results to conclude on the quality of their metadata as well as on their availability, and how it evolves over these harvesting rounds. The list of working services was decreasing every month almost constantly, and less than half of the initial services continued working at the end.

Nevertheless, the OAI services did not have a solid behavior, which make any assumption and conclusion harder to reach. In [4] Kapidakis explored the behavior of the information services over a small period of time, so that no permanent changes to their behavior were expected: during 21 harvesting rounds in three days intervals for a period of two months. He classified the services into five classes, according to their reliability in their behavior, and examined each class separately. He found that the service failures are quite a lot, and many unexpected situations are formed.

We have to address a fundamental issue: when a harvesting task should be considered successful. This issue is not trivial because (a) each harvesting task is a complex process, with many points of failure – not all of them of equal importance and (b) each harvesting task produces an outcome, that is either normal completion or an error message, but this outcome does not coincide with other task characteristics like the number of records returned: tasks that return without an error message do sometimes return records and also tasks that declare that complete without an error message sometimes return no records.

Therefore, on this work we studied the error messages and the patterns in which they appear and also the tasks that returned no records, to decide which is the most essential condition to characterise task as successful. For this reason, we performed and used a large number of harvesting rounds and examined in detail all harvesting error messages, to better understand them and the causes that triggered them.

The rest of the paper is organised as follows: In Sect. 2 we describe our methodology and the data we collected for that purpose. In Sect. 3 we study the error messages that the failed harvesting services returned and try to discover what they represent. In Sect. 4 we try to detect when the harvesting tasks time out and/or return records, to find patterns of behavior. We conclude on Sect. 5.

2   Methodology and Collected Data

It is difficult to understand, analyze or predict the behavior of network services, because it depends on many factors, many of which may be external to the service and unknown. Nevertheless, there may be some significant factors of the service configuration or maintenance, or their environment (including the accessing network) that can be considered.

To reveal common behavior patterns, we created an OAI client using the oaipy library and used several harvesting rounds, where on each one we asked each of the 2138 OAI-PMH sources listed in the official OAI Registered Data Providers (https://www.openarchives.org/Register/BrowseSites) on January of 2016 for a similar task: to provide 1000 (valid – non deleted) metadata records. Such tasks are common for the OAI-PMH services, which periodically satisfy harvesting requests for the new or updated records, and involve the exchange of many requests and responses, starting from the a negotiation phase for the supported features of the two sides. Our sequential execution of all these record harvesting tasks from the corresponding specific services usually takes more then a day to complete. Sometimes the tasks time out resulting to abnormal termination of the task: we set a timeout deadline to 1 h for each task.

We repeated a new harvesting round with a task for each service in constant intervals, asking the exact same requests every 36 h for a period of 9 months (June of 2016 to March of 2017). In the following, we further analyze the errors as permanent or temporary, and examine the error distribution per service.

Ideally, a task will complete normally, returning all requested metadata records – 1000, or all available records (even zero) if fewer are only available on that service. Other behaviors are also possible - and of interest when studying the behavior of each service. A task may return an error, declaring that the final task goal cannot be satisfied. We also consider as error the situation of the abnormal termination of a timed out task and such a task may return some records.

Finally, it is also possible to have a situation that the service actually returns less records than the ones available and requested, but reports that the task completes normally.

3   The Errors of the Services and Their Semantics

Our 214163 (54%) harvesting tasks did not complete normally. The reasons for such failures may be attributed to either the intermediate network or to the service – and can be either temporary or permanent. The fact that in all 185 rounds the 30155 tasks of 163 services consistently completed normally leads as to believe that the network connectivity of our OAI client is not a primary reason for the task failures in the remaining services, although we cannot exclude it, as the harvesting expands over many hours of operation.

A significant number of tasks fail on each round, and this rate is slowly increasing. The average number of failed tasks per round is 1158, with minimum 1108 and maximum 1222. Some rounds have an instantaneous increase to the number of failed tasks (round 175 is the maximum), but not too high to indicate a network problem on the OAI client side.

Our data confirms that the failed tasks are already surprisingly many and increase slowly, as was first discovered in [3]. Figure 1a shows how these task failures are mapped to the harvesting services, as each service includes 185 of these tasks, one for each round. We observe that 253 services (shown first) had no failures at all, the 1046 (shown last) services failed on all 185 of their tasks, and the remaining 839 services failed some times, most of them only a few, but others much more. Therefore, we have good, reliable services, a lot of constantly failing services and many that their outcome is affected by the environment.

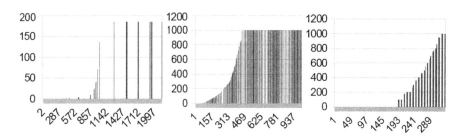

Fig. 1. (a) The number of failed tasks per service for the 2138 services. (b) The distribution of the number of records returned by a service and its frequency for the 1092 services. (c) The distribution of the number of returned records in 334 tasks that timed out.

Each OAI task consists of many actions, including mainly a negotiation in communication and many record requests. Therefore, its outcome may not be trivially classified as full success or failure, but may be something in between, with contradicting indicators, like when retrieving records and getting an error status, or when retrieving no records and completing normally. Therefore, the success of a task has to be clearly defined, so that it can be used later on. In this work, we decided to initially adopt the status returned from the task in order to characterise a success or failure, not considering the number of records actually returned. This is what each task declares anyway. This approach has the advantage that on the failures we can interpret the returned error status to further explore the reason for the failure.

The error messages reported by the failed tasks, as returned by each service and processed by the client, are briefly listed in the first column of Table 1, with the number of tasks they report them in the third column. They are service specific and the exact semantics of each error message is not globally defined.

Table 1. Service errors and the number of tasks they appear, outcome patterns they are contained and services that always return only them

Error short name	Error Category	Tasks	Outcomes	Services
URLError	Communication	68941	106	316
HTTPError	Communication	67035	96	160
BadVerbError	OAI-PMH protocol	33264	44	64
XMLSyntaxError	OAI-PMH protocol	23501	55	36
NoRecordsMatchError	OAI-PMH protocol	6082	26	4
DatestampError	OAI-PMH protocol	3835	11	6
BadArgumentError	OAI-PMH protocol	3220	10	6
error	Communication	3179	68	6
Error	Communication	1302	25	2
UnicodeEncodeError	OAI-PMH protocol	1267	5	
BadResumptionTokenError	OAI-PMH protocol	632	8	1
SSLError	Communication	610	5	1
BadStatusLine	OAI-PMH protocol	565	36	
CannotDisseminateFormatError	OAI-PMH protocol	317	3	
IncompleteRead	Communication	79	17	
Timeout		334	48	
Normal Completion		181367		253
Total		**395530**	**563**	**855**

Apart from the 181367 normal responses and the timed-out responses, we present the remaining errors by splitting them into two categories, the ones related to the network communication and the ones that are specific to the OAI-PMH operation, and its data interpretation and exchange and we show it on the second column of Table 1. We observe that the communication errors, and more specifically the *HTTPError* and *URLError*, appear much more often than the OAI-PMH protocol errors.

There are services that failed on each of their 185 tasks. There seems to be no coincidence behind this: there must be permanent issues that do not permit the service to complete normally (including possible incompatibility with our client), requiring human intervention (in the data or the software) to correct them. But these issues are not revealed easily, especially when the service does not fail in the same way each time. There may be some temporary issues, possibly on some tasks only, that apply in an earlier stage than the permanent ones, and force the task to terminate prematurely, reporting one of these issues instead, and hiding the permanent issues. Thus, the analysis of each behavior is usually complex.

In order to discover the semantics of the reported errors and the issues they imply, we have to examine their presence in many failed tasks and in the formed outcome patterns, especially since we have no access to the environment of a service (like the documentation for its operation) for further investigation.

In an effort to further clarify the semantics of the error messages, we investigated the combinations of outcomes that appear in the tasks of each service: either the error-less normal completion or an error message. By counting only the existence of the outcomes in all tasks on each service, and not their order or frequency, we reached 210 distinct outcome (completion/error) pattern combinations. The 106 of them were followed by only one service, and the rest 104 matched 2 or more services. The fifth column of Table 1 lists the number of services that follow each of the 12 service patterns that consist of only a single outcome (the same error message or normal completion) in all its 185 tasks.

In the remaining 198 outcome patterns with 2 or more (up to a maximum of 7) outcomes in their 185 tasks, the most common patterns consists of the *HTTPError* and *URLError* outcomes (93 services), and also one of these two and the normal completion (87 services). The other patterns are not followed by more than 8 services. The error messages that appear in the fifth column of Table 1 appear mostly as permanent errors, preventing any task to do any more progress.

Some error messages are more common in these patterns than others. The fourth column of Table 1 displays the error messages by frequency in the remaining 198 outcome patterns, a total of 563 error messages. There are 5 error messages that do not appear in the fourth column of Table 1: *CannotDisseminateFormatError*, *UnicodeEncodeError*, *BadStatusLine*, *IncompleteRead* and *Timeout*. These have a higher probability that they are temporary errors, but we cannot conclude that by their participation in the outcome patterns. Only *IncompleteRead* and *Timeout* are found with certainty to be temporary.

No error message seems to be part of very few error patterns. The frequency of all error messages in the patterns is more or less proportional to their overall frequency in all tasks.

4   Collective Behavior of the Services

Below, we examine the behavior of the participating services collectively, to better understand the possible outcomes, and their likelihood.

The 181367 tasks (46%) completed normally with average response time 56.9 s, although 69 of them returned no records. For tasks that returned records, the minimum response time was less than a second and the maximum was 3584 s, in a case that the information service returned 1000 records. Because of the short interval between our harvesting rounds, in most cases neither the service software nor the records in each service were modified between rounds.

The number of services that competed normally on each of the 185 rounds range from 916 to 1030, with an average of 980 and standard deviation 19. As there are no rounds that had much higher or lower normally completed task rates, we believe that there were no special conditions (possibly on our network) that affect the outcome of our tasks and contaminate our analysis, that should be excluded.

We study the distribution of the number of records returned by each service for common values, other than the requested records, 1000. Figure 1b presents the distribution of the number of records returned by each of the 1092 services that completed normally on each of the 185 rounds. 839 of the services completed normally only on some rounds, and 90 of them completed normally on all rounds, but did not provide the same number of records on each round. In these cases, the maximum number of records returned was considered. The remaining 163 services always returned the same number records. The more rounds a service completed normally – especially when their maximum is repeated in many rounds, the safer it is to assume that we have found the number of records that should be returned on each task.

Only 646 of the 1092 services returned 1000 records, and may actually hold more records. On the other hand, 69 services completed normally although always returned 0 records. The remaining 377 services returned from 2 to 991 records (270 of them returned the same records consistently in all 185 rounds – the others occasionally returned less) and with no obvious pattern. The number of returned records seems to vary unpredictably, so it should be the number of records that the service can provide, and not a number dictated by the software API..

The other 214163 (54%) tasks ended with an error: in 206138 (96%) of these tasks, no records were returned at all, while in the remaining 8025 tasks some records were indeed returned.

Among the failed tasks, 334 tasks did not terminate normally but timed out after an hour of trying. With an average service response time less than 57 s, and a standard deviation close to 137 s in the 181367 tasks that did not completed normally, our 3600 s time out deadline seems reasonable. The 188 of the timed out tasks (56%) returned no records, while 14 such tasks returned all 1000 records, but afterwards did not terminate normally, but timed out. The remaining 132 timed out tasks returned from 50 to 998 records, with an average of 442 but with no obvious concentration of values and with very high standard deviation (268).

The tasks that timed out and returned zero records are between 1 and 12 (1.7 on average) per round from the 188 tasks on 110 rounds, while the tasks that timed out and returned some records but not all 1000 are between 1 and 4 (1.3 on average) per round from the 132 tasks on 100 rounds. It seems that the tasks that timed out and returned zero records coexist more often, and the network conditions on the harvesting round have a small correlation to the time out. Figure 1c shows how the distribution of the number of returned records for the 334 timed out tasks.

Only 121 of the 2138 services ever had a time out, up to 43 ones, with an average of 2.8 and standard deviation of 6.4, in the 334 total tasks. Only 7 services timed out in more than 5 rounds, while most services only timed out in one (90 services) or two (17 services) rounds. Most of them return 0 records, but the rest of the timed out tasks return a number of records that has an almost uniform distribution, up to the maximum 1000. Our data also indicate that the time out behavior do change (on numbers and returned results) on each round.

The 8025 tasks that failed but retuned some records do failed with all 15 different error messages (the majority of them, 5030, failed with the *XMLSyntaxError* message).

Excluding the tasks that timed out, 7879 other tasks also returned records and 21 of them (with 4 different error messages) returned all 1000 requested records. The remaining 7858 tasks returned from 5 to 997 records. Thus the errors can occur on all stages of the service requests, and may not affect the task in an important way.

The 7879 tasks that failed, but not with a time out, and retuned some records had a maximum response time of 2374 s and an average of 51 s, which is even less that the maximum response time of 3585 s and the average of 60 s of the tasks that completed normally and also returned some records. Thus these failed tasks have a smaller response time, failing in an earlier stage.

The 8797 tasks that completed normally but returned no records, they all completed after a maximum of 930 s, and on the average on 3.5 s, and were the fastest of all tasks. Most of these tasks belong to 17 services that returned no records on all 185 rounds and on 16 services that returned no records on 184 rounds. Also, 34 services had one such task each, and fewer services had from 2 to 182 such tasks. We conclude that a few services consistently return no records, while others may do so temporarily.

5   Conclusions and Future Work

The OAI tasks were executed in multiple rounds, which were in small intervals to avoid significant changes on the services. The behavior of a service over multiple rounds of tasks is not always consistent, and often changes between two rounds, because of temporary problems.

OAI is a protocol that works unattended but needs site administration, maintenance and monitoring tools. OAI and its implementations are vulnerable to many network conditions and often return less records than those requested.

Most errors can appear as both permanent and temporary. The OAI servers and clients should use more fine grained error messages, and each one of them would cover with fewer semantics that will also indicate permanent and temporary unrecoverable issues.

The time out behavior seems to be strongly correlated to the specific service, as some services are much more prone to time out than others.

It seems appropriate to consider a task successful when it returns any records, even when it ended with an error or time out. Furthermore, the number of records returned, when less than the ones available by the service, does not seem biased for services either completing normally or with an error message. This consideration contradicts the reasonable alternative to only consider the return outcome for the task success, as the two considerations often disagree: when a task returns no records on its normal completion and when a task returns records while also returning an error message.

In the future, we plan combine the response time of the tasks and services with the conditions of the time out, of the incomplete read and of the return of zero records to see how they are related.

References

1. Fuhr, N., Tsakonas, G., Aalberg, T., Agosti, M., Hansen, P., Kapidakis, S., Klas, P., Kovács, L., Landoni, M., Micsik, A., Papatheodorou, C., Peters, C., Sølvberg, I.: Evaluation of digital libraries. Int. J. Digit. Libr. **8**(1), 21–38 (2007)
2. Kapidakis, S.: Rating quality in metadata harvesting. In: Proceedings of the 8th ACM International Conference on PErvasive Technologies Related to Assistive Environments (PETRA 2015), Corfu, Greece, 1–3 July 2015. ACM International Conference Proceeding Series (2015). ISBN 978-1-4503-3452-5
3. Kapidakis, S.: Exploring metadata providers reliability and update behavior. In: Fuhr, N., Kovács, L., Risse, T., Nejdl, W. (eds.) TPDL 2016. LNCS, vol. 9819, pp. 417–425. Springer, Cham (2016). doi:10.1007/978-3-319-43997-6_36
4. Kapidakis, S.: Exploring the consistent behavior of information services. In: CSCC 2016, Corfu, 13–16 July (2016)
5. Lagoze, C., Krafft, D., Cornwell, T., Dushay, N., Eckstrom, D., Saylor, J.: Metadata aggregation and automated digital libraries: a retrospective on the NSDL experience. In: Proceedings of the 6th ACM/IEEE-CS Joint Conference on Digital Libraries (JCDL 2006), pp. 230–239 (2006)
6. Moreira, B.L., Goncalves, M.A., Laender, A.H.F., Fox, E.A.: Automatic evaluation of digital libraries with 5SQual. J. Informetrics **3**(2), 102–123 (2009)
7. Ward., J.: A quantitative analysis of unqualified dublin core metadata element set usage within data providers registered with the open archives initiative. In: Proceedings of the 3rd ACM/IEEE-CS Joint Conference on Digital Libraries (JCDL 2003), pp. 315–317 (2003). ISBN 0-7695-1939-3
8. Zhang, Y.: Developing a holistic model for digital library evaluation. J. Am. Soc. Inf. Sci. Technol. **61**(1), 88–110 (2010)

Semantic Enrichment of Web Query Interfaces to Enable Dynamic Deep Linking to Web Information Portals

Arne Martin Klemenz[✉] [iD] and Klaus Tochtermann

ZBW – Leibniz-Information Centre for Economics, Kiel, Germany
{a.klemenz,k.tochtermann}@zbw.eu

Abstract. This article addresses how to improve the automated accessibility and visibility of information from *Web Information Portals* and in particular virtual library systems. Information from web information portals could provide great value to satisfy information needs. But most of this information stays hidden in data silos which are part of that section of the web that is not indexable by common search engines and is therefore called *Deep Web*. Shared vocabularies like *Schema.org* helped to increase machine readability of structured information on the web in general, but markup vocabularies didn't increase the accessibility and visibility of information from data silos. This article addresses the limitations regarding the accessibility of information from data silos on the Deep Web and proposes an extension to Schema.org to fill the identified gaps. The extension improves the automated accessibility and visibility of information provided in web information portals by providing *Dynamic Deep Linking* capabilities to Deep Web data silos by lifting web forms of web information portals to the level of machine understandable semantic *Web Query Interfaces*.

Keywords: Dynamic Deep Linking · Schema.org extension · Web query interface · Web information portals · Virtual library systems

1 Introduction

The Web is a continuously growing diverse set of information. Various kinds of web-accessible data sources like search or information portals provide access to vast amounts of information and can be classified into service-oriented web services on the one hand and user-oriented web portals on the other hand. Service-oriented web services like web APIs (Application Programming Interfaces) provide information in a machine readable and accessible way to ensure its retrievability. In contrast to this, user-oriented web portals like information portals and in particular virtual library systems provide information in a way that is machine readable in terms of displaying purpose but primarily intended to be human readable. Information in user-oriented web portals – hereafter referred to as *Web Information Portals* – is usually retrieved based on user interactions like user-initiated web form submissions. These web form submissions become the focal point of interest in this article as it makes automated access and retrieval of information difficult and therefore causes this information to be hidden in *data silos* of the so called *Deep Web* [1].

© Springer International Publishing AG 2017
J. Kamps et al. (Eds.): TPDL 2017, LNCS 10450, pp. 553–559, 2017.
DOI: 10.1007/978-3-319-67008-9_45

The Deep Web refers to the part of the web that is not indexable by common search engines due to its limited accessibility. Web crawlers basically rely on hyperlinks to discover new information on the web [6]. The limited automated accessibility and therefore limited visibility of information from web information portals is caused by the likewise high complexity of web form submissions that rely on user input.

In contrast, the *Surface Web* [1] consists of information that can be easily accessed by common search engines. Most users even rely on search engines when searching information from known information sources like a well-known subject portal [9]. In contrast to classical library *OPACs* (Online Public Access Catalogues) which provide a service for local users, modern virtual library systems aim to provide global services. Whereas classical OPACs often exclude search engine crawlers on purpose [2], virtual libraries rely on search engine optimizations to reach their targeted user group.

This article proposes the semantic enrichment of web forms based on an extension to the open vocabulary standard *Schema.org*. The extension investigates the potential of semantic annotations for web forms to provide *Dynamic Deep Linking* capabilities and therefore increase the automated accessibility of information from web information portals with a focus on the special characteristics of virtual library systems.

2   Related Work

The above described user behavior in information gathering corresponds with the search engines intention to refine their search results based on the usage of structured data. For example, this applies to rich snippets [11] as well as embedded additional information in search results pages. This was one of the major reasons for launching the Schema.org initiative. Schema.org schemata supported the semantic description of static entities with a focus on optimized ranking and rich snippet generation. The schemata had a wider range and less specialization in contrast to previous vocabularies and extremely improved machine readability of information that is covered by the provided schemata. But this just applies to information that was already accessible by search engines and so far just machine readable in terms of displaying purpose.

As the web is not just about static descriptions of entities, the Schema.org community announced Schema.org *Actions*[1] to describe actions performed on entities. One of these indicates search actions based on form submissions to ease access to annotated websites for web crawlers and provide *Sitelinks Search*[2] functionality. As lots of information on the web still cannot be reached due to the semantic complexity of web forms, this work investigates further potential of semantic annotations for web forms with the application of *Linked Data* principles. These principles refer to "best practices for publishing and connecting structured data [...] to create typed links between data from different sources" [3] and led to improvements regarding the retrievability of information on the web, e.g. based on *Semantic Search* capabilities.

---

[1] Schema.org Actions: http://blog.schema.org/2014/04/announcing-schemaorg-actions.html
[2] Sitelinks Search: https://developers.google.com/structured-data/slsb-overview

The challenge of improving the automated accessibility of information on the web has been addressed from several perspectives in the past. On the one hand, Semantic Web Services play an increasingly important role in web data integration processes. In particular, *Hydra* a lightweight vocabulary has been published aiming to create hyper-media-driven Web APIs [8]. In contrast to this article, the Hydra W3C community puts its focus on service-oriented web services. Moreover, the *OpenURL* framework for open reference linking in the scholarly information environment provides linking capabilities to library services going beyond the classic notion of a reference link [10]. It "provides a standardized format for transporting bibliographic metadata [...] between information services" [10]. The format is mainly used for link resolvers and has the basic concept of deep linking information in target services in common with this work. But in contrast to the service-oriented OpenURL format this work provides a more generic vocabulary with Dynamic Deep Linking capabilities to user-oriented web information portals.

On the other hand, previous research focuses on the extraction of information from Deep Web data sources. Special emphasis has been placed on surfacing information respectively the information extraction from the Deep Web [4] based on automated web form discovery, understanding and classification approaches [12, 13]. Additionally, general Deep Web harvesters have been developed [5, 14]. While most of these approaches address the Deep Web data integration challenges from the retrieving services perspective and have to deal with either strict limitations regarding their application domain or their efficiency, this work switches the perspective and addresses these challenges from the information providing services perspective [7].

3 Semantic Deep Search Extension

Semantic annotations provide further potential for the improved access to Deep Web data silos and in particular web information portals. Virtual library systems rely on expert generated bibliographic metadata to describe their provided content. This bibliographic metadata is usually based on authority data and vocabularies defined as thesauri which play a significant role in the targeted search. Whereas Linked Data principles have a widespread use in service-oriented information provision and access, they still lack recognition when accessing user-oriented web information portals. We investigate the further potential of expert generated bibliographic metadata based on additional semantic annotations for web forms following the Linked Data principles. Web forms are hereafter referred to as *web query interfaces* to indicate their relevancy for the automated accessibility of information in web information portals.

The *Semantic Deep Search Extension* to Schema.org should meet the following challenges to enable better automated access to Deep Web data silos:

1. Identify web query interfaces of web information portals (*Service Discovery*)
2. Select web query interfaces for specific information need (*Service Classification*)
3. Generate service-specific query URLs (*Dynamic Deep Linking*)

According to the introduced switch in perspective in contrast to approaches from the retrieving services perspective, *Deep Web Service Endpoints* (underlying retrieval

service of a web query interface) need to be self-describing in terms of general classification purpose and in terms of their detailed URL parameter assignment. As a result,
service endpoints serve as semantic APIs to the primarily just user-accessible search
functionality of web information portals. The overall objective of this approach is a
framework for *Semantic Deep Search* based on *Dynamic Deep Linking*.

In contrast to Schema.org's syntactical *SearchAction* markup, the extension provides
a vocabulary that is capable of describing the detailed web query interface semantics.
General information for the discovery and classification of the service is provided based
on so called *content* properties: e.g. content domain, language, content type and licensing
information. Furthermore, this extension specifies detailed *WebFormElement* properties
which correspond to the service parameters or sets of related parameters of a web query
interface to describe *semantic parameter constraints* like input domain (valid parameter
values) and output range (output restrictions triggered by input values) as well as *structural* or *semantic parameter dependencies*.

This model is based on the formalization of a web query interface. A web query
interface can be formally described as a service with a set of input variables X which
correspond to the service parameters. The result set of a query submission is based on
several restrictions to the underlying dataset defined by each input variable $x_i \in X$. Each
variable $x_i \in X$ has a specified input domain inp_i. An assigned variable value specifies
a restriction out_i regarding the result dataset. The input domain inp_i and output restriction
out_i of x_i can be formalized as graph patterns. Furthermore, there might exist semantic
dependencies between related variables or sets of variables. For example, the value of
a variable $x_i \in X$ (e.g. select field) might restrict or redefine the input domain inp_j and
output restriction out_j of a related variable $x_j \in X$ (e.g. input field). The altered input
domain inp_j' is a restricted subset or redefinition of inp_j and the output restriction out_j' is
a restricted subset or redefinition of out_j. Therefore, the overall result of a query submission for a valid variable assignment $X = \{x_1, \dots, x_n\}$ can be specified by a conjunctive

query: $CQ(X) = T_0 \cap \left(\bigcap_{x_i \in X} T_{out_i} \right) = T_r.$

The result set can described as graph pattern T_r which is the conjunctive intersection
of all graph subsets T_{out_i} defined by formal restrictions out_i of variables $x_i \in X$ and the
provided dataset of a web information portal described by graph pattern T_0.

The Semantic Deep Search Extension for Schema.org provides the expressiveness
to define these introduced formal parameter constraints and semantic dependencies. The
full vocabulary extension is publicly available on the extension website[3].

In the following, selected aspects of a prototype annotation utilizing the implemented
extension will be introduced to illustrate some of the main concepts like semantic
parameter dependencies and parameter constraints. The prototype implementation is
based on a web query interface from EconBiz[4]. The described example web query interface consists of three parameters: input field, select field and checkbox. The specified

---

[3] Semantic Deep Search Extension website: http://semdeepsearch.vocab-ext.appspot.com/
[4] EconBiz, subject portal for economics and business studies: https://www.econbiz.de/

select field defines restrictions to the whole range of attributes of the specified input field depending on its selected value. As an example, the restriction caused by the selection of the predefined value "Author" can be described based on the following annotation (in Notation 3; _:elem1 references input field):

```
@prefix schema: <http://schema.org/> .
@prefix gndo: <http://d-nb.info/standards/elementset/gnd#> .
... a schema:WebFormParameterValue ;
      schema:webFormParameterValue "Author" ;
      schema:definesWebFormElementRestriction
        [ a schema:WebFormElementRestriction ;
          schema:restrictedWebFormElement _:elem1 ;
          schema:webFormParameterVocabulary
            "http://d-nb.info/gnd/" ;
          schema:webFormParameterInputDomain
            [ a schema:ValueInputDomain ;
              schema:valueInputDomainClass "gndo:Person" ;
              schema:valueInputDomainProperty
                "gndo:preferredNameForThePerson" ] ;
          schema:webFormParameterOutputRange "schema:author" ] ...
```

With ?val_elem1 as value of the input field element, the corresponding output restriction can be formalized as RDF statement:

```
?person a gndo:Person .
?person gndo:preferredNameForThePerson ?val_elem1 .
?publication a schema:CreativeWork .
?publication schema:author ?person .
```

The Deep Search URL which can be generated based on the semantic annotation is: https://econbiz.de/Search/Results?lookfor=[?val_elem1]&type=Author.

In addition, search operators have special relevance for virtual library systems. It is common to refine a search query based on classifications, descriptors or authority data. To be capable of describing specifications like these, the extension introduces *WebForm-PrefixSearchOperator*s as shown in the following example annotation:

```
... a schema:WebFormPrefixSearchOperator ;
      schema:prefixSearchOperatorPrefix "gnd" ;
      schema:prefixSearchOperatorPrefixNamespace
        "http://d-nb.info/gnd/" ;
      schema:prefixSearchOperatorVocabulary "http://d-nb.info/gnd/" ;
      schema:prefixSearchOperatorInputDomain "gndo:Person" ;
      schema:prefixSearchOperatorRange "schema:author" . ...
```

This annotation will allow parametrized Linked Data based Deep Links like: https://econbiz.de/Search/Results?lookfor=gnd:[GND-IDENTIFIER]

Overall, automated systems are capable of generating service-specific query URLs for a specific information need to perform a *Semantic Deep Search* based on *Dynamic Deep Linking*. This means that automated systems, which understand the web query interface markup, are able to provide links to any search results page for any specific information need. The entire markup example applying the full vocabulary range of the extension is available for review on the vocabulary extension website.

4   Summary and Future Work

This article introduced the *Semantic Deep Search Extension* to Schema.org to improve the automated accessibility of web information portals and in particular virtual library systems. These are part of the so called Deep Web and thus not indexable by common search engines. The extension adds further expressiveness to Schema.org *SearchActions* to provide a semantic markup for *WebForms*. As a result, the information providing web information portal is able to provide a self-describing semantic annotation that enables automated access. In contrast, previous research focused these challenges from the retrieving services perspective, e.g. based on form understanding. The semantic markup allows automated systems to lead users to the search results page of an annotated web information portal that is the most expedient according to their information need. The generation of query URLs is introduced as *Dynamic Deep Linking* and has great potential in combination with *Semantic Search* strategies.

Our future work will focus on the distribution of the introduced extension to the Schema.org community with the intention to enter the routine for official extension candidates. The widespread acceptance of the extension is fundamental to achieve its full potential and finally its acceptance by the search engines that are part of the Schema.org initiative. In addition, future work will concern evaluation studies.

References

1. Bergman, M.K.: White paper: the deep web: surfacing hidden value. J. Electron. Publish. **7**(1), 1–17 (2001)
2. Blandford, A.: Google, public libraries, and the deep web. Dalhousie J. Interdiscip. Manage. **11** (2015)
3. Bizer, C., Heath, T., Berners-Lee, T.: Linked data-the story so far. In: Semantic Services, Interoperability and Web Applications: Emerging Concepts, pp. 205–227 (2009)
4. Ferrara, E., De Meo, P., Fiumara, G., Baumgartner, R.: Web data extraction, applications and techniques: a survey. Knowl.-Based Syst. **70**, 301–323 (2014)
5. Furche, T., Gottlob, G., Grasso, G., Guo, X., Orsi, G., Schallhart, C.: The ontological key: automatically understanding and integrating forms to access the deep Web. VLDB J. **22**(5), 615–640 (2013)
6. Henzinger, M.R.: Hyperlink analysis for the web. IEEE Internet Comp. **5**(1), 45–50 (2001)
7. Klemenz, A.M., Tochtermann, K.: Semantification of Query Interfaces to Improve Access to Deep Web Content. SDA, pp. 104–111 (2013)
8. Lanthaler, M., Gütl C.: Hydra: a vocabulary for hypermedia-driven web apis. In: LDOW, vol. 996 (2013)
9. Purcell, K., Brenner, J., Rainie L.: Search engine use 2012 (2012)
10. Van de Sompel, H., Beit-Arie, O.: Open linking in the scholarly information environment using the OpenURL framework. New Rev. Inf. Netw. **7**(1), 59–76 (2001)
11. Steiner, T., Troncy, R., Hausenblas, M.: How Google is using linked data today and vision for tomorrow. In: Proceedings of Linked Data in the Future Internet, vol. 700 (2010)

12. Wang, L., Hawbani, A., Wang, X.: Focused deep web entrance crawling by form feature classification. In: Wang, Yu., Xiong, H., Argamon, S., Li, X., Li, J. (eds.) BigCom 2015. LNCS, vol. 9196, pp. 79–87. Springer, Cham (2015). doi:10.1007/978-3-319-22047-5_7

13. Zhang, Z., He, B., Chen-Chuan Chang, K.: Understanding web query interfaces: best-effort parsing with hidden syntax. In: Proceedings of the 2004 ACM SIGMOD International Conference on Management of Data, pp. 107–118. ACM (2004)

14. Zhao, F., Zhou, J., Nie, C., Huang, H., Jin, H.: SmartCrawler: a two-stage crawler for efficiently harvesting deep-web interfaces. IEEE Trans. Serv. Comput. 9(4), 608–620 (2016)

A Complete Year of User Retrieval Sessions in a Social Sciences Academic Search Engine

Philipp Mayr[(✉)] and Ameni Kacem

GESIS - Leibniz Institute for Social Sciences, Cologne, Germany
{philipp.mayr,ameni.sahraoui}@gesis.org

Abstract. In this paper, we present an open data set extracted from the transaction log of the social sciences academic search engine sowiport. The data set includes a filtered set of 484,449 retrieval sessions which have been carried out by sowiport users in the period from April 2014 to April 2015. We propose a description of interactions performed by the academic search engine users that can be used in different applications such as result ranking improvement, user modeling, query reformulation analysis, search pattern recognition.

Keywords: Whole session retrieval · Information behavior · Session log analysis · User session data · Social sciences users

1 Introduction

Every Digital Library (DL) system generates huge amounts of usage data and DL operators often face the problem of not being able to report about the real usage on an expressive level that is moreover understandable for laymen. Reporting average statistics like number of unique sessions, page impressions, amount of actions and even click-through rates is not enough because these numbers cannot represent and explain the underlying pattern of the information behavior of DL users. Exploratory search in DLs and academic search engines [1] is a rewarding research environment for interactive IR researchers because evolving searches with complex search tasks can be observed much easier compared to web search where searchers often jump into different websites. In DLs, users typically stay in the system and work with the variety of facilities it offers. This is due to the fact that state-of-the-art DLs offer dozens of possibilities to navigate and interact with the search system [2,3]. Our motivation in proposing this data set is grounded in the observation that in the field very few open data sets which support whole session investigation exist. To the best of our knowledge there is no open data set available from academic search engines or DLs with full coverage of whole session information. Among the available data sets, we find the most famous evaluation campaign TREC (Text REtrieval Conference) which proposed TREC Session[1] [4] and Interactive[2] tracks. In fact, one way to

---

[1] http://trec.nist.gov/data/session.html.

[2] http://trec.nist.gov/data/interactive.html.

© Springer International Publishing AG 2017
J. Kamps et al. (Eds.): TPDL 2017, LNCS 10450, pp. 560–565, 2017.
DOI: 10.1007/978-3-319-67008-9_46

enhance the development and evaluation of information-seeking systems is to propose shareable data sets in order to facilitate the collaboration within an interdisciplinary team including developers, computer scientists, and behavioral experts who work together in order to explore new ideas and propose improvements [5].

Consequently, with the proposed data set we want to support DL developers and IR researchers to work on the analysis of whole retrieval sessions. These practitioners need such data sets to propose methods and techniques which allow us to examine search steps, analyze usage data, understand the underlying information behavior covered in search sessions that are performed by geographically distributed persons.

2   Related Work

Interactive information retrieval (IIR) refers to a research discipline that studies the interaction between the user and the search system. In fact, researchers have moved from considering only the current query to consider the user's past interactions. Research approaches aim to understand the user search behavior in order to improve the ranking of results after submitting a query and enhance the user experience with an IR system. Thus, they study concepts such as search strategies [1,6], search term suggestions [7], communities' detection [8], personalization of search results, recommendation's impact [7], users information needs frequency and change. Many interactive IR models have been proposed in the literature (e.g. [9]) that describe the user's behavior by different steps (stages) of information seeking and interacting with an information retrieval system. In order to evaluate and analyze such models and approaches log analysis has been introduced. In [10], the authors proposed a detailed overview of the history and development of transaction log analysis by examining possible applications and features analysis. Jones et al. [11] investigated transaction logs for the Computer Science Technical Reports Collection of the New Zealand DL. The authors analyzed query complexity, query terms change, sessions frequency and length.

3   Dataset

Sowiport[3] is a DL for the Social Sciences that contains more than nine million records, full texts and research projects included from twenty-two different databases whose content is in English and German [2]. This data set **Sowiport User Search Sessions Data Set (SUSS)**[4] [12] contains individual search sessions extracted from the transaction log of sowiport. The data was collected over a period of one year (between 2nd April 2014 and 2nd April 2015). The web server log files and specific JavaScript-based logging techniques were, first, used to capture the user behavior within the system. Then, the log was heavily filtered to

---

[3] http://www.sowiport.de.
[4] To download the dataset: http://dx.doi.org/10.7802/1380.

exclude transactions performed by robots and short interactions limited to one action per session. After that, all transaction activities are mapped to a list of 58 different user actions which cover all types of activities and pages that can be carried out/visited within the system (e.g. typing a query, visiting a document, selecting a facet, exporting a document, etc.). For each action, a session id, the date stamp and additional information (e.g. query terms, document ids, and result lists) are stored. Based on the session id and date stamp, the step in which an action is conducted and the length of the action is included in the data set as well. The session id is assigned via browser cookies and allows tracking user behavior over multiple search sessions. Session boundaries were specified after a threshold period indicating a period of inactivity and thus the end of the session. In our data set this threshold is equal to 20 min. Thus, in the data set we find 484,449 individual search sessions and a total of 7,982,427 log entries.

4   Preliminary Analysis

In this section, we present first descriptive analysis of the SUSS data set regarding sessions, users and searches. These analyses are not following concrete research questions but are intended to show the richness of this open data set.

4.1   Description of Actions

Searching sowiport can be performed through an *All fields* search box (default search without specification), or through specifying one or more field(s): title, person, institution, number, keyword or year. The users' main actions are described in Table 1. We grouped the main actions into two categories: "Query"-related and "Document"-related actions. Another categorization of actions was proposed in [7] by specifying search interactions and successive positive actions.

4.2   Users and Sessions

Given the data set described in Sect. 3, we first analyze the user types. A user can perform a search and submit a query to sowiport without signing up. Registered users can keep the search history, add a document to favorites and create favorite lists according to their interests. We found 1,509 registered users who performed 3,372 unique sessions (0.69%). The rest of the sessions in sowiport were performed by non-registered users (99.31%).

4.3   Investigation of Actions

Main user actions as described before can be categorized into actions regarding either search queries or documents. These actions are used in different scales in the data set. Query-related actions represent 29.84% while document-related actions represent 35.79% of the total amount of actions. The rest of actions contain navigational interactions such as logging in the system, managing favorites, and accessing the system pages.

Table 1. Main actions performed by users in sowiport

Category	Action	Description	Frequency
Query	query_form	Formulating a query	179,964
	search	A search result list for any kind of search	848,556
	search_advanced	A search with the advanced settings that can limit the search fields, information type, etc	103,432
	search_keyword	A search for a keyword	43,608
	search_thesaurus	Usage of the thesaurus system	71,599
	search_institution	A search for an institution	13,104
	search_person	A search for a specific person (author/editor)	93,083
Document	view_record	Displaying a record in the result list after clicking on it	1,344,361
	view_citation	View the document's citation(s)	24,994
	view_references	View the document's references	2,086
	view_description	View the document's abstract	86,752
	export_bib	Export the document through different formats	27,229
	export_cite	Export the document's citations list	27,385
	export_mail	Send the document via email	10,987
	to_favorites	Save the document to the favorite list	5,431

Figure 1 shows the frequencies of the top six most used actions by the users in the data set. We notice that the actions *"view_record"* and *"search"* are the most used ones before *"query_form"* and *"search_keyword, person, institution"*.

In Table 2, we show a specific session, the user's ID and the actions' label and length in seconds. In this session, the user with ID *41821* started with logging into the system and then submitted a query describing his/her information need (*query_form*). After getting the result list, labeled as *resultlistids* and viewing a document, the user performed additional searches (*searchterm_2*), and displayed some results' content (*view_record*). Finally, he/she checked the external availability of a result (*goto_google_scholar*). We notice that the user spent more than 40% of the time reading documents' content.

In Fig. 2, we display the number of actions per session. We note that the average number of actions per session is 16 and only sessions with a minimum of one action are considered in this data set. We conclude, from this figure, that the number of sessions with less than 16 actions (n = 384,087) is much larger than the number of sessions having over 16 actions (n = 100,360).

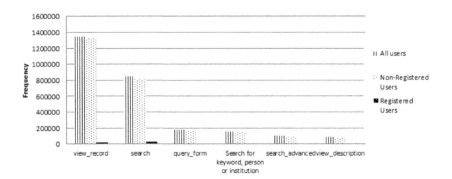

Fig. 1. Frequency distribution of the six most performed action groups

Table 2. Sample of a session search for a specific user

User ID	Date	Action label	Action length (s)
41821	2014-10-28 16:08:46	goto_login	1
	2014-10-28 16:09:13	query_form	22
	2014-10-28 16:09:35	search	10
	2014-10-28 16:09:35	resultlistids	10
	2014-10-28 16:09:45	view_record	31
	2014-10-28 16:09:45	docid	31
	2014-10-28 16:10:16	view_record	392
	2014-10-28 16:16:48	search	10
	2014-10-28 16:16:48	searchterm_2	10
	2014-10-28 16:16:48	resultlistids	10
	2014-10-28 16:16:58	view_record	9
	2014-10-28 16:17:07	goto_google_scholar	0

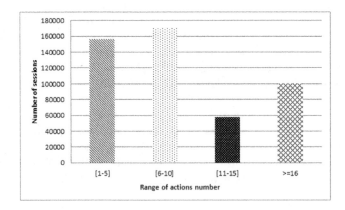

Fig. 2. Distribution of the Number of actions contained in a session

5   Future Work

For academia there is a need for open data sets which provide information about the variety of retrieval sessions and help to study and understand the abstract information behavior and common scan paths of academic users in a DL. In fact, session log provision and investigation open opportunities to enhance DLs' systems and to offer new services. Some possible future work based on our proposed data set can be outlined as follows: finding and studying abstract user groups like exhaustive or effective users; modeling academic users; analyzing reformulation and refining strategies; identifying various search phases like starting; chaining, browsing and differentiating; task characterization and prediction; personalization of search results according to the user behavior within search sessions.

Acknowledgement. This work was funded by Deutsche Forschungsgemeinschaft (DFG), grant no. MA 3964/5-1; the AMUR project at GESIS together with the working group of Norbert Fuhr. The AMUR project aims at improving the support of interactive retrieval sessions following two major goals: improving user guidance and system tuning.

References

1. Carevic, Z., Lusky, M., van Hoek, W., Mayr, P.: Investigating exploratory search activities based on the stratagem level in digital libraries. Int. J. Dig. Libr. (2017). https://link.springer.com/article/10.1007/s00799-017-0226-6
2. Hienert, D., Sawitzki, F., Mayr, P.: Digital Library Research in Action Supporting Information Retrieval in Sowiport. D-Lib Mag. **21**(3/4) (2015). doi:10.1045/march2015-hienert, http://www.dlib.org/dlib/march15/hienert/03hienert
3. Fuhr, N., et al.: Evaluation of digital libraries. Int. J. Dig. Libr. **8**(1), 21–38 (2007)
4. Kacem, A., Boughanem, M., Faiz, R.: Emphasizing temporal-based user profile modeling in the context of session search. In: SAC, pp. 925–930. ACM (2017)
5. Kelly, D., Dumais, S.T., Pedersen, J.O.: Evaluation challenges and directions for information-seeking support systems. IEEE Comput. **42**(3), 60–66 (2009)
6. Carevic, Z., Mayr, P.: Survey on high-level search activities based on the stratagem level in digital libraries. In: Fuhr, N., Kovács, L., Risse, T., Nejdl, W. (eds.) TPDL 2016. LNCS, vol. 9819, pp. 54–66. Springer, Cham (2016). doi:10.1007/978-3-319-43997-6_5
7. Hienert, D., Mutschke, P.: A usefulness-based approach for measuring the local and global effect of IIR services. In: Proceedings of CHIIR 2016, pp. 153–162. ACM (2016)
8. Akbar, M., Shaffer, C.A., Fox, E.A.: Deduced social networks for an educational digital library. In: Proceedings of JCDL 2012, pp. 43–46. ACM (2012)
9. Ellis, D.: A behavioural approach to information retrieval system design. J. Documentation **45**(3), 171–212 (1989)
10. Peters, T.A.: The history and development of transaction log analysis. Libr. Hi Tech. **11**(2), 41–66 (1993)
11. Jones, S., Cunningham, S.J., McNab, R.: An Analysis of Usage of a Digital Library. In: Nikolaou, C., Stephanidis, C. (eds.) ECDL 1998. LNCS, vol. 1513, pp. 261–277. Springer, Heidelberg (1998). doi:10.1007/3-540-49653-X_16
12. Mayr, P.: Sowiport User Search Sessions Data Set (SUSS) (2016)

Social Dendro: Social Network Techniques Applied to Research Data Description

Nelson Pereira[✉], João Rocha da Silva, and Cristina Ribeiro

INESC TEC, Faculdade de Engenharia, Universidade do Porto,
Rua Dr. Roberto Frias, 4200-465 Porto, Portugal
nelsonpereira1991@gmail.com, joaorosilva@gmail.com, mcr@fe.up.pt

Abstract. Research data management has become an integral part of the research workflow. Currently, concern with data appears mainly at the very last stages of projects, rather than being present from the moment of data creation. The goal of this work is to make data easier to find, share and reuse through early metadata production and in-group review. The approach proposed in this paper, Social Dendro, introduces social network concepts such as posts, shares and comments, in Dendro, our research data management platform. The implementation follows the ontology-based architecture of the platform. Results of a preliminary user test have provided insights for future improvements.

Keywords: Data repository · Data curation · Research data management · Social networks · User interfaces · Ontologies

1 Introduction

Research data management (RDM) is a very complex problem involving a multitude of stakeholders and issues ranging from the social to the technical [2]. Furthermore, it is becoming an essential part of research workflows, as funding institutions are either recommending [4] or requiring [5] the inclusion of Data Management Plans (DMPs) in research project proposals.

A series of surveys conducted by the Data Archiving and Networked Services (DANS) showed that large amounts of information created during the research process, from 70 to 90%, is not stored outside of the lab context [3]. It is also clear that researchers need to be involved in the data curation process, because they know much about the data, and such information is crucial for producing the quality metadata required to interpret datasets. However, their engagement relies on the availability (existence) of tools that can handle the details and formalities involved in the practice of data curation, so that their focus can remain on the research activity.

Dendro[1] is a collaborative data management platform currently in use at the University of Porto. Its goal is to provide a friendly, "Dropbox"-like

---

[1] Link: http://github.com/feup-infolab/dendro.

© Springer International Publishing AG 2017
J. Kamps et al. (Eds.): TPDL 2017, LNCS 10450, pp. 566–571, 2017.
DOI: 10.1007/978-3-319-67008-9_47

interface combined with data description features built over ontologies [7]. Dendro's main focus is the description of the files and folders created by research groups. Dendro is designed to assist researchers during the data production phase and complements existing repository platforms, supporting researchers in the organization and description of the data before they are deposited in repositories such as Zenodo, Figshare, the EUDAT B2Share or CKAN.

In this paper, we present the design and implementation of Social Dendro, an extension that draws inspiration from the Science Repositories 2.0 concept [1]. Data management is regarded there as a social process, involving researchers right from the moment of data creation—covering deposit, reuse and replication of experiments. We carried out a preliminary analysis for best suited ontologies, allowing us to build the data model for Social Dendro and integrate it in Dendro's data model [6]. We then carried out a small usability test covering data description in this collaborative setting.

2   Social Dendro

Being aware of work by others is one of the cornerstones of collaboration systems [8]. Given the collaboration features that are already in place and its graph data model, Dendro is a good base for the implementation of the vision presented in Science 2.0 repositories.

This approach advocates that the sharing and reuse of information should occur as early as possible in the research cycle [1], and goes on to propose the introduction of posts, ratings, comments, and likes as a way to review the research activity and its products. The nurturing of tacit knowledge, which is mostly shared by discussion with other individuals, is also quite important in the context of the research activity, as it can influence the reproducibility of results in experiments. Social Dendro is expected to preserve this very valuable type of knowledge, via social network techniques, during and after the research activity.

2.1   Data Model

Following the principles of data sharing and reuse proposed by the Semantic Web initiative[2], we have identified ontologies and classes that match the concepts of like, share, and comment. The schema.org ontology[3] already has a set of classes for these concepts, namely `SocialMediaPosting`, `ShareAction`, `CommentAction` and `LikeAction`. They were adopted in Social Dendro.

As an example, consider a user filling in descriptor "Creator", from the Dublin Core Terms ontology[4], for a folder. A new `SocialMediaPosting` instance is created as a result (see 1 in Fig. 1).

---

[2] Link: https://www.w3.org/standards/semanticweb/.
[3] Link: http://schema.org/.
[4] Available at http://bloody-byte.net/rdf/dc_owl2dl.

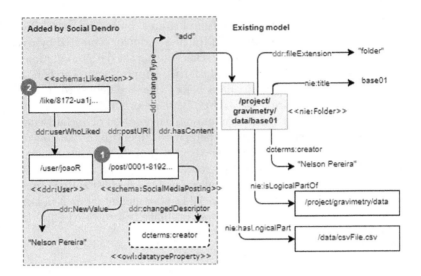

Fig. 1. Social Dendro creates a post as a descriptor is added

In the "Added by Social Dendro" section of Fig. 1, we can see the data changes triggered by this Social Dendro event in the Dendro graph. Properties newValue, changedDescriptor, hasContent and changeType are defined in the Dendro ontology, represented here with the ddr prefix. The Social Dendro event generates instances for these properties as depicted.

Following this interaction, another user liked the SocialMediaPosting. This is represented in the graph as a LikeAction instance (see **2**) associated to this SocialMediaPosting object, with the properties userWhoLiked and postUri identifying the user who liked the post and the post that was liked.

3   Usability Tests

The implementation of Social Dendro was validated through usability tests with a set of 8 users. The subjects included researchers, students and software developers. Each experiment consisted in a sequence of real-time interactions between two users in the Dendro platform under the close monitoring of two evaluators.

Dendro is a web application, so we considered the five quality components specified by Jacob Nielsen to test website interfaces: learnability, efficiency, memorability, errors and satisfaction[5]. Memorability was left out due to the short time span of the evaluation.

We requested the participants to fill a short questionnaire before the start of the experiment. The questions are in Table 1; some of them have an open-ended nature, while others ask the participants to rate their experience with a specific concept in a scale of 1 to 5, low to high.

---

[5] Link: https://www.nngroup.com/articles/usability-101-introduction-to-usability/.

Table 1. Questions of the preliminary questionnaire

ID	Description
QI1	Do you have any previous experience with the Dendro platform?
QI2	Please designate your degree of experience with research data management (1–5)
QI3	What do you usually do with the data from your projects?
QI4	Please rate how frequently you use social networks (1–5)

The results for question **QI1** showed that 6 out of 8 people had previously interacted with Dendro. Note however that, for all participants, this was the first interaction with the Social Dendro extension.

Results for **QI2** showed that 3 out of the 8 individuals acknowledge to have a very low experience level in research data management (level 2). Also 3 in 8 claim to have level 4 experience in this field. The highest level of experience was only associated with 1 of the evaluated individuals. We had therefore a set of users with balance between experts and non-experts in research data management.

The answers to **QI3** reveal that storing research data in personal computers or in external hard drives is still very common, a fact that reveals the need for RDM tools.

Finally, **QI4** shows that most of the participants use social networks very frequently.

3.1   The Tasks for the Usability Tests

The tasks designed for the experiments are abbreviated in Table 2. To simulate a creator and a collaborator interacting on a Dendro project, each task requires two users (**A** and **B**), on two separate computers.

Table 2. Description of the evaluated tasks

ID	Description
TA1	Find the "Social Dendro" section
TA2A	As user A, create a project, add user B as collaborator, add a folder and upload a file
TA2B	As user B, wait to be added to the project, then create a folder, and upload a file
TA3	Add two metadata descriptors to the created folder
TA4	Check the posts generated from the changes
TA5	Like a post
TA6	Comment a post
TA7	Share a post
TA8	Identify the notifications section

Although **TA1** seemed a very simple task, it had an average time for completion of more than 2 min. Most of the users commented that the Social Dendro timeline section was not easily accessible.

Tasks **TA2A** and **TA2B** had the highest average completion time, with values above 5 min. One of the required steps for user A was to add user B as a collaborator. As the input for this step required writing the full user URI of the collaborator, it was observed that it took some users several tries to succeed.

The average completion time for **TA4** was above one minute, perhaps because the default Dendro interface includes an area that shows the recent changes on the metadata for the selected file or folder. The Social Dendro timeline, on the other hand, shows the changes made to resources in all the projects where the user participates and this similarity was confusing. As for the remaining tasks, the average completion times were quite low as expected.

3.2   Post-experiment Questionnaire

To collect feedback and assess user satisfaction with Social Dendro, users were required to answer a post-experiment questionnaire. In some questions they were requested feedback on each of the social components—like, comment, share, as well as on the notifications. In others, they where asked to rate each social component on various properties relevant to the research activity, in a scale of 1 to 5, low to high: the **utility** in the context of RDM; the **visibility of research**; the **quality of metadata**; and the reduction of the **time for adaptation** to Dendro.

Users considered that the like component was the least useful feature, giving it the lowest average degree of utility (3.5) in the context of RDM. As it was also the social component with the highest standard deviation value (1.41), we may conjecture that the usefulness of this component for RDM is less obvious than the others. All base data of this study will be available at our institutional repository.

4   Conclusions and Future Work

Social Dendro adds social features to Dendro, allowing users in a research project to describe data and like, share and comment on each other's descriptors. Its user interface needs improvement to be easily used and understood, as seen by the time some users took to complete some of the tasks in the user experiments. Specifically, the like component was shown to lack usefulness in contrast to the share and comment components.

In research contexts, the ability to give both positive and negative ratings is essential, since criticism and peer review are at the basis of the scientific process. At this stage, we are planning to introduce an upvote/downvote system for descriptors, files, and folders.

The script given to the users about the tasks proved to be confusing at some points, making the process more complex than anticipated. The evaluation was focused mainly on usability and could not cover the impact of Social Dendro on RDM workflows. Future experiments will address this issue, by surveying researchers as they interact with their own data in real scenarios, as well as introducing a larger set of participants.

Acknowledgements. This work is financed by the ERDF- European Regional Development Fund through the Operational Programme for Competitiveness and Internationalization - COMPETE 2020 Programme and by National Funds through the Portuguese funding agency, FCT - Fundação para a Ciência e a Tecnologia within project POCI-01-0145-FEDER-016736.

References

1. Assante, M., et al.: Science 2.0 repositories: time for a change in scholarly communication. In: D-Lib Magazine 21.1/2 (2015). doi:10.1045/january2015-assante
2. Borgman, C.L.: The conundrum of sharing research data. J. Am. Soc. Inf. Sci. Technol. **63**(6), 1059–1078 (2012). doi:10.1002/asi.22634
3. Dillo, I., Doorn, P.K.: The Dutch data landscape in 32 interviews and a survey. Reporting year: 2011. Data Archiving and Networked Services (DANS) (2011). ISBN: 978-94-90531-10-2
4. European Commission. Guidelines on Open Access to Scientific Publications and Research Data in Horizon 2020 (2013)
5. National Science Foundation. Grants.Gov Application Guide A Guide for Preparation and Submission of NSF Applications via Grants.gov (2011)
6. Rocha, J., Ribeiro, C., Lopes, J.C.: Ontology-based multi-domain metadata for research data management using triple stores. In: Proceedings of the 18th International Database Engineering & Applications Symposium (2014). doi:10.1145/2628194.2628234
7. Rocha, J., et al.: The Dendro research data management platform: applying ontologies to long-term preservation in a collaborative environment. In: iPres Conference Proceedings (2014)
8. Yang, S.J.H., Chen, I.Y.L.: A social network-based system for supporting interactive collaboration in knowledge sharing over peer-to-peer network. Int. J. Hum. Comput. Stud. **66**(1), 3650 (2008). doi:10.1016/j.ijhcs.2007.08.005

Incidental or Influential? - Challenges in Automatically Detecting Citation Importance Using Publication Full Texts

David Pride[✉] and Petr Knoth

The Knowledge Media Institute, The Open University, Milton Keynes, UK
{david.pride,petr.knoth}@open.ac.uk

Abstract. This work looks in depth at several studies that have attempted to automate the process of citation importance classification based on the publications' full text. We analyse a range of features that have been previously used in this task. Our experimental results confirm that the number of in-text references are highly predictive of influence. Contrary to the work of Valenzuela et al. (2015) [1], we find abstract similarity one of the most predictive features. Overall, we show that many of the features previously described in literature are not particularly predictive. Consequently, we discuss challenges and potential improvements in the classification pipeline, provide a critical review of the performance of individual features and address the importance of constructing a large scale gold-standard reference dataset.

1 Introduction

The three largest citation databases; Google Scholar, Web of Science (WoS) and Scopus all give prominence to citation counts. However, it has been long established that treating all citations with equal weight is counterintuitive. Garfield, the original proponent of the JIF [2], proposed a range of 15 different reasons a paper may be cited.

In this paper, we address the problem of identifying influential citations based on publications' full text. The rest of the paper is organised as follows. In Sect. 2, we introduce key studies on which our work is based. We then discuss the approach for detecting influential citation, providing a critical analysis of features previously applied in this task in Sect. 3, selecting a set of three key features for further analysis. We present a comparative study of the identified features in Sect. 4, together with the challenges inherent in this task.

2 Related Work

There have been several different methodologies applied to this task, Hou et al. (2011) [3] first suggest the idea of using an internal citation count based on the full text of a research paper rather than just the bibliography to determine influence.

© Springer International Publishing AG 2017
J. Kamps et al. (Eds.): TPDL 2017, LNCS 10450, pp. 572–578, 2017.
DOI: 10.1007/978-3-319-67008-9_48

They demonstrate a positive correlation between the number of times a citation occurs and its overall influence on the citing paper. Zhu et al. [4] suggests a range of 40 classification features including both semantic and metric features to determine influence. Most recently, Valenzuela et al. (2015) [1] made significant efforts to construct a reference set which was publicly released and which this study relies heavily on. They suggest a range of 12 features, many of which show similarity with those of [4].

All of the studies under consideration use a range of different features and test them on different datasets. Consequently, getting a deeper understanding of which of the previously suggested features are most effective at this task is needed.

3   Methodology

The typical workflow for classifying citation types involves extracting the full text of the manuscript, parsing the text to detect document structure and then applying a classifier trained using machine learning approaches.

In the rest of this section, we describe this workflow concentrating on the selection of features used in the citation type classification task.

3.1   Classification Features Used by Prior Studies

One of the overriding aims of this work is to establish which of the previously identified classification features perform most strongly as predictors of citation importance and to use this as a baseline from which to build future work.

We consider the features presented in the two most recent studies. In [4] we first see an expansion of the features into a rich range that move beyond simple counting of in-text citations;

We analysed the 40 features used by [4] and 12 features used in the study of [1]. Of the 40 features suggested by [4], a combination of just 4 features resulted in the best performance of the model. Adding features beyond this actually lowered the performance. Out of the 12 features of [1], we found three features irreproducible (F3, F5[1], F12), we were unable to reliably replicate two features due to PDF extraction issues (F2, F6) and we elected not to use two features as they rely on external and potentially changing evidence (F10, F11). Two features we tested (F7, F8) did not produce any significant correlation with the gold standard.

Of the three remaining features of [1], we found a complete overlap of two features between [1] and [4] (F1-countsInPaperWhole, F4-aux_SelfCite) and a close match on the third (F9-simTitleCore). These three selected features correspond to the best (F1-countsInPaperWhole) feature of Zhu, the worst feature of Valenzuela (F9-simTitleCore) and a third where the opinion regarding the usefulness of this feature was divided between the two studies (F4-aux_SelfCite).

---

[1] We attempted to reproduce this feature, but failed due to Valenzuela's dictionary of cue words not being available.

3.2   Classification

Using the identified features, we perform a binary incidental/influential classification. WEKA 3 [5] was selected as the machine learning toolset in our study.

4   Results

4.1   Dataset

The dataset released by [1] contains incidental/influential human judgments on
465 citing-cited paper pairs for articles drawn from the 2013 ACL anthology,
the full texts of which are publicly available. The judgment for each citation
was determined by two expert human annotators and each citation was assigned
a label. Using the author's binary classification, 396 citation pairs were ranked
as incidental citations and 69 (14.3%) were ranked as influential (important)
citations.

4.2   Analysis and Comparison of Selected Features

Our experiments tested a range of features and their efficacy as predictors of citation influence. We achieved the best results using the Random Forests Classifier.
We tested the model using bagging with 100 iterations and a base learner, using
a 10-fold cross-validation methodology. The WEKA toolset was used to generate
P/R curves for each of the individual features as well as the combination of all
the features (Table 1).

Table 1. Interpolated precision at different recall levels for all features for the random
forest classifier.

Feature	P@R = 0.05	P@R = 0.1	P@R = 0.3	P@R = 0.5	P@R = 0.7	P@R = 0.9
F1	0.4	0.34	0.33	0.3	0.26	0.21
F4	0.27	0.35	0.14	0.15	0.14	0.14
F9	0.46	0.49	0.21	0.2	0.18	0.16
All	0.5	0.38	0.37	0.37	0.29	0.23

We also measured the correlation between each of the individual features and
the classification given by the human annotators. Valenzuela et al. [1] present
their results in terms of P/R values for each feature whereas [4] shows the Pearson
correlation with their gold standard. We therefore present the results of our
experiments in both formats to allow for accurate comparison. Our work confirms
the earlier findings reported in [1,4] that the number of direct instances of a
citation within a paper is a clear indicator of citation influence. We also find
that author overlap, or self-citation, does have value as a classification feature.

Table 2. Comparison of results by feature

Feature	Precision@Recall = 0.9		Pearson r	
	Valenzuela et al. [1]	Our results	Zhu et al. [4]	Our results
Direct citations	**0.30**	**0.21**	**0.330**	0.281
Abstract similarity	0.14	0.14	N/A	**0.373**
Author overlap	0.22	0.16	0.020	0.132

Contrary to the work of [1] we find that the similarity between abstracts is more predictive of citation influence than previously shown.

The correlation of this feature with the reference set ($r = 0.373$, $p < 0.01$, 2-tailed) was the highest of all the features we tested. It is our contention that testing all features using P/R values, at $R0.90$ masks some of the predictive value of those features when the dataset contains only a small number of instances of the influential class. Table 2 shows the precision of the random forests classifier at various recall levels. It can been seen from these results that the classifier initially performs quite well and identifies many of the influential cases, however it has difficulty identifying the last few instances which substantially decreases the classifier's performance at $R0.90$. Using Mean Average Precision (MAP) or a similar metric that provides a single-figure measure of quality across recall levels would be a better choice in this case.

4.3   Results for Individual Features

F1 - Number of Direct Citations: This feature is rated as the highest value in terms of predictive ability by [4] and the second highest by [1]. The latter shows P0.30 at R0.90, however our results demonstrate a slightly lower P value, P0.21 at R0.90. [4] lists the equivalent 'countsinPaper_Whole' as the most significant feature of their classifier, with a Pearson correlation coefficient of P0.35. We find a Pearson correlation of P0.28 (significant at the 0.01 level, 2-tailed) for this feature with our dataset. The small difference in this result is likely caused by the differences in the two datasets. Our results therefore confirm that the number of times a citation appears is a strong indicator of that citation's influence.

F4 - Author Overlap: The results from the two earlier studies for this feature vary considerably. In the results for [1] this is the third ranked 'most significant feature' with P0.22 for R0.90. We find slightly less precision than [1] for this feature; P0.16 at R0.90. [4]'s results show little correlation with their gold standard for the similar feature aux_selfCite (Pearson 0.02). Interestingly, despite the low correlation, this feature was the fourth one selected by their model and did indeed improve the performance of the classifier, albeit only slightly. The experiments with our dataset show a far stronger positive correlation, P0.132 (significant at the 0.01 level, 2-tailed), than that found by [4].

F9 - Abstract Similarity: Whilst [4] generated many similarity-based features, they did not compare citing abstract and cited abstract. This is somewhat surprising as we consider it to be an interesting feature and one that also seems innately logical. The abstract similarity is calculated as the cosine similarity of the tf-idf scores of the two abstracts. By ensuring that the dataset only contains valid data, i.e. the abstract is available for both citing and cited paper, a direct comparison can be made for this feature with [1] who rank this as the lowest of their twelve features, P0.14 at R0.90.

Here our results are the same as [1], with P0.14 at R0.90. However, the Pearson correlation with the gold standard dataset for this feature is the highest of the three features tested in our experiments. We find a Pearson correlation of 0.373 (significant at the 0.01 level, 2-tailed). This feature was not tested by any of the other earlier studies covered in this work. Our results demonstrate that abstract similarity between citing and cited paper is more predictive of citation influence that previously shown.

4.4   The Value of Complex Features

Many of the complex features tested by previous studies have been shown to have little predictive ability in regards to classifying citation function or importance. Some of the most basic features have been shown to offer the strongest potential in identifying important or influential citations. Our research confirms that one of the most simplistic features, i.e. the number of times a citation appears in a paper, is highly predictive of influence.

Replicating complex features is a non-trivial task unless exact details of how the values for these features were calculated or source code are provided by the original study. We believe that it is essential that the types and values of all features should be provided as part of the research dataset (as opposed to providing just source prior to feature extraction) to serve as a roadmap in replicating them.

5   Discussion

One of the major limitations of this and previous studies is the size of the publicly available, annotated, datasets. The study by [1] uses 465 citing/cited paper pairs. The study by [4] uses just 100 papers by 40 authors. Due to the unbalanced split between the incidental and influential classes, our complete dataset contained only 61 examples of the positive (influential) class. We argue that due to the relative sparsity of influential citations a much larger reference set is required. This is equally true for negative citations, which have been shown to be even rarer. Training a classifier when the dataset contains so few instances of the non-neutral classes is problematic and we will address this in future work. The construction of a gold standard dataset containing many thousands of annotated citations, rather than a few hundred, is a significant undertaking but we believe this is a vital step in improving the abilities of the classification models.

There is a noticeable difference between the datasets used by [1,4] which warrants further study. The [1] dataset annotation was undertaken by two independent annotators and finds significant value in using author overlap as a classification feature. However, the [4] reference set is annotated by the authors themselves and this study ranks author overlap/self-citation as being of very low importance. It may be that is demonstrates shyness or reticence on behalf of authors to regard their own, earlier, work as being a significant influence.

Finally we argue that if a citation is considered influential, this original influence remains regardless of external factors or the environment. Therefore, classification features which rely on external and potentially fluid information should be used somewhat cautiously. In future work we will address this issue in greater detail.

6   Conclusions

Of the features we tested, we find the feature *Abstract Similarity* shows the strongest positive correlation for predicting citation influence. We find *Number of Direct Citations* to also be highly predictive and we find *Author Overlap/Self-Citation* to be less predictive but still valuable as a classification feature. It is important to note that many of the features suggested by earlier studies have been shown to have little predictive ability.

There is scope for further work surrounding the efficacy and in particular the reproducibility of some of the previously tested classification features. Many of the earlier studies in this domain present results based on sometimes complex and irreproducible features. We contest that this is detrimental to this area of study as a whole and, whilst earlier studies have identified several effective features, having the ability to reproduce them is fundamental to further development in the area of citation classification.

Whilst it may be a relatively easy task for a human being to identify important or influential citations, building a model to automatically classify these citations with any degree of accuracy is a non-trivial task. A larger scale reference set than those used in this and previous studies is essential, particularly due to the inevitably skewed nature of any dataset of citations annotated according to influence or importance.

Acknowledgements. This work has been funded by Jisc and has also received support from the scholarly communications use case of the EU OpenMinTeD project under the H2020-EINFRA-2014-2 call, Project ID: 654021.

References

1. Valenzuela, M., Ha, V., Etzioni, O.: Identifying meaningful citations. In: AAAI Workshops (2015)
2. Garfield, E., et al.: Citation analysis as a tool in journal evaluation, American Association for the Advancement of Science (1972)
3. Hou, W.R., Li, M., Niu, D.K.: Counting citations in texts rather than reference lists to improve the accuracy of assessing scientific contribution. BioEssays **33**(10), 724–727 (2011)
4. Zhu, X., Turney, P., Lemire, D., Vellino, A.: Measuring academic influence: not all citations are equal. J. Assoc. Inf. Sci. Technol. **66**(2), 408–427 (2015)
5. Witten, I.H., Frank, E., Hall, M.A., Pal, C.J.: Data Mining: Practical Machine Learning Tools and Techniques. Morgan Kaufmann, San Francisco (2016)

User Interactions with Bibliographic Information Visualizations

Athena Salaba[1](✉) and Tanja Merčun[2]

[1] Kent State University, Kent, OH 44242, USA
asalaba@kent.edu
[2] University of Ljubljana, Ljubljana, Slovenia
tanja.mercun@ff.uni-lj.si

Abstract. The paper presents preliminary results on a study testing user interactions with five prototype systems, including four different visualizations of FRBR-based bibliographic information and one more typical bibliographic information system. Performance and perceptions findings using the same tasks across the different visualizations are reported with a discussion on the implications for the design of future bibliographic information interfaces.

Keywords: Information visualization · User interfaces · Bibliographic information · User study · FRBR-based displays · Evaluation

1 Introduction

Bibliographic information retrieval systems, such library catalogs, have a long history of being described as difficult to use, time-consuming, producing long results lists users need to shift through to find those most relevant to their needs. With the emergence of the IFLA Functional Requirements models (FRBR) [1] several studies examined the potential improvements in system design, display of results, and hierarchical browsing of work-based information, including exploring relationships among variations of a work and relationships between works represented in the bibliographic data [2–4] but most importantly the potential to enhance users' experience searching and browsing bibliographic data. A few prototypes were developed and tested using FRBR-based systems [5] but very few examined FRBR-based implementations of visualizing bibliographic information [6].

This paper presents a continuation of a study (2012-Slovenia) that examined user performance and preferences between a typical bibliographic information system and four hierarchical visualizations of work-family based bibliographic information.

2 Study Background

The study reported here (2016-United States), utilizes the same prototype designs first introduced in a similar, 2012 study [6], using a similar methodology. Five different prototype designs (Fig. 1) were implemented and tested. Four of them are using

© Springer International Publishing AG 2017
J. Kamps et al. (Eds.): TPDL 2017, LNCS 10450, pp. 579–584, 2017.
DOI: 10.1007/978-3-319-67008-9_49

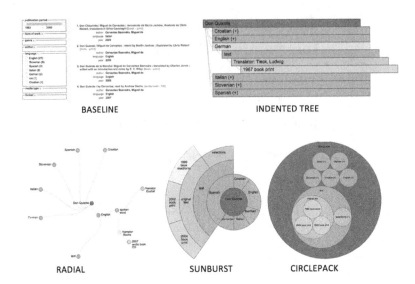

Fig. 1. Four hierarchical visualizations and one baseline prototype displays

different hierarchical visualization techniques to create displays of the same information using the work-family concept: Circlepack, Radial tree, Intended tree, and Sunburst.

The fifth design implements a typical bibliographic information system with faceted navigation, using edition-based displays of information. This design, Baseline, serves as a typical interface users currently experience, for comparisons to the visualization interfaces.

The bibliographic data included in this study are FRBR-based work family data sets of three works, varying in the degree of a work's expression and manifestation complexity. For example, one work has only two expressions in English (*Destiny of the Republic*) and another has multiple expressions, including translations in several languages, other variations, and related works, contributing to the workset's complexity (*Don Quixote*), with the third being a children's fictional work, offering a variety of expressions (*Histoire de Babar, le petit éléphant*). Only one common work family data set, *Don Quixote*, was used in both the 2012 and 2016 studies in order to provide comparative data. One of the goals for repeating the study using one common work, was to test the feasibility of designing FRBR-based visualization displays that appeals to users in different cultural environments. To expand the original study, the 2016 study used two new work family sets, one non-fiction with limited complexity and one fictional children's work in order to test if work complexity has an effect on user's interactions and perceptions of FRBR-based visualizations of bibliographic data.

The data collection took place in April 2016. The 79 participants were undergraduate students of a large university in the United States, representing different areas of academic study. Each participant interacted with four interfaces, the Baseline and three visualizations. The same tasks for each work (total 10 tasks for each) were randomly assigned to each system interaction. Tasks ranged from simple finding and identifying to more exploratory and understanding questions. Tasks were grouped as

Table 1. Number of tests performed per visualization interface

BASELINE	RADIAL	CIRCLEPACK	INDENTED	SUNBURST	TOTAL
69	52	51	52	51	275

work-set related (labeled in graphs as versions), other related works (related), and author-related (author). Sequencing of the interface, work, and task group for each test was designed to avoid order bias. In total, there were 275 valid tests, distributed among interfaces as indicated in Table 1.

Participants were asked to assign a difficulty score for each task they completed (1 = very easy, 5 = very difficult) and rank the interface designs at the completion of the study from their most favorite to their least favorite. In addition to these user-reported measures, researchers recorded time-on-task, completion success, and navigation success for each task.

3 Findings

The study looked at several measures and factors. This paper is limited to summary findings on a number of performance and perception measures. Preliminary comparison findings between the two studies were reported in 2016 [7].

3.1 Performance

Three performance measures are reported here: time it took for each participant to complete each task, how successful each participant was to complete each task, and how successful each participant was in navigating each system, whether through result lists and facets for the Baseline system or the hierarchical navigation features of the visualization systems.

Time on Task. Overall, participants needed considerably more time to complete tasks in the Baseline system than any of the four visualization systems (Fig. 2). On average, participants needed the least amount of time when using the Indented tree.

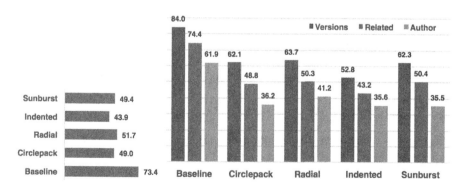

Fig. 2. Average (mean) time on task

When comparing the time on task by task group, participants needed more time for work-related tasks than related works or author-based tasks. Again, we see that Baseline time on task being highest for all three task groups among all systems (Fig. 2). Indented tree had the lowest time on task for work-related tasks and related works tasks, and Sunburst for author-related tasks.

Task Success. Participants were least successful in completing their tasks when using the Baseline system. They were most successful when using Indented tree with a close second when using Sunburst (Fig. 3).

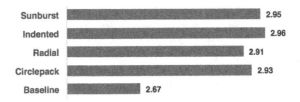

Fig. 3. Average (mean) task success for each design

Relationships to other works is one of the areas where work-based bibliographic information system design offers improvements over the traditional edition-based bibliographic information displays.

Navigation Success. Each design offers navigation features to aid users in their tasks. For example, Baseline offers faceted navigation and visualizations offer different hierarchical navigation features for expanding or collapsing displays.

When examining how participants used these navigation features to complete their tasks, we see that they were least successful using the Radial tree followed by the Baseline system (Fig. 4).

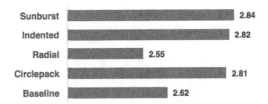

Fig. 4. Overall average (mean) navigation success

3.2   Perception

Participant perceptions on ease of task and their system preference ranked from the most favorite to their least favorite are reported in this section.

Ease of Task. At the completion of each task, participants were asked to rate how easy or difficult the individual task was, with "1" being "very easy" and "5" being "very difficult."

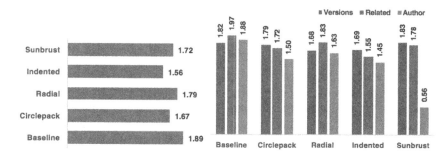

Fig. 5. Average task difficulty

Overall, participants rated the tasks more difficult when using the Baseline system and easiest when using the Indented tree (Fig. 5). When comparing task groups, author-related tasks using Sunbursts were rated considerably easier than using any other system, in addition to being rated the easiest among all task groups across systems.

System Raking. At the completion of their tests, participants were asked to rank the systems they interacted with from their most favorite to their least favorite, based on their experiences using each system to complete their tasks. Looking at the interfaces ranked as their most favorite (ranked #1), participants favored Circlepack the most with a close second the Indented tree and Radial the least. According to Fig. 6, when combining the #1 and #2 rankings for each system, Indented tree is ranked considerably higher (67.31%) than other systems, with Baseline ranked lowest (44.93%).

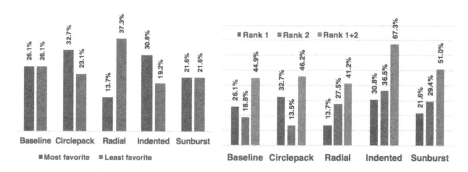

Fig. 6. Participant system ranking

4   Discussion and Conclusion

The study examined if there are any differences on performance and perceptions between the Baseline prototype and the four visualizations systems. Overall, the study findings show that the Baseline prototype performed least favorably among all systems on a number of measures, including longest time-on-task, least successful task completion, highest task difficulty, and second to last on navigation success.

This later finding is interesting, given the fact that the majority of current systems offer faceted navigation. Further examination of the factors for the low performance on navigation success is warranted.

The study shows significant difference between the Baseline system and the visualizations on the time it took participants to complete their tasks and successful completion of tasks with visualizations performing higher overall. Participants ranked three of the four visualizations (Indented, Sunburst, and Circlepack) higher than the Baseline system based on their experiences. Also, when it comes to more complex tasks of exploring relationships, the Intended tree and Sunburst outperformed the other systems.

The results suggest that, generally, hierarchical organization and visual display of bibliographic information enables users to better navigate bibliographic work families and complete their tasks successfully. Although it was their first interactions with FRBR-based systems, participants expressed their preference for these visualization systems over the Baseline system. These findings are similar to the 2012 studies, although a comparison between the 2012 and 2016 studies show that the was a significant difference in the time it took to complete the tasks, with the US group average time of task completion being longer.

A detailed analysis between the repeated work family in the two studies and the newly added work families in the 2016 study will examine whether some of the differences between the prototype design scores derived due to the different environments or the different work sets. In addition, further analysis is needed to explore the differences in performance and perceptions based on work complexity and task complexity. Findings from these experimental studies of prototype systems will inform better visualization systems design for bibliographic information and improve user experience.

Acknowledgements. Prototypes developed in cooperation with Dr. Trond Aalberg (NTNU, Norway).

References

1. IFLA Study Group on the Functional Requirements for Bibliographic Records: Functional Requirements for Bibliographic Record, final report. K.G. Saur, Munich (1998)
2. Mimno, D., Crane, G., Jones, A.: Hierarchical catalog records: implementing a FRBR catalog. D-Lib Mag. **11**(10) (2005). doi:10.1045/october2005-crane
3. Zhang, Y., Salaba, A.: What do users tell us about FRBR catalogs? Cataloging Classif. Q. **50** (5–7), 705–723 (2012)
4. Carlyle, A.: Understanding FRBR as a conceptual model: FRBR and the bibliographic universe. Libr. Resour. Tech. Serv. **50**(4), 264–273 (2006)
5. Zhang, Y., Salaba, A.: Implementing FRBR in Libraries: Key Issues and Future Directions. Neal-Schuman, New York (2009)
6. Merčun, T., Žumer, M., Aalberg, T.: Presenting bibliographic families using information visualization: evaluation of FRBR-based prototype and hierarchical visualizations. J. Assoc. Inf. Sci. Technol. **68**(2), 392–411 (2017)
7. Merčun, T., Salaba, A., Žumer, M.: User testing of prototype systems in two different environments: preliminary results. In: Morishima, A., et al. (eds.) ICADL 2016, LNCS, vol. 10075, pp. 104–109. Springer, Cham (2016)

Towards Building Knowledge Resources from Social Media Using Semantic Roles

Diana Trandabăţ[✉]

University "Al. I. Cuza" of Iasi, Iaşi, Romania
dtrandabat@info.uaic.ro

Abstract. Text semantics is a well-hidden treasure, whose deciphering requires deep understanding. Artificial Intelligence enhances computers with human-like judgments, thus decoding the covered message and sharing it between machines is one of the main challenges that the computational linguistics domain faces nowadays. In an attempt to learn how humans communicate, computers use language models derived from human knowledge. While still far from completely understanding insinuated messages in political discourses, computer scientists and linguists have joined efforts in modeling a human-like linguistic behavior. This paper aims to introduce the *VoxPopuli* platform, an instrument to collect user generated content, to analyze it and to generate a map of semantically-related concepts to capturing crowd intelligence.

Keywords: Semantic roles · Knowledge resources · Social media

1 Introduction

Language technology is generally acknowledged today as one of the key growth areas in information technology. The META-NET White Paper series "Europe's Languages in the Digital Age" [15], warned that languages may find it difficult to survive in the digital age, if the support for language technologies will not receive a boost. Building machine-readable knowledge bases takes a huge amount of time and resources, both financial and human (trained experts). Since today we found ourselves in an era in which software learns from its users and all of the users are connected, this paper proposes a natural language processing application which explores the social web in a new and innovative way, based on semantic frames, in order to extract the wisdom of crowds captured within.

With such knowledge bases, easily and dynamically created for different users, contexts or time frames, a gap will be filled between where we are now and where we could be in artificial intelligence: computers could be engaged in "intellectual" cooperation (with humans, or even more futuristic, with each other) in order to foster creativity, innovation and inventiveness.

Social media refers in fact to Web 2.0 applications which support user content-creation and collaboration. People seek and share ideas, information, experiences, expertise, opinions, and emotion with both acquaintance and strangers on the Internet, based on the effect of the Wisdom of Crowds [13]. Over the last few years, the use of Social Media

© Springer International Publishing AG 2017
J. Kamps et al. (Eds.): TPDL 2017, LNCS 10450, pp. 585–591, 2017.
DOI: 10.1007/978-3-319-67008-9_50

has increased tremendously all over the world. Through *VoxPopuli*, people's contribution can reach a much wider audience than their small group of friends, by contributing to a "universal" knowledge base. The huge popularity of social networks provides an ideal environment for scientists to test and simulate new models, algorithms and methods to process knowledge and *VoxPopuli* provides a platform to do precisely this job.

The paper is structured as follow: Sect. 2 gives a short overview of the current state of the art in analyzing user generated content and semantic roles, while Sect. 3 discusses the proposed methodology. Section 4 briefly discusses the evaluation of our platform before drawing some conclusions in the last section.

2 State-of-the-Art

Since the emerging of user generated contents (UGC), researchers have tried to automatically understand the opinions and sentiments that people are communicating [10, 11]. However, most analyses over social media were so far limited to identify user profiles or group behavior, extract sentiments expressed in specific posts, or identify topics in order to adapt recommendation systems. This paper is a position paper proposing the extraction of structured knowledge from social media using semantic frames, a direction yet unexplored.

Semantic roles [4] allow to identify when, where, why or how an event takes place, by clarifying the context of a sentence in terms of relations between the predicational word [3] and its semantic roles. The SRL system we propose: (a) is adapted for UGC (as opposed to existing systems, trained on news date); (b) incorporates, besides syntactic information, named entity recognition and topic information.

For extracting events and relations from texts, worth noticing is the work done by the Watson group at IBM on relationship extraction and snippets evaluation with applications to question answering [12], but also the work in [2], where semantic roles and event extraction are considered structurally identical tasks. Our approach extracts relations between concepts using semantic roles, similar up to some extend to the work in [2], but tailored for UGC.

3 Methodology

The architecture of the *VoxPopuli* platform involves 4 distinct modules (see Fig. 1), each of them specialized on a specific task and corresponding to a different objective.

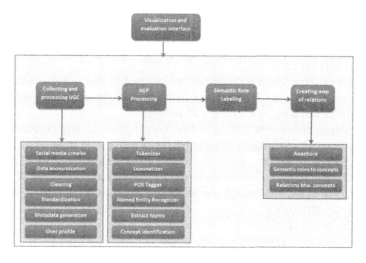

Fig. 1. Architecture of the *VoxPopuli* platform

3.1 Collecting and Pre-processing UGC

User-generated content (UGC) is defined as: *"any form of content (…) of media that was created by users of an online system or service, often made available via social media websites"* [1].

The *VoxPopuli* platform collects textual UGC for analysis of individual and collective behavior, without accessing any personal data of users[1]. Privacy and copyright in using social media data is an open issue. Hoser and Nitschke [8] discuss the ethics of mining social networks, suggesting that researchers should not access personal data that users did not share for research purpose, even when they are publicly available. On the other side, from a pure technical point of view, if for using the private data on social networks the user's agreement is needed, public postings, such as Facebook walls, Tweets, YouTube or Flickr comments, blogs and wikis count as public behavior. Furthermore, specialized APIs exist, allowing collection of social media data.

This module performs the following consecutive tasks:

- Identify User Generated Content (UGC) sources;
- Apply a social web crawler;
- Anonymyzing, cleaning and standardization;
- Metadata generation;
- Identify user profiles and classify user generated content types.

---

[1] According to the Directive 95/46/EC of the European Parliament and of the Council, personal data is defined as: "'personal data' shall mean any information relating to an identified or identifiable natural person ('data subject'); an identifiable person is one who can be identified, directly or indirectly, in particular by reference to an identification number or to one or more factors specific to his physical, physiological, economic, cultural or social identity".

After identifying UGC sources (social media site, folksonomies, blogs and review sites, etc.), a collective data crawler uses a set of concurrent processes to query the social web using specialized search or streaming APIs, such as Archivist; YouTube Developer Page or Flickr API Gardens.

VoxPopuli platform ensures no relation to a natural person is made from collected data, no personal data are stored or used, and that all texts are properly shuffled and anonymized, before being cleaned and standardized.

The standardization step is focuses on noisy content: social media content often has unusual spelling (e.g. 2moro), irregular capitalization (e.g. all capital or all lowercase letters), emoticons (e.g. :-P), and idiosyncratic abbreviations (e.g. ROFL, ZOMG). Spelling and capitalization normalization methods have been developed [7], coupled with location-based linguistic variations in shortening styles in microtexts [6].

The last step identifies of user types, depending on the most frequent concepts in a user's content and its writing style, based on the ontology proposed in [9].

3.2   NLP Processing

Once the collection of texts from UGC is created, data is explored using a series of NLP processes: tokenization, lemmatization, part-of-speech tagging, named entity recognition, sentiment analysis, topics and concept identification.

We tested our platform for the Romanian language, therefore specific NLP tools were applied, but the Platform is built to be language independent, allowing the inclusion of different language specific tools.

For sentiment analysis, *VoxPopuli* uses regular expressions to: convert the texts to lowercase; discard words shorter than two characters; remove special diacritic signs, URLs, as well as unsupported symbols (such as "?" or "@"); remove duplicated vowels in the middle of the words (e.g. cooooool). Subsequently, the sentiment is extracted by combining two methods: (1) a Naïve Bayes classifier, trained on Semeval 2016 data, using the following features: tokenized unigrams, emoticons, hash tags, similar to [5] and (2) the AlchemyAPI, applies to tweets to extract the expressed opinion.

The Topic Extraction module identifies different types of significant events using named entities, concepts extracted from the Romanian WordNet using hypernyms and hash tags. After the identification of topics, they are classified using a hybrid text classification model, combining statistical classification with rule-based filtering, identifying the thematic area of the message (transport, economy, daily life), alert situations (road accidents, fires, street violence), specific locations (building, means of transport) or events to which the text refers (cultural or sport events).

3.3   Semantic Role Labeling

The next module of *VoxPopuli* applies a semantic role labeling system. We adapted the SRL parser [14] developed for news texts, in order to cope with social media input, due to a set of challenging social media characteristics, syntactically and semantically different than the ones of the texts the role labeler is trained on: short messages, noisy content, temporal and social context, multilingual.

Since it is time-consuming to annotate UGC with semantic roles in a large enough corpus to be used for training a classifier, our technique was to alter the training set, by including broken language, typing errors, limiting the number of words/characters in sentences, etc.) and run the machine learning algorithms again. The major shortcoming of this method is that it is not based on a real, naturally occurring language. Therefore, we decided to also use the initial SRL parser, improved with a set of post-processing patterns. The two methods are combined in a voting algorithm, which decides statistically on the semantic roles to apply for the user generated content.

3.4   Creating a Semantically-Related Map of Concepts

The most challenging module of *VoxPopuli* extracts individual and collective intelligence from UGC based on the semantic annotation. This modules populates a knowledge resource in three steps: (1) first, it extracts from each UGC the predicational words, for which semantic roles are annotated; (2) semantic frame analysis is used to extract relations between concepts (semantic roles) collocated with the specific predicate; and (3) the concepts found in the UGC in relation to a specific predicational word are mapped to the ones already introduced in the knowledge base (if any) using anaphora resolution and/or the WordNet (5) hyponymy hierarchy.

The concepts and their references are linked using a simple anaphora resolution method, based on a set of reference rules.

The created knowledge base, stored in RDF format, can be validated through a specialized interface. At this stage, we are still fine-tuning *VoxPopuli* platform, so we only validated a small number of relations (2000 relations).

4   Evaluation

For the evaluation of *VoxPopuli* platform, we analyses a set of 2000 relations extracted from user-generated content using semantic roles. The distribution of the total number of annotated roles per sentence is: 14% sentences with 3 annotated roles; 62% sentences with 2 roles; 24% sentences with 1 role. For the evaluation we only considered sentences having at most 3 annotated semantic roles, a limitation imposed in the testing version of our system which can be removed latter.

In this first stage of analyzing our platform, we focused on six semantic roles: Entity, Item, Manner Duration, Place and Time.

Figure 2 presents a distribution of the types of semantic roles. The overall accuracy for the identified relations is over 86%. Most error cases were introduced by: (1) incorrect mapping of semantic roles to their predicational word, in cases when more than one word appeared in the sentence; (2) partial annotation of the semantic role, i.e. only the head of the constituent, not the whole constituent was selected; (3) errors in generalization using WordNet, e.g. the pronoun he is generalized as helium.

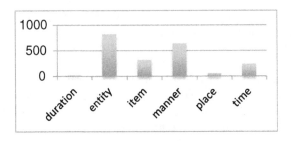

Fig. 2. Distribution of semantic roles

5 Discussion

The major contribution of this paper is a method for building a knowledge base from user generated content. Our results suggest that semantic role information can be used to automatically generate a knowledge resource. This pilot study needs to be extended to a larger scale, including different types of semantic roles.

The next obvious stage is to merge our resource to existing linked open data repositories. We also intend to expand the *VoxPopuli* platform in order to extract similar map of relations from scientific literature in order to predict future "hot" research topics.

References

1. Chua, T.-S., Li, J., Moens, M.-F.: Mining User Generated Content. Chapman and Hall/CRC, Boca Raton (2014)
2. Chen, C.-M., Chen, L.-H.: A novel approach for semantic event extraction from sports webcast text. Multimed. Tools Appl. **71**(3), 1937–1952 (2014)
3. Curteanu, N.: Contrastive meanings of the terms "predicative" and "predicational" in various linguistic theories (i, ii). Comput. Sci. J. Moldova **11**(4), 2003 (2003)
4. Daniel, G., Jurafsky, D.: Automatic labeling of semantic roles. Comput. Linguist. **28**(3), 245–288 (2002)
5. Go, A., Bhayani, R., Huang, L.: Twitter Sentiment Classification using Distant Supervision, Technical report (2009)
6. Gouws, S., Metzler, D., Cai, C., Hovy, E.: Contextual bearing on linguistic variation in social media. In: Proceedings of Workshop on Languages in Social Media, LSM-2011, pp. 20–29 (2011)
7. Han, B., Baldwin, T.: Lexical normalisation of short text messages: makn sens a #twitter. In: Proceedings of the 49th ACL-HLT 2011, pp. 368–378 (2011)
8. Hoser, B., Nitschke, T.: Questions on ethics for research in the virtually connected world. Soc. Netw. **32**(3), 180–186 (2010). doi:10.1016/j.socnet.2009.11.003
9. Macovei, A., Gagea, O., Trandabăţ, D.: Towards creating an ontology of social media texts. In: Trandabăţ, D., Gîfu, D. (eds.) RUMOUR 2015. CCIS, vol. 588, pp. 18–31. Springer, Cham (2016). doi:10.1007/978-3-319-32942-0_2
10. Nakov, P., Ritter, A., Rosenthal, S., Stoyanov, V., Sebastiani, F.: SemEval-2016 task 4: sentiment analysis in Twitter. In: Proceedings of SemEval 2016 (2016)

11. Russell, M.A.: Mining the Social Web: Data Mining Facebook, Twitter, LinkedIn, Google+, GitHub, and More (2013)
12. Schlaefer, N., Chu-Carroll, J., Nyberg, E., Fan, J., Zadrozny, W., Ferrucci, D.: Statistical source expansion for question answering. In: Proceedings of CIKM (2011)
13. James, S.: The wisdom of crowds. Doubleday (ed.) (2005). ISBN: 0-385-50386-5
14. Diana, T.: Mining Romanian texts for semantic knowledge. In: Proceedings of ISDA 2011, Cordoba, Spain, pp. 1062–1066 (2011)
15. Trandabăţ, D., Irimia, E., Barbu, M.V., Cristea, D., Tufis, D.: The Romanian language in the digital age. In: White Paper Series, p. 87. Springer (2012). ISBN: 978-3-642-30702-7

Poster and Demonstration Paper

Towards Finding Animal Replacement Methods

Nadine Dulisch and Brigitte Mathiak[(✉)]

GESIS - Leibniz Institute for the Social Sciences,
Unter Sachsenhausen 6-8, Cologne, Germany
{nadine.dulisch,brigitte.mathiak}@gesis.org

Abstract. Protecting animal rights and reducing animal suffering in experimentation is a globally recognized goal in science. Yet numbers have been rising, especially in basic research. While most scientists agree that they would prefer to use less invasive methods, studies have shown that current information systems are not equipped to support the search for alternative methods. In this paper, we outline our investigations into the problem. We look into supervised and semi-supervised methods and outline ways to remedy the problem. We learned that machine assisted methods can identify the documents in question, but they are not perfect yet and in particular the question about gathering sufficient training data is unsolved.

Keywords: Classification · Animal welfare

1 Introduction

Researchers from Life Sciences that contemplate to use animal testing are motivated by ethical, financial and often legal incentives to try and find alternatives that enable them to answer the same research questions, but with less animal involvement.

Unfortunately, current strategies for literature search do not support search for animal test alternatives very well. Dutch animal welfare officers have been asked for their strategies in finding alternative methods [2]. None found this task easy and reported that the most successful way of finding good alternatives was word of mouth.

When interviewing experts in finding such documents, it becomes clear that there are two criteria. *Similarity* is how close the document is to the experiment we want to replace. *Relevance* measures how likely it is that the document describes a method that causes less animal suffering. In order to give a comprehensive list of candidate documents, we need to take both criteria into account.

2 Related Work

Our attempt at solving the animal test replacement problem is not the first one. Go3R[1] is a semantic search engine based in PubMed and ToxNet[2] prioritizing

---

[1] http://www.gopubmed.org/web/go3r/.
[2] http://toxnet.nlm.nih.gov.

© Springer International Publishing AG 2017
J. Kamps et al. (Eds.): TPDL 2017, LNCS 10450, pp. 595–598, 2017.
DOI: 10.1007/978-3-319-67008-9_51

3R and toxicology. While the toxicology use case is interesting and useful, the focus on recall makes it hard to handle basic research questions, which typically do not fit semantic categories as neatly. AltBib[3] is not an independent search engine, but rather suggests query term expansions that, among other factors, utilize the MeSH classification for animal testing alternatives. This classification is not systematically used to tag all methods that are developed as alternatives, but seems to focus on documents about animal testing alternatives on a meta level. In 2015 there were less than 3000 documents with that MeSH term.

Table 1. Overview over # of positive, negative and not classified instances for the different use cases.

Use case	Reference document (PMID)	Relevance			Similarity			Animal test		
		+	-	/	+	-	/	+	-	/
1	16192371	13	85	2	15	77	8	13	70	17
2	11932745	21	70	9	13	78	9	47	39	14
3	11489449	13	80	7	4	75	21	37	55	8

3   Corpus

To our knowledge, there is no corpus available to use as training data, a problem that has been plaguing Information Retrieval from the very beginning [1]. We structured our corpus around individual use cases, mimicking the application process for getting a permission to conduct a specific animal experiment. The starting point was a document describing an animal test (reference document), which was chosen by the domain expert. To search for possible alternative methods, we used the PubMed functionality to find similar documents based on substring similarity[4].

The first 100 hits of documents similar to the reference document were downloaded and then assessed by the domain expert according to criteria the expert set down beforehand. Additionally to the aforementioned dimensions of result similarity and replacement relevance, we also asked the expert to give us information on whether the document described animal experiments or not.

Non-classification occurred when there was not enough information to make a sensible decision, e.g. missing abstract, missing relevant information, non-available full text or, in rare cases, when the expert felt not knowledgeable enough about the domain to make a judgement call. In the following experiments, we only used the classified documents.

---

[3] http://toxnet.nlm.nih.gov/altbib.html.

[4] http://www.ncbi.nlm.nih.gov/books/NBK3827/#pubmedhelp.
   Computation_of_Similar_Articl.

For each document we collected the following metadata: title, abstract, PubMed Central ID, URLs, journal name, availability status of the full text and MeSH term information.

Table 1 gives an overview over our created use cases. The table shows the number of positive (+) and negative (−) instances included in the datasets. The datasets include only few positive instances, leading to strong bias in learning.

4   Experiments

4.1   Classification

What we are most interested in, is if trained algorithms are able to distinguish between positive and negative instances. In this experiment, we compared the prediction performance of standard data mining algorithms, which included J48 (C4.5 decision tree), JRIP (Propositional rule learner), SMO (Sequential minimal optimization), Naïve Bayes, Bayes Net and LWL(Locally weighted learning). We conducted the experiments applying the data mining software Weka[5] (version 3.6) and used Weka's implementation of the aforementioned algorithms. For classification we used Weka's "FilteredClassifier", applying the "StringToWord-Vector" filter to handle string attributes. This filter transforms string attributes into an attribute set that represents word occurrence information[6]. We used leave-one-out evaluation for all experiments, based on the original dataset. For the results see Table 2.

Table 2. Average F-Score over all three use cases, differentiated after algorithm and metric. Note that the F-Score is calculated from the point of view of the positive instances, therefore the expected value for random choice is very low due to the bias.

Target attribute	Unbalanced dataset					
	J48	JRIP	Naïve Bayes	Bayes Net	SMO	LWL
Relevance	0.51	0.36	0.44	0.66	0.52	0.42
Similarity	0.14	0.08	0.36	0.28	0.18	0.07
Animal test	0.79	0.87	0.91	0.87	0.94	0.83

We immediately discovered that the unbalancedness of the datasets, with as little as 4 positive examples were creating serious problems as especially relevant and similar documents were only rarely correctly classified (cf. Table 2). This is particularly devastating for similarity, where most results are worse than or close to random. Results for relevance are not ideal, but clearly better than random. Animal tests can be detected quite reliably, but positive and negative instances are much better distributed, as you can see in Table 1.

---

[5] http://www.cs.waikato.ac.nz/ml/weka/.

[6] http://weka.sourceforge.net/doc.dev/weka/filters/unsupervised/attribute/
StringToWordVector.html.

Table 3. F-Score for semi-supervised learning. Averaged over all use cases. Number in parentheses is the original value.

Target attribute	Naïve Bayes	SMO
Relevance	0.56 (0.44)	0.53 (0.52)
Similarity	0.38 (0.36)	0.43 (0.18)

Semi-supervised Learning. As discussed before, we had only comparably few labeled documents available, and in real life, we might have even less. What we have not leveraged so far is that we had a high number of unlabeled documents available to us that fit the general topic. Following the self-training methodology laid out by [3] we used a semi-supervised learning approach in which unlabeled documents were used to counteract the scarcity of training data.

The semi-supervised approach improves results for relevance compared to the original values for the unbalanced dataset. As Table 3 shows, the F-Score value for relevance increases for both top algorithms, but only moderately. The SMO F-Score for similarity, however, raises more significantly, which seems to indicate that the lack of training data impacted the classifiers ability to successfully predict similarity.

5   Conclusions and Future Work

While we do not have a workable prototype for up-ranking animal replacement methods yet, we believe we have made important inroads and identified some roadblocks. On the bright side, we have shown that relevant documents can be found with machine learning given enough training data. Methods to reduce the need for training data have been tested and were found to be successful.

A more direct approach would be to use un-supervised methods of finding similar documents. Bibliometric methods seem hopeful, but positive and negative effects overlap and cancel each other out. We tried using classifiers across use cases, but without any improvements. On all fronts, it becomes clear that more training data is needed.

References

1. Jones, K.S., van Rijsbergen, C.J.: Information retrieval test collections. J. Documentation **32**(1), 59–75 (1976)
2. van Luijk, J., Cuijpers, Y., van der Vaart, L., de Roo, T.C., Leenaars, M., Ritskes-Hoitinga, M.: Assessing the application of the 3rs: a survey among animal welfare officers in The Netherlands. Lab. Anim. **47**(3), 210–219 (2013)
3. Zhu, X.: Semi-Supervised Learning Literature Survey. Technical Report 1530, Computer Sciences, University of Wisconsin-Madison (2005)

Environmental Monitoring of Libraries with MonTreAL

Marcel Großmann[1(✉)], Steffen Illig[2], and Cornelius Matějka[1]

[1] Computer Networks Group, University of Bamberg, 96047 Bamberg, Germany
marcel.grossmann@uni-bamberg.de,
cornelius-lucian.matejka@stud.uni-bamberg.de
[2] University Library of Bamberg, 96047 Bamberg, Germany
steffen.illig@uni-bamberg.de

Abstract. An ever-increasing amount of devices connected over the Internet pave the road towards the realization of the 'Internet of Things' (IoT) idea. With IoT, endangered infrastructures can easily be enriched with low-cost, energy-efficient monitoring solutions, thus alerting is possible before severe damage occurs. We developed a library wide humidity and temperature monitoring framework MonTreAL, which runs on commodity single board computers. In addition, our primary objectives are to enable flexible data collection among a computing cluster by migrating virtualization approaches of data centers to IoT infrastructures.

We evaluate our prototype of the system MonTreAL at the University Library of Bamberg by collecting temperature and humidity data.

Keywords: IoT · Single board computer · Container · Virtualization · Monitoring · ARM · Sensor

1 Introduction

The environment in archives and libraries should be maintained within specified tolerances in order to guarantee cultural heritage, e.g. books, magazines, electronic media to be conserved and prevented from serious damage [1]. Our university library maintains several depots to store stocks and equipment, which are partly hosted in buildings under monumental protection. Occurring problems range from simple things, such as broken lights, to more problematic ones, like water damages which could lead to mold formation. In that case a regular supervision of a depot is absolutely essential. To overcome severe issues and to avoid health risks while doing manual data collection, the endangered depot was equipped with temperature and humidity sensors attached to a single board computer (SBC). The relation of both, humidity and temperature, is absolutely necessary for an appropriate climate control in depots [1].

These requirements lead to the idea of creating a distributed sensor environment that covers more areas within a single depot and can be distributed over multiple depots. The data captured by those sensors should then be aggregated

© Springer International Publishing AG 2017
J. Kamps et al. (Eds.): TPDL 2017, LNCS 10450, pp. 599–602, 2017.
DOI: 10.1007/978-3-319-67008-9_52

to a single point, where it can be processed and used, e.g., for creating graphs and analyzing trends. Also the captured data should be made accessible via a web interface once the sensor network is fully operational.

2   IoT Monitoring Solutions in Other Areas of Application

Already existing monitoring solutions are provided in several other areas. For instance, Lewis *et al.* [2] propose an environmental monitoring in a quality-controlled calibration laboratory. In the e-health sector, Jassas *et al.* [3] implemented a prototype with e-health sensors attached to the RPi. Medical sensors measure patients' physical parameters and the RPi collects and transfers them to the cloud environment, as real-time data. However, the former prototypes miss an architectural concept and are not considering the privacy of sensor data. Instead, they send, process, and store the data on remote cloud servers. Only Hentschel *et al.* [4] introduce a concept with local supernodes, which are sensor enhanced RPis. In their model the nodes behave autonomously and carry out simple tasks like processing data or communication with other devices. Unfortunately, each supernode needs to be managed and thus, attaching new devices increases maintenance complexity.

In contrast to those approaches, we focus on transferring data center technology to energy-efficient devices, to enable an easily updatable, scalable, and manageable framework. Moreover, obtained sensor data are processed locally and are saved on the RPis without the need to send them to cloud services.

3   Realisation of MonTreAL

We built MonTreAL (Monitoring Treasures of all libraries) to measure temperature and humidity in our library. Due to our initial goal to perform the measurements with SBCs, MonTreAL takes advantage of Docker[1] and Docker Swarm as underlying technologies.

The system consists of two main components as depicted in Fig. 1. Devices that act as *Workers* are connected to sensors and are distributed within a building. Their tasks is a regular interaction with their environment to gather and process information. Therefore, a *sensor* container running an application to interact with the sensors is responsible to correctly process and send the gathered data to a messaging *queue* container. *Sensor* containers are able to address several different sensor types, which are connected via *GPIO* or *USB* interfaces to a SBC and will port the diversity of manufacturers' dependent data formats into a more uniform format for further processing. Every *sensor* container is configured to add a unique ID to every package it sends to the queue to distinguish and match the data with their origin later. The sensors we are using are simple temperature and humidity sensors like the *DHT-22*[2], which is directly

---

[1] https://www.docker.com/.
[2] https://www.adafruit.com/product/385.

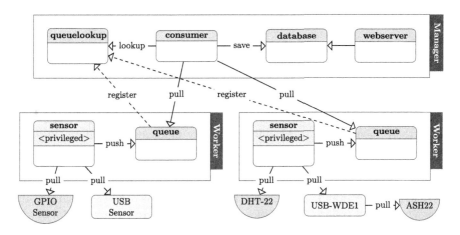

Fig. 1. Architectural overview of MonTreAL

connected to the GPIO interface of the RPi. Or, a more complex sensor system, which consists of a receiver (*USB-WDE1*[3]) that is connected via USB and several autonomous sensor devices (*ASH22*[4]), which regularly send their data to the receiver by radio.

The second main component of MonTreAL is one (or possibly more) device(s) representing the *Manager*. Its task is to accumulate and store the gathered data and provide interfaces to allow user interaction. To fetch the gathered sensor data from the queues, a *consumer* container running a dedicated consumer implementation periodically requests the addresses of all running queues from a *queuelookup* container and fetches all available data. The consumer then stores the data in a *database* container for long-term storage. MonTreAL also provides the user with the possibility to view the gathered data in a suitable way. It runs a simple *webserver* container, which allows the user to view gathered data of every single sensor in a graph and to filter it by date. Furthermore, the front end indicates, which sensors are currently online or offline, or when problems with devices occur, e.g., inoperable devices, broken sensors, empty batteries, etc.

To summarize, MonTreAL allows to set up a distributed system of IoT devices equipped with specific sensors and to gather, accumulate and store the collected data without the need of a powerful backend. It also features a simple front end to view the data. MonTreAL relies on the existence of a network infrastructure (LAN, WLAN) and operates with only a few needed configuration options. We constantly evaluate our prototype at the University Library of Bamberg by collecting temperature and humidity data with the capacity of three RPis.

---

[3] https://www.elv.de/output/controller.aspx?cid=74&detail=10&detail2=44549.
[4] https://www.elv.de/elv-funk-aussensensor-ash-2200-fuer-z-b-usb-wde-1-ipwe-1.html.

4   Conclusion and Future Work

Our IoT prototype MonTreAL implements a distributed temperature and humidity monitoring framework, which is delivered by Docker images for several Docker supported SBC architectures like ARM, AARCH64 and x86_64. It is made publicly available at the Github[5] repository *unibaktr/MonTreAL* containing the source code. By defining those images, virtualization is achieved at IoT level and by these means MonTreAL offers an easily manageable solution. Furthermore, deploying the system to a container enabled cluster achieves reliability and resilience through the underlying Docker Swarm paradigm. Even, if services are unavailable or are scaling up or down, collected data remain in the distributed message queue and is processed as soon as the consumer recognizes the queue again. Data collected from message queues are processed and stored in a database for long term evaluations. However, if sensors, services, the underlying network, or in the worst case SBCs fail, the consumer notifies the user by alerts.

By now, the RPis, more precisely the plugged in SD cards, provide a bottleneck for MonTreAL. Their limited lifetime is evoked by a relatively small number of write cycles and stands in contradiction to a persistent storage solution. This might be a less relevant problem for a stateless container that only produces runtime data. However, for the database we plan to evaluate the possibility to persist data on a reliable storage solution, which is attached to the database container. Moreover, MonTreAL currently supports only one depot with several sensors attached to multiple RPis. In the future, more containers should enhance the framework to remotely manage multiple depots within one web service.

Acknowledgement. The authors would like to thank the Hypriot team (https://blog.hypriot.com/crew/) for their great effort to port Docker to ARM platforms and for providing Hypriot OS.

References

1. Glauert, M.: Klimaregulierung in Bibliotheksmagazinen. In: Hauke, P., Werner, K.U. (eds.) Bibliotheken bauen und ausstatten. Institut fü Bibliotheks- und Informationswissenschaft (2009). http://edoc.hu-berlin.de/docviews/abstract.php?id=30204
2. Lewis, A., Campbell, M., Stavroulakis, P.: Performance evaluation of a cheap, open source, digital environmental monitor based on the Raspberry pi. Measurement 87, 228–235 (2016). http://www.sciencedirect.com/science/article/pii/S0263224116001871
3. Jassas, M.S., Qasem, A.A., Mahmoud, Q.H.: A smart system connecting e-Health sensors and the cloud. In: 2015 IEEE 28th Canadian Conference on Electrical and Computer Engineering (CCECE), pp. 712–716, May 2015
4. Hentschel, K., Jacob, D., Singer, J., Chalmers, M.: Supersensors: Raspberry pi devices for smart campus infrastructure. In: 2016 IEEE 4th International Conference on Future Internet of Things and Cloud (FiCloud), pp. 58–62, August 2016

---

[5] https://github.com.

Introducing Solon: A Semantic Platform for Managing Legal Sources

Marios Koniaris[1]($\boxtimes$), George Papastefanatos[2], Marios Meimaris[2],
and Giorgos Alexiou[2]

[1] KDBS Lab, School of ECE,
National Technical University of Athens, Athens, Greece
mkoniari@dblab.ece.ntua.gr
[2] Athena Research Center, Athens, Greece

Abstract. In this paper we introduce *Solon*, a legal document management platform aiming to improve access to legal sources by offering advanced modelling, managing and mining functions. It utilizes a novel method for extracting semantic representations of legal sources from unstructured formats, interlinking and enriching them with advanced classification features. Also, it provides refined search results utilizing the structure and specific features of legal sources, allowing users to connect and explore legal resources according to individual needs.

Keywords: Digital libraries · Information retrieval · Legal informatics

1 Introduction

As a consequence of many open data initiatives, a plethora of publicly available portals and datasets provide legal resources to citizens and legislation stakeholders. However, legal resources are mostly disseminated in a semantically poor, human-readable textual representation [1], mainly PDF, which can not capture the structure and the legal semantics of the data, making it impossible to reuse and establish an interoperability layer among repositories in the Semantic Web.

To address these issues, we introduce *Solon*[1], an advanced system architecture, aiming to assist users locate and retrieve legal and regulatory documents within the exact context of a conceptual reference. It consists of several different components, exposed as REST services. It operates on unstructured legal sources, capturing the internal organisation of the textual structure and the legal semantics, interlinking them based on discovered references and classifying them according to a set of rules. It exploits the semantic representation of legal sources, offering, among others fine-grained search results and enabling users to organize legal information according to individual needs. Recent efforts have also addressed the need for semantic representation of greek legal information [2],

---

[1] Solon was an Athenian lawmaker, credited with having laid the foundations for Athenian democracy.

© Springer International Publishing AG 2017
J. Kamps et al. (Eds.): TPDL 2017, LNCS 10450, pp. 603–607, 2017.
DOI: 10.1007/978-3-319-67008-9_53

whereas *Solon* has been successfully deployed in a public sector production environment[2]. In this paper, we initially demonstrate the main features offered, we present the main architectural components and discuss future work aspects.

2   Architecture

Requirements and General Characteristics. The main requirements for *Solon*, are focused on (i) support for automatic and manual import of unstructured legal sources from predefined repositories, (ii) automatic structural analysis and semantic representation of legal sources, (iii) automatic discovery and resolution of legal citations, (iv) automatic classification of legal sources based on custom rules, (v) support for manual content curation, (vi) multi criteria and multi faceted search using all metadata identified in documents, and (vii) support for user-defined collections of legal resources around a topic.

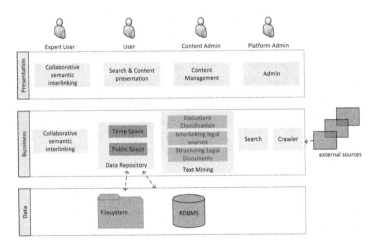

Fig. 1. High-level view of *Solon* logical and conceptual architecture

Architecture. The system's architecture, Fig. 1, is composed of different components exposed to the remaining platform as REST services. The core business layer consists of (i) the Document Repository, which provides functionality for storing and managing complex legal sources, (ii) the Crawler module, which harvests remote information sources as input data, (iii) the Text Mining module, which transforms the Crawler's input into a semantically rich data structure, (iv) the Search module, which is responsible for the efficient indexing and retrieval of legal information, and (v) the Collaborative Semantic Interlinking module. Complementary to the aforementioned modules is the presentation layer, which offers UI and Administration functionality.

---

[2] http://www.publicrevenue.gr/elib/.

Data Model. *Solon* utilizes the Akoma Ntoso (AKN) schema [3] to model legal documents, an OASIS standard, XML schema for modelling parliamentary, legislative, and judiciary documents. To accommodate for the structure and metadata of Greek legal and administrative documents we provide extensions to the schema in terms of Dublin Core and FRBR vocabularies. In short, our model maps all sections of a legal document, such as articles and paragraphs, into semantically rich legal resources, identified by a URI.

Legal Document Repository. Our legal document repository was build on top of Flexible Extensible Digital Object Repository Architecture, Fedora[3]. Since a legal resource may be accompanied by several files e.g., manifestation format (word, pdf), XML based representation (AKN schema), accompanying material (images), objects are stored in digital containers utilizing a directed acyclic graph of resources where edges represent a parent-child relation. Digital Container management (CRUD) operations are exposed through a RESTful HTTP API, implementing the W3C Linked Data Platform specification.

Crawler. The crawler is based on a distributed architecture, and consists of (a) a Crawler manager and (b) various crawler implementations, extending a common Crawler interface, delivering relevant data to the Crawler manager. The Crawler manager interacts with the Legal Document Repository, ensuring consistent data validation and storage, avoiding also duplication of content. It is also responsible for the periodic scheduling and performance monitoring of all crawling activities within the system.

Text Mining. A pipeline strategy invokes a list of transformers in sequence, that acquire a structured semantic representation of legal sources, interlink legal sources and perform advanced classification based upon tailor made rules that map text to semantic components.

- **Parser.** *Solon* employs the automatic structuring and semantic indexing approach for legal documents, presented in [4].

- **Interlinking legal sources.** Given the abundant use of citations between legal sources, *Solon* employs automated methods for identification and resolution of legal citations, between Greek and EU legal sources, following the methodology utilized in [5]. This component first discovers citations, and then resolves them by creating their respective URIs, following ELI, an EU proposed standard for a European Legislation Identifier. The legal corpus can then be modelled using a graph model as presented in [6].

- **Document Classification.** The Document Classification mechanism is based on a custom developed rule engine following a deterministic approach. Rules are defined through the administration UI module and executed against the legal sources using priorities. Rules can be simple or combined, forming complex chains of operation, acting upon the textual data or metadata of the legal sources.

Collection Management. *Solon* employs a linked data enabled collaborative semantic interlinking mechanism, that allows users to create legal

---

[3] http://fedorarepository.org.

collections, provide custom semantic annotations to legal resources, and collaborate on shared resources, utilizing the model presented in [7].

Search. *Solon*'s information retrieval component, has been build on top of Solr[4], integrated with our repository component. Since legal documents tend to be quite long, covering multiple topics, we follow structured retrieval techniques, utilizing the legal sources hierarchical structure of nested elements.

3   Evaluation and Demonstrator

Solon has been successfully deployed in a public sector production environment (see footnote 2), under the supervision of the Independent Authority for Public Revenue[5], aiming to provide semantic access to Greek tax legislation. It currently hosts more than 4000 legal and regulatory documents. In the demo, we will showcase the main functionality, addressing the needs of both: (a) the public users e.g., browse the legal knowledge base, search, cite legal resources and (b) the authenticated users e.g., select a legal document and upload it to the repository, curate the structure/metadata, publish it, create legal collections.

4   Conclusion and Future Work

In this paper, we presented Solon a platform suitable for modelling, managing and mining legal sources. As future work, we are investigating the adoption of the recently proposed ELI extension as an OWL ontology and the temporal management of legal sources. Additionally, we plan to employ search result diversification methods, as a means of improving user satisfaction by increasing the variety of information shown to user, based on our previous work where we performed an exhaustive evaluation of several state of the art methods [8].

References

1. Inter-Parliamentary Union: World e-Parliament Report (2016). http://www.ipu.org/pdf/publications/eparl16-en.pdf
2. Chalkidis, I., Nikolaou, C., Soursos, P., Koubarakis, M.: Modeling and querying greek legislation using semantic web technologies. In: Blomqvist, E., Maynard, D., Gangemi, A., Hoekstra, R., Hitzler, P., Hartig, O. (eds.) ESWC 2017. LNCS, vol. 10249, pp. 591–606. Springer, Cham (2017). doi:10.1007/978-3-319-58068-5_36
3. Barabucci, G., Cervone, L., Palmirani, M., Peroni, S., Vitali, F.: Multi-layer markup and ontological structures in Akoma Ntoso. In: Casanovas, P., Pagallo, U., Sartor, G., Ajani, G. (eds.) AICOL -2009. LNCS, vol. 6237, pp. 133–149. Springer, Heidelberg (2010). doi:10.1007/978-3-642-16524-5_9
4. Koniaris, M., Papastefanatos, G., Vassiliou, Y.: Towards automatic structuring and semantic indexing of legal documents. In: Proceedings of the 20th Pan-Hellenic Conference on Informatics, PCI 2016. ACM (2016)

---

[4] http://lucene.apache.org/solr/.
[5] http://www.aade.gr/, formerly known as General Secretariat of Public Revenue.

5. Opijnen, M.v., Verwer, N., Meijer, J.: Beyond the experiment: the extendable legal link extractor (2015). https://ssrn.com/abstract=2626521
6. Koniaris, M., Anagnostopoulos, I., Vassiliou, Y.: Network analysis in the legal domain: A complex model for european union legal sources. In: Physics and Society, Cornell University Library, arxiv (2015). http://arXiv.org/abs/1501.05237
7. Meimaris, M., Alexiou, G., Papastefanatos, G.: LinkZoo: a linked data platform for collaborative management of heterogeneous resources. In: Presutti, V., Blomqvist, E., Troncy, R., Sack, H., Papadakis, I., Tordai, A. (eds.) ESWC 2014. LNCS, vol. 8798, pp. 407–412. Springer, Cham (2014). doi:10.1007/978-3-319-11955-7_57
8. Koniaris, M., Anagnostopoulos, I., Vassiliou, Y.: Evaluation of diversification techniques for legal information retrieval. Algorithms **10**(1), 22 (2017)

Towards a Semantic Search Engine for Scientific Articles

Bastien Latard[1,2]([✉]), Jonathan Weber[1], Germain Forestier[1],
and Michel Hassenforder[1]

[1] MIPS, University of Haute-Alsace, Mulhouse, France
bastien.latrad@uha.fr
[2] MDPI AG, Basel, Switzerland

Abstract. Because of the data deluge in scientific publication, finding relevant information is getting harder and harder for researchers and readers. Building an enhanced scientific search engine by taking semantic relations into account poses a great challenge. As a starting point, semantic relations between keywords from scientific articles could be extracted in order to classify articles. This might help later in the process of browsing and searching for content in a meaningful scientific way. Indeed, by connecting keywords, the context of the article can be extracted. This paper aims to provide ideas to build such a smart search engine and describes the initial contributions towards achieving such an ambitious goal.

1 Introduction

Keeping up-to-date in a specific research field is a tedious and complex task. This is mandatory as it allows researchers to increase their knowledge on a domain and acquire latest ideas. Hence, choosing the correct approach is the first step of any research work. Despite—*or because of*—the data deluge in scientific publication, researchers spend a significant amount of time searching for articles related to their scientific interests.

An editorial from *Nature* [1] clearly expressed the continued frustration of the scientific community concerning the incredible potential that text mining of scientific literature represents. However, text miners often face the barrier of publishers' legal restrictions (i.e., closed access). The average growth of scientific literature is estimated to be 3 million new articles per year from journals and conferences over the last 4 years, with 3.3 million articles produced in 2016 (http://www.scilit.net). This massive amount of data is published by more than 6000 publishers in around 47,000 scientific journals. These de-centralised and separated platforms further complicate the research process because scientists are unable to go through them all in order to search for relevant articles. Thus, they have to rely on big databases or indexing companies which provide either an incomplete corpus due to selection criteria or only display articles from their own platforms. Moreover, their search engines often offer very limited search functionalities, and this is the problem we want to tackle.

J. Kamps et al. (Eds.): TPDL 2017, LNCS 10450, pp. 608–611, 2017.
DOI: 10.1007/978-3-319-67008-9_54

To tackle this problem, our approach consists in using semantic relations between keywords to extract the main categories of the articles. This approach simultaneously validates both the context of the article and the context of the word, thus providing the correct category. Effendy and Yap [2] discussed the potential of using semantic mining tools to extract the best category of a conference. This is exactly what our framework aims to do.

2   Method

Our approach uses BabelNet [3] which is a multilingual lexicographic and encyclopaedic database based on the smart superposition of semantic lexicons (WordNet, VerbNet) together with other collaborative databases (Wikipedia and other Wiki data). A query for a term through BabelNet returns "dictionary entries", synonyms, categories or domains. Each synset S contains the relative categories C, domains D and synonyms syn within the specific concept:

$$S = \{C, D, syn\} \tag{1}$$

Assuming that synonyms of keywords might be an interesting way to connect several articles, BabelNet is the knowledge database on which our framework will rely. However, BabelNet lacks specificity and searching for one word can return synsets from various different contexts. For example, "flight" returns 36 synsets, from a South Korean movie to the verb 'to fly'. Consequently, a method to filter out unrelated synsets is mandatory.

Because synonyms are too specific, and domains are too general, categories have been naturally chosen in order to identify overlapping between synsets from different keywords. Indeed, if several keywords share the same category, then this is potentially the correct category in regards to the article context. In addition, the greater the number of keywords sharing the same category, the higher the confidence. Thus, connecting the returned synsets based on their categories is an interesting way to naturally filter out all of the unrelated synsets.

This approach does filter some content, but still returns "living people; English-language films; celestial mechanics; American films" as the main categories for keywords "nonlocal gravity; celestial mechanics; dark matter". Constant noise (*_singer, *_album, etc.), meaningless in our scientific context, has been identified. A parameter can now be set in order to force the automatic filtering of identified noise. Most of the remaining noise is finally naturally filtered out, and "celestial mechanics" is finally returned as the main category.

Our final goal is to apply this valuable added knowledge to all articles from the scientific literature database, Scilit (http://www.scilit.net), developed by MDPI (http://www.mdpi.com). To validate our approach, a manual analysis on a subset of 595 articles from seven journals (six about Physical Science and one about Pediatrics) has been conducted. We evaluated the correctness of the categories based on the connection of keywords by their synsets. This approach provides good precision—from 96% to 100%—depending on the threshold which identify the data as correct not. Indeed, strictly selecting only categories shared by three different keywords or more leads to a high degree of confidence (100% precision), but a recall of 9%. By being more tolerant and considering all categories shared by at least two keywords, precision slightly decreased (96%) but we significantly gain in recall (47%). Moreover, similar proportions are observed for *Children*, the journal about Pediatrics (from 100% to 92%). This validates that our approach may be used in several domains.

The main drawback of our approach is that correct categories have been identified for only 22% of the articles within the subset. Figure 1 illustrates the reason for the law recall and coverage of our approach.

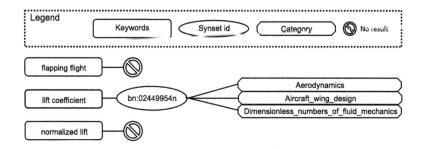

Fig. 1. Limits of the exact search: only one keyword from three return data.

One of the reasons for this law coverage is that BabelNet often returns no result for composed keywords (multi-word keywords), as shown in Fig. 1, where no data is returned for two of three keywords. In our approach—proposing only categories shared by at least two keywords—the degree of confidence is not high enough to return the categories. We will investigate further a way to propose some categories from these composed keywords in our future work. In doing so, we aim to significantly gain in recall, and cover many more articles.

3   Perspectives

Making scientific recommender systems smarter is crucial in order to help scientists in their mandatory and tedious bibliographical research phase. The approach proposed could be the first step in building such a smart system. Indeed, analysing the correctness of the main category based on the overlapping of the keywords category confirms the logic of our approach. In the future, we plan to extend the search in order to extract categories from composed keywords. Splitting on spaces would provide some data for sub-keywords. Then, applying the same logic as described in our approach (i.e., connect by common category) will filter out unrelated items, and categories from connected items might be used for the global category connection. By taking the example from Fig. 1, splitting "flapping flight" on spaces will return 3 and 25 synsets, respectively for "flapping" and "flight" (Fig. 2):

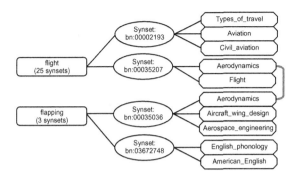

Fig. 2. The category "Aerodynamics" is returned as the main category of "flapping flight". Other categories are filtered out by this connection.

This further search will successfully identify "Aerodynamics" as the main category of "flapping flight". Thus, our approach would connect "Aerodynamics" based on both keywords. Extracting the part-of-speech (with a syntactical analyser like SyntaxNet [4] or CoreNLP [5]) from long keywords could be an interesting extra source of information for refining requests on BabelNet. Finally, Fig. 3 shows the main logic of our next contribution: to process in a smarter way the keywords that do not return any satisfactory results. Later, we might also generate a graph inherited from the BabelNet's synsets as in [6].

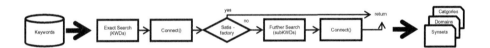

Fig. 3. Illustration of the general logic of our approach in a future work

References

1. (Editorial), N.: Gold in the text? Nature **483**(7388), 124 (2012)
2. Effendy, S., Yap, R.H.C.: The problem of categorizing conferences in computer science. In: Fuhr, N., Kovács, L., Risse, T., Nejdl, W. (eds.) TPDL 2016. LNCS, vol. 9819, pp. 447–450. Springer, Cham (2016). doi:10.1007/978-3-319-43997-6_41
3. Navigli, R., Ponzetto, S.P.: BabelNet: the automatic construction, evaluation and application of a wide-coverage multilingual semantic network. Artif. Intell. **193**, 217–250 (2012)
4. Andor, D., Alberti, C., Weiss, D., Severyn, A., Presta, A., Ganchev, K., Petrov, S., Collins, M.: Globally normalized transition-based neural networks. In: ACL (2016)
5. Manning, C.D., Surdeanu, M., Bauer, J., Finkel, J.R., Bethard, S., McClosky, D.: The Stanford CoreNLP natural language processing toolkit. In: ACL, pp. 55–60 (2014)
6. Franco-Salvador, M., Cruz, F.L., Troyano, J.A., Rosso, P.: Cross-domain polarity classification using a knowledge-enhanced meta-classifier. Knowl. Based Syst. **86**, 46–56 (2015)

Development of an RDF-Enabled Cataloguing Tool

Lucy McKenna[1]([✉]), Marta Bustillo[2], Tim Keefe[3], Christophe Debruyne[1], and Declan O'Sullivan[1]

[1] ADAPT Centre, Trinity College Dublin, Dublin, Ireland
lucy.mckenna@adaptcentre.ie
[2] UCD Library, University College Dublin, Dublin, Ireland
[3] Digital Resources and Imaging Services, Trinity College Dublin, Dublin, Ireland

Abstract. By generating bibliographic records in RDF, libraries can publish and interlink their metadata on the Semantic Web. However, there are currently many barriers which prevent libraries from doing this. This paper describes the process of developing an RDF-enabled cataloguing tool for a university library in an attempt to overcome some of these obstacles.

Keywords: Semantic web · Linked Data · MODS · RDF · Library · Interface design · Usability testing

1 Introduction

The Digital Resources and Imaging Services (DRIS) department of the Library of Trinity College Dublin (TCD) hosts the Digital Collections Repository of the university. This repository provides open access to TCD's collection of digitised cultural heritage materials which includes manuscripts, letters, books, images, and other archival materials. DRIS aims to publish the bibliographic data of its collections as RDF in order for these materials to be discoverable on the SW, increasing the visibility and use of the library's resources. Additionally, RDF metadata published by DRIS could be interlinked with Linked Data (LD) emerging from other institutions, facilitating library users to access a web of related data from a single information search [1].

2 Libraries and Linked Data

Although not yet widely used, libraries are publishing bibliographic metadata as RDF in increasing quantities [1,2]. However, librarians have reported a number barriers in using LD to its full potential including that LD software is not tailored to the specific needs and expertise of librarians but rather technical experts. Other reported challenges included a lack of authority control on the SW, difficulties establishing interlinks, and few examples of useful applications of LD

© Springer International Publishing AG 2017
J. Kamps et al. (Eds.): TPDL 2017, LNCS 10450, pp. 612–615, 2017.
DOI: 10.1007/978-3-319-67008-9_55

in the library domain that would justify the allocation of time and resources to its generation [3, 4]. These challenges were experienced by DRIS and prevented the library from publishing its metadata to the SW. As such a bespoke RDF-enabled cataloguing interface was developed for DRIS. The aim of the interface was to explore whether such a tool could be used by DRIS to successfully generate MODS-RDF records for a small sample of records thus demonstrating the potential for LD software specifically designed for library use.

3   MODS and MADS

The Metadata Object Description Schema (MODS) is an XML schema for a bibliographic element set that can be used for the purpose of cataloging digital resources [5]. The full schema consists of 20 top-level elements, for example TitleInfo and Name, which are used to provide information on the title and creator of a work. The majority of MODS elements contain subelements, such as title, subtitle, and namePart, as well as attributes which describe the metadata itself, for example, the authority source from which a title or name was taken, or the language used when cataloguing.

MODS was selected as the output schema for the tool as it was sufficiently detailed for DRIS's cataloguing purposes and a MODS-RDF ontology was already available [6]. Additionally, a set of MODS implementation guidelines was developed by the Digital Library Federation's (DLF) Aquifer Initiative thus allowing for the standardisation of MODS records [7].

The Metadata Authority Description Schema (MADS) [8] can serve as a companion to MODS to provide metadata regarding the authority sources used in a record when describing names, organisations, genres, or subjects for example. Like MODS, a MADS-RDF ontology already exists [9]. Both MODS and MADS share a number of subelements, such as those in TitleInfo, Name and Subject. The schemas also share all attributes. Interestingly the MODS-RDF ontology excludes all elements it has in common with MADS. As such, in order to generate a full MODS record in RDF, both ontologies must be used.

4   Interface Design and Testing

A semi-structured interview was carried out with the DRIS metadata cataloguer in order to establish a set of tool requirements, and a mock-up of the cataloguing interface was subsequently developed. User requirements included:

- Facilitating cataloguing efficiency by automating input where possible.
- Publishing MODS records that meet DLF-Aquifer requirements by forcing data entry for certain fields and constraining data entry options for others.
- Further constraining data entry options as per the specific needs of DRIS.
- Providing additional administrative data entry fields.

The completed interface was programmed to initially constrain data entry options to only those elements and subelements which were identified as required fields by the DLF. This was done to ensure that the minimal data requirements for each record were met prior to the addition of supplementary metadata. Once these fields were complete, data entry options expanded to include recommended and optional fields.

Data entry fields and dropdown menu options were programmed to dynamically alter based on prior selections made during the cataloguing process. This ensured that data entry options were restricted to DLF recommendations. For example, in the Name element, DLF require that the resource creator's name should be taken from the Name and Title Authority Source Codes maintained by the Library of Congress (LOC). Thus the list of options in the authority menu was constrained to these sources, this was then further constrained to display only the sources used by DRIS. Data entry fields also self-populated based on prior selections allowing for a more efficient cataloguing process. For example, again in Name, after selecting an authority source the Authority-URI field self-populated. This also highlights how the tool was capable of accepting URIs to other LD datasets - a first step in the LD interlinking process.

The interface was tested by observing the DRIS metadata cataloguer using the tool to create a bibliographic record. Although results indicated some issues with the interface layout, the librarian felt that the tool would be useful for creating more authoritative RDF datasets and that it could facilitate increased LD generation by librarians rather than technical experts alone.

5   Record Generation

Data from the interface was stored in a relational database. In order to uplift this data to RDF an R2RML mapping was developed based on the MODS and MADS RDF ontologies. R2RML is a W3C Recommendation for declaring mappings from relational databases to RDF datasets [10]. In the process of adding MADS to the mappings it was noted that, unlike MODS-RDF where properties are represented individually, some MADS-RDF properties were grouped in collections including the subelements in TitleInfo and Name. Collections are a special RDF construct used to represent lists. This grouping allows for labels, such as title and subtitle, or first and last names, to be reconstructed with all elements in the correct order. However, at the time of the project, R2RML did not support the mapping of RDF collections, thus some metadata, such as subtitle, and more than one namePart were omitted. Despite this setback, semi-complete RDF records were generated for a small sample of DRIS's materials. A number of SPARQL (RDF query language) queries were successfully run over the RDF dataset including typical searches by author, date, and genre, as well as more interesting and detailed searches by ISO Language and Country Codes, authority sources, controlled vocabulary terms, and URIs.

This issue inspired a separate project in which an R2RML expansion supporting the mapping of RDF Collections (and Containers) was developed [11].

This expansion facilitated the uplift of all metadata in the database to RDF, allowing for the publication of complete MODS records.

6   Conclusions and Future Directions

Providing librarians with bespoke LD tools would allow for increased publication of rich LD datasets. It is likely that LD generated by librarians would be treated with increased credibility and thus used more frequently as libraries are viewed as trustworthy and authoritative sources of information. LD created by librarians will follow specific and standardised bibliographic schemas, and use long established authorities and controlled vocabularies to describe resources. This would increase the level of authority control on the SW, allowing for similar entities to be identified consistently across the SW leading to richer search results.

Future research will explore how to engage librarians in the process of interlinking with LD datasets published by other libraries and related institutions rather than just large scale authorities (LOC) and LD datasets (DBpedia). This would allow library users to access larger amounts of related data from single information search.

Acknowledgments. This study is supported by the Science Foundation Ireland (Grant 13/RC/2106) as part of the ADAPT Centre for Digital Content Platform Research (http://www.adaptcentre.ie/) at Trinity College Dublin.

References

1. Hastings, R.: Linked data in libraries: status and future direction. Comput. Libr. **35**, 12–16 (2015)
2. Mitchell, E.T.: Library linked data: early activity and development. Libr. Technol. Rep. **52**, 5–33 (2016)
3. Hallo, M., Lujan Mora, S., Trujillo Mondejar, J.C.: Transforming library catalogs into Linked Data. In: ICERI (2013)
4. OCLC: Linked Data Survey (2017). http://www.oclc.org/research/themes/data-science/linkeddata.html
5. Library of Congress: MODS (2017). http://www.loc.gov/standards/mods/
6. Library of Congress: MODS-RDF (2012). http://www.loc.gov/standards/mods/modsrdf/v1/modsrdf.owl
7. Digital Library Federation: MODS Implementation Guidelines (2009). https://wiki.dlib.indiana.edu/download/attachments/24288/DLFMODS_Implementation-Guidelines.pdf
8. Library of Congress: MADS (2017). http://www.loc.gov/standards/mads/
9. Library of Congress: MADS-RDF (2017). http://www.loc.gov/standards/mads/rdf/mads-ontology-20101119.owl
10. W3C: R2RML (2012). https://www.w3.org/TR/r2rml/
11. Debruyne, C., McKenna, L., O'Sullivan, D.: Extending R2RML with support for RDF collections and containers to generate MADS-RDF datasets. In: TPDL (2017)

Towards Semantic Quality Control of Automatic Subject Indexing

Martin Toepfer[1]([envelope]) and Christin Seifert[2]

[1] ZBW – Leibniz Information Centre for Economics,
Düsternbrooker Weg 120, 24105 Kiel, Germany
m.toepfer@zbw.eu
[2] University of Passau, Innstraße 33, 94032 Passau, Germany
christin.seifert@uni-passau.de

Abstract. Automatic subject indexing is a key technology for digital libraries, however, factors like concept drift hinder its success in practice. Releasing high-quality results into productive retrieval systems may still be possible when thorough quality control is applied, which may support algorithmic improvements and allow to create high precision filters. Since errors and their relevance can depend on characteristics of concepts and their relations, evaluations should take semantic aspects into account. For this reason, we present the prototype of a web-based reviewing tool which especially aims at fostering semantic analysis and visualization, that is, considering relations, properties and semantic categories of concepts, algorithms and reviews. The tool uses techniques of the Semantic Web. Its application is demonstrated by example.

Keywords: Quality control · Automatic subject indexing · Semantics

1 Introduction

Accurate indexing of documents with subject headings (descriptors, concepts) of controlled vocabularies enables high-quality semantic access to digital libraries. Automation of this task has been addressed by many researchers, for instance, in the field of machine learning and multi-label classification. In practice, different factors hinder the success of automatic methods, thus libraries apply them either only as assistants [1,3], or as autonomous agents restricted to special types of documents [7]. In particular if predictions are passed to productive retrieval systems without human intervention, continuous testing and control becomes crucial to ensure high-quality results over time. In this paper, we present a web-based application for reviewing automatically predicted subject headings. In order to recognize semantic patterns in errors, integration of background knowledge from thesauri is desirable. We build upon technology from the Semantic Web for data modelling to foster analysis and visualization of relations, properties and semantic categories of concepts, indexing approaches and ratings.

© Springer International Publishing AG 2017
J. Kamps et al. (Eds.): TPDL 2017, LNCS 10450, pp. 616–619, 2017.
DOI: 10.1007/978-3-319-67008-9_56

2   Background

Put briefly, subject indexing aims to determine the most relevant subjects of documents comprehensively, precisely and concisely. Controlled vocabularies are used to reduce ambiguity and enable further semantic applications. In this work, we use the STW thesaurus 9.02[1] [2], which addresses economics and related subject areas. It has more than 6,000 concepts with links between broader (BT), narrower (NT), and semantically related (RT) concepts. Descriptors are additionally linked to semantic categories. Regarding automatic subject indexing, we assume that there is a main system under review, for instance, a fusion system [6] that combines lexical approaches, which use keyword matching, and associative approaches, which learn synonymous expressions from examples.

Common approaches for evaluation of automatic methods leverage corpora with documents that have already been indexed professionally and compare them with subjects predicted by algorithms. Several classification metrics, like precision, recall, and F1, or ranking metrics can be computed and different averaging techniques may be used, for instance, aggregation by concept or by instance. Beyond these evaluation approaches, subject-specific analysis and fine-grained ratings are used as well, for instance, at the German National Library [7].

Since cleansing and evaluation tasks are crucial but often costly parts of projects, general purpose tools like OpenRefine[2] and various specialized user interfaces for annotation and evaluation tasks have been developed in different domains, for instance, ontology alignment and object-vocabulary automatic linking, which also have to deal with fuzzy matching problems.

3   Reviewing Subject Headings

The main view for reviewing subject headings is depicted in Fig. 1. On the top, meta-data (title, author keywords, abstract) is shown ⬚1 for determining relevant subjects manually. A table ⬚2 summarizes the concepts (rows) that have been proposed by different indexing approaches (columns). Each concept can be rated individually ⬚3. Missing concepts can be added ⬚4. When finished reviewing the concepts, the reviewer enters a final decision for the document on a 3-point scale ⬚5, which especially determines if the automatically generated descriptors must be rejected because the proposed subjects would be misleading. A graph visualization[3] depicts relations between proposed concepts (direct RT relations and paths of BT) and their semantic categories ⬚6. Some decisions may be subtle, like disambiguation between Germany, Germans and German (language).

The tool especially targets the precision of automatic indexing, thus the most relevant role of concept-level ratings is to prevent misleading concepts ($-$) in the output. The other levels (0, $+$, $++$) denote increasing preciseness and relevance. Detailed information on the rating guidelines can be accessed by a dialog ⬚7.

---

[1] www.zbw.eu/en/stw-info/ (accessed: 10.04.2017).
[2] www.openrefine.org/ (accessed: 15.06.2017).
[3] The graph visualization is below the table in the user interface, but depicted next to it due to space constraints.

Fig. 1. User interface for reviewing subject headings.

4   Semantic Quality Control

In contrast to plain applications of precision, recall and F1, the tool proposed here aims at semantic quality control, that is, taking semantic categories and relations between concepts as specified in thesauri into account. In the context of measuring inter-indexer consistency, disregarding semantic relationships between concepts has been criticised, for instance, by Medelyan and Witten [5] who proposed a measure that incorporates RT and BT/NT relations. Some thesauri provide further structure beyond these relations, for instance, the STW (cf. Sect. 2). At the top level, it has seven semantic categories (row names in Fig. 2). In order to leverage this background knowledge, we build upon Semantic Web technology for data modeling which can be used by digital libraries to expose Linked Open Data [4]. We utilize well-known schemas: dublin core[4], SKOS[5], rev[6], and MUTO[7]. With this representation, queries on reviews can be formulated in SPARQL, accessing semantic properties and relations.

To illustrate a simple case of semantic quality control, Fig. 2 shows an analysis of artificial concept ratings by rating value, indexing algorithm (Agent) and semantic category. It can be seen that severe errors are imbalanced, e.g. fusion does not make any errors for geographic names. Such insights can help to improve the overall indexing system by weighting assignments for each category dependent on the algorithm. Also rules for filtering can be developed. For instance, geographic names proposed by knn may be blocked.

---

[4] dublincore.org/documents/dcmi-terms/ (accessed: 10.04.2017).

[5] www.w3.org/2004/02/skos/ (accessed: 10.04.2017).

[6] vocab.org/review/ (accessed: 10.04.2017).

[7] muto.socialtagging.org/core/v1.html (accessed: 10.04.2017).

rating value	(rating 1)				(rating 2)				(rating 3)				(rating 4)			
Agent / Semantic Category	fusion	ranking	knn	binary relev.	fusion	ranking	knn	binary relev.	fusion	ranking	knn	binary relev.	fusion	ranking	knn	binary relev.
Economics	1											1				1
Business economics	1											1				1
Economic sectors	1											1	2	1	2	1
Commodities	1		1													
Related subject areas													3	2	3	2
Geographic names	1	3	1					1					3	2		1
General descriptors	1							1								
Total	6	3	2					2				3	8	5	5	6

Fig. 2. Contingency table of ratings aggregated by method and semantic category.

5 Discussion and Future Work

The software is under active development. In particular, we plan to support confidence information and quality estimation. Experiments have to be conducted to evaluate and improve the system. Some aspects of the implementation are currently tuned to the STW, especially regarding semantic properties that are beyond the scope of the SKOS specification, and thus differ among thesauri.

Acknowledgements. We would like to thank the indexing professionals who set the reviewing guidelines for valuable feedback, and thank the anonymous reviewer, who pointed us to related work in other domains, for constructive advice.

References

1. Berrios, D.C., Cucina, R.J., Fagan, L.M.: Methods for semi-automated indexing for high precision information retrieval. JAMIA **9**, 637–652 (2002)
2. Gastmeyer, M., Wannags, M., Neubert, J.: Relaunch des Standard-Thesaurus Wirtschaft - Dynamik in der Wissensrepräsentation. Inf. Wiss. Praxis **67**(4), 217–240 (2016)
3. Hinrichs, I., Milmeister, G., Schäuble, P., Steenweg, H.: Computerunterstützte Sacherschließung mit dem Digitalen Assistenten (DA-2). o-bib. Das offene Bibliotheksjournal **3**(4), 156–185 (2016)
4. Latif, A., Borst, T., Tochtermann, K.: Exposing data from an open access repository for economics as linked data. D-Lib Mag. **20**(9/10), 7 (2014)
5. Medelyan, O., Witten, I.H.: Measuring inter-indexer consistency using a thesaurus. In: Proceedings of Joint Conference on Digital Libraries, pp. 274–275. ACM (2006)
6. Toepfer, M., Seifert, C.: Descriptor-invariant fusion architectures for automatic subject indexing. In: Proceedings of Joint Conference on Digital Libraries (2017, accepted)
7. Uhlmann, S.: Automatische Beschlagwortung von deutschsprachigen Netzpublikationen mit dem Vokabular der Gemeinsamen Normdatei (GND). Dialog mit Bibliotheken **25**(2), 26–36 (2013)

Doctoral Consortium Paper

Research Data in Scholarly Practices: Observations of an Interdisciplinary Horizon2020 Project

Madeleine Dutoit[✉]

Oslo and Akershus University College of Applied Sciences, Oslo, Norway
madeleine.dutoit@hioa.no

Keywords: Research data · Sharing · Interdisciplinary · Scholarly practices · Scholarly communication · Horizon2020

1 Research Topic and Questions

My research project belongs within Library and Information Science in the area of Scholarly Communication. It started in October 2016 and will continue for three years. The study focuses on researchers and research data - more specifically on research data sharing in the scholarly practices of an interdisciplinary research project mandated by a data policy. A Horizon2020-project including four different disciplines will be investigated. These EU-projects are by default obligated to develop data management plans (DMPs). Few studies have been done on this subject and the questions to answer are many. I have chosen to focus on the following questions: what does data mean to the different participating disciplines? How do the researchers work with the data of the project and how do they share the data between the represented disciplines? What are the effects of the data policy on the daily research work?

2 Introduction

Within the area of Scholarly Communication, many things are changing rapidly today. One of these changes concerns how research data is viewed and valued. Different stakeholders demonstrate an unprecedented level of interest in how researchers communicate their findings (Jubb 2013). Governments, universities and research funders, among them actors as the OECD and the EU, are currently formulating digital data policies that requires granted research projects to develop plans for their data management. Researchers receiving funding are expected to develop data management plans where data storage, handling and access is specified. Within scholarly communication research data as new actors are predicted to become "recognized as significant scholarly contributions in their own right" (Hey et al. 2009). Other authors terms this development as an institutionalization of open access to research data (e.g. Mauthner and Parry 2013) but it is perhaps early to use this term, since policies have not yet been in place or applied consistently long enough. In either case, researchers sharing data with one another is per se nothing new, but the external demands on researchers to do this sharing are. I see

© Springer International Publishing AG 2017
J. Kamps et al. (Eds.): TPDL 2017, LNCS 10450, pp. 623–627, 2017.
DOI: 10.1007/978-3-319-67008-9_57

these policies as constituting a prominent example of this new interest coming from outside the researchers closest community, directing towards openness to research data. These policies are changing the conditions for and practices of data sharing.

Then what does research data mean and what are the drivers behind this development? To begin with, the concept of research data is complex and research seems only to agree in that no single definition is sufficient. Data have many kinds of value that varies widely over place, time and context (Borgman 2015). She, like other authors, agrees on that the sometimes more interesting question than what are data is *when* are data, claiming data to be "emergent, relational, and shaped by their use" (Haider and Kjellberg 2016). However, in order to analyse data in the context of scholarly communication, Borgman decides that a narrower approach to the concept will suffice which is why I settle with her definition of data as "representations of observations, objects, or other entities used as evidence of phenomena for the purposes of research or scholarship" (Borgman 2015). The general underlying motive for the aspirations of opening up access to data is the idea that accessible research data can contribute to benefit both research itself and the society in general. The driving arguments are mainly political, quality improvement of research through facilitation of transparency, and economical, enabling researchers and other interested to utilise the data for further research or innovation (NordForsk 2016).

I would here like to clarify what open data means in this discussion, since what is meant by openness and the degree of accessibility varies. Open can mean free and accessible data posted on a researchers personal website or published alongside a scholarly journal article as well as data deposited in a repository accessed only after registering and requesting it. In a report by The UK Royal Society speaks of "qualified openness" in the meaning of open data as "accessible, useable, assessable and intelligible" data (Royal Society 2012), implying that not all data are equally interesting or important. Though different approaches to openness, stakeholders agree that not all data can be made open, certain data will remain confidential for commercial, safety, privacy or security reasons.

3   Contributions

The results of this study will have both theoretical and practical implications. Learning more on how researchers from different disciplines collaborate on research data issues and how the data is shared with other project members, will increase our knowing on how knowledge is created jointly in temporary research projects. I hope to clarify the process of shaping common decisions on questions on research data and factors influencing this process. Additionally, with more knowledge of the epistemological bases of data in scholarly practices, the researchers' daily data "doings" and negotiations, and what data means to them, these behaviours can be better understood. Practically this new knowledge can be used when developing and forming services adapted to support the needs of research groups.

Work remains to be done in order to reach the economic and political goals set up for open data sharing. There is a gap between the aspirations of openness of research data and the actual sharing being done. The above mentioned data policies rarely describe data as the complex heterogeneous phenomenon it actually is, moreover varying from discipline to discipline, but rather simplifies without recognizing the many obstacles present before a realization of these demands is possible. Some problems that have to be resolved are technical issues, others are infrastructural. Many authors indicate that these problems are the easier ones to solve. What is needed and is crucial, but more difficult to achieve, is to change the scholarly cultures and practices of researchers (Hey et al. 2009). And studies on disciplinary cultures show that we are facing multiple cultures (e.g. Kim and Stanton 2016).

In order to support the process of changing scholarly practices and develop infrastructure supporting data sharing, further research on how research data in scholarly practices looks like within interdisciplinary collaborations. Studies on the practices of scholars related to research data will "be imperative to improve our understanding of both the epistemological bases and the actual practices that arise from new forms of collaboration and novel approaches to data management" (Palmer and Cragin 2008).

4   Theoretical Framework

For this project the theoretical approach is based on practice theory. Although there is no such thing as a unified practice theory or single practice-based approach, its origin coming from different intellectual backgrounds, the way of how to view organizational knowing is central and unifying. This challenges today's assumptions of knowledge based on rationalistic and cognitivist learning (Cox 2012) in considering knowing as "situated in the system of ongoing practices of action, as relational, mediated by artefacts, and always rooted in a context of interaction" (Nicolini et al. 2003). The philosopher Schatzki describes a practice as an "array of human activity" and as "bodily doings and sayings" (Schatzki et al. 2001). They are routine-based activities and things said or unsaid, and they materialize un-reflected knowledge. For studies in LIS, this approach can therefore serve as a useful instrument when examining the social aspects of scholarly practices, the tacit knowledge of researchers and their approach in information use or information sharing.

Based in practice theory, moving away from the individualistic focus, the scholar will be studied mainly as a member of a community. She will be seen as a carrier of a practice, neither autonomous nor dependent of social culture (Cox 2012). The results will thus be analysed seeing data activities less governed by individual needs and more as having a social nature: as an information activity that is woven through social practices.

5   Methodology

Three different qualitative methods for studying research data in scholarly practices have been chosen since qualitative methods are well suited to describe phenomena in context and provide an interpretation that leads to a greater understanding of the phenomena (Justesen and Mik-Meyer 2012). The methods I find most suitable are interviews, focus groups and participant observations. Researchers within an identified research project funded by the EU Research and Innovation programme Horizon 2020 will be studied. This project has been chosen firstly since projects within Horizon 2020 from January 2017 by default are part of the Open Data Pilot and thereby "must deposit your data in a research data repository where they will be findable and accessible for others" (European Commission 2016). In order to specify data collection, handling, sharing and curation, the projects are suggested to develop data management plans. Secondly, the project was chosen since it is constituted of researchers representing four different disciplines. The researchers thus have different traditions of handling data and their data will differ. Different scholarly practices will meet when project participants will discuss and agree on data management issues.

I have chosen to observe one single Horizon2020-project for this study in order to make it fit my time-frame and to be able to go deep into the subject; I want to search for "thick descriptions". The disadvantage with this choice is that it is difficult at this stage to say how representative my study will be. However, it would be quite difficult to find projects identically constituted in order to make an exact comparison. Naturally it would be very interesting to investigate several projects of this kind, to make a largely scaled qualitative study of this kind. Unfortunately, this will have to be for others to realize.

Semi-structured individual interviews as well as focus groups will be conducted with a yet unknown number of the researchers in order to find information of data in their scholarly practices, of cultures and norms, that can answer my research questions. Focus groups will be used as a "parallel force" to the interviews allowing me to observe the process of the participants managing their role both as an individual as a representative of the collective (Barbour 2013). Additionally, groups can prompt talk and interacting in offering other audience than the researcher (Macnaghten and Myers 2007) and points of view are argued for or defended in dialogue with the other participants (Tenopir et al. 2011). This information hopefully allows me to identify or grasp the data in the scholarly practices of the group and what is considered general behavior and unusual.

References

Barbour, R.S.: Analysing focus groups. In: Flick, U. (ed.) The SAGE Handbook of Qualitative Data Analysis. SAGE Publications, London (2013)

Borgman, C.L.: Big Data, Little Data, No Data: Scholarship in the Networked World. The MIT Press, Cambridge (2015)

European Commission: What is the Open Research Data Pilot? (2016). https://www.openaire.eu/opendatapilot. Accessed 2 June 2017

Cox, A.M.: An exploration of the practice approach and its place in information science. J. Inf. Sci. **38**, 176–188 (2012)

Haider, J., Kjellberg, S.: Data in the making. In: Rekers, J.V., Sandell, K. (eds.) New Big Science in Focus: Perspectives on ESS and MAX IV. Lunds universitet, Lund (2016)

Hey, A.J.G., Tansley, S., Tolle, K.M.: The Fourth Paradigm: Data-Intensive Scientific Discovery. Microsoft Research, Redmond (2009)

Jubb, M.: Introduction: scholarly communications – disruptions in a complex ecology. In: Judd, M., Shorley, D. (eds.) The Future of Scholarly Communication. Facet Publishing, London (2013)

Justesen, L., Mik-Meyer, N.: Qualitative research methods in organisation studies. Hans Reitzels Forlag, København (2012)

Kim, Y., Stanton, J.M.: Research Article. J. Assoc. Inf. Sci. Technol. **67**, 776–799 (2016)

Macnaghten, P., Myers, G.: Focus groups. In: Seale, C., Gobo, G., Gubrium, J.F., Silverman, A.D. (eds.) Qualitative Research Practice. Sage, London (2007)

Mauthner, N.S., Parry, O.: Open access digital data sharing: principles, policies and practices. Soc. Epistemol. **27**, 47–67 (2013)

Nicolini, D., Gherardi, S., Yanow, D.: Knowing in organizations: a practice-based approach. M.E. Sharpe, Armonk (2003)

Nordforsk. Open Access to Research Data: Status, Issues and Outlook, Oslo (2016)

Palmer, C.L., Cragin, M.H.: Scholarship and disciplinary practices. Ann. Rev. Inf. Sci. Technol. **42**, 165–212 (2008)

Schatzki, T.R., Knorr-Cetina, K., Savigny, E.V.: The practice turn in contemporary theory. Routledge, New York (2001)

Royal Society: Science as an open enterprise. The Royal Society Science Policy Center (2012)

Tenopir, C., Allard, S., Douglass, K., Aydinoglu, A.U., Wu, L., Read, E., Manoff, M., Frame, M.: Data sharing by scientists: practices and perceptions. PLoS One **6**, e21101 (2011)

Research Data in Norway:
How Do Expectations, Demands and Solutions Correspond in the Knowledge Infrastructure for Research Data?

Live Kvale[(⊠)] [iD]

Oslo and Akershus University College of Applied Sciences, Oslo, Norway
live.kvale@hioa.no

Abstract. Amounts of digital data combined with incentives for open research challenges researchers to share their research data (RD). Researchers are meeting requirements for data management plans (DMPs), and in Norway the Research Council has made it a priority to develop an infrastructure for making RD available. This project investigates how these constraints influence the researchers' choices and how the solutions created fit with the expectations from other stakeholders. Is there a satisfying dialog between the research environments and the service providers? Do the researchers experience that their voices are heard by the infrastructure providers and the research funders? The work aims to strengthen the dialog between the stakeholders in the knowledge infrastructure (KI) for RD and contribute to an improvement of this.

Keywords: Research data management · Knowledge infrastructure

1 Background

Edwards [1] defines KI as "robust networks of people, artifacts, and institutions that generate, share and maintain specific knowledge about the human and natural worlds". Borgman [2] elaborates on this: "[knowledge] infrastructures are not engineered of fully coherent processes. Rather they are best understood as ecologies or complex adaptive systems" and "these networks include technology, intellectual activities, learning, collaboration, and distributed access to human expertise and to documented information". By this she includes, in the context of RD, not only the technical solutions we often associate with the use of the term infrastructure, but also human recourses and competence. Sharing of RD can be viewed from different perspectives; organizational, political, technical, and ethical. Different stakeholders typically weight each of these differently, yet they are all important to keep in mind in order to get the whole picture of the KI for RD.

Norwegian KI stakeholders include funders, researchers, infrastructure providers and institutional research support functions such as libraries, research administration, and IT. If we look into the terms technical solutions, human resources, and competence, there appears to be a clear connection between these. It is, however, not evident that the stakeholders related to the different functions act with a full understanding of

© Springer International Publishing AG 2017
J. Kamps et al. (Eds.): TPDL 2017, LNCS 10450, pp. 628–631, 2017.
DOI: 10.1007/978-3-319-67008-9_58

the matter as a whole and that they are aware of the other stakeholders' positions. A better understanding of the different stakeholders expectations in regards of roles and responsibilities will make it possible to improve the workflow within the KI.

The KI for RD in Norway is undergoing big changes in order to follow requirements of open science coming from both the EU [3] and several research environments. The research is expected to give input for how the KI can be improved. As an approach for investigating science and technology Bruno Latour's [4] seven rules of method for studying "science in action" will be applied.

2 Research Question and Method

The main research question *How do expectations, demands and solutions correspond in the KI around RD that is being established in Norway?* reflects on the KI as a unity where the researcher, the technical infrastructure and its providers, the funders, and the political control exist as separate but connected entities. In turn, we can divide the research question into two sub-areas:

1: What expectations do the different stakeholders have of each other?
2: What expectations do the different stakeholders have of sharing RD (making RD available)?

This leads to the following sub-questions: How do funders influence the researcher's choice regarding RD? How do funders wish to influence the researcher's choice regarding RD? How do the different funding programs influence the services and tools developed? How do the funders wish services and tools to be developed? To what extent do the researchers find the solutions available for RD to be satisfactory? To what extent do the service providers find their solutions to be satisfactory? How do research support services adapt to new demands and needs? What role do the research support services[1] at the universities have, from the perspective of the other stakeholders? (in relation to knowledge spread, choices/political influence, and other possible roles). What roles and functions do the research support services see for themselves?

A Delphi inspired mixed methods study is applied with data collection through interviews with the different stakeholders. Two rounds of interviews will be conducted in order to identify possible agreement on different roles and responsibility between stakeholders. Interviewees will be key-persons from the two main data storage facilities representing the infrastructure, the research council and the ministry of knowledge representing the funders, researchers with H2020 project requiring DMPs representing researchers, and from the research support services individuals engaged in DM services will be interviewed. Researchers and research support staff from the universities of Bergen, Oslo, Trondheim, and Tromsø will be interviewed. These universities (BOTT) have a long tradition of collaboration on administrative support systems, IT, and infrastructure. In addition, a questionnaire will be sent out to a larger group of

---

[1] Research support services at the universities include IT, library and research office.

researchers with questions based on preliminary analysis. Relevant documents and policies will be analyzed.

3 Challenges

At this stage, the questions I ask myself relate much to definitions, theoretical perspective and methodology. I find that the challenges here often are interconnected and it will be my focus for the autumn of 2017 to sharpen and pin down what "my way" is when it comes to theoretical perspective and method. Below are some of the questions I tangle with.

3.1 What Are Expectations?

According to Merriam Webster expectation is "The act or state of expecting: anticipation in expectation of what would happen" [5], but it can also be expectations for an economic recovery such as prospects of inheritance. Even if the anticipation in expectation of what would happen is not stated as a positive, the term can still give the impression that the anticipation is of something positive that will come. In my usage, I do not wish to enforce this positive sentiment but rather for the interviewees to express their thoughts on the subject, independent of these being positive or negative. In other words, I wish to use "Expectation" as a neutral term.

3.2 How Do I Identify and Measure Expectations?

Clearly stating expectations is one thing; it will be more difficult to identify what might not be reflected upon by the interviewees. I hope that doing two rounds of interviews can be useful, since it makes it possible to follow up on issues that emerge from my analysis of the first interview.

3.3 The National Perspective

The selection of the national perspective might not be obvious. My reasons for such a restrain are based current organization and traditions within the higher education system in Norway. Many solutions are national or shared by several universities. Further juridical issues are grounded in national legislation, which for data that is openly shared is not an issue, but there are still a lot of data with access restrictions for which national storage facilities are necessary. At the same time, European requirements, international partners and the research as a global network of knowledge makes the national perspective redundant, and somewhat "old school" in a global world.

3.4 My Role

Too much reflection upon one's own role in a PhD project might turn the focus off the project itself. I do however sometimes find myself on the other side of the table as I am both writing a PhD thesis on the subject and working with it from the University of

Oslo Library (UiOL). Being a visible agitator for sharing of RD in Norway for some years, and outspoken on my personal views in my position at UiOL, I risk that this influences how my respondents see me and how they respond to me. I do not believe that I can be completely objective, but by revising and reflecting upon the decisions I make along the way and in particular during data analysis, combined with discussing with peers, I will aim at making my research process as transparent as possible.

3.5   Pinning Down the Methodology

Many of the issues above relate to what methodological approach I end up with, both theoretical perspectives and data collection methods will somehow be a response to the issues above. I would therefore be most grateful on feedback that helps me make decisions regarding my methodological approach.

4   Path Forward

During the autumn of 2017 I will plan my first round of data collection, define, and describe my methodological approach. In the beginning of 2018 the first data collection will take place, followed by a stay at the University of Illinois Urbana-Champagne were I hope to be discuss findings data collection with fellows. In the last part of 2018 a second round of data collection will take place. After this the focus will be analysis, writing, presenting, and publication, my deadline is in primo 2021.

References

1. Edwards P.: A Vast Machine : Computer Models, Climate Data, and the Politics of Global Warming. MIT Press (2010)
2. Borgman, C.L.: Big Data, Little Data, No Data: Scholarship in the Networked World. MIT Press (2015)
3. European Commission. H2020 Programme - Guidelines on Open Access to Scientific Publications and Research Data in Horizon 2020. European Commission (2016)
4. Latour, B.: Science in Action: How to Follow Scientists and Engineers Through Society. Open University Press, Milton Keynes (1987)
5. Expectation - Merriam-Webster Dictionary, http://www.merriam-webster.com/dictionary/expectation. Accessed 25 May 2017

Top-Down and Bottom-up Approaches to Identify the Users, the Services and the Interface of a 2.0 Digital Library

Elina Leblanc[✉]

LUHCIE and LIG Laboratory, Université Grenoble-Alpes, Grenoble, France
elina.leblanc@univ-grenoble-alpes.fr

Abstract. In spite of the existence of theoretical models for digital libraries (DL), studies and guidelines about the identification of the DL users, the users and their needs in terms of interface and services are still not well known, especially in the context of a 2.0 DL. Yet, this type of DL has become crucial for the projects that want to be more anchored in the Web environment and want to better fulfil the expectations of their potential users. Our work "Enriched digital libraries: users and their interfaces" aims to better understand the needs of the users in terms of interaction and participation in a DL called *Fonte Gaia Bib*, through the development of an enriched and participative DL. This paper will first present the challenges of the elaboration of this type of interface, from both the DL and user points of view. Then, it will focus on the method chosen to achieve it, which is a combination of a top-down approach (state of the art of the DLs' services) and a bottom-up one (identification of users profiles and requirements through user studies). This method has already produced good results and has highlighted common practices, services and roles, that can constitute the basis for the development of an interactive and participative DL.

Keywords: Digital library · User studies · Interfaces · Bottom-up and top-down approaches · Services · Collaboration

1 The User, the Great Unknown of the Digital Libraries?

Digital Libraries' (DL) theoretical models, such as the Digital Library Reference Model (DLRM) or the Interaction Triptych Framework (ITF), make the user one of the pillars of this type of resources. They insist on the need to take the user into account during the development of a DL, to achieve well-balanced interfaces [15, 32]. However, DL interfaces are often based on a set of presuppositions and false beliefs, and are ultimately the reflection of the needs of the creators of DLs themselves [12, 13]. Instead of being a bridge between the users and the digital contents, interfaces appear as a place of conflict between the users as they are imagined by the creators, and the *real* users.

To balance this situation and to better know these specific users and their practices, studies have been carried out since the late 1990s [4, 5, 7, 18, 19, 29–31, 33]. Some DLs,

© Springer International Publishing AG 2017
J. Kamps et al. (Eds.): TPDL 2017, LNCS 10450, pp. 632–639, 2017.
DOI: 10.1007/978-3-319-67008-9_59

such as *Gallica*[1] or *Europeana*[2], have also launched extensive user studies to adapt their interfaces to the evolution of their users. The success of those studies has led to the publication of guidelines that help and encourage similar projects to do the same [2, 3, 6, 8, 11, 14, 25]. However, despite these recommendations and studies, DLs' users are still not well known. On one hand, the studies focus on one facet of users' practices (usually the search and reading practices). On the other hand, they are linked to a specific project and were launched several years after the release of the first interface. Therefore, they only provide a biased view of the DLs public, i.e. the behaviours of a public for one specific DL, and general studies are still too few.

Moreover, the DLs are currently at a turning point. They gradually move towards so-called 2.0 interfaces, where the users become actors, instead of being passive readers. These new-generation libraries are no longer mere data silos, where the data are static and only available for viewing, but spaces for sharing and collaborating, where the data are renewed, enriched and dynamic, mainly through users' activities such as the addition of annotations or links. If many studies on user participation have been done for other humanities projects [1, 9, 16, 17, 26–28], they are still rare for DLs. The needs of the users in terms of interaction with DLs contents and collaboration are not clearly identified.

2   Who Are the Real Users of Digital Libraries? Objectives and Methods

On the basis of these studies and through the development of a specific DL called *Fonte Gaia Bib*[3], a 2.0 DL for the Italian studies, the project "Enriched Digital Libraries: Users and their Interfaces" aims to lay the foundations of the needs of users in the context of an enriched and participative DL. Several research questions are at the base of this work: Who are the intended and the real end-users of 2.0 scientific DLs: is it us (i.e. academics folk with a high digital literacy), someone like us, or someone else? What does it mean in terms of self-perception and of expectations to be a user of a 2.0 DL? What kind of interface do we have to build in order to match those expectations and to shape implicit needs? Can we imagine a personalized and user-adjustable interface that remains intuitive and effective? By extension, this work focuses on the formation of a mixed community, made of academics and members of the larger public, both groups being the intended public of the library we are building. It also investigates which factors are likely to encourage users to be engaged with a DL: is it the content, the interface, the infrastructure or the community gathered around a DL, that lead them to use a resource rather than another?

This work aims to propose a method that goes beyond the principle of "If we build, they will come", which is particularly persistent in digital humanities projects [12, 13,

---

[1] http://gallica.bnf.fr/accueil [Accessed 28/05/2017].

[2] http://www.europeana.eu/portal/fr [Accessed 28/05/2017].

[3] Presentation of the digital library's project: http://fontegaia.hypotheses.org [Accessed 28-05-2017]. *Fonte Gaia Bib*: http://www.fontegaia.eu [Accessed 28-05-2017].

22, 34]. This latter principle seems to be based on a top-down approach pushed to its extreme, i.e. the creators develop an interface only from their own perspective. This hegemony of the top may lead to unsuitable interfaces, as we mentioned earlier. This is why we have chosen to adopt a twofold approach: a bottom-up approach, based on the needs of the potential users, *and* a top-down approach, based not only on the objectives of the creators but also on projects that already exist. In merging these two approaches, the goal is to produce a user-centred interface, made up of innovative functionalities identified with the top-down approach, while remaining consistent with what users do and want via the bottom-up approach. This twofold approach corresponds to the first step of this work. The results obtained will be a base for the development of prototypes, which will then be tested with a panel of users to analyse their activities and to identity new needs.

The *Fonte Gaia Bib* case will help us to offer a set of recommendations for the elaboration of a 2.0 DL and to contribute to the definition of 2.0 DLs users. These reflections about users and their interfaces will also bring knowledge about the engage-ment of the public with digital written heritage and about the factors that lead to the constitution of mixed communities of users[4].

3   Preliminary Results: Users and Services

3.1   From a Digital Library Point of View: The Top-Down Approach

The top-down approach was based on a state of the art of DLs and digital humanities projects, which have some similarities with *Fonte Gaia Bib*. After the analysis of these projects, it appears that the interfaces of DLs take the form of a mosaic of services. These services act like the key components of the relationship between users and digitized contents. This relationship can be defined as interdependent. Indeed, when users access a digitized content, the interface provides one or several services that the users then use. The nature of a service is dependent both on the type of the content chosen and on the profile of the users. However, the features of a particular content can be influenced by the service used to interact with it. Finally, if the profile of the users does not change, their methods of work can be modified by the services offered (Fig. 1).

Let us take the example of the downloading service. This service is based on proposing different formats of export. However, these formats depend on the nature of the resource: if a text offers many formats such as .pdf, .epub, .jpeg, .tiff or .xml, an image has limited possibilities, mostly .tiff and .jpeg. The form of the service itself is then influenced by a particular resource. But, the aspect of a resource can also be modi-fied by the export format chosen by users. A resource in PDF or in XML will not have the same form and the same features. In return, the users have new possibilities to see and analyse the resource. For example, they can annotate a resource with the PDF format, but a XML format is rather for automatic processing of texts, data extraction etc. Thus, the downloading service can enrich the experience of users by providing different ways to apprehend a resource.

---

[4] Presentation of the thesis: http://fontegaia.hypotheses.org/1050 [Accessed 23-06-2017].

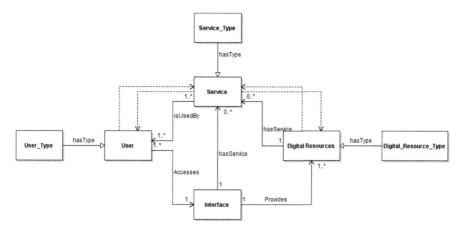

Fig. 1. UML representation of the User-Resource-Service relationship

This model leads to the categorisation of the services already present in DLs, and their modelling in a UML ontology. This categorisation underlines that DLs waver between fidelity to the traditional missions of libraries by offering services that copy the one of the latter (consulting of documents, advice services, communication services), and innovation, under the influence of Web 2.0 technologies (user-generated-content services). These different types of services tend to change the relationships between DLs and their users. From a unilateral relationship, where DLs provide tools and their expertise to their users, DLs move little by little towards a multilateral relationship, where users collaborate with DLs and other users. This ontology gives us then an insight into the internal functions of a DL and helps to envisage potential services for *Fonte Gaia Bib*.

3.2   From a User Point of View: The Bottom-up Approach

The needs of users have been identified using a bottom-up approach, via dissemination of a questionnaire (67 answers, May 2016) [21] and conducting interviews (8 participants, February–April 2017). The objective of this first user study was (1) to better know the profiles, the practices and the needs of users of DLs; (2) to identify their high-priority services; (3) to measure their degree of engagement with interfaces by presenting interactive and participative services. Compared to other similar endeavours, such as the ones of *Gallica* and *Europeana*, this study had the distinctive characteristic to precede the public release of *Fonte Gaia Bib* and then to attempt to define a general profile for the users of specialised DLs[5]. Moreover, it gave an important insight into the notions of interaction and collaboration.

The questionnaire allowed us to identify the core stakeholders of DLs, i.e. a public made of researchers, PhD students and GLAMs' professionals. It appears that these

---

[5] "Specialised DLs" are projects that focus on one discipline, type of contents, period or author, by opposition to "generalist DLs", that aggregate large types of contents and do not make any distinction between them [10, 24].

different groups have similar needs and practices, namely the way they read a digital document (scanning, use of the table of contents and of the full-text search to select interesting parts), the way they search (dominance of the simple search) or the way they want to contribute (addition of tags, comments and bibliographical references) [20]. These results have been confirmed by interviews with eight representatives of the core stakeholders of the DLs (2 academics, 2 PhD students, 2 students and 2 librarians).

However, this study has also highlighted several users' patterns that are independent from the professional backgrounds, and that represent distinctive roles within a DL:

- The passer-by reader: These "ephemeral" users favour the reading locally on their own machine to reading on the Web. They glean contents from library to library and have a superficial knowledge of the functionalities provided by DLs. They do not have a preferred DL, but rather a preferred portal or search engine that helps them to find contents of interest. They put the emphasis on the contents rather than on the interface or on the services of DLs.
- The active reader: These users are engaged in DLs and favour online reading. They have a good knowledge of the services offered, but focus their attention on select functionalities, such as zoom, full-text search and interactive table of contents. For them, a DL is a reading space, where they can manage their selections, their searches and their downloaded material.
- The expert reader: These users have a high degree of engagement with DLs. They explore all services for the purpose of analysis and reuse of content. They consider DLs as workspaces, where they can compare several documents, make critical annotations or work in group.

These roles underline that DLs have a utilitarian function: they fulfil research objectives that have been carefully thought through by the users. These latter do not come to a DL for their own enjoyment, but with a specific purpose in mind [14, 23]. The services expected by the participants of the study are thus essentially linked to search, export and analysis activities, which are oriented towards the use and reuse of content.

The same assessment can be made for collaboration. While all the participants declare a willingness to collaborate with other users, the way they see this collaboration depends on their profession: researchers and PhD students are oriented towards scientific collaboration; students towards a pedagogical collaboration; librarians towards a professional collaboration, where the contributors become citizen librarians. The volunteers are thus interested by participative working tools, that allow them to share their knowledge and skills and that can help the progress of their research, their studies or their professional tasks.

4   When the Two Approaches Meet: Conclusions and Perspectives

After comparing the results of the top-down and bottom-up phases, it was possible to define a set of services that have the specificity to engage users in each step of a digital document's lifecycle. From the suggestions of which work to digitize, to the dissemination of contents (social networks, Web 2.0 tools), to the involvement in the pre-processing of documents (transcription, OCR correction) and their analysis (annotations, easy-reading tools, personal workspace), users can contribute to the improvement and the renew of the digitized documents of DLs.

A second categorisation can be overlaid on the first one: users' roles in DLs that have been modelled during the user studies. The first categorisation corresponds to the DL point of view; the second, to the users' point of view. In this second classification, each service that has been identified previously corresponds to a role within a DL and to the needs and expectations of the users that follow this role. Services connected with reading, search and export correspond to the passer-by reader role. Services for the description and the improvement of the content, as well as connected with the management of reading (history, personal annotations) are typical to the active reader role. Services related to advanced reading (comparison, text and image manipulation) and to critical annotations are in the realm of expert reader role.

The definition of these roles helps the on-going modelling of the DL's interface, via the realisation of mock-ups and prototypes. The challenge will be to be able to standardize these roles and relevant services into one interface, and to allow users to move easily from one role to the next in a flexible way. The objective is to avoid the forms of role segregation between users, and to consider the potential evolutions of their practices.

In the meantime, the bottom-up phase enters a new stage: the identification of the potential users coming from a larger public, who remained in the shade during the first study. After this new study (May–June 2017), the aim will be to compare the practices of the academics and of the amateurs, in order to identify common patterns or differences that will help to understand how we can create a diversified and welcoming community of users. These results will be also taken into account during the development of an enriched and participative DL.

References

1. Alam, S.L., Campbell, J.: Crowdsourcing motivations in a not-for-profit GLAM context: the Australian newspapers digitisation program. In: ACIS 2012: Location, Location, Location: Proceedings of the 23rd Australasian Conference on Information Systems 2012, Geelong, pp. 1–11 (2012)
2. Assadi, H., Beauvisage, T., Lupovici, C., Cloarec, T.: Users and uses of online digital libraries in France. In: Koch, T., Sølvberg, I.T. (eds.) ECDL 2003. LNCS, vol. 2769, pp. 1–12. Springer, Heidelberg (2003). doi:10.1007/978-3-540-45175-4_1
3. Beaudouin, V., Denis, J.: Observer et évaluer les usages de Gallica. Réflexion épistémologique et stratégique. Telecom ParisTech, BnF (2014)
4. Bishop, A.P.: Making digital libraries go: comparing use across genres. In: Proceedings of the Fourth ACM Conference on Digital Libraries, pp. 94–103. ACM, New York (1999)
5. Blandford, A., et al.: Use of multiple digital libraries: a case study. In: Proceedings of the ACM/IEEE-CS Joint Conference on Digital Libraries, 1st, Roanoke, Virginia, 24–28 June 2001, pp. 179–188. ACM, New York (2001)
6. Bonneau, J.: Enquête auprès des usagers de la bibliothèque numérique Gallica. Bibliothèque nationale de France, Paris (2017)

7. Bryan-Kinns, N., Blandford, A.: A survey of user studies for digital libraries. RIDL Working Paper (2000)
8. Caffo, R., et al.: Hanbook on Cultural Web User Interaction. Minerva, Rome (2008)
9. Causer, T., Wallace, V.: Building a volunteer community: results and findings from Transcribe Bentham. Digit. Humanit. Q. **6**, 2 (2012)
10. Claerr, T., Westeel, I. (eds.): Manuel de constitution de bibliothèques numériques. Éditions du Cercle de la librairie, Paris (2013)
11. Dierickx, B., Natale, M.T.: Report on the user needs and requirements. AthenaPlus (2013)
12. Dinet, J.: Pour une conception centrée-utilisateurs des bibliothèques numériques. Commun. Lang. **161**, 59–74 (2009)
13. Dobreva, M., et al. (eds.): User studies for digital library development. Facet Publishing, London (2012)
14. GMV: Évaluation de l'usage et de la satisfaction de la bibliothèque numérique Gallica et perspectives d'évolution. BnF, Paris (2012)
15. Gonçalves, M.A., et al.: Streams, structures, spaces, scenarios, societies (5 s): a formal model for digital libraries. ACM Trans. Inf. Syst. **22**(2), 270–312 (2004)
16. Holley, R.: Crowdsourcing and social engagement in libraries: the state of play (2011). http://aliasydney.blogspot.fr/2011/06/crowdsourcing-and-social-engagement-in.html
17. Holley, R.: Many Hands Make Light Work: Public Collaborative OCR Text Correction in Australian Historic Newspapers. National Library of Australia (2009)
18. Jones, S., et al.: An analysis of usage of a digital library. University of Waikato, Department of Computer Science (1998)
19. Kimani, S., et al.: Digital library requirements: a questionnaire-based study. In: Theng, Y.-L., et al. (eds.) Digital Libraries. Design, Development, and Impact. pp. 287–297. IGI Global, Hershey (2009)
20. Leblanc, E.: À la découverte des utilisateurs de bibliothèques numériques: Les résultats du questionnaire Fonte Gaia. http://fontegaia.hypotheses.org/1902
21. Leblanc, E.: La parole est à vous: premier questionnaire de Fonte Gaia/Primo questionario di Fonte Gaia. http://fontegaia.hypotheses.org/1673
22. Markus, M.L., Keil, M.: If we build it, they will come: designing information systems that people want to use. Sloan Manage. Rev. **35**(4), 11–25 (1994)
23. Matharan, J., et al.: Rapport d'étude sur les usages communautaires et collaboratifs, sur place et à distance, des ressources numérisées de la BnF. Bibliothèque nationale de France (2008)
24. Mion Mouton, F.: Bibliothèques numériques et coopération: comparaisons internationales. Enssib (2012)
25. Rasmussen, K.G., et al.: Recommendations for Conducting User Tests. EuropeanaConnect (2011)
26. Research Information Network: If you build it, will they come? How researchers perceive and use Web 2.0. Research Information Network, London (2010)
27. Ridge, M.: Crowdsourcing our cultural heritage. Ashgate, Farnham, Surrey (2014)
28. Romeo, F., Blaser, L.: Bringing citizen scientists and historians together. Presented at the Museums and the Web 2011: Proceedings, Toronto March 31 (2011)
29. Sfakakis, M., Kapidakis, S.: User behavior tendencies on data collections in a digital library. In: Agosti, M., Thanos, C. (eds.) ECDL 2002. LNCS, vol. 2458, pp. 550–559. Springer, Heidelberg (2002). doi:10.1007/3-540-45747-X_41
30. Tammaro, A.M.: User perceptions of digital libraries: a case study in Italy. Perform. Meas. Metr. **9**(2), 130–137 (2008)

31. Theng, Y.L., Duncker, E., Mohd-Nasir, N., Buchanan, G., Thimbleby, H.: Design guidelines and user-centred digital libraries. In: Abiteboul, S., Vercoustre, A.-M. (eds.) ECDL 1999. LNCS, vol. 1696, pp. 167–183. Springer, Heidelberg (1999). doi:10.1007/3-540-48155-9_12
32. Tsakonas, G., Papatheodorou, C.: Exploring usefulness and usability in the evaluation of open access digital libraries. Inf. Process. Manag. **44**(3), 1234–1250 (2008)
33. Warwick, C., et al.: If you build it will they come? the LAIRAH study: quantifying the use of online resources in the arts and humanities through statistical analysis of user log data. Lit. Linguist. Comput. **23**(1), 85–102 (2008)
34. Wilson, L.A.: If we build it, will they come? library users in a digital world. In: Lee, S.H. (ed.) Improved Access to Information: Portals, Content Selection, and Digital Information, pp. 19–28. The Haworth Information Press, Binghamton, New York (2003)

Cross-Language Record Linkage Across Humanities Collections Using Metadata Similarities Among Languages

Yuting Song[(⊠)]

Ritsumeikan University, Kusatsu, Shiga 5258577, Japan
gr0260ff@ed.ritsumei.ac.jp

Abstract. This paper proposes a method for cross-language record linkage across digital humanities collections by exploiting similarities between metadata values in different languages without using any translation method. Our method represents metadata values in Japanese and English as vectors by using monolingual word embeddings. Then, we calculate similarity between metadata value vectors by learning a mapping between vector spaces that represent Japanese and English. The proposed method could help users to acquire multilingual information of the objects in digital collections. We evaluate the effectiveness of our method on Japanese Ukiyo-e print databases in Japanese and English.

Keywords: Cross-language record linkage · Word embeddings · Digital humanities collections

1 Introduction

Over the past decade, more and more libraries, museums and galleries around the world have been digitalizing their collections and making them accessible online. It opens up new opportunities to acquire valuable knowledge from vast amounts of information about these digital collections. The metadata, which are used to provide information about the records in digital collections, are created independently by heterogeneous institutions using different natural languages. For instance, Japanese Ukiyo-e woodblock prints[1] have been digitized by many museums in Japan and Western countries and described by the metadata values in their native languages. As a consequence, identical records that refer to the same object could be described in different languages. Given that there is multilingual information in metadata of identical records, it is important to provide technologies for finding these identical records in order to aggregate multilingual knowledge about objects.

Record linkage [1] is a task of finding record pairs that refer to the same object across multiple data sources, which has been studied for many years. Our research focuses on a new field of cross-language record linkage, where records are from the data sources with metadata in different languages. In particular, we aim for cross-language record linkage across digital humanities collections in Japanese and English by using textual metadata

---

[1] Ukiyo-e is a type of Japanese traditional woodblock print, which is known as one of the popular arts of the Edo period (1603–1868).

© Springer International Publishing AG 2017
J. Kamps et al. (Eds.): TPDL 2017, LNCS 10450, pp. 640–643, 2017.
DOI: 10.1007/978-3-319-67008-9_60

values. It is challenging due to several reasons: (1) metadata values are expressed in different languages, therefore similarity measures cannot be employed directly; (2) even if the machine translation system can be used to translate metadata values into the same language in order to calculate similarities between them, machine translation systems have poor performance on specific domains [2] due to the difficulty of obtaining a domain-specific bilingual corpus for training system.

2 Proposed Method

Our proposed method focuses on the similarity matching phase of cross-language record linkage, which is an important phase that determines whether two records represent the same object. In this section, we first introduce our approach of representing textual metadata values. Then, a method of learning a mapping between vector spaces that represent Japanese and English is provided for calculating the similarity between metadata values.

2.1 Representations of Metadata Values

We represent textual metadata values as vectors by using word embeddings [3], which are dense, low-dimensional and real-valued vectors for representing words. Through these embedded word representations, the words with a similar meaning have closer distances in a vector space, e.g. *vector*("storm") is close to *vector*("hurricane"), which means the semantic relationships between words can be captured. Moreover, the semantic relationships between words can be expressed as linear operations in a vector space, e.g., *vector*("Berlin") - *vector*("Germany") + *vector*("France") is close to *vector*("Paris").

Our method of representing textual metadata values is inspired by the characteristics of word embeddings. More specifically, we firstly learn Japanese and English word embeddings by using Word2Vec toolkit. Then, we represent textual metadata values in Japanese and English by additive combination of the vector embeddings of words that compose the metadata values. Fig. 1 illustrates our method of representing metadata values.

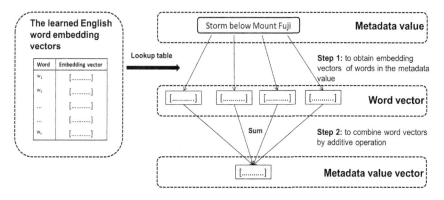

Fig. 1. The process of our method of representing metadata values

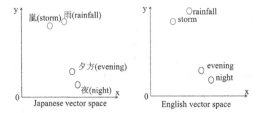

Fig. 2. The word vector representations of weathers and times in Japanese and English

2.2 Similarity Calculation Between Metadata Value Vectors

We calculate the similarity between metadata value vectors by learning a mapping between vector spaces that represent Japanese and English. Our proposed method is motivated by the idea in [4] that the same concepts have similar geometric arrangements across the vector spaces that represent different languages, which is illustrated in Fig. 2. Taking the concept of weather as an example, the relative positions of "雨 (rainfall)" and "嵐 (storm)" in the vector space that represents Japanese (the graph on the left) are similar to the relative positions of "rainfall" and "storm" in the English vector space (the graph on the right). What is more important is that the relationship between vector spaces that represent these two languages can possibly be captured by learning a mapping between them, e.g. a liner mapping. If we know some word pairs in Japanese and English, e.g. "雨" and "rainfall", "嵐" and "storm", we can learn a mapping that can help us to transform other words in the Japanese vector space to the English vector space.

Similar to the idea above, we learn a liner mapping between vector spaces that represent Japanese and English in order to transform the Japanese metadata value vectors to the vector space that represents English. Suppose we are given a set of textual metadata value pairs and their associated vector representations $\{x_i, z_i\}_{i=1}^{n}$, where $x_i \in \mathbb{R}^1$ is the vector representation of Japanese metadata value i, and $z_i \in \mathbb{R}^2$ is the vector representation of its corresponding English metadata value that is obtained by our method in Sect. 2.1. Our goal is to find a mapping matrix W such that Wx_i approximates z_i. In practice, W can be learned by the following optimization problem shown in Eq. (1), which can be solved with stochastic gradient descent.

$$\min_w \sum\nolimits_{i=1}^{n} \| Wx_i - z_i \|^2 \tag{1}$$

At the time of similarity calculation, for any given new Japanese metadata value vector x, we transform it into a vector space that represents English by computing $z = Wx$. Then, we can calculate the similarity between metadata values in Japanese and English by comparing the transformed vectors of Japanese metadata with other English metadata value vectors.

3   Experiment

In this section, we show the preliminary results of our proposed method in finding the identical Ukiyo-e prints across databases in Japanese and English.

In the experiments, the titles of Ukiyo-e prints are used to calculate similarities between Ukiyo-e prints. We train the Japanese and English word vectors on Japanese and English Wikipedia articles using Word2Vec toolkit. In the process of learning the mapping between the vector spaces that represent Japanese and English, we use 600 Japanese-English parallel short sentence pairs for pre-training. In order to make this mapping more accurate to transform Ukiyo-e titles, we further use 74 pairs of Japanese and English Ukiyo-e titles to optimize this mapping, in which each pair of titles refers to the same Ukiyo-e prints. The similarities between the Ukiyo-e titles are calculated by cosine similarity metric.

We use 173 pairs of Japanese and English Ukiyo-e titles as the test data to evaluate our method. The precision at top-n are used to evaluate the experimental results. In order to verify the effectiveness of using Ukiyo-e titles to optimize the mapping in the phase of pre-training, we show the results of both conditions of using Ukiyo-e titles and without using them. The experimental results are shown in Table 1.

These results show that the precisions can be improved by using Japanese and English Ukiyo-e titles to optimize the mapping between Japanese and English vector spaces. The experimental results also confirm the usefulness of our proposed method for linking the same Ukiyo-e prints in Japanese and English.

Table 1. The experimental results

	Precision			
	In top-1	Within top-5	Within top-10	Within top-15
Without using Ukiyo-e titles	2.3%	12.2%	17.4%	22.7%
Using Ukiyo-e titles	29.1%	41.9%	50.0%	54.7%

4 Conclusions

In this paper, we proposed a method of cross-language record linkage by measuring the similarity between textual metadata values without using any translation methods. In the future, we plan to explore different embedded vector representations of metadata values. Besides we will evaluate our method on data sources in other languages.

References

1. Ahmeh, K.E., Panagiotis, G.I., Vassillios, S.V.: Duplicate record detection:a survey. IEEE Trans. Knowl. Data Eng. **9**(1), 1–16 (2007). IEEE
2. Hua, W., Haifeng, W., Chengqing, Z.: Domain adaptation for statistical machine translation with domain dictionary and monolingual corpora. In: 22nd International Conference on Computational Linguistics, pp. 993–1000. ACL, Manchester (2008)
3. Mikolov, T., Chen, K., Corrado, G., Dean, J.: Efficient estimation of word representations in vector space. arXiv preprint arXiv:1301.3781 (2013)
4. Mikolov, T., Le, Q.V. and Sutskever, I.: Exploiting similarities among languages for machine translation. arXiv preprint arXiv:1309.4168 (2013)

Machine Learning Architectures for Scalable and Reliable Subject Indexing
Fusion, Knowledge Transfer, and Confidence

Martin Toepfer[✉]

ZBW – Leibniz Information Centre for Economics,
Düsternbrooker Weg 120, 24105 Kiel, Germany
m.toepfer@zbw.eu

Abstract. Digital libraries desire automatic subject indexing as a scalable provider of high-quality semantic document representations. The task is, however, complex and challenging, thus many issues are still unsolved. For instance, certain concepts are not detected accurately, and confidence estimates are often unreliable. Accurate quality estimates are, however, crucial in practice, for example, to filter results and ensure highest standards before subsequent use. The proposed thesis studies applications of machine learning for automatic subject indexing, which faces considerable challenges like class imbalance, concept drift, and zero-shot learning. Special attention will be paid to architecture design and automatic quality estimation, with experiments on scholarly publications in economics and business studies. First results indicate the importance of knowledge transfer between concepts and point out the value of so-called fusion approaches that carefully combine lexical and associative subsystems. This extended abstract summarizes the main topic and status of the thesis and provides an outlook on future directions.

Keywords: Automatic subject indexing · Machine learning · Quality control

1 Introduction

By subject indexing, libraries create concise yet comprehensive descriptions of documents with terms of controlled vocabularies like MeSH[1], LCSH[2], GND[3], or STW[4]. These structured semantic document representations are highly valuable for digital libraries since they support services like semantic browsing, multilingual information retrieval and recommendation, or trend detection. According to the TPDL-2017 theme "Part of the machine: turning complex into scalable"[5],

---

[1] www.nlm.nih.gov/mesh/.
[2] id.loc.gov/authorities/subjects.html.
[3] www.dnb.de/gnd.
[4] www.zbw.eu/stw.
[5] www.tpdl.eu/tpdl2017/.

© Springer International Publishing AG 2017
J. Kamps et al. (Eds.): TPDL 2017, LNCS 10450, pp. 644–647, 2017.
DOI: 10.1007/978-3-319-67008-9_61

digital libraries demand for solutions that ease access to large amounts of heterogeneous data, thus making accurate automatic subject indexing a key technology for their infrastructure. The challenges that need to be solved are, however, considerable: subject indexing is a complex cognitive task which involves several aspects of the human mind, such as natural language processing and semantic reasoning under uncertainty. Its automation thus requires artificial intelligence and machine learning, having many options to model the task and encode it in terms of input, output, features, dependencies, and objectives. In the past, such architectures have been realized in different ways.

For instance, several researchers regarded subject indexing as a multi-label classification task [2,9], which differs from standard classification in that multiple possibly interrelated classes have to be assigned per document. Since the number of different classes is remarkably large, often exceeding several thousands of concepts, system engineers have to be very careful. Complex models with many variables can quickly suffer from too few training examples to estimate parameters reliably, and in order to recognize previously unseen concepts, zero-shot learning [8] and exploitation of external knowledge sources have to be regarded.

For these reasons, researchers have made several assumptions and simplifications. For instance, Medelyan and Witten [7] made a specific *invariance assumption* that allows to learn parameters shared by all concepts, *transferring knowledge* between them. This reduces the minimum amount of documents necessary for training and enables zero-shot learning, however, it requires extensive linguistic knowledge from thesauri. Other approaches make "naive" independence assumptions for the sake of efficiency. Very common are binary relevance approaches like the work of Wilbur and Kim [11], using independent classifiers for each subject heading, thus learning associations between terms and concepts. More complex architectures have been constructed by *combining different approaches* using meta-learning [4] or learning to rank [3]. Yet, research on the design of machine learning architectures with heterogeneous modules for automatic subject indexing is limited. Their composition and setup has numerous parameters, configuration options, and design choices.

Interestingly, although in practice accurate *quality estimates* are fundamental to ensure highest standards of generated meta-data, there have been negative experiences with confidence values provided by systems[6], hence, it becomes an attractive research topic. Finally, note that evaluation in automatic indexing is non-trivial and may require more detailed analysis than standard metrics like precision and recall.

2 Contributions and Research Questions

The proposed thesis aims to make contributions in the fields of digital libraries and computer science, gaining insights into applications of machine learning

---

[6] For instance, due to experiments at the ZBW and correspondence with the German National Library at a recent workshop on "Computer-assisted Subject Cataloguing", 2017 in Stuttgart, Germany.

techniques, and exploring different approaches, their effectiveness and efficiency. Related to Manning's thoughts on computational linguistics and deep learning [5], the thesis strives for thorough problem analysis and meaningful composition of machine learning architectures, which may provide more general insights.

In this regard, the thesis will be directed by the following questions:

1. Architectures:
   (a) How do aspects of *architecture design* (modelling: encoding of input, features, output, dependencies, objectives) and requirements of the environment (properties of thesauri, characteristics and availability of training data, dynamics of the domain) relate to each other? In particular, which effects does architecture design have according to concept drift and zero-shot learning?
   (b) What is the role of *invariance* assumptions, which enable to share parameters in learning? Where do they apply? How are they modeled and integrated into systems?
   (c) Can different approaches be "meaningfully" combined, that is, can we leverage individual advantages effectively?
2. How can (reliable) confidence estimates be computed for automatic subject indexing? What are relevant aspects of "confidence" in subject indexing? How do the terms "quality" and "confidence" relate to each other in the field of automatic subject indexing? How do confidence and quality estimation fit into encompassing architectures?

Finally, many research activities focused on the medical domain where subject-specific solutions are available and can be incorporated. This thesis will investigate a less-studied domain, namely scholarly publications related to economics, where comparable solutions do not exist. Differing challenges may emerge, with the prospect of novel insights.

3   Approach

As part of the thesis, different automatic subject indexing architectures and confidence estimation approaches are analysed, designed, implemented, and evaluated.

The project looks at approaches like dictionary matching, ranking and associative methods, and determines how they fit into encompassing architectures, especially with respect to concept drift.

Regarding confidence estimation, techniques similar to the DeepQA architecture of IBM Watson [1] will be considered. The project will first collect ideas for implementation: confidence estimation features for automatic indexing must be developed, and meaningfully grouped into evidence profiles. Finally, experimental evaluation will be performed.

Theoretical analysis will be taken into account, however, it may be limited due to the fuzzy nature of the tasks [9]. Therefore, experiments will be conducted to test systems and justify hypothesis empirically. Commonly used metrics are precision, recall and F_1 [2,9], although they may be too shallow to assess quality reasonably. Semantic relations between concepts [6] and graded ratings may therefore be taken into account. Confidence estimation methods may be evaluated using ranking metrics. The data that will mainly be used, has the following properties: scholarly publications related to economics, written in English, only descriptive metadata (short texts), indexed with descriptors of the STW.

4   Status Summary

A major contribution has already been accomplished by analysis of architectures, development of specific fusion systems, and experiments on short texts (titles and author keywords) [10]. This work will be supplemented by certain extensions, their analysis, and experiments. For instance, different approaches to apply learning components for fusion will be explored. Work on confidence estimation is at an early stage and thus may profit from exchange with the digital libraries community.

References

1. Ferrucci, D.A., Brown, E.W., Chu-Carroll, J., Fan, J., Gondek, D., Kalyanpur, A., Lally, A., Murdock, J.W., Nyberg, E., Prager, J.M., Schlaefer, N., Welty, C.A.: Building watson: an overview of the DeepQA project. AI Mag. **31**(3), 59–79 (2010)
2. Gibaja, E., Ventura, S.: A tutorial on multilabel learning. ACM Comput. Surv. **47**(3), 52: 1–52: 38 (2015)
3. Huang, M., Névéol, A., Lu, Z.: Recommending MeSH terms for annotating biomedical articles. JAMIA **18**(5), 660–667 (2011)
4. Jimeno-Yepes, A., Mork, J.G., Demner-Fushman, D., Aronson, A.R.: A one-size-fits-all indexing method does not exist: automatic selection based on meta-learning. JCSE **6**(2), 151–160 (2012)
5. Manning, C.D.: Computational linguistics and deep learning. Comput. Linguist. **41**(4), 701–707 (2015)
6. Medelyan, O., Witten, I.H.: Measuring inter-indexer consistency using a thesaurus. In: Proceedings of Joint Conference on Digital Libraries, pp. 274–275. ACM (2006)
7. Medelyan, O., Witten, I.H.: Domain-independent automatic keyphrase indexing with small training sets. J. Am. Soc. Inf. Sci. Technol. **59**(7), 1026–1040 (2008). http://dx.doi.org/10.1002/asi.20790
8. Palatucci, M., Pomerleau, D., Hinton, G.E., Mitchell, T.M.: Zero-shot learning with semantic output codes. In: Bengio, Y., Schuurmans, D., Lafferty, J.D., Williams, C.K.I., Culotta, A. (eds.) Advances in Neural Information Processing Systems 22, pp. 1410–1418. Curran Associates, Inc. (2009). http://papers.nips.cc/paper/3650-zero-shot-learning-with-semantic-output-codes.pdf
9. Sebastiani, F.: Machine learning in automated text categorization. ACM Comput. Surv. **34**(1), 1–47 (2002)
10. Toepfer, M., Seifert, C.: Descriptor-invariant fusion architectures for automatic subject indexing. In: Proceedings of Joint Conference on Digital Libraries (2017). Accepted
11. Wilbur, W.J., Kim, W.: Stochastic gradient descent and the prediction of MeSH for PubMed records. Proc. AMIA Ann. Symp. **2014**, 1198–1207 (2014)

Explaining Pairwise Relationships Between Documents

Nils Witt[(✉)]

ZBW-Leibniz Information Centre for Economics, Kiel, Germany
n.witt@zbw.eu

Abstract. Current methods in automatic text summarization only take a single document into account. In contrast, the proposed research in this paper aims at summarizing the discrepancy between two documents. We approach this by transforming documents into a representation where mathematical operations have a semantical correspondence. This allows us to recombine documents and draw a summary from the recombined vector. A discrepancy summary can briefly convey what a reader can additionally learn from reading an unknown document with respect to a document that the reader is already familiar with.

1 Introduction

In natural language processing(NLP) applications documents are usually transformed into a new reference system which allows applying mathematical measure to determine their differences. We want to investigate these differences between two documents with the goal of acquiring human-interpretable knowledge. To achieve this goal we developed an abstract model that captures and formalizes automatic dissimilarity summarization(ADS). This model will be accompanied by an evaluation framework that objectively compares different implementations of this model. Afterwards we will evaluate the performance of different implementations and investigate enhancements. To the best of our knowledge this question has not been studied, albeit the potential benefits in NLP applications. Word2Vec [4,7] and its increments (e.g. [4,9]) are promising candidates for a suitable data representation for ADS, as they comprise a shallow understanding of language. Summaries generated by ADS are more beneficial to users than plain summaries as they take the history of documents that a user has already read into account. These information can support users by choosing the next document to read with respect to previously read documents.

2 Research Questions

There is a vast amount of text representations available to determine the difference of two documents mathematically. These representations have important practical applications but fall short on depicting their insights to humans directly, which is the motivation for the proposed research. We have identified the following research questions:

© Springer International Publishing AG 2017
J. Kamps et al. (Eds.): TPDL 2017, LNCS 10450, pp. 648–651, 2017.
DOI: 10.1007/978-3-319-67008-9_62

- What kind of text representation is most appropriate for automatic discrepancy summarization?
- Is the information that is contained in the order of words better encoded by a sequence of word representations or by a document-wide representation?
- What is the most suitable way to convey the discrepancy between documents (text-based, graphical or other representations)?
- How can we determine the quality of a summary?

These questions are governed by two areas of research in natural language processing: *(1) Information Representation:* Often simple word count models like bag of words [3] and tf-idf [5] are used to represent text. In contrast, topic models like LDA [2] represent entire documents by uncovering their latent topic structure and encoding each document in a collection as a composition of those topics. Word embedding models [1] map each word in a vocabulary onto a vectors of real numbers. These vectors are arranged such that they encode semantical and syntactical properties of the language. *(2) Automatic Text Summarization:* Automatic text summarization is the process of shortening a text document while preserving salient aspects of it. There are two different approaches to the problem. *(1) Extraction:* Extraction methods try to find a subset of sentences from the original document that cover salient aspects of the original document. Which is what TextRank [6] does, by constructing a graph where every sentences is a vertex that has an edge to sentences that are sufficiently equal to the sentence. Subsequently, PageRank is used to determine the importance of each sentence. *(2) Abstraction:* Abstraction methods build semantical representation of text documents an derive a summary from that representation, which mimics what people do when they summarize a document.

3   Approach

The proposed research wants to represent the content differences of two documents such that they are human-interpretable, easily comprehensible and quickly to compute. At first an estimator (e.g. Recurrent Neural Network (RNN)) is trained to recover a text that generated the corresponding data representation. Secondly, a recombination takes place that generates a new data representation. And lastly, the trained estimator is used to find a coherent interpretation of this newly created data representation (see Fig. 1). The model is required to learn the language of the documents to a degree that allows it to generate correct language which is plausible as Sutskever et al. [8] have been able to create a RNN that generates English text despite the fact that it was trained on character level. Since there is no proper baseline available that we can compare against, the initial step is to create a simple implementation of the model described in Fig. 1. This includes a sentence-wise tf-idf encoding and a simple Gaussian naive Bayes classifier. Afterwards we will create and apply an evaluation framework that reflects the desired properties:

- **Correctness**: The summary produced is coherent given the two documents that it was derived from. A simple approach to test this is to consider the

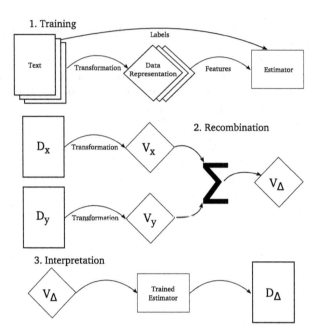

Fig. 1. Abstract model of the automatic discrepancy summarization. During the training phase, an estimator learns to recreate a text from its data representation. In the recombination phase, a new vectors gets created by applying some function (e.g. subtraction) onto those two vectors (The Ds refer to a text unit whereas the Vs are the corresponding data representations). The Interpretation phase transforms the newly created vector into a human-understandable form.

extremes: What is the result of comparing a document with itself? Since there is no discrepancy, the summary should be empty. What happens when two topically orthogonal documents are examined? The summary should be dominated by one of the documents.

- **Interpretable**: Is a user able to comprehend the results produced and draw conclusion from them? For example, if the summary is a sentence, the sentence must be syntactically, grammatically and semantically correct.
- **Generalizable**: The technique must work on text documents across different domains. Moreover, the model must produce coherent results on domains that were not covered in the set of training documents.
- **Deterministic**: The system always produces the same output when it is given the same inputs.

Subsequently, more sophisticated implementations will be investigated and compared against the baseline results. Currently, two experiments are projected:

1. **Topic Models**: A topic model will be used to represent the text units. Since topic models preserve more structure of the text, it is expected to obtain better results.

2. **Text Embeddings**: In this experiment we want to investigate the question whether the sequential nature of text documents should be captured by the representation of the text or by the estimator. This leads to two different settings: (1) Sequence-aware model: Sentences will be transformed into a sequence of word vectors. The estimator will be a RNN. (2) Sequence-aware data representation: Sentences will be transformed into a single paragraph vector. The estimator will be a support vector machine.

4   Preliminary Results

In [10] we have derived words describing the difference between two documents from their document embeddings. That is, given two document vectors d_0 and d_1, we compiled a set of word vectors $w_0 \ldots w_n$ such that $d_0 \approx d_1 \circ w_0 \circ w_1 \circ \ldots \circ w_n$ ($\circ$ was either the summation or the subtraction operator). We approached that task with an iterative process that generated a word at each iteration. It began by finding the word that maximizes $similarity(d_0, d_1 \circ w_0)$ and continued by maximizing $similarity(d_0 \circ w_0, d_1 \circ w_0 \circ w_1)$ etc. We found that this leads to a list of words that is descriptive of the content difference of d_0 and d_1. However, we realized that words are not the correct text unit for the given task, as they provide to few information and, lacking context, lead to ambiguous and false conclusions. Hence, we want to investigate alternative methods that may overcome the drawbacks while preserving the beneficial properties of our previous solution.

References

1. Bengio, Y., Ducharme, R., Vincent, P., Jauvin, C.: A neural probabilistic language model. J. Mach. Learn. Res. **3**, 1137–1155 (2003)
2. Blei, D.M., Ng, A.Y., Jordan, M.I.: Latent dirichlet allocation. J. Mach. Learn. Res. **3**, 993–1022 (2003)
3. Harris, Z.S.: Distributional structure. Word **10**(2–3), 146–162 (1954)
4. Le, Q.V., Mikolov, T.: Distributed representations of sentences and documents. arXiv preprint arXiv:1405.4053 (2014)
5. Luhn, H.P.: A statistical approach to mechanized encoding and searching of literary information. IBM J. Res. Dev. **1**(4), 309–317 (1957)
6. Mihalcea, R., Tarau, P.: Textrank: Bringing order into texts. In: AFCL (2004)
7. Mikolov, T., Sutskever, I., Chen, K., Corrado, G.S., Dean, J.: Distributed representations of words and phrases and their compositionality. In: Advances in Neural Information Processing Systems 26, pp. 3111–3119. Curran Associates, Inc. (2013)
8. Sutskever, I., Martens, J., Hinton, G.E.: Generating text with recurrent neural networks. In: Proceedings of the 28th International Conference on Machine Learning (ICML-11), pp. 1017–1024 (2011)
9. Trask, A., Michalak, P., Liu, J.: sense2vec-a fast and accurate method for word sense disambiguation. arXiv preprint arXiv:1511.06388 (2015)
10. Witt, N., Seifert, C., Granitzer, M.: Explaining topical distances using word embeddings. In: Database and Expert Systems Applications, 2016 27th International Workshop on Text-based Information Retrieval, pp. 212–217. IEEE (2016)

Studying Conceptual Models for Publishing Library Data to the Semantic Web

Sofia Zapounidou(⊠)

Department of Archives, Library Science and Museology,
Ionian University, Corfu, Greece
l12zapo@ionio.gr

Abstract. This thesis studies the library data and the way that linked data technologies may affect libraries. The thesis aims to contribute to the research regarding the development and implementation of a framework for the integration of bibliographic data in the semantic web. It seeks to make sound propositions for the interoperability of conceptual bibliographic models, as well as for future library systems and search environments integrating bibliographic information.

Keywords: BIBFRAME · Conceptual models · EDM · FRBR · FRBRoo · Interoperability · Linked data

1 Research Area

Libraries generate and preserve bibliographic data by describing the resources they provide access and the items they keep in their collections. These bibliographic descriptions have been specified by standards, such as the ISBD family of standards and the AACR2 rules. Codification and sharing of bibliographic data is currently realized using the MARC21 and UNIMARC standards. The MARC structure and format originates in the 1960s and was developed in pace with the technology of the time. Technology has evolved facilitating the delivery of services using structured data from one or more domains. Integration of library data into the semantic web (SW) demands a shift in conceptual data models and data format according to the SW principles and standards. Thus, the library community has started either (re)using existing data models, such as the Functional Requirements for Bibliographic Records (FRBR) and FRBR Object-Oriented (FRBRoo) or developing new models, such as the British Library Data model and the Bibliographic Framework (BIBFRAME) for representing bibliographic information according to the new needs and formats. There are currently projects applying these models and publishing bibliographic data in linked data format (e.g. British National Bibliography, National Library of Spain Catalog). There are also other efforts considering the description of bibliographic entities, broader in scope or related to other domains, such as the Bib.schema.org for the web community, the CIDOC-CRM, the Europeana Data Model and the DPLA Metadata Application Profile for the cultural heritage domain. Other projects related to different aspects of the transition of library data into the SW are the Bibflow project examining future cataloguing workflows, the Linked Data for Production (LD4P) project

© Springer International Publishing AG 2017
J. Kamps et al. (Eds.): TPDL 2017, LNCS 10450, pp. 652–655, 2017.
DOI: 10.1007/978-3-319-67008-9_63

investigating both changes in workflows and extension of BIBFRAME to serve scholarly communication and description of special materials needs. All these initiatives call 'library data' the universe of bibliographic data and their relationships and agree that the shift from legacy to library linked data needs a new framework for bibliographic data definition, representation and interoperability. The existing relationships between the bibliographic data and how they should be preserved in the SW environment is an issue that this research effort focuses on. In general the research is initiated into the context of the following research questions:

- How could library data and their relationships be modeled respecting library domain principles and linked data prerequisites?
- What changes are expected regarding existing tools, practices and services? How will internal workflows and end-user services (existing or potential) be affected?
- How shall libraries collaborate and interoperate with each other, and with third parties, such as archives, museums, publishers, information resource providers?

2 Aim and Objectives

The aim of the thesis is to contribute to the evolution and integration of library data into the SW. The thesis' objectives lie on a framework defined by three axes:

- **New data models.** The thesis seeks to participate in the identification of library data representations that may support enhanced information services within the SW. Specific library data use cases and the way they are represented by library data models shall be examined. Specific element sets and value vocabularies shall be also identified for the expression of certain types of bibliographic information.
- **Workflows.** Adoption of linked data shall affect the overall operation and management of a library. What new services shall or may libraries offer? What kind of data (internal of from third parties) may facilitate the delivery of these services? The thesis seeks to study the requirements for these services in a SW environment by focusing on tools, new workflows and interoperability of data shared between libraries, or between libraries and other organizations.
- **Testbed.** To propose the design and development of a testbed within the scope of assessing the thesis' findings. On the basis of good practices [1, 2] this testbed shall include: (a) test scenarios, (b) test data, (c) workflows and use of data in the scenarios, (d) evaluation criteria regarding the data expressiveness, ease of implementation and use for both libraries and end-users, (e) proposed software tools to be used in implementing the thesis' findings and in managing the test data.

3 Approach and Preliminary Results

The methodology's first step has been a literature survey regarding new library data models and their semantics. The well-known bibliographic conceptual models FRBR, FRBRoo, BIBFRAME, as well as the Europeana Data Model (EDM) have been studied so far. The expressiveness of the models with regard to specific use cases has been the thesis' first object of investigation. The cases are of varying complexity, from single-volume monographs to aggregates, involving also the representation of content relationships (derivation, adaptation, equivalence, whole-part) and large bibliographic families. The term bibliographic family, as defined in [3], refers to a group of related bibliographic works that somehow derive from a common progenitor. To cover all these cases, selected records from library catalogs have been used.

The research begun with an initial study [4] that revealed similarities and divergences between the models. More specifically, it exhibited that there is "more common ground between FRBR and FRBRoo, and between EDM and BIBFRAME". The similarity between FRBR and FRBRoo was expected, since the latter extends the semantics of the former. The common ground between EDM and BIBFRAME seemed interesting for interoperability reasons. Earlier in the same year the EDM-FRBRoo application profile was presented [5]. Therefore, the research proceeded [6, 7] focusing on a possible BIBFRAME-EDM application profile. The study presented in [7] used four different approaches for the representation of monographs in EDM: according to the library alignment report, using the *ore:Proxy* class, using the *edm:InformationResource* class, and using both *ore:Proxy* and *edm:InformationResource* classes. All three studies [4, 6, 7] revealed that for the case of monographs semantic interoperability between the models is desired and possible.

This finding was further studied in the four models (FRBR, FRBRoo, BIBFRAME, EDM) using more bibliographic records as test data. The bibliographic records were selected to study content relationships also. The content relationships studied are derivation, equivalence and the whole-part relationship. This study [8] identified that for the description of each case, there can be more than one alternative representations enabled by each model's semantics. The research has also revealed similarities and divergences between the models that may facilitate or hinder interoperability and sharing of library data. This last finding is currently tested in mappings from FRBR to BIBFRAME using bibliographic records as test data. All bibliographic records describe works from the same bibliographic family. The mappings are also studied for the preservation of the explicit and the implicit relationships between members of a bibliographic family. Control of bibliographic families and explicit linkages between their members would enhance navigation in a linked data environment. During these tests, it has been discovered that there are conditions enabling mappings, e.g. the existence of a specific attribute of a class or a specific value to an attribute, and that different mappings may be required for the preservation of content relationships. The findings may influence future cataloguing policies, workflows, software interfaces, as well as prospective mappings for sharing or integration purposes.

4 Planned Work and Future Directions

The research plans to develop more mappings to transform data from FRBR and BIBFRAME and vice versa, and to evaluate them using records from library catalogs. The consolidated FRBR model (FRBR-LRM) shall be approved in 2017 and mappings will take under consideration any changes with regard to Group 1 entities, on which this thesis focuses. Since preservation of bibliographic families is an important issue, probably separate mapping algorithms are going to be developed for each bibliographic case. At this stage, it will also be evaluated whether existing MARC21 records include explicit statements of content relationships that may enable preservation of bibliographic families after transformation of data. The mapping algorithms may be expressed in a mapping language, such as the x3ml [9] or the RML [10]. Another issue to be investigated is whether available software tools for automatic conversion of MARC21 data in linked data, produce conversions that ensure preservation of content relationships and bibliographic families.

References

1. Ogle, V., Wilensky, R.: Testbed development for the Berkeley Digital Library Project. D-Lib Mag. **2**, 1–6 (1996)
2. Strodl, S., Rauber, A., Rauch, C., Hofman, H., Debole, F., Amato, G.: The DELOS testbed for choosing a digital preservation strategy. In: Sugimoto, S., Hunter, J., Rauber, A., Morishima, A. (eds.) ICADL 2006. LNCS, vol. 4312, pp. 323–332. Springer, Heidelberg (2006). doi:10.1007/11931584_35
3. Smiraglia, R., Leazer, G.: Derivative bibliographic relationships: The work relationship in a global bibliographic database. J. Am. Soc. Inf. Sci. **50**, 493–504 (1999)
4. Zapounidou, S., Sfakakis, M., Papatheodorou, C.: Highlights of library data models in the era of linked open data. In: Garoufallou, E., Greenberg, J. (eds.) MTSR 2013. CCIS, vol. 390, pp. 396–407. Springer, Cham (2013). doi:10.1007/978-3-319-03437-9_38
5. Doerr, M., Gradmann, S., Le Boeuf, P., Aalberg, T., Bailly, R., Olensky, M.: Final Report on EDM–FRBRoo Application Profile Task Force. (2013)
6. Zapounidou, S., Sfakakis, M., Papatheodorou, C.: Integrating library and cultural heritage data models: the BIBFRAME - EDM case. In: Panhellenic Conference on Informatics Proceedings. ACM, Athens (2014)
7. Zapounidou, S., Sfakakis, M., Papatheodorou, C.: Library data integration: towards bibframe mapping to EDM. In: Closs, S., Studer, R., Garoufallou, E., Sicilia, M.-A. (eds.) MTSR 2014. CCIS, vol. 478, pp. 262–273. Springer, Cham (2014). doi:10.1007/978-3-319-13674-5_25
8. Zapounidou, S., Sfakakis, M., Papatheodorou, C.: Representing and integrating bibliographic information into the Semantic Web: a comparison of four conceptual models. J. Inf. Sci. (2016). doi:10.1177/0165551516650410
9. Kondylakis, H., Doerr, M., Plexousakis, D.: Mapping Language for Information Integration. Technical report 385. Institute of Computer Science, FORTH-ICS, Heraklion, Crete, Greece (2006)
10. Dimou, A., Vander Sande, M., Colpaert, P., Verborgh, R., Mannens, E., de Walle, R.: RML: A Generic Language for Integrated RDF Mappings of Heterogeneous Data. In: Bizer, C., Heath, T., Auer, S., Berners-Lee, T. (eds.) Proceedings of the 7th Workshop on Linked Data on the Web, co-located with the 23rd International World Wide Web Conference (WWW 2014), Seoul, Korea, 8 April 2014

Tutorials

Putting Historical Data in Context:
How to Use DSpace-GLAM

Andrea Bollini[(⊠)] and Claudio Cortese

4Science S.r.l., Milan, Italy
{fandrea.bollini,claudio.corteseg}@4science.it

1 Aims, Scope and Learning Objectives

The proposed tutorial introduced attendees to DSpace-GLAM (Galleries, Libraries, Archives, Museums), the Digital Library Management System based on DSpace (http://www.dspace.org) and DSpace-CRIS (https://wiki.duraspace.org/display/DSPACECRIS/DSpace-CRIS+Home), developed by 4Science for the management, analysis and preservation of digital cultural heritage, covering its functional and technical aspects.

DSpace-GLAM is an additional open-source configuration for the DSpace platform. It extends the DSpace data model providing the ability to manage, collect and expose data about every entity important for the cultural heritage domain, such as persons, organizations, projects, events, places, concepts and so on. The extensible data model was explained in depth, through examples and discussions with participants. Other main topics were DSpace-GLAM "components", relationships management and network analysis.

Finally, 4Science new add-ons for digital cultural resources fruition and analysis (the IIIF-International Image Interoperability Framework-Image Viewer, the Audio/Video Streaming Module, the OCR Module and the CKAN integration) were illustrated. At the end of the tutorial the participants were able to understand the DSpace-GLAM data model, to adapt it to their needs and to evaluate if DSpace GLAM fits the needs of their institution.

2 Relevance to TPDL 2017 and Significance for the Research Field

In the last years humanities are witnessing a growth of available data, thanks to the increasing use of databases, electronic journals, digitization of cultural heritage and tools for data extraction and analysis. In this context, scholars and cultural heritage professionals have to be able to correlate different data sources, to better investigate the articulation of historical phenomena and of the transformation processes that affected human history and culture. In the analysis of the digital data, indeed, it is essential that they are not considered in isolation but in conjunction with all the contextual information needed to answer the research questions. Historical data are often fragmentary,

© Springer International Publishing AG 2017
J. Kamps et al. (Eds.): TPDL 2017, LNCS 10450, pp. 659–660, 2017.
DOI: 10.1007/978-3-319-67008-9

partial and biased, so frequently we can understand a content only in a contextual framework, analyzing its relationships within a global and multidimensional approach.

Since almost two decades Digital Libraries have been managing a variety of objects, like texts, audios, videos, blogs and so on, and they are still evolving to scale into Virtual Research Environments (VREs), integrating the entire extent of an institution's scholarship, including articles, theses, dissertations, journals and also research datasets.

In the Big Data Age, Digital Library Management Systems have to forcefully enter the research process to manage both qualitative and quantitative aspects of digital cultural heritage and allow researchers also to analyze data, highlighting and enhancing their relationships at different scales and to explain their interpretations about the important dimensions of variation and about the network of contextual links that affect the historical sources. Therefore, the flexibility of the data model and the availability of tools for analysis and interpretation become fundamental features for such systems. DSpace-GLAM has been developed to face these challenges.

Using DSpace-GLAM, institutions will be able to manage, analyze and preserve digital objects. together with historical, archaeological or other cultural datasets, relating them with other entities (persons, events, places, concepts, contexts, etc.) for describing the context of cultural objects and data, according to different granularity levels.

3 Target Audience and Expected Prerequisite Knowledge of the Audience

The level of this tutorial was introductory. It was addressed to librarians, archivists, historians, archaeologists, researchers and to those who wanted to build their own digital library, but did not want to write their own software, nor buy a proprietary solution. No programming ability was required. Basic knowledge of digital libraries and repositories architectures and of the relational model, though not mandatory, can guarantee a better learning experience.

Innovation Search

Michail Salampasis[⊠]

Department of Informatics,
Technological Educational Institute of Thessaloniki, Sindos, Greece
msa@it.teithe.gr

1   Aims, Scope and Learning Objectives of the Tutorial

Innovation search presents many challenges to the research community and also to professional searchers and search solution providers. Patents are complex technical documents, whose content appears in many languages and contains images, chemical and genomic structures and other forms of data, intermixed and cross-referring with the text material. Further much innovation search involves the search of other forms of technical information such as scientific papers, or the integration of open linked data and so on with the patent data. Finally, the realistic presentation of search and analysis results to often non-technical and time-poor audiences for purpose of strategic decision making presents particular challenges.

The course reviewed the state of the art and pointed out where the key challenges are, especially for early stage researchers and innovation professionals in patent search and related disciplines. The objectives of the tutorial were:

- Understand the international patent system, patent searching and the relevant state-of-the-art.
- Understand the key limitations and challenges for the research community in the development of patent retrieval and innovation search systems in general.
- Understand how recent developments in information retrieval, multilingual and interactive information access may be applied to patent searching research.

2   Relevance and Significance to TPDL2017

The patent related business is worth billions, but still many problems of innovation search are solved on a daily basis by the patent community, often in somewhat inadequate labor-intensive ways. The challenge for the research and scientific community is to provide better solutions without increasing the already heavy burden on relevant technical, legal and language experts.

© Springer International Publishing AG 2017
J. Kamps et al. (Eds.): TPDL 2017, LNCS 10450, pp. 661–662, 2017.
DOI: 10.1007/978-3-319-67008-9

3 Target Audience

Intellectual property and patents in particular have been extremely present in research, public discussion as well as in the media, in the last ten years. Innovation is very important in today's society and heavily reflected on the economic analyses, calls for collaborative research etc., therefore it demands for a serious consideration also by researchers and experts from related disciplines. We hence expected the target audience to consist mainly of two groups. First, postgraduate students and post-doc researchers from academia engaging in studies related to information retrieval, professional search systems and natural language processing. Second, researchers from other related disciplines and professionals (e.g. search solutions providers) who will be given the opportunity to enhance their expertise towards the area innovation search.

Since the class introduced foundations and basic concepts of patents and innovation search, it was also accessible to individuals not familiar with the field of information retrieval. We did not rely on any particular prior knowledge.

Enabling Precise Identification and Citability of Dynamic Data

Recommendations of the RDA Working Group on Data Citation

Andreas Rauber[✉]

Information and Software Engineering Group (IFS),
Institute of Software Technology and Interactive Systems (ISIS), Vienna
University of Technology, Vienna, Austria
rauber@ifs.tuwien.ac.at

1 Introduction

"Sound, reproducible scholarship rests upon a foundation of robust, accessible data. For this to be so in practice as well as theory, data must be accorded due importance in the practice of scholarship and in the enduring scholarly record. In other words, data should be considered legitimate, citable products of research. Data citation, like the citation of other evidence and sources, is good research practice and is part of the scholarly ecosystem supporting data reuse." (Data Citation principles, [1])

While the importance of these Data Citation Principles is by now widely accepted, several challenges persist when it comes to actually providing the services needed to support precise identification and citation of data, particularly in dynamic environments. In order to repeat an earlier study, to apply data from an earlier study to a new model, we need to be able to precisely identify the very subset of data used. While verbal descriptions of how the subset was created (e.g. by providing selected attribute ranges and time intervals) are hardly precise enough and do not support automated handling, keeping redundant copies of the data in question does not scale up to the big data settings encountered in many disciplines today. Conventional approaches, such as assigning persistent identifiers to entire data sets or individual subsets or data items, are not sufficient to meet these requirements. This problem is further exacerbated if the data itself is dynamic, i.e. if new data keeps being added to a database, if errors are corrected or if data items are being deleted.

Starting from the Data Citation Principles we reviewed the challenges identified above and discussed the solutions and recommendations that have been elaborated within the context of a Working Group of the Research Data Alliance (RDA) on Data Citation: Making Dynamic Data Citeable. These approaches are based on versioned and time-stamped data sources, with persistent identifiers being assigned to the time-stamped queries/expressions that are used for creating the subset of data.

© Springer International Publishing AG 2017
J. Kamps et al. (Eds.): TPDL 2017, LNCS 10450, pp. 663–664, 2017.
DOI: 10.1007/978-3-319-67008-9

We reviewed examples of how these can be implemented for different types of data, including SQL-style databases, comma-separated value files (CSV) and others, and took a look at operational implementations in a variety of data centers.

Reference

1. Data Citation Synthesis Group: Joint Declaration of Data Citation Principles. In: Martone. M. (ed.) FORCE11. San Diego, CA (2014)

Enriching Digital Collections Using Tools for Text Mining, Indexing and Visualization

Riza Batista-Navarro[1([⊠])], Axel J. Soto[1], Nhung T.H. Nguyen[1],
William Ulate[2], and Sophia Ananiadou[1]

[1] School of Computer Science, University of Manchester, Manchester, UK
{friza.batista,axel.soto,nhung.nguyen,
sophia.ananiadoug}@manchester.ac.uk
[2] Missouri Botanical Garden, St. Louis, MO, USA
william.ulate@mobot.org

1 Aims, Scope and Learning Objectives

The tutorial demonstrated a suite of tools for text mining, semantic indexing and visualization that facilitated enhanced searching and exploration of digital collections. Specifically, we aimed to provide:

- an introduction to modular text mining and indexing workflows developed using the Argo platform (http://argo.nactem.ac.uk);
- an overview of the Elasticsearch indexing engine and the Kibana visualization platform;
- the know-how on building and visualizing semantic indexes over digital collections without any programming effort.

The tutorial will cover the end-to-end automatic generation and visualization of a semantically enabled search index over digital collections. By the end of the tutorial, the audience will have gained knowledge on:

- exemplar digital collections (e.g., the Biodiversity Heritage Library, British Medical Journal) enhanced with text-mined semantic metadata and visualization tools;
- information extraction methods for generating semantic metadata over textual collections;
- employing Argo to construct text mining workflows that generate Elastic-search indexes for searching over digital collections;
- using the Kibana platform to generate dashboards and visually explore digital collections indexed with Elasticsearch.

J. Kamps et al. (Eds.): TPDL 2017, LNCS 10450, pp. 665–666, 2017.
DOI: 10.1007/978-3-319-67008-9

2   Relevance to TPDL 2017 and Significance to the Field

With the vast amounts of heterogeneous data that many digital libraries hold, finding information relevant to users has become a challenge. One of the most complex types of data is text written in natural language, whose unstructured and ambiguous nature poses a barrier to the accessibility and discovery of information. Furthermore, the volume of available data makes the exploration and discovery of meaningful content difficult. This can be alleviated by means of a combination of semantic indexing and interactive visualization. Firstly, documents can be indexed with semantic metadata, e.g., by tagging them with terms that indicate their "aboutness". As manually indexing these documents is impractical, automatic tools capable of generating semantic meta-data and building search indexes have become attractive solutions. Secondly, users of digital libraries need to be provided with the necessary tools to explore collections and be able to quickly answer analytical questions based on the data. Information visualization, therefore, represents a valuable asset as it aims at showing summarized information in an intuitive manner.

In this tutorial, we aimed to demonstrate how digital library developers and managers (who do not necessarily have the expertise on natural language processing, text mining and visualization) can make their digital collections easier to search and explore. To this end, we showed how Argo can facilitate the development of their own customized, modular workflows for automatic semantic metadata generation and search index construction. Moreover, we showed how the Kibana platform can be used to slice and dice different views of the data and facilitate their visual exploration.

In this way, the tutorial provided digital library practitioners with the necessary technical know-how on building and visualizing semantic search indexes without any programming effort. We believe that this in turn will allow various digital libraries to build search systems that enable users to find and discover information of interest in a more scalable and efficient manner.

3   Target Audience

This tutorial did not require any prerequisite knowledge and the concepts which will be presented were at the introductory level. Although its aim was to enable the audience to build a technical artifact, no knowledge of programming was required, owing to the Argo platform's graphical interface for workflow construction as well as Kibana's ready-to-use visualizations based on Elasticsearch queries. We wished to reach out to attendees who were developers and managers of digital libraries interested in enhancing their systems with capabilities that facilitate semantic searching and visualization over digital collections.

Workshops

NKOS 2017 – 17th European Networked Knowledge Organization Systems Workshop

Philipp Mayr[1(✉)], Douglas Tudhope[2], Koraljka Golub[3],
Christian Wartena[4], and Ernesto William De Luca[5]

[1] GESIS - Leibniz Institute for the Social Sciences, Cologne, Germany
philipp.mayr@gesis.org
[2] Hypermedia Research Group, Faculty of Computing, Engineering and Science,
University of South Wales, Pontypridd, UK
douglas.tudhope@southwales.ac.uk
[3] Department of Library and Information Science,
Linnaeus University, Växjö, Sweden
Koraljka.golub@lnu.se
[4] Hochschule Hannover, Abteilung Information
und Kommunikation, Hannover, Germany
christian.wartena@hs-hannover.de
[5] Georg-Eckert-Institut - Leibniz-Institut für internationale
Schulbuchforschung, Braunschweig, Germany
deluca@gei.de

1 Introduction

The 17th NKOS workshop at TPDL2017 explored the potential of Knowledge Organization Systems, such as classification systems, taxonomies, thesauri, ontologies, and lexical databases in the context of current developments and possibilities. These tools help to model the underlying semantic structure of a domain for purposes of information retrieval, knowledge discovery, language engineering, and the semantic web. The workshop provided an opportunity to discuss projects, research and development activities, evaluation approaches, lessons learned, and research findings.

2 Themes

Main workshop themes were:

- KOS Alignment: KOS alignment or terminology mapping plays a vital role in NKOS for many years. We wanted to sort out the needs (use cases) of KOS alignments in the new environment of Linked Open Data. We planned to collect methodologies, best practices, guidelines and tools. This included manual and automatic alignments.
- KOS Linked Open Data: Recent years have seen an increasing trend to publication of KOS as Linked Data vocabularies. We needed discussion of practical initiatives to link between congruent vocabularies and to provide effective web services and APIs so that applications can build upon them.

© Springer International Publishing AG 2017
J. Kamps et al. (Eds.): TPDL 2017, LNCS 10450, pp. 669–670, 2017.
DOI: 10.1007/978-3-319-67008-9

- KOS and Document Retrieval: Documents or parts of documents are nowadays not only accessible via their metadata, but their abstracts and in many cases the full texts are electronically available. Thus, these documents also can be found by search engines. Given this possibility of full text search the role of classification and annotation had to be redefined. Questions like the following ones arise: can traditional knowledge organization and document annotation improve full text retrieval? Are classification, categorisation, annotation, tagging, and full text retrieval complementary, or how can they be made complementary? What should be the focus of annotation, if full text retrieval is available?

3   Website

More information can be found at https://at-web1.comp.glam.ac.uk/pages/research/hypermedia/nkos/nkos2017/.

MDQual – (Meta)-Data Quality Workshop

Dimitris Gavrilis[1(✉)] and Christos Papatheodorou[2]

[1] Department of Electrical and Computer Engineering,
University of Patras, Patras, Greece
gavrilis@gmail.com
[2] Department of Archives, Library Science and Museology,
Ionian University, Corfu, Greece
papatheodor@ionio.gr

1 Introduction

It is well known that we are rapidly moving towards a data driven world, where all aspects in our everyday lives are data driven. In all domains, from healthcare to retail and finance, data is collected, analyzed and used to make decisions, usually utilizing machine learning techniques. Data Science involves collecting, cleansing and integrating data prior of analysis. The quality of this data is critical and directly affects the outcome of all data science related tasks. Moreover, metadata is used to annotate data and facilitate data organization and retrieval. Metadata quality also directly affects retrieval and other operations (such as data integration) and workflows that are metadata driven.

Although various metrics have been proposed to measure metadata and data quality, in most cases they are highly subjective and/or domain specific. Moreover, they are directly related to the intended use of the data, meaning that a dataset could be of high quality for one use and of low quality for another. In all cases, (meta)data quality has a tremendous impact on data science related tasks and ultimately in everyday life. The proposed workshop aims at exploring the various quality issues found in people working with both data and metadata across domains. An inter-disciplinary workshop where data scientists across different domains will meet and:

- share and exchange experiences regarding (meta)data quality;
- identify patterns in (meta)data quality;
- share methodologies and metrics that will help to measure (meta)-data quality;
- share/propose tools that can be used effectively in improving (automatically) (meta)-data quality.

This initiative aimed at bringing together a community of data scientists that have expertise in a diverse set of domains, such as archives and libraries, healthcare, biology, humanities, computer science and engineering, environment, agriculture, economics, etc. Apart from sharing metrics and methods to identify and resolve quality issues and evaluate datasets, the workshop aimed at promoting the use of tools and services for the automatic measurement and improvement of (meta)data quality. Although few such

© Springer International Publishing AG 2017
J. Kamps et al. (Eds.): TPDL 2017, LNCS 10450, pp. 671–672, 2017.
DOI: 10.1007/978-3-319-67008-9

tools are available in the market, a good number of standalone micro-services are available and can be used to automatically improve (meta)data quality.

We welcomed position papers expressing the data and metadata quality needs from content providers (libraries, archives, museums, public and private sector organizations that manage multimedia content). Moreover, we welcomed research papers that described methods, metrics, services and tools for measuring and ensuring quality. The workshop provided a session for demonstrating implemented systems and services in order to trigger discussions on real world needs and running systems.

2 Topics

Indicative topics of the Workshop were:

- Data and metadata quality measurement methods
- Data and metadata quality requirements for e-research, health, education and digital humanities, etc.
- Metrics for data quality measurement in for e-research, health, education and digital humanities, etc.
- Metrics for metadata quality measurement in for e-research, health, education and digital humanities, etc.
- Tools and services for measuring quality
- Tools and services for improving quality
- Services for automatic data and metadata enrichment

3 Website

More information can be found at http://qualitics.org/mdqual2017.

TDDL 2017 – 1st International Workshop on Temporal Dynamics in Digital Libraries

Annalina Caputo[1], Nattiya Kanhabua[2(✉)], Pierpaolo Basile[3],
and Séamus Lawless[4]

[1] ADAPT Centre, School of Computer Science and Statistics,
Trinity College, Dublin, Ireland
caputoa@tcd.ie

[2] Database, Programming and Web Technologies (DPW) Group,
Department of Computer Science, Aalborg University, Aalborg, Denmark
nattiya@cs.aau.dk

[3] SWAP Group, Department of Computer Science,
University of Bari Aldo Moro, Bari, Italy
pierpaolo.basile@uniba.it

[4] Knowledge and Data Engineering Group, School of Computer Science
and Statistics, Trinity College, Dublin, Ireland
seamus.lawless@scss.tcd.ie

1 Introduction

In Digital Libraries, which can often span several epochs, time is a critical factor. It is the means by which understanding, searching, and exploring these collections of data. Temporal dynamics, i.e. time-based patterns and trends, underpin language usage, entity references, and cultural and economic trends. Users accessing the information contained in Digital Libraries have to deal with their partial knowledge of these phenomena (word meaning variation, entity temporal ambiguity, specific events and time-related trends), as well as their own temporal evolution, i.e. their change in interests, preferences, and goals over time. Intercepting, representing, and predicting these dynamics is fundamental to the intelligent information access in Digital Libraries.

This workshop proposed to bring together researchers and practitioners from different backgrounds in order to identify and discuss research trends, challenges, and new opportunities related to the time-aware intelligent access to Digital Libraries.

2 Topics of Interest

We invited papers that pertain to the workshop theme including, but not limited, to:

- Diachronic analysis of language
- Time-aware Information Retrieval for Digital Libraries
- Time-aware Recommender Systems for Digital Libraries
- Timeline Summarization

J. Kamps et al. (Eds.): TPDL 2017, LNCS 10450, pp. 673–674, 2017.
DOI: 10.1007/978-3-319-67008-9

- Time-aware User Modeling for Digital Libraries
- Event detection
- Time-aware entity disambiguation
- Topic detection and tracking
- Temporal clustering
- Timeline interfaces
- Temporal queries
- Historical studies and computational history
- Topic and entity evolution
- Opinion changes over time
- Web archive-related topics

3   Website

More information can be found at http://tddl2017.github.io/

FUTURITY 2017 – Workshop on Modeling Societal Future

Daniela Gîfu[1,2(✉)] and Diana Trandabăț[1]

[1] "Alexandru Ioan Cuza" University of Iași, Romania, Iasi, Romania
{daniela.gifu,dtrandabat}@info.uaic.ro
[2] Romanian Academy - Iași Branch, Iasi, Romania

1 Introduction

People seek and share ideas, information, experiences, expertise, opinions, and emotion with both acquaintance and strangers on the Internet, based on the effect of the Wisdom of Crowds. Over the last few years, the use of Social Media has increased tremendously all over the world. The huge popularity of social networks provides an ideal environment for scientists to test and simulate new models, algorithms and methods to process knowledge. Structured social knowledge can be used by different actors (companies, public institutions, researchers and scholars interested in formal and empirical analysis of social trends) to understand the behaviors in users or groups.

As recent advances in information and communication technologies continue to reshape the relationship between governments and citizens, opportunities emerge at both ends. Citizens route their voices through new electronic channels, hoping to have their opinions heard at any time from any place. At the same time, companies are willing to identify user's opinion and perceived contexts about their products.

Taking advantage of this huge knowledge "repository", and the new search and extraction methods, the scientific program of FUTURITY-2017 invited papers focusing on the following (and related) topics:

- Extracting knowledge from social web;
- Collaborative and interactive search;
- Conversational search interaction;
- Community behavioral analysis;
- Intelligent personal assistants;
- Semantics in digital libraries;
- Extracting and mining forum data;
- Social media and linked data methodologies in real-life scenarios;
- Collaborative tools and services for citizens, organizations, communities;
- Creating and using structured social media-based resources through social web mining;
- Exploring crowdsourcing and user communities;
- Strategic early warning systems and detection of week signals;

© Springer International Publishing AG 2017
J. Kamps et al. (Eds.): TPDL 2017, LNCS 10450, pp. 675–676, 2017.
DOI: 10.1007/978-3-319-67008-9

- Using the social web to foster innovation;
- Exploring the digital cultural heritage;
- Interaction with the web as a mental, social and physical extension of people.

In this context, the specific aim of FUTURITY-2017 was to establish a consolidated community of internationally appreciated language technology practitioners from different backgrounds, with interests in real-life applications, bridging the gap between research and innovation in order to make sense of crowdsourced knowledge and foreseen future societal challenges.

2 Activities

The workshop intended to be a half-day workshop, tailored around the following schedule:

- Opening session and ice-breaking team building activities, meant to familiarize participants with each other;
- Presentations of papers focusing on challenging research questions;
- Poster presentations during coffee break/lunch;
- A two hours active brainstorming activity (see below)
- A final round table, summarizing ideas and enhancing collaborations.

FUTURITY-2017 had a brainstorming session on three societal innovation scenarios (the topics list was open, participants were asked to propose discussion topics when registering): (1) multilingual collaborative and interactive search; (2) innovative conversational agents for the social web; (3) "intellectual" cooperation between humans and computers.

The organizers acted as facilitators, making sure all participants were engaged in discussion, by actively working is small groups, using creative instruments (from classical mind maps, to Round-Robin brainstorming or Six Thinking Hats techniques). The output of the brainstorming sessions will be at least one viable research project draft.

3 Website

More information can be found at https://profs.info.uaic.ro/~futurity/

Author Index

Alexiou, Giorgos 603
Ameri, Shirin 3
Ananiadou, Sophia 665
Arampatzis, Avi 274, 394
Auer, Sören 315, 328, 342

Balke, Wolf-Tilo 169
Basile, Pierpaolo 673
Batista-Navarro, Riza 665
Benevenuto, Fabrício 537
Bollini, Andrea 659
Bönisch, Thomas 140
Borbinha, José 128
Bozzon, Alessandro 86
Bustillo, Marta 612

Canuto, Sérgio 103
Caputo, Annalina 673
Carevic, Zeljko 459
Charalampous, Aristotelis 181
Clos, Jérémie 498
Clough, Paul 207, 434
Correia, Nuno 525
Cortese, Claudio 659
Cox, Simon 74
Crestani, Fabio 473

da Silva, Camila Wohlmuth 525
da Silva, João Rocha 566
de Siqueira, Gustavo Oliveira 103
De Luca, Ernesto William 669
Debruyne, Christophe 531, 612
Demidova, Elena 116
Devaraju, Anusuriya 74
Dimitropoulos, Harry 355
Dores, Wellington 537
Douligeris, Christos 407
Dulisch, Nadine 595
Dutoit, Madeleine 623

Effrosynidis, Dimitrios 394

Fafalios, Pavlos 261
Fathalla, Said 315, 342
Forestier, Germain 608
Foster, Jonathan 434
Foufoulas, Yannis 355
Fourkioti, Olga 274
Fragkeskos, Kyriakos 86
Fraser, Ryan 74
Freire, Nuno 220
Fujii, Atsushi 485

Gavrilis, Dimitris 671
Gîfu, Daniela 675
Golub, Koraljka 669
Gonçalves, Marcos André 103
Gooch, Phil 287
Goodale, Paula 207
Gossen, Gerhard 116
Großmann, Marcel 599

Hall, Mark 434
Hassenforder, Michel 608
Heaven, Rachel 49
Hill, Timothy 207
Houben, Geert-Jan 86
Howard, John B. 220
Hui, Kit-Ying 49

Illig, Steffen 599
Ioannidis, Yannis 355
Iosifidis, Vasileios 261, 369
Isaac, Antoine 220

Jack, Kris 287

Kacem, Ameni 560
Kanhabua, Nattiya 673
Kapidakis, Sarantos 544
Karalis, Apostolos 407
Keefe, Tim 612
Kern, Dagmar 446
Kippar, Jaagup 421

Klemenz, Arne Martin   553
Klump, Jens   74
Knoth, Petr   181, 287, 572
Koniaris, Marios   603
Krestel, Ralf   40
Kvale, Live   628
Kyprianos, Konstantinos   407

Laender, Alberto H.F.   103, 537
Lange, Christoph   3, 315, 328, 342
Latard, Bastien   608
Lawless, Séamus   673
Leblanc, Elina   632
Lofi, Christoph   86
Lusky, Maria   446

Manguinhas, Hugo   220
Markert, Katja   61
Massie, Stewart   49
Matějka, Cornelius   599
Mathiak, Brigitte   595
Mayr, Philipp   560, 669
McKenna, Lucy   531, 612
Meimaris, Marios   603
Merčun, Tanja   579
Mesbah, Sepideh   86
Mets, Õnne   421
Mountantonakis, Michalis   155
Müller, Mark-Christoph   300

Na, Jin-Cheon   382
Nam, Le Nguyen Hoai   511
Nejdl, Wolfgang   61
Nguyen, Nhung T.H.   665
Nkisi-Orji, Ikechukwu   49
Ntoutsi, Eirini   261, 369

O'Sullivan, Declan   531, 612
Oelschlager, Annina   369

Pankowski, Tadeusz   27
Papadakis, Ioannis   407
Papastefanatos, George   603
Papatheodorou, Christos   15, 671
Paramita, Monica Lestari   207
Pereira, Nelson   566
Petras, Vivien   233
Pinto, José María González   169
Pride, David   572
Proença, Diogo   128

Quoc, Ho Bao   511

Rafailidis, Dimitrios   473
Rauber, Andreas   663
Ribeiro, Cristina   566
Risch, Julian   40
Risse, Thomas   116, 246
Robson, Glen   220

Sadeghi, Afshin   328
Salaba, Athena   579
Salampasis, Michail   661
Schembera, Björn   140
Seifert, Christin   193, 616
Sfakakis, Michalis   15
Soares, Elias   537
Soleman, Sidik   485
Song, Yuting   640
Soto, Axel J.   665
Stamatogiannakis, Lefteris   355
Stefanidis, Kostas   261
Stiller, Juliane   233
Symeonidis, Symeon   274, 394

Tahmasebi, Nina   246
Tan, Sang-Sang   382
Tey, Victor   74
Tochtermann, Klaus   553
Toepfer, Martin   616, 644
Trandabăţ, Diana   585, 675
Tudhope, Douglas   669
Tzitzikas, Yannis   155

Ulate, William   665

Vahdati, Sahar   3, 315, 342
Van Canh, Tran   61
van Hoek, Wilko   459
Vidal, Maria-Esther   328
Vieira, Ricardo   128

Wacker, Dirk   446
Walsh, David   434
Wartena, Christian   669
Weber, Jonathan   608
Wiratunga, Nirmalie   49, 498
Witt, Nils   193, 648
Wyborn, Lesley   74

Zapounidou, Sofia   15, 652

Printed in the United States
By Bookmasters